Lecture Notes in Artificial Intelligence 3698

Edited by J. G. Carbonell and J. Siekmann

Subseries of Lecture Notes in Computer Science

W0235418

Ulrich Furbach (Ed.)

KI 2005:
Advances in
Artificial Intelligence

28th Annual German Conference on AI, KI 2005
Koblenz, Germany, September 11-14, 2005
Proceedings

 Springer

Series Editors

Jaime G. Carbonell, Carnegie Mellon University, Pittsburgh, PA, USA
Jörg Siekmann, University of Saarland, Saarbrücken, Germany

Volume Editor

Ulrich Furbach
Universität Koblenz-Landau
Fachbereich Informatik
Universitätsstrasse 1, 56070 Koblenz, Germany
E-mail: uli@uni-koblenz.de

Library of Congress Control Number: 2005931931

CR Subject Classification (1998): I.2

ISSN   0302-9743
ISBN-10   3-540-28761-2 Springer Berlin Heidelberg New York
ISBN-13   978-3-540-28761-2 Springer Berlin Heidelberg New York

Springer is a part of Springer Science+Business Media

springeronline.com

© Springer-Verlag Berlin Heidelberg 2005
Printed in Germany

Typesetting: Camera-ready by author, data conversion by Scientific Publishing Services, Chennai, India
Printed on acid-free paper   SPIN: 11551263   06/3142   5 4 3 2 1 0

Preface

This volume contains the research papers presented at *KI 2005*, the 28th German Conference on Artificial Intelligence, held September 11–14, 2005 in Koblenz, Germany.

KI 2005 was part of the *International Conference Summer Koblenz 2005*, which included conferences covering a broad spectrum of topics that are all related to AI: tableau-based reasoning methods (TABLEAUX), multi-agent systems (MATES), automated reasoning and knowledge representation (FTP), and software engineering and formal methods (SEFM).

The Program Committee received 113 submissions from 22 countries. Each paper was reviewed by three referees; after an intensive discussion about the borderline papers during the online meeting of the Program Committee, 29 papers were accepted for publication in this proceedings volume.

The program included three outstanding keynote talks: Ian Horrocks (University of Manchester, UK), Luc Steels (University of Brussels and Sony) and Sebastian Thrun (Stanford University), who covered topics like logical foundation, cognitive abilities of multi-agent systems and the DARPA Grand Challenge.

KI 2005 also included two excellent tutorials: *Techniques in Evolutionary Robotics and Neurodynamics* (Frank Pasemann, Martin Hülse, Steffen Wischmann and Keyan Zahedi) and *Connectionist Knowledge Representation and Reasoning* (Barbara Hammer and Pascal Hitzler). Many thanks to the tutorial presenters and the tutorial chair Joachim Hertzberg. Peter Baumgartner, in his role as a workshop chair, collected 11 workshops from all areas of AI research, which also includes a meeting of the German Priority Program on *Kooperierende Teams mobiler Roboter in dynamischen Umgebungen*.

I want to sincerely thank all the authors who submitted their work for consideration and the Program Committee members and the additional referees for their great effort and professional work in the review and selection process. Their names are listed on the following pages.

Organizing a conference like *KI 2005* is not possible without the support of many individuals. I am truly grateful to the local organizers for their hard work and help in making the conference a success: Gerd Beuster, Sibille Burkhardt, Ruth Götten, Vladimir Klebanov, Thomas Kleemann, Jan Murray, Ute Riechert, Oliver Obst, Alex Sinner, and Christoph Wernhard.

September 2005   Ulrich Furbach

Organization

Program and Conference Chair

Ulrich Furbach   University of Koblenz-Landau, Germany

Program Committee

Slim Abdennadher	The German University in Cairo
Elisabeth André	University of Augsburg
Franz Baader	University of Dresden
Peter Baumgartner	Max-Planck-Institute for Informatics (Workshop Chair)
Michael Beetz	Technical University of Munich
Susanne Biundo	University of Ulm
Gerhard Brewka	University of Leipzig
Hans-Dieter Burkhard	Humboldt University Berlin
Jürgen Dix	University of Clausthal
Christian Freksa	University of Bremen
Ulrich Furbach	University of Koblenz-Landau (General Chair)
Enrico Giunchiglia	Genova University
Günther Görz	Friedrich-Alexander University of Erlangen-Nürnberg
Andreas Günther	University of Hamburg
Barbara Hammer	University of Clausthal
Joachim Hertzberg	University of Osnabrück (Tutorial Chair)
Otthein Herzog	TZI University of Bremen
Steffen Hölldobler	University of Dresden
Werner Horn	University of Vienna
Stefan Kirn	University of Hohenheim
Gerhard Kraetzschmar	University of Applied Sciences, Bonn-Rhein-Sieg
Rudolf Kruse	Otto-von-Guericke-University of Magdeburg
Nicholas Kushmerick	University College Dublin
Gerhard Lakemeyer	University of Aachen
Hans-Hellmut Nagel	University of Karlsruhe
Bernhard Nebel	University of Freiburg
Bernd Neumann	University of Hamburg
Ilkka Niemelä	Helsinki University of Technology
Frank Puppe	University of Würzburg

Martin Riedmiller	University of Osnabrück
Ulrike Sattler	University of Manchester
Amal El Fallah Seghrouchni	University of Paris 6
Jörg Siekmann	German Research Center for Artificial Intelligence
Steffen Staab	University of Koblenz-Landau
Frieder Stolzenburg	University of Applied Studies and Research, Harz
Sylvie Thiébaux	The Australian National University, Canberra
Michael Thielscher	University of Dresden
Toby Walsh	National ICT Australia
Gerhard Weiss	Technical University of Munich

Additional Referees

Vazha Amiranashvili	Andreas Heß	Stephan Otto
Martin Atzmüller	Jörg Hoffmann	Günther Palm
Sebastian Bader	Jörn Hopf	Dietrich Paulus
Volker Baier	Lothar Hotz	Yannick Pencole
Thomas Barkowsky	Tomi Janhunen	Tobias Pietzsch
Joachim Baumeister	Yevgeny Kazakov	Marco Ragni
Brandon Bennett	Rinat Khoussainov	Jochen Renz
Sven Bertel	Phil Kilby	Kai Florian Richter
Gerd Beuster	Alexandra Kirsch	Jussi Rintanen
Thomas Bieser	Dietrich Klakow	Michael Rovatsos
Alexander Bockmayr	Alexander Kleiner	Marcello Sanguineti
Peter Brucker	Birgit Koch	Ferry Syafei Sapei
Andreas Brüning	Klaus-Dietrich Kramer	Bernd Schattenberg
Frank Buhr	Kai Lingemann	Sebastian Schmidt
Wolfram Burgard	Christian Loos	Lars Schmidt-Thieme
Nachum Dershowitz	Bernd Ludwig	Martin Schuhmann
Klaus Dorfmüller-Ulhaas	Carsten Lutz	Bernhard Schüler
Phan Minh Dung	Stefan Mandl	Holger Schultheis
Manuel Fehler	Hans Meine	Inessa Seifert
Alexander Ferrein	Erica Melis	John Slaney
Felix Fischer	Ralf Möller	Freek Stulp
Arthur Flexer	Reinhard Moratz	Iman Thabet
Lutz Frommberger	Armin Mueller	Bernd Thomas
Alexander Fuchs	Jan Murray	Jan Oliver Wallgrün
Sharam Gharaei	Matthias Nickles	Diedrich Wolter
Axel Großmann	Malvina Nissim	Stefan Wölfl
Hans Werner Guesgen	Peter Novak	Michael Zakharyaschev
Matthias Haringer	Andreas Nüchter	Yingqian Zhang
Pat Hayes	Oliver Obst	
Rainer Herrler	Karel Oliva	

Organization

KI 2005 was organized by the Artificial Intelligence Research Group at the Institute for Computer Science of the University of Koblenz-Landau.

Sponsoring Institutions

University of Koblenz-Landau
City of Koblenz
Government of Rhineland-Palatine
Griesson - de Beukelaer GmbH

Table of Contents

Diagnosis

Neural Networks

Planning

Robotics

Cognitive Modelling / Philosopy / Natural Language

Hierarchy in Fluid Construction Grammars

Joachim De Beule[1] and Luc Steels[1,2]

[1] Vrije Universiteit Brussel, Belgium
[2] Sony Computer Science Laboratory, Paris, France
{joachim, steels}@arti.vub.ac.be
http://arti.vub.ac.be/~{joachim,steels}/

Abstract. This paper reports further progress into a computational implementation of a new formalism for construction grammar, known as Fluid Construction Grammar (FCG). We focus in particular on how hierarchy can be implemented. The paper analyses the requirements for a proper treatment of hierarchy in emergent grammar and then proposes a particular solution based on a new operator, called the J-operator. The J-operator constructs a new unit as a side effect of the matching process.

1 Introduction

In the context of our research on the evolution and emergence of communication systems with human language like features[5], we have been researching a formalism for construction grammars, called Fluid Construction Grammar (FCG) [6]. Construction grammars (see [3], [2], [4] for introductions) have recently emerged from cognitive linguistics as the most successful approach so far for capturing the syntax-semantics mapping. In FCG (as in other construction grammars), a construction associates a semantic structure with a syntactic structure. The semantic structure contains various units (corresponding to lexical items or groupings of them) and semantic categorizations of these units. The syntactic structure contains also units (usually the same as the semantic structure) and syntactic categorizations.

FCG uses many techniques from formal/computational linguistics, such as feature structures for representing syntactic and semantic information and unification as the basic mechanism for the selection and activation of rules. But the formalism and its associated parsing and production algorithms have a number of unique properties: All rules are bi-directional so that the same rules can be used for both parsing and production, and they can be flexibly applied so that ungrammatical sentences or meanings that are only partly covered by the language inventory, can be handled without catastrophic performance degradation. We therefore prefer the term templates instead of rules. Templates have a score which reflects how 'grammatical' they are believed to be. The score is local to an agent and based solely on the agent's interaction with other agents. Our formalism is called *Fluid* Construction Grammar as it strives to capture the fluidity by which new grammatical constructions can enter or leave a language inventory.

U. Furbach (Ed.): KI 2005, LNAI 3698, pp. 1–15, 2005.

The inventory of templates can be divided depending on what role they play in the overall language process. Morph(ological) templates and lex(ical)-stem templates license morphology and map lexical stems into partial meanings. Syncat and sem-cat templates identify syntactic or semantic features preparing or further instantiating grammatical constructions. Gram(matical) templates define grammatical constructions.

So far FCG has only dealt with single layered structures without hierarchy. It is obvious however that natural languages make heavy use of hierarchy: a sentence can contain a noun phrase, which itself can contain another noun phrase etc. This raises the difficult technical question how hierarchy can be integrated in FCG without loosing the advantages of the formalism, specifically the bidirectionality of the templates. This paper presents the solution to this problem that has proven effective in our computational experiments.

2   Hierarchy in Syntax

It is fairly easy to see what hierarchy means on the syntactic side and there is a long tradition of formalisms to handle it. To take an extremely simple example: the phrase "the big block" combines three lexical units to form a new whole. In terms of syntactic categories, we say that "the" is an article, "big" is an adjective, and "block" is a noun and that the combination is a noun phrase, which can function as a unit in a larger structure as in "the blue box next to the big block". Traditionally, this kind of phrase structure analysis is represented in graphs as in figure 1. Using the terminology of (fluid) construction grammar,

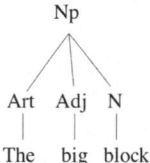

Fig. 1. Hierarchical syntactic structure for a simple noun phrase

we would say that "the big block" is a noun phrase construction of the type Art-Adj-Noun. It consists of a noun-phrase (NP) unit which has three subunits, one for the Article, one for the Adjective and one for the Noun. "The blue box next to the big block" is another noun phrase construction of type NP-prep-NP. So constructions come in different types and subtypes. The units in a construction participate in particular grammatical relations (such as head, complement, determiner, etc.) but the relational view will not be developed here.

Various syntactic constraints are associated with the realization of a construction which constrain its applicability. For example, in the case of the Art-Adj-Noun construction, there is a particular word order imposed, so that the

Article comes before the Adjective and the Adjective before the Noun. Agreement in number between the article and the noun is also required. In French, there would also be agreement between the adjective and the noun, and there would be different Art-Adj-Noun constructions with different word order patterns depending on the type of adjective and/or its semantic role ("une fille charmante" vs. "un bon copain").

When a parent-unit is constructed that has various subunits, there are usually properties of the subunits (most often the head) that are carried over to the parent-unit. For example, if a definite article occurs in an Art-Adj-Noun construction, then the noun phrase as a whole is also a definite noun phrase. Or if the head noun is singular than the noun phrase as a whole is singular as well. It follows that the grammar must be able to express (1) what kind of units can be combined into a larger unit, and (2) what the properties are of this larger unit.

The first step toward hierarchical constructions consists in determining what the syntactic structure looks like before and after the application of the construction. Right before application of the construction in parsing, the syntactic structure is as shown in figure 2. It corresponds to the left part of figure 6. Application of the Art-Adj-Noun construction to this syntactic structure should

```
((top
  (syn-subunits (determiner modifier head))
  (form ((precedes determiner modifier)
         (precedes modifier head))))
 (determiner
  (syn-cat ((lex-cat article) (number singular)))
  (form ((stem determiner "the"))))
 (modifier
  (syn-cat ((lex-cat adjective)))
  (form ((stem modifier "big"))))
 (head
  (syn-cat ((lex-cat noun) (number singular)))
  (form ((stem head "block")))))
```

Fig. 2. Syntactic structure for the phrase "The big block" as it is represented in FCG before the application of an NP construction of the type Art-Adj-Noun. The structure contains four units, one of them (the top unit) has the other three as subunits. The lexicon contributes the different units and their stems. The precedes-constraints are directly detectable from the utterance. This structure should trigger the NP construction.

result in the structure shown in figure 3. It corresponds to the right part of figure 6. As can be seen a new NP-unit is inserted between the top-unit and the other units. The NP-unit has the determiner, modifier and head-unit as subunits and also contains the precedence constraints involving these units. The lexical category of the new unit is NP and it inherits the number of the head-unit.

```
((top
    (syn-subunits (np-unit)))
 (np-unit
    (syn-subunits (determiner modifier head))
    (form ((precedes determiner modifier)
           (precedes modifier head)))
    (syn-cat ((lex-cat NP) (number singular))))
 (determiner
    (syn-cat ((lex-cat article) (number singular)))
    (form ((stem determiner "the"))))
 (modifier
    (syn-cat ((lex-cat adjective)))
    (form ((stem modifier "big"))))
 (head
    (syn-cat ((lex-cat noun) (number singular)))
    (form ((stem head "block")))))
```

Fig. 3. Syntactic structure for the phrase "The big block" as it is represented in FCG after the application of an NP construction of the type Art-Adj-Noun. A new NP-unit is inserted between the top-unit and the other units, which now covers the article, adjective and noun units.

Let us now investigate what the constructions look like that license this kind of transformation.

In FCG a construction is defined as having a semantic pole (left) and a syntactic pole (right). The application of a template consists of a matching phase followed by a merging phase. In parsing, the syntactic pole of a construction template must match with the syntactic structure after which the semantic pole may be merged with the semantic structure to get the new semantic structure. In production, the semantic pole must match with the semantic structure and the syntactic pole is then merged with the syntactic structure. Merging is not always possible. This section considers first the issue of syntactic parsing.

Our key idea to handle hierarchy is to construct a new unit as a side effect of the matching and merging processes. Specifically, we can assume that, for the example under investigation, the syntactic pole should at least contain the units shown in figure 4. This matches with the syntactic structure given in 2 and can therefore be used in parsing to test whether a noun-phrase occurs. But it does not yet create the noun-phrase unit.

This is achieved by the J-operator.[1] Units marked with the J-operator are ignored in matching. When matching is successful, the new unit is introduced and bound to the first argument of the J-operator. The second argument should already have been bound by the matching process to the parent unit from which the new unit should depend. The third argument specifies the set of units that will be pulled into the newly created unit. The new unit can contain additional

---

[1] J denotes the first letter of Joachim De Beule who established the basis for this operator.

```
((?top
    (syn-subunits (== ?determiner ?modifier ?head))
    (form (== (precedes ?determiner ?modifier)
              (precedes ?modifier ?head))))
 (?determiner
    (syn-cat (== (lex-cat article) (number singular)))))
 (?modifier
    (syn-cat (== (lex-cat adjective))))
 (?head
    (syn-cat (== (lex-cat noun) (number singular))))))
```

Fig. 4. The part of the NP-construction that is needed to match the structure in figure 2. Symbols starting with a question-mark are variables that get bound during a successful match or merge. The ==, or *includes symbol*, specifies that the following list should at least contain the specified elements but possibly more.

```
((?top
    (syn-subunits (== ?determiner ?modifier ?head))
    (form (== (precedes ?determiner ?modifier)
              (precedes ?modifier ?head))))
 (?determiner
    (syn-cat (==1 (lex-cat article) (number ?number))))
 (?modifier
    (syn-cat (==1 (lex-cat adjective))))
 (?head
    (syn-cat (==1 (lex-cat noun) (number ?number))))
 ((J ?new-unit ?top (?determiner ?modifier ?head))
    (syn-cat (np (number ?number))))))
```

Fig. 5. The pole of figure 4 extended with a J-operator unit. The function of the ==1 or *includes-uniquely* symbol will be explained shortly, in matching it roughly behaves like the == symbol.

slot specifications, specified in the normal way, and all variable bindings resulting from the match are still valid. An example is shown in figure 5.

Besides creating the new unit and adding its features, the overall structure should change also. Specifically, the new-unit is declared a subunit of its second argument (i.e. ?top in figure 5) and all feature values specified in this parent unit are moved to the new unit (i.e. the syn-subunits and form slots in figure 5.) This way the precedence relations in the form-slot of the original parent or any other categories that transcend single units (like intonation) can automatically be moved to the new unit as well. Thus the example syntactic structure of figure 2 for "the big block" is indeed transformed into the target structure of figure 3 by applying the pole of figure 5. The operation is illustrated graphically in figure 6. Note that now the np-unit is itself a unit ready to be combined with others if necessary.

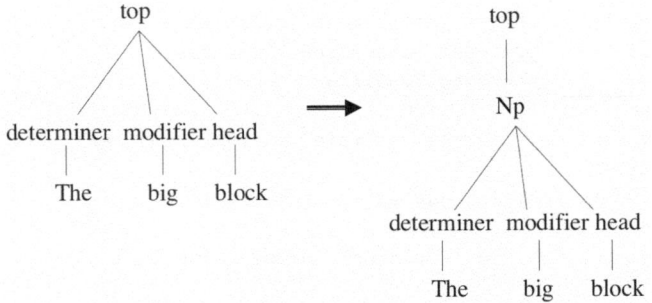

Fig. 6. Restructuring performed by the J-operator

Formally, the J-operator is a tertiary operator and is written as (variable names are arbitrary):

(J ?new-unit ?parent (?child-1 ... ?child-n))

The second argument is called the parent argument of the J-operator and the third the children-argument. Any unit-name in the left or right pole of a template may be governed by a J-operator. The parent and children-arguments should refer to other units in the same pole.

Matching a pattern containing a J-operator against a target structure is equivalent to matching the pattern *without* the unit marked by the J-operator.

Assume a set of bindings for the parent and children in the J-operator, then merging a pattern containing a J-operator with a target is defined as follows:

1. Unless a binding is already present, the set of bindings is extended with a binding of the first argument ?new-unit of the J-operator to a new constant unit-name N.
2. A new pattern is created by removing the unit marked by the J-operator from the original pattern and adding a unit with name N. This new unit has as slots the union of the slots of the unit marked by the J-operator and of the unit in the original pattern as specified by the parent argument. This pattern is now merged with the target structure.
3. The feature values of the parent specified in the J-pattern are removed from the corresponding unit in the target structure and the new unit is made a subunit of this unit.
4. Finally, all remaining references to units specified by the children are replaced by references to the new unit.

This is illustrated for the target structure in figure 2 and the pattern of figure 5. The pattern figure 5 matches with figure 2 because the J-unit is ignored. The resulting bindings are:

[?top/top, ?determiner/determiner, ?modifier/modifier,
?head/head, ?number/singular]   (B1)

The new pattern is given by

```
((?top
    (syn-subunits (== ?determiner ?modifier ?head))
    (form (== (precedes ?determiner ?modifier)
              (precedes ?modifier ?head))))
 (?determiner
    (syn-cat (==1 (lex-cat article) (number ?number))))
 (?modifier
    (syn-cat (==1 (lex-cat adjective))))
 (?head
    (syn-cat (==1 (lex-cat noun) (number ?number))))
 (np-unit
    (syn-cat (np (number ?number)))
    (syn-subunits (== ?determiner ?modifier ?head))
    (form (== (precedes ?determiner ?modifier)
              (precedes ?modifier ?head)))))
```

Before merging, these bindings are extended with a binding [?new-unit/np-unit].
Merging this pattern with figure 2 given the bindings (B1) results in

```
((top
    (syn-subunits (determiner modifier head))
    (form ((precedes determiner modifier)
           (precedes modifier head)))
 (np-unit
    (syn-subunits (determiner modifier head))
    (form ((precedes determiner modifier)
           (precedes modifier head)))
    (syn-cat ((lex-cat NP) (number singular))))
 (determiner
    (syn-cat ((lex-cat article) (number singular)))
    (form ((stem determiner "the"))))
 (modifier
    (syn-cat ((lex-cat adjective)))
    (form ((stem modifier "big"))))
 (head
    (syn-cat ((lex-cat noun) (number singular)))
    (form ((stem head "block")))))
```

The last two steps remove the syn-subunits and form slots from the top unit and
result in the structure of figure 3.

Note that the application of a template now actually consists of three phases:
first the matching of a pole against the target structure, next the merging of the
same structure with the altered pole as specified above and finally the normal
merging phase between the template's other pole and the corresponding target
structure (although here too J-operators might be involved).

```
((top
    (syn-subunits (determiner modifier head))
  (determiner
    (form ((stem determiner "the"))))
  (modifier
    (form ((stem modifier "big"))))
  (head
    (form ((stem head "block")))))))
```

Fig. 7. Syntactic structure while producing "the big block", before the application of an NP-construction

In syntactic production the syntactic structure after lexicon-lookup and before application of the Art-Adj-Noun construction is as shown in figure 7. As will be discussed shortly, in production the left (semantic) pole of the construction is matched with the semantic structure and its right (syntactic) pole (shown in figure 5) has to be *merged* with the above syntactic structure. Here again the J-operator is used to create a new unit and link it from the top, yielding again the structure shown in figure 3.

It is important to realise that merging may fail (and hence the application of the construction may fail) if there is an incompatibility between the slot-specification in the template's pattern and the target structure with which the pattern is merged. Indeed this is the function of the *includes-uniquely* symbol ==1. It signals that there can only be one clause with the predicates that follow. Whereas if the slot-specification starts with a normal includes symbol ==, it means that the clause(s) can simply be added. The different forms of a slot-specification and their behavior in matching and merging are summarised in the table below:

Notation	Read as	Matching and merging behavior
$(a_1...a_n)$	equals	target must contain *exactly* the given elements
$(== a_1..a_n)$	includes	target must contain *at least* the given elements
$(== 1\ a_1...a_n)$	includes uniquely	target must contain *at least* the given elements and may not contain duplicate predicates

For example, the syn-cat slot of the determiner unit in the NP-construction's syntactic pole (see figure 5) is specified as (==1 (lex-cat article) (number ? number)). This can be merged only if the lex-cat of the target unit is none other than article and if the ?number variable is either not yet bound (in which case it becomes bound to the one contained in the target unit) or else only if it is bound to the number value in the target unit.

Note also that no additional mechanisms are needed to implement agreement or the transfer from daughter to parent. Indeed, in figure 5 ?number is the same for the determiner and modifier units. If these units would have different number values in the target structure then the merge is blocked. This implements a test on agreement in number. And because the new-unit has its number category specified by the same variable ?number, it will inherit the number of its constituents as specified.

provided, in which Art-Adj-Noun specifies that

3   Hierarchy in Semantics

The mechanisms proposed so far implement the basic mechanism of syntactically grouping units together in more encompassing units. Any grammar formalism supports this kind of grouping. However constructions have both a syntactic and a semantic pole, and so now the question is how hierarchy is to be handled semantically in tight interaction with syntax.

A construction such as "the big block" not only simply groups together the meanings associated with its component units but also adds additional meaning, including what should be done with the meanings of the components to obtain the meaning of the utterance. FCG uses a Montague-style semantics (see [1] for an introduction), which means that individual words like "ball" or "big" introduce predicates and that constructions introduce second order semantic operators that say how these predicates are to be used (see [7] for a sketchy description of the system IRL generates this kind of operators and how it is used).

For example, we can assume a semantic operator find-referent-1, which is a particular type of a find-referent operator that uses a quantifier, like [the], a property, like [big], and a prototype of an object, like [ball], to find an object in the current context. It would do this by filtering all objects that match the prototype, then filtering them further by taking the one who is biggest, and then taking out the unique element of the remaining singleton.

The initial semantic structure containing the meaning of the phrase "the big block" now looks as in figure 8. Lexicon-lookup introduces units for each of

```
((top
    (meaning ((find-referent-1 obj det1 prop1 prototype1)
             (quantifier det1 [the])
             (property prop1 [big])
             (prototype prototype1 [ball]))))))
```

Fig. 8. The initial semantic structure for producing the phrase "the big block"

the parts of the meaning that are covered by lexical entries and transforms this structure into the one in figure 9: The challenge now is to apply a construction which pulls out the additional part of the meaning that is not yet covered and constructs the appropriate new unit. Again the J-operator can be used with exactly the same behavior as seen earlier. The J-unit is ignored during matching but introduces a new unit during merging as explained before: the new unit is linked to the parent and the parent slot specifications contained in the template are moved to the new unit.

```
((top
    (sem-subunits (determiner modifier head))
    (meaning ((find-referent-1 obj det1 prop1 prototype1))))
 (determiner
    (referent det1)
    (meaning ((quantifier det1 [the]))))
 (modifier
    (referent prop1)
    (meaning ((property prop1 [big]))))
 (head
    (referent prototype1)
    (meaning ((prototype prototype1 [ball])))))
```

Fig. 9. The semantic structure for producing "the big block" after lexicon lookup. Three lexical entries were used, respectively covering the parts of meaning (quantifier det1 [the]), (property prop1 [big]) and (prototype prototype1 [ball]).

Hence, the semantic pole of the Art-Adj-Noun construction looks as follows:

```
((?top
    (sem-subunits (== ?determiner ?modifier ?head))
    (meaning
       (== (find-referent-1 ?obj ?det1 ?prop1 ?prototype1))))
 (?determiner (referent ?det1))
 (?modifier (referent ?prop1))
 (?head (referent ?prototype1))
 ((J ?new-unit ?top) (referent ?obj)
  (?determiner ?modifier ?head)))
```

Application of this semantic pole to the semantic structure of figure 9 yields:

```
((top
    (sem-subunits (np-unit)))
 (np-unit
    (referent obj)
    (sem-subunits (determiner modifier head))
    (meaning ((find-referent-1 obj det1 prop1 prototype1))))
 (determiner
    (referent det1)
    (meaning ((quantifier det1 [the]))))
 (modifier
    (referent prop1)
    (meaning ((property prop1 [big]))))
 (head
    (referent prototype1)
    (meaning ((prototype prototype1 [ball])))))
```

As on the syntactic side, various kinds of selection restrictions could be added. On the semantic side they take the the form of semantic categories that may have

```
Art-Adj-Noun Construction:
    ((?top
        (sem-subunits (== ?determiner ?modifier ?head))
        (meaning
          (==
            (find-referent-1 ?obj ?quantifier ?property ?prototype))))
      (?determiner (referent ?quantifier))
      (?modifier
        (referent ?property)
        (sem-cat (==1 (property-domain ?property physical))))
      (?head
        (referent ?prototype)
        (sem-cat (==1 (prototype-domain ?prototype physical))))
      ((J ?new-unit ?top)
        (referent ?obj)
        (sem-cat (==1 (object-domain ?obj physical)))))
    <-->
    ((?top
        (syn-subunits (== ?determiner ?modifier ?head))
        (form (== (precedes ?determiner ?modifier)
                  (precedes ?modifier ?head))))
      (?determiner
        (syn-cat (==1 (lex-cat article) (number ?number))))
      (?modifier
        (syn-cat (==1 (lex-cat adjective))))
      (?head
        (syn-cat (==1 (lex-cat noun) (number ?number))))
      ((J ?new-unit ?top)
        (syn-cat (np (number ?number))))))
```

Fig. 10. The complete Art-Adj-Noun construction. The double arrow $<-->$ seperates the semantic pole of the construction (above the arrow) form the syntactic pole (below the arrow) and suggests the bi-directional application of the construction during parsing and producing.

been inferred during re-conceptualization (i.e. using sem-templates.) These selection restrictions would block the application of the construction while matching in production and while merging in parsing. For example, we could specialise the construction's semantic pole to be specific to the domain of physical objects (and hence physical properties and prototypes of physical objects) as follows:

```
((?top
    (sem-subunits (== ?determiner ?modifier ?head))
    (meaning
      (==
        (find-referent-1 ?obj ?quantifier ?property ?prototype))))
  (?determiner (referent ?quantifier))
  (?modifier
```

```
    (referent ?property)
    (sem-cat (==1 (property-domain ?property physical))))
  (?head
    (referent ?prototype)
    (sem-cat (==1 (prototype-domain ?prototype physical))))
  ((J ?new-unit ?top)
    (referent ?obj)
    (sem-cat (==1 (object-domain ?obj physical)))))
```

The complete Art-Adj-Noun construction as a whole is shown in figure 10.
    syntactic parsing. syntactic pole. This unit is top unit are
    As an example of semantic parsing, suppose the semantic structure after
lexicon-lookup and semantic categorisation is as in figure 11.

```
((top
    (sem-subunits (determiner modifier head)))
  (determiner (referent ?quant1)
    (meaning ((quantifier ?quant1 [the]))))
  (modifier
    (referent ?prop1)
    (meaning ((property ?prop1 [big])))
    (sem-cat ((property-domain ?prop1 physical-object))))
  (head (referent ?prototype1)
        (meaning ((prototype ?prototype1 [ball])))
        (sem-cat ((prototype-domain ?prototype1 physical-object)))))
```

Fig. 11. Semantic structure after lexicon lookup and semantic categorisation while
parsing the phrase "the big block"

Merging now takes place and succeeds because all semantic categories are
compatible. The final result is:

```
((top
    (sem-subunits (np-unit)))
  (np-unit
    (referent ?obj)
    (sem-subunits (determiner modifier head))
    (meaning ((find-referent-1 ?obj ?quant1 ?prop1 ?prototype1)))
    (sem-cat ((object-domain ?obj physical))))
  (determiner
    (referent ?det1)
    (meaning ((quantifier ?quant1 [the]))))
  (modifier
    (referent ?prop1)
    (meaning ((property ?prop1 [big])))
    (sem-cat ((property-domain ?prop1 physical-object))))
```

```
(head
   (referent ?prototype1)
   (meaning ((prototype ?prototype1 [ball])))
   (sem-cat ((prototype-domain ?prototype1 physical-object)))))
```

4   The J-Operator in Lexicon Look-Up

In the previous sections it was assumed that the application of lex-stem templates resulted in a structure containing separate units for every lexical entry. For example, it was assumed that the lex-stem templates transform the flat structure of figure 8 into the one in figure 9. However an additional transformation is needed for this, and in this section we show how the the J-operator can be used to accomplish this.

In production the initial semantic structure is as in figure 8. Now consider the following lexical entry:

```
(def-lex-template [the]
   ((?top (meaning (== (quantifier ?x [the]))))
    ((J ?new-unit ?top)))
 <-->
   ((?top (syn-subunits (== ?new-unit))
          (form (== (string ?new-unit "the"))))
    ((J ?new-unit ?top)
          (form (== (stem ?new-unit "the")))
          (syn-cat (== (lex-cat article)))))),
```

and similar entries for [big] and [block].

Matching the left pole of the [the]-template with the structure in figure 8 yields the bindings:

```
[?top/top, ?x/det1].
```

As explained, merging this pole with figure 8 extends these bindings with the binding [?new-unit/[the]-unit] and yields:

```
((top
    (sem-subunits ([the]-unit))
    (meaning ((find-referent-1 obj det1 prop1 prototype1)
              (property prop1 [big])
              (prototype prototype1 [ball]))))
 ([the]-unit
    (referent det1)
    (meaning ((quantifier det1 [the])))))
```

Note that, because of the working of the J-operator, a new unit is created which has absorbed the (quantifier det1 [the]) part of the meaning from the top unit. Applying also the templates for [big] and [block] finally results in the desired structure shown in figure 9.

In production, the initial syntactic structure is empty and simply equal to
((top)). But merging the right pole of the [the]-template given the extended
binding yields:

```
((top
    (syn-subunits ([the]-unit)))
 ([the]-unit
    (form ((string [the]-unit "the") (stem [the]-unit "the")))
    (syn-cat ((lex-cat article)))))
```

Similarly applying the templates for [big] and [block] results in figure 7.

As a final example, consider that the initial semantic structure in parsing is
also empty and equal to ((top)). Merging of the left poles of the lexical entries
results in

```
((top
    (sem-subunits ([the]-unit [big]-unit [block]-unit)))
 ([the]-unit (referent ?x-1)
    (meaning ((quantifier ?x-1 [the]))))
 ([big]-unit
    (referent ?x-2)
    (meaning ((property ?x-2 [big]))))
 ([block]-unit
    (referent ?x-3)
    (meaning ((prototype ?x-3 [ball])))))
```

which is, apart from the sem-cat features which have to be added by sem-
templates (re-conceptualization), indeed equivalent with figure 11.

5   Conclusions

This paper introduced the J-operator and showed that it is a very elegant solution to handle hierarchical structure, both in the syntactic and semantic domain.
All the properties of FCG, and particularly the bi-directional applicability of
templates, could be preserved. There are of course many related topics and remaining issues that could not be discussed in the present paper. One issue is the
percolation of properties from the 'head' node of a construction to the governing
node. Although this can be regulated explicitly (as shown in the Art-Adj-Noun
construction where the number of the noun percolates to the number of the
governing NP), it has been argued (in HPSG related formalisms) that there are
default percolations. This can easily be implemented in FCG as will be shown in
a forthcoming paper. Another issue is the handling of optional components, i.e.
elements of a construction which do not obligatory appear. This can be handled
by another kind of operator that regulates which partial matches are licenced
as will also be shown in another paper. Finally, we have already implemented
learning operators that invent or acquire the kind of constructions discussed
here, but their presentation goes beyond the scope of the present paper.

Acknowledgement

This research has been conducted at the Sony Computer Science Laboratory in Paris and the University of Brussels VUB Artificial Intelligence Laboratory. It is partially sponsored by the ECAgents project (FET IST-1940). We are indebted to stimulating discussions about FCG with Benjamin Bergen, Joris Bleys, and Martin Loetzsch.

References

1. Dowty, D., R. Wall, and S. Peters (1981) Introduction to Montague Semantics. D. Reidel Pub. Cy., Dordrecht.
2. Goldberg, A.E. 1995. *Constructions.: A Construction Grammar Approach to Argument Structure.* University of Chicago Press, Chicago
3. Kay, P. and C.J. Fillmore (1999) Grammatical Constructions and Linguistic Generalizations: the What's X Doing Y? Construction. Language, may 1999.
4. Michaelis, L. 2004. *Entity and Event Coercion in a Symbolic Theory of Syntax* In J.-O. Oestman and M. Fried, (eds.),Construction Grammar(s): Cognitive Grounding and Theoretical Extensions. Constructional Approaches to Language series, volume 2. Amsterdam: Benjamins.
5. Steels, L. (2003) Evolving grounded communication for robots. Trends in Cognitive Science. Volume 7, Issue 7, July 2003 , pp. 308-312.
6. Steels, L. (2004) Constructivist Development of Grounded Construction Grammars Scott, D., Daelemans, W. and Walker M. (eds) (2004) Proceedings Annual Meeting Association for Computational Linguistic Conference. Barcelona. p. 9-1
7. Steels, L. (2000) The Emergence of Grammar in Communicating Autonomous Robotic Agents. In: Horn, W., editor, Proceedings of European Conference on Artificial Intelligence 2000. Amsterdam, IOS Publishing. pp. 764-769.

Description Logics in Ontology Applications
(Abstract)

Ian Horrocks

Faculty of Computer Science,
University of Manchester, Great Britain

Abstract. Description Logics (DLs) are a family of logic based knowledge representation formalisms. Although they have a range of applications (e.g., configuration and information integration), they are perhaps best known as the basis for widely used ontology languages such as OWL (now a W3C recommendation). This decision was motivated by a requirement that key inference problems be decidable, and that it should be possible to provide reasoning services to support ontology design and deployment. Such reasoning services are typically provided by highly optimised implementations of tableaux decision procedures. In this talk I will introduce both the logics and decision procedures that underpin modern ontology languages, and the implementation techniques that have enabled state of the art systems to be effective in applications in spite of the high worst case complexity of key inference problems.

U. Furbach (Ed.): KI 2005, LNAI 3698, p. 16, 2005.
© Springer-Verlag Berlin Heidelberg 2005

175 Miles Through the Desert
(Abstract)

Sebastian Thrun

Artificial Intelligence Lab (SAIL),
Stanford University, United States of America

Abstract. The DARPA Grand Challenge is one of the biggest open challenges for the robotics community to date. It requires a robotic vehicle to follow a given route of up to 175 miles across punishing desert terrain, without any human supervision. The Challenge was first held in 2004, in which the best performing team failed after 7.3 miles of autonomous driving. The speaker heads one out of 195 teams worldwide competing for the 2 Million Dollar price. Thrun will present the work of the Stanford Racing Team, which is developing an automated car capable of desert driving at up to 50km/h. He will report on research in areas as diverse as computer vision, control, fault-tolerant systems, machine learning, motion planning, data fusion, and 3-D environment modeling.

U. Furbach (Ed.): KI 2005, LNAI 3698, p. 17, 2005.

A New n-Ary Existential Quantifier in Description Logics

Franz Baader, Eldar Karabaev, Carsten Lutz[1], and Manfred Theißen[2]

[1] Theoretical Computer Science, TU Dresden, Germany
[2] Process Systems Engineering, RWTH Aachen, Germany

Abstract. Motivated by a chemical process engineering application, we introduce a new concept constructor in Description Logics (DLs), an n-ary variant of the existential restriction constructor, which generalizes both the usual existential restrictions and so-called qualified number restrictions. We show that the new constructor can be expressed in $\mathcal{ALCQ}$, the extension of the basic DL $\mathcal{ALC}$ by qualified number restrictions. However, this representation results in an exponential blow-up. By giving direct algorithms for $\mathcal{ALC}$ extended with the new constructor, we can show that the complexity of reasoning in this new DL is actually not harder than the one of reasoning in $\mathcal{ALCQ}$. Moreover, in our chemical process engineering application, a restricted DL that provides only the new constructor together with conjunction, and satisfies an additional restriction on the occurrence of roles names, is sufficient. For this DL, the subsumption problem is polynomial.

1 Introduction

Description Logics (DLs) [2] are a class of knowledge representation formalisms in the tradition of semantic networks and frames, which can be used to represent the terminological knowledge of an application domain in a structured and formally well-understood way. DL systems provide their users with inference services (like computing the subsumption hierarchy) that deduce implicit knowledge from the explicitly represented knowledge. For these inference services to be feasible, the underlying inference problems must at least be decidable, and preferably of low complexity. This is only possible if the expressiveness of the DL employed by the system is restricted in an appropriate way. Because of this restriction of the expressive power of DLs, various application-driven language extensions have been proposed in the literature (see, e.g., [3,9,22,16]), some of which have been integrated into state-of-the-art DL systems [15,13].

The present paper considers a new concept constructor that is motivated by a process engineering application [23]. This constructor is an n-ary variant of the usual existential restriction operator available in most DLs. To motivate the need for this new constructor, assume that we want to describe a chemical plant that has a reactor with a main reaction, and *in addition* a reactor with a main and a side reaction. Also assume that the concepts *Reactor_with_main_reaction* and *Reactor_with_main_and_side_reaction* are defined such that the first concept

U. Furbach (Ed.): KI 2005, LNAI 3698, pp. 18–33, 2005.

subsumes the second one. We could try to model this chemical plant with the help of the usual existential restriction operator as

$$Plant \sqcap \exists has\_part.Reactor\_with\_main\_reaction \sqcap$$
$$\exists has\_part.Reactor\_with\_main\_and\_side\_reaction.$$

However, because of the subsumption relationship between the two reactor concepts, this concept is equivalent to

$$Plant \sqcap \exists has\_part.Reactor\_with\_main\_and\_side\_reaction,$$

and thus does *not* capture the intended meaning of a plant having *two* reactors, one with a main reaction and the other with a main and a side reaction. To overcome this problem, we consider a new concept constructor of the form $\exists r.(C_1, \ldots, C_n)$, with the intended meaning that it describes all individuals having n *different* r-successors $d_1, \ldots, d_n$ such that d_i belongs to C_i ($i = 1, \ldots, n$). Given this constructor, our concept can correctly be described as

$$Plant \sqcap \exists has\_part.(Reactor\_with\_main\_reaction,$$
$$Reactor\_with\_main\_and\_side\_reaction).$$

The situation differs from other application-driven language extensions in that the new constructor can actually be expressed using constructors available in the DL $\mathcal{ALCQ}$, which can be handled by state-of-the-art DL systems (Section 3). Thus, the new constructor can be seen as syntactic sugar; nevertheless, it makes sense to introduce it explicitly since this speeds up reasoning. In fact, expressing the new constructor with the ones available in $\mathcal{ALCQ}$ results in an exponential blow-up. In addition, the translation introduces many "expensive" constructors (disjunction and qualified number restrictions). For this reason, even highly optimized DL systems like RACER [13] cannot handle the translated concepts in a satisfactory way. In contrast, the direct introduction of the new constructor into $\mathcal{ALC}$ does not increase the complexity of reasoning (Section 4). Moreover, in the process engineering application [23] mentioned above, the rather inexpressive DL $\mathcal{EL}^{(n)}$ that provides only the new constructor together with conjunction is sufficient. In addition, only concept descriptions are used where in each conjunction there is at most one n-ary existential restriction for each role. For this restricted DL, the subsumption problem is polynomial (Section 5). If this last restriction is removed, then subsumption is in coNP, but the exact complexity of the subsumption problem in $\mathcal{EL}^{(n)}$ is still open (Section 6). Because of space constraints, some of the technical details are omitted: they can be found in [5].

2   The DL $\mathcal{ALCQ}$

Concept descriptions are inductively defined with the help of a set of *constructors*, starting with a set N_C of *concept names* and a set N_R of *role names*. The

Table 1. Syntax and semantics of $\mathcal{ALCQ}$

Name	Syntax	Semantics
conjunction	$C \sqcap D$	$C^{\mathcal{I}} \cap D^{\mathcal{I}}$
negation	$\neg C$	$\Delta^{\mathcal{I}} \setminus C^{\mathcal{I}}$
at-least qualified number restriction	$\geqslant n\,r.C$	$\{x \mid card(\{y \mid (x,y) \in r^{\mathcal{I}} \wedge y \in C^{\mathcal{I}}\}) \geq n\}$

constructors determine the expressive power of the DL. In this section, we restrict the attention to the DL $\mathcal{ALCQ}$, whose concept descriptions are formed using the constructors shown in Table 1. Using these constructors, several other constructors can be defined as abbreviations:

- $C \sqcup D := \neg(\neg C \sqcap \neg D)$ (disjunction),
- $\top := A \sqcup \neg A$ for a concept name A (top-concept),
- $\exists r.C := \geqslant 1\,r.C$ (existential restriction),
- $\forall r.C := \neg \exists r.\neg C$ (value restriction),
- $\leqslant n\,r.C := \neg(\geqslant (n+1)\,r.C)$ (at-most restriction).

The semantics of $\mathcal{ALCQ}$-concept descriptions is defined in terms of an *interpretation* $\mathcal{I} = (\Delta^{\mathcal{I}}, \cdot^{\mathcal{I}})$. The domain $\Delta^{\mathcal{I}}$ of $\mathcal{I}$ is a non-empty set of individuals and the interpretation function $\cdot^{\mathcal{I}}$ maps each concept name $A \in N_C$ to a subset $A^{\mathcal{I}}$ of $\Delta^{\mathcal{I}}$ and each role $r \in N_R$ to a binary relation $r^{\mathcal{I}}$ on $\Delta^{\mathcal{I}}$. The extension of $\cdot^{\mathcal{I}}$ to arbitrary concept descriptions is inductively defined, as shown in the third column of Table 1. Here, the function *card* yields the cardinality of the given set.

A *general $\mathcal{ALCQ}$-TBox* is a finite set of general concept inclusions (GCIs) $C \sqsubseteq D$ where C, D are $\mathcal{ALCQ}$-concept descriptions. The interpretation $\mathcal{I}$ is a model of the general $\mathcal{ALCQ}$-TBox $\mathcal{T}$ iff it satisfies all its GCIs, i.e., if $C^{\mathcal{I}} \subseteq D^{\mathcal{I}}$ holds for all GCIs $C \sqsubseteq D$ in $\mathcal{T}$.

We use $C \equiv D$ as an abbreviation of the two GCIs $C \sqsubseteq D$, $D \sqsubseteq C$. An *acyclic $\mathcal{ALCQ}$-TBox* is a finite set of *concept definitions* of the form $A \equiv C$ (where A is a concept name and C an $\mathcal{ALCQ}$-concept description) that does not contain multiple definitions or cyclic dependencies between the definitions. Concept names occurring on the left-hand side of a concept definition are called *defined* whereas the others are called *primitive*.

Given two $\mathcal{ALCQ}$-concept descriptions C, D we say that C *is subsumed by* D *w.r.t. the general TBox* $\mathcal{T}$ $(C \sqsubseteq_{\mathcal{T}} D)$ iff $C^{\mathcal{I}} \subseteq D^{\mathcal{I}}$ for all models $\mathcal{I}$ of $\mathcal{T}$. Subsumption w.r.t. an acyclic TBox and subsumption between concept descriptions (where $\mathcal{T}$ is empty) are special cases of this definition. In the latter case we write $C \sqsubseteq D$ in place of $C \sqsubseteq_{\emptyset} D$. The concept description C is *satisfiable* (w.r.t. the general TBox $\mathcal{T}$) iff there is an interpretation $\mathcal{I}$ (a model $\mathcal{I}$ of $\mathcal{T}$) such that $C^{\mathcal{I}} \neq \emptyset$.

The complexity of the subsumption problem in $\mathcal{ALCQ}$ depends on the presence of GCIs. Subsumption of $\mathcal{ALCQ}$-concept descriptions (with or without acyclic TBoxes) is PSPACE-complete and subsumption w.r.t. a general $\mathcal{ALCQ}$-

TBox is EXPTIME-complete [24].[1] These results hold both for unary and binary coding of the numbers in number restrictions, but in this paper we restrict the attention to unary coding (where the size of the number n is counted as n rather than $\log n$).

3   The New Constructor

The general syntax of the new constructor is

$$\exists r.(C_1, \ldots, C_n)$$

where $r \in N_R$, $n \geq 1$, and $C_1, \ldots, C_n$ are concept descriptions. We call this expression an n-ary existential restriction. Its semantics is defined as

$$\exists r.(C_1, \ldots, C_n)^{\mathcal{I}} := \{x \mid \exists y_1, \ldots, y_n. \ (x, y_1) \in r^{\mathcal{I}} \wedge \ldots \wedge (x, y_n) \in r^{\mathcal{I}} \wedge \\ y_1 \in C_1^{\mathcal{I}} \wedge \ldots \wedge y_n \in C_n^{\mathcal{I}} \wedge \bigwedge_{1 \leq i < j \leq n} y_i \neq y_j\}.$$

We call the DL whose concept descriptions are formed using the constructors conjunction, negation, and n-ary existential restriction $\mathcal{EL}^{(n)}\mathcal{C}$. It is an immediate consequence of the semantics of n-ary existential restrictions that the at-least restriction $\geqslant n\, r.C$ can be expressed by the n-ary existential restriction $\exists r.(C, \ldots, C)$.[2] Consequently, all of $\mathcal{ALCQ}$ can be expressed within $\mathcal{EL}^{(n)}\mathcal{C}$.

Conversely, can we express n-ary existential restrictions within $\mathcal{ALCQ}$? We have seen in the introduction that, in general, $\exists r.(C_1, \ldots, C_n)$ cannot be replaced by the conjunction $\exists r.C_1 \sqcap \ldots \sqcap \exists r.C_n$ since this conjunction does not ensure the existence of n different r-successors. However, $\mathcal{ALCQ}$ provides us with the more expressive qualified number restriction constructor. Let us first consider the case $n = 2$. We claim that $\exists r.(C_1, C_2)$ can be expressed by the $\mathcal{ALCQ}$-concept description

$$D := (\geqslant 1\, r.C_1) \sqcap (\geqslant 1\, r.C_2) \sqcap (\geqslant 2\, r.(C_1 \sqcup C_2)).$$

It is clear that any individual belonging to $\exists r.(C_1, C_2)$ also belongs to D. Conversely, assume that x belongs to D. Then x has two distinct r-successors y_1, y_2, both belonging to $C_1 \sqcup C_2$. If one of them belongs to C_1 and the other to C_2, then we are done. Otherwise, we have two cases: (i) both belong to $C_1 \sqcap \neg C_2$, or (ii) both belong to $\neg C_1 \sqcap C_2$. We restrict our attention to the first case (since the second is symmetric). Due to the conjunct $\geqslant 1\, r.C_2$ in D, x has an r-successor in C_2, which is different from y_1 since y_1 does not belong to C_2. Consequently, there are two distinct r-successors of x, one belonging to C_1 and the other belonging to C_2, which shows that x belongs to $\exists r.(C_1, C_2)$.

This result can be extended to arbitrary n.

---

[1] In [24], acyclic TBoxes are not considered, but it is easy to show that the usual approach for handling acyclic TBoxes without using exponential space [18] extends to $\mathcal{ALCQ}$ (see [6]).

[2] Since we assume unary coding of numbers in number restrictions, this translation is linear. Otherwise, it would be exponential.

Theorem 1. *The n-ary existential restriction constructor can be expressed within $\mathcal{ALCQ}$, and thus $\mathcal{ALCQ}$ and $\mathcal{EL}^{(n)}\mathcal{C}$ have the same expressive power.*

To prove this theorem we show that $\exists r.(C_1, \ldots, C_n)$ can be expressed by the $\mathcal{ALCQ}$-concept description

$$D_n := \bigsqcap_{\{i_1,\ldots,i_k\} \subseteq \{1,\ldots,n\}} (\geqslant k\, r.(C_{i_1} \sqcup \ldots \sqcup C_{i_k})).$$

It is again clear that any individual belonging to the concept $\exists r.(C_1, \ldots, C_n)$ also belongs to D_n. The other direction is an easy consequence of Hall's theorem [14]. Let $F = (S_1, \ldots, S_n)$ be a finite family of sets. This family has a *system of distinct representatives (SDR)* iff there are n distinct elements $s_1, \ldots, s_n$ such that $s_i \in S_i$ $(i = 1, \ldots, n)$.

Theorem 2 (Hall). *The family $F = (S_1, \ldots, S_n)$ has an SDR iff $\mathrm{card}(S_{i_1} \cup \ldots \cup S_{i_k}) \geq k$ for all $\{i_1, \ldots, i_k\} \subseteq \{1, \ldots, n\}$, where $i_1, \ldots, i_k$ are distinct.*

Now, assume that the individual x belongs to D_n. For $i = 1, \ldots, n$, let S_i be the set of r-successors of x that belong to C_i. By the definition of D_n, the family $(S_1, \ldots, S_n)$ satisfies the condition of Hall's theorem, and thus it has an SDR. This SDR obviously shows that x belongs to $\exists r.(C_1, \ldots, C_n)$.

The proof of Theorem 1 shows that the subsumption problem in $\mathcal{EL}^{(n)}\mathcal{C}$ can be reduced to the subsumption problem in $\mathcal{ALCQ}$, and thus DL systems like RACER that can handle $\mathcal{ALCQ}$ can in principle be used to compute subsumption in $\mathcal{EL}^{(n)}\mathcal{C}$. However, the translation from $\mathcal{EL}^{(n)}\mathcal{C}$ into $\mathcal{ALCQ}$ described above is obviously exponential. In addition, the constructs it introduces (disjunctions and qualified number restrictions) are hard to handle for tableau-based subsumption algorithms like the one used by RACER. In fact, faced with the $\mathcal{ALCQ}$-translations of the $\mathcal{EL}^{(n)}\mathcal{C}$-concept descriptions

$$C := \exists r.(A_1 \sqcap B_1, A_2 \sqcap B_2, A_3 \sqcap B_3, A_4 \sqcap B_4),$$
$$D := \exists r.(A_1, A_2, A_3, A_4),$$

it takes RACER[3] 57 minutes to find out that $C \sqsubseteq D$. For the 5-ary variant of this example, RACER did not finish its computation within 4 hours.

This problem can be due either to the inherently higher complexity of reasoning in $\mathcal{EL}^{(n)}\mathcal{C}$, or to the translation. We will see in the next section that the latter is the culprit.

4   Complexity of Reasoning in $\mathcal{EL}^{(n)}\mathcal{C}$

The exponential translation of $\mathcal{EL}^{(n)}\mathcal{C}$-concepts into $\mathcal{ALCQ}$-concepts together with the known complexity of the subsumption problem in $\mathcal{ALCQ}$ (see Section 2) yields the following complexity upper-bounds for the subsumption problem in $\mathcal{EL}^{(n)}\mathcal{C}$: EXPSPACE for subsumption of concept descriptions and 2EXPTIME for

define procedure $\mathcal{EL}^{(n)}\mathcal{C}$-World$(\Delta, \Gamma)$

   if Δ is not a type for Γ **then**

      return false

   for all $r \in \mathsf{rol}_\exists(\Delta)$ **do**

      non-deterministically choose an $n \leq N_r(\Gamma)$ and sets $\Psi_0, \ldots, \Psi_{n-1} \subseteq r\text{-}\mathsf{cl}(\Delta)$

      if $\Psi_0, \ldots, \Psi_{n-1}$ is not a successor candidate for Δ w.r.t. Γ **then**

         return false

      for all $i < n$ **do**

         if $\mathcal{EL}^{(n)}\mathcal{C}$-World$(\Psi_i, r\text{-}\mathsf{cl}(\Delta)) = $ false **then**

            return false

  return true

Fig. 1. The procedure $\mathcal{EL}^{(n)}\mathcal{C}$-World

subsumption w.r.t. a general TBox. The next theorem shows that these upper-bounds are not optimal.

Theorem 3. *The subsumption problem in* $\mathcal{EL}^{(n)}\mathcal{C}$ *is* PSPACE-*complete for subsumption between concept descriptions and* EXPTIME-*complete for subsumption w.r.t. a general TBox.*

The hardness results are an immediate consequence of the corresponding hardness results [11] for the subsumption problem in $\mathcal{ALC}$ (which allows for conjunction, negation, and existential restrictions). Since $\mathcal{EL}^{(n)}\mathcal{C}$ is closed under negation, it is enough to prove the upper bounds for the satisfiability problem. To show the PSPACE-upper bound, we adapt the "witness algorithm" (also called **K**-worlds algorithm) commonly used in modal logics to show that satisfiability in the modal logic **K** is in PSPACE (see, e.g., [7]). The EXPTIME-upper bound is proved by an adaptation of Pratt's "elimination of Hintikka sets" approach to show that satisfiability in propositional dynamic logic (PDL) is in EXPTIME (see also [7]). But first, we must introduce some notation.

In the following, we assume that all concept descriptions are built using only the constructors conjunction, negation, and n-ary existential restriction. We use $\mathsf{sub}(C)$ to denote the set of all *subconcepts* of C, $\mathsf{sub}(\mathcal{T})$ to denote $\bigcup_{C \sqsubseteq D \in \mathcal{T}}(\mathsf{sub}(C) \cup \mathsf{sub}(D))$, and define the *closure* of C and $\mathcal{T}$ as

$$\mathsf{cl}(C, \mathcal{T}) := \mathsf{sub}(C) \cup \mathsf{sub}(\mathcal{T}) \cup \{\neg D \mid D \in \mathsf{sub}(C) \cup \mathsf{sub}(\mathcal{T})\}.$$

We use $\mathsf{cl}(C)$ as an abbreviation for $\mathsf{cl}(C, \emptyset)$. Let Γ be a set of concept descriptions. A set $\Psi \subseteq \Gamma$ is a *type for* Γ iff it satisfies the following conditions:

- for all $C \sqcap D \in \Gamma$: $C \sqcap D \in \Psi$ iff $\{C, D\} \subseteq \Psi$;
- for all $\neg(C \sqcap D) \in \Gamma$: $\neg(C \sqcap D) \in \Psi$ iff $\{\neg C, \neg D\} \cap \Psi \neq \emptyset$;
- for all $\neg C \in \Gamma$: $\neg C \in \Psi$ iff $C \notin \Psi$.

---

[3] RACER Version 1.7.23; on a Pentium 4 machine, 2 Ghz, 2 GB memory; under Redhat Linux.

define procedure $\mathcal{EL}^{(n)}\mathcal{C}$-Elim$(C, \mathcal{T})$

> Set $i := 0$ and $\mathfrak{T}_0$ to the set of all types for C and $\mathcal{T}$
>
> **repeat**
>> $\mathfrak{T}_{i+1} := \{\Gamma \in \mathfrak{T}_i \mid \Gamma \text{ is not moribund in } \mathfrak{T}_i\}$
>>
>> $i := i + 1$
>
> **until** $\mathfrak{T}_i = \mathfrak{T}_{i-1}$
>
> **if** there is a $\Gamma \in \mathfrak{T}_i$ with $C \in \Gamma$ **then**
>> **return** true
>
> **return** false

Fig. 2. The procedure $\mathcal{EL}^{(n)}\mathcal{C}$-Elim

Intiuitively, a type for $\mathsf{cl}(C, \mathcal{T})$ can be used to describe to which subconcepts of $C, \mathcal{T}$ an individual of a given interpretation belongs or not. Individuals having identical types behave the same w.r.t. subconcepts of $C, \mathcal{T}$, and thus, in the algorithms, types can be used to represent the relevant properties of individuals. Basically, the EXPTIME-upper bound is due to the fact that there are only exponentially many types for $\mathsf{cl}(C, \mathcal{T})$. In case $\mathcal{T}$ is empty, there are still exponentially many types, but the way one goes through them is such that only polynomially many of them need to be held in memory at the same time.

Let Γ be a set of concept descriptions, and r a role name. Then $\mathsf{rol}_\exists(\Gamma)$ denotes the set of role names r such that $\exists r.(C_1, \ldots, C_k) \in \Gamma$ for some sequence of concept descriptions $C_1, \ldots, C_k$; moreover, for every role name r we set

$$r\text{-}\mathsf{con}(\Gamma) := \{C_1, \ldots, C_k \mid \exists r.(C_1, \ldots, C_k) \in \Gamma \text{ or } \neg \exists r.(C_1, \ldots, C_k) \in \Gamma\},$$
$$r\text{-}\mathsf{cl}(\Gamma) := \{D, \neg D \mid D \in \mathsf{sub}(E) \text{ for some } E \in r\text{-}\mathsf{con}(\Gamma)\},$$
$$N_r(\Gamma) := \sum\nolimits_{\exists r.(C_1, \ldots, C_k) \in \Gamma} k.$$

Finally, let $\Psi \subseteq \Gamma$, $\Phi_0, \ldots, \Phi_{n-1}$ a (possibly empty) sequence of subsets of Γ, and r a role name. Then $\Phi_0, \ldots, \Phi_{n-1}$ is a *successor candidate* for Ψ w.r.t. r and Γ if, for all $\exists r.(C_1, \ldots, C_k) \in \Gamma$, we have $\exists r.(C_1, \ldots, C_k) \in \Psi$ iff there are $i_1, \ldots, i_k < n$ such that $C_j \in \Phi_{i_j}$ for $1 \le j \le k$ and $i_j \ne i_\ell$ for $1 \le j < \ell \le k$.

The following lemma, whose proof can be found in [5], states that the procedure introduced in Fig. 1 decides satisfiability of $\mathcal{EL}^{(n)}\mathcal{C}$-concept descriptions.

Lemma 1. *The $\mathcal{EL}^{(n)}\mathcal{C}$-concept description C is satisfiable iff there exists a set $\Psi \subseteq \mathsf{cl}(C)$ with $C \in \Psi$ such that $\mathcal{EL}^{(n)}\mathcal{C}$-World$(\Psi, \mathsf{cl}(C))$ returns* true.

In [5] it is also shown that $\mathcal{EL}^{(n)}\mathcal{C}$-World is a non-deterministic algorithm that runs in polynomial space. Because of Savitch's theorem, which says that PSPACE = NPSPACE, this yields the desired PSPACE upper-bound.

Let us now turn to the case of satisfiability w.r.t. a general TBox. Let C be a concept and $\mathcal{T}$ a TBox. A set $\Psi \subseteq \mathsf{cl}(C, \mathcal{T})$ is a *type for C and $\mathcal{T}$* if it is a type for $cl(C, \mathcal{T})$ and additionally satisfies the following property: for all $D \sqsubseteq E \in \mathcal{T}$, $D \in \Psi$ implies $E \in \Psi$.

A type Γ is called *moribund* w.r.t. a set of types $\mathfrak{T}$ if there exists a role name r such that there is no sequence $\Phi_0, \ldots, \Phi_{n-1} \in \mathfrak{T}$ with $n \leq N_r(\Gamma)$ that is a successor candidate for Γ w.r.t. r and $\mathsf{cl}(C, \mathcal{T})$.

Lemma 2. *The procedure $\mathcal{EL}^{(n)}\mathcal{C}$-Elim introduced in Fig. 2 decides satisfiability of C w.r.t. $\mathcal{T}$ in exponential time.*

5 A Tractable Sublanguage

In the chemical process engineering application mentioned above [23], the full expressive power of $\mathcal{EL}^{(n)}\mathcal{C}$ is actually not needed. This application is concerned with supporting the construction of mathematical models of process systems by storing building blocks for such models in a class hierarchy. In order to retrieve building blocks, one can then either browse the hierarchy or formulate query classes. In both cases, the existence of efficient algorithms for computing subsumption between class descriptions is an important prerequisite.

The frame-like formalism for describing classes of such building blocks introduced in [23] can be expressed in the *sublanguage $\mathcal{EL}^{(n)}$ of $\mathcal{EL}^{(n)}\mathcal{C}$*, which allows for conjunction, n-ary existential restrictions, and the top concept. Moreover, since in each frame a given slot-name can be used only once, it is sufficient to consider *restricted $\mathcal{EL}^{(n)}$-concept descriptions* where in each conjunction there is at most one n-ary existential restriction for each role: an $\mathcal{EL}^{(n)}$-concept description is *restricted* iff it is of the form

$$A_1 \sqcap \ldots \sqcap A_n \sqcap \exists r_1.(B_{1,1}, \ldots, B_{1,\ell_1}) \sqcap \ldots \sqcap \exists r_m.(B_{m,1}, \ldots, B_{m,\ell_m}),$$

where $A_1, \ldots, A_n$ are concept names, $r_1, \ldots, r_m$ are *distinct* role names, and $B_{1,1}, \ldots, B_{m,\ell_m}$ are restricted $\mathcal{EL}^{(n)}$-concept descriptions.

For example, the $\mathcal{EL}^{(n)}$-concept description $\exists r.(A, \exists r.(B, C)) \sqcap \exists s.(A, A)$ is restricted whereas the description $\exists r.(A, \exists r.(B, C)) \sqcap \exists r.(A, A)$ is not.

As in the case of $\mathcal{EL}$ [4], the fragment of $\mathcal{EL}^{(n)}$ admitting only unary existential restrictions, restricted $\mathcal{EL}^{(n)}$-concept descriptions can be translated into $\mathcal{EL}^{(n)}$-*description trees*, where the nodes are labeled with sets of concept names and the edges are labeled with role names. For example, the restricted $\mathcal{EL}^{(n)}$-concept descriptions

$$A \sqcap \exists r.(A, B \sqcap \exists r.(B, A), \exists r.(A, A \sqcap B)) \quad \text{and} \quad A \sqcap \exists r.(A, B, \exists r.(A, A))$$

yield the description trees depicted in Fig. 3. Given a restricted $\mathcal{EL}^{(n)}$-concept description C, we denote the corresponding description tree by T_C. Formally, this tree is described by a tuple $T_C = (V, E, v_0, \ell)$, where V is the finite set of nodes, $E \subseteq V \times N_R \times V$ is the set of N_R-labeled edges, $v_0 \in V$ is the root, and $\ell : V \longrightarrow 2^{N_C}$ is the node labeling function.

In [4], it was shown that subsumption between $\mathcal{EL}$-concept descriptions corresponds to the existence of a homomorphism between the corresponding description trees. In $\mathcal{EL}^{(n)}$, we must additionally require that the homomorphism is injective.

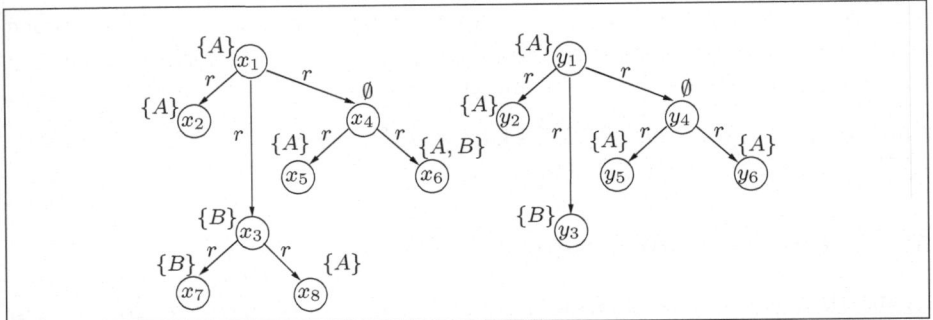

Fig. 3. Two $\mathcal{EL}^{(n)}$-description trees

Definition 1. *Given two $\mathcal{EL}^{(n)}$-description trees $T_1 = (V_1, E_1, v_{0,1}, \ell_1)$ and $T_2 = (V_2, E_2, v_{0,2}, \ell_2)$, a* homomorphism *$\varphi : T_1 \longrightarrow T_2$ is a mapping $\varphi : V_1 \longrightarrow V_2$ such that*

- *$\varphi(v_{0,1}) = v_{0,2}$,*
- *$\ell_1(v) \subseteq \ell_2(\varphi(v))$ for all $v \in V_1$, and*
- *$(\varphi(v), r, \varphi(w)) \in E_2$ for all $(v, r, w) \in E_1$.*

This homomorphism is an embedding *iff the mapping $\varphi : V_1 \longrightarrow V_2$ is injective.*

For example, mapping y_i to x_i for $i = 1, \ldots, 6$ yields an embedding from the description tree on the right-hand side of Fig. 3 to the description tree on the left-hand side. If we changed the label of x_6 to $\{B\}$, then there would still exist a homomorphism between the two trees (mapping both y_5 and y_6 onto x_5), but not an embedding.

The following theorem can be shown similarly to the proof of the corresponding result for $\mathcal{EL}$ [4] (see [5] for details).

Theorem 4. *Let C, D be restricted $\mathcal{EL}^{(n)}$-concept descriptions and T_C, T_D the corresponding description trees. Then $C \sqsubseteq D$ iff there exists an embedding from T_D into T_C.*

To show that subsumption between restricted $\mathcal{EL}^{(n)}$-concept descriptions is a polynomial-time problem, it remains to be shown that the existence of an embedding can be decided in polynomial time. First, let us recall the well-known bottom-up approach for testing for the existence of a homomorphism [21,4].

Let $T_1 = (V_1, E_1, v_{0,1}, \ell_1)$ and $T_2 = (V_2, E_2, v_{0,2}, \ell_2)$ be two $\mathcal{EL}^{(n)}$-description trees, and assume that we want to check whether there is a homomorphism from T_1 to T_2. The idea underlying the polynomial time test is to compute, for each $v \in V_1$, the set $\delta(v)$ of all nodes $w \in V_2$ such that there is a homomorphism from the subtree of T_1 with root v to the subtree of T_2 with root w. Once these sets δ are computed for all nodes of T_1, we can simply check whether $v_{0,2}$ belongs to $\delta(v_{0,1})$. The sets $\delta(v)$ are computed in a bottom-up fashion, where a node is treated only after all its successor nodes have been considered:[4]

---

[4] For example, one can use a postorder tree walk [10] of the nodes of T_1 to realize this.

1. If v is a leaf of T_1, then $\delta(v)$ simply consists of all the nodes $w \in V_2$ such that $\ell_1(v) \subseteq \ell_2(w)$.
2. Let v be a node of T_1 and let $(v, r_1, v_1), \ldots, (v, r_k, v_k)$ be all the edges in E_1 with first component v. Since we work bottom up, we know that the sets $\delta(v_1), \ldots, \delta(v_k)$ have already been computed. The set $\delta(v)$ consists of all the nodes $w \in V_2$ such that
   (a) $\ell_1(v) \subseteq \ell_2(w)$ and
   (b) for each $i, 1 \leq i \leq k$ there exists a node $w_i \in \delta(v_i)$ such that $(w, r_i, w_i) \in E_2$.

It is easy to show that this indeed yields a polynomial-time algorithm for checking the existence of a *homomorphism* between two $\mathcal{EL}^{(n)}$-description trees.

If we want to test for the existence of an *embedding*, we must modify Step 2 of this algorithm. In fact, we must ensure that distinct r-successors of v can be mapped to distinct r-successors of w. This can be achieved as follows:

2′. Let v be a node of T_1, and for each role r let $(v, r, v_{1,r}), \ldots, (v, r, v_{k_r, r})$ be the edges in E_1 with first component v and label r. Since we work bottom up, we know that the sets $\delta(v_{1,r}), \ldots, \delta(v_{k_r, r})$ have already been computed. The set $\delta(v)$ consists of all the nodes $w \in V_2$ satisfying the following two properties:
   (a) $\ell_1(v) \subseteq \ell_2(w)$,
   (b) for all roles r, the family $F_r(w) := (S_{1,r}(w), \ldots, S_{k_r, r}(w))$ has an SDR, where the members of this family are defined as

$$S_{i,r}(w) := \{ w' \in \delta(v_{i,r}) \mid (w, r, w') \in E_2 \}.$$

Obviously, the existence of an SDR for $F_r(w)$ allows us to map the r-successors of v to *distinct* r-successors of w, and thus construct an embedding. For this algorithm to be polynomial, it remains to be shown that the existence of an SDR can be decided in polynomial time. Note that Hall's characterization of the existence of an SDR obviously does not yield a polynomial-time procedure. However, checking for the existence of an SDR is basically the same as solving the maximum bipartite matching problem, which can be done in polynomial time since it can be reduced to a network flow problem [10].

To be more precise, let $(L \cup R, E)$ be a bipartite graph, i.e., $L \cap R = \emptyset$ and $E \subseteq L \times R$. A *matching* is a subset M of E such that each node in $L \cup R$ occurs at most once in M. This matching is called *maximum* iff there is no other matching having a larger cardinality. As shown in [10], such a maximum matching can be computed in time polynomial in the cardinality of V and E.

Let $F = (S_1, \ldots, S_n)$ be a finite family of finite sets, and let $L := \{1, \ldots, n\}$ and $R = S_1 \cup \ldots \cup S_n$.[5] We define the set of edges of the bipartite graph $G_F = (L \cup R, E)$ as follows:

$$E := \{(i, s) \mid s \in S_i\}.$$

---

[5] Without loss of generality we can assume that $L \cap R = \emptyset$.

It is easy to see that the family F has an SDR iff the corresponding bipartite graph G_F has a maximum matching of cardinality n. In fact, $(1, s_1), \ldots, (n, s_n)$ is a maximum matching iff $s_1, \ldots, s_n$ is an SDR.

Thus, we have shown that the existence of an embedding can be decided in polynomial time. Together with Theorem 4, this yields the following tractability result:

Corollary 1. *Subsumption between restricted $\mathcal{EL}^{(n)}$-concept descriptions can be decided in polynomial time.*

A first implementation of this polynomial-time algorithm behaves much better than the translation approach on the example concept descriptions C, D from Section 3 and their obvious extensions to larger n. For small n, the subsumption relationship is found immediately (i.e., with no measurable run-time), and even for $n = 100$, the runtime (of our unoptimized implementation) is just 1 second. One could argue that the comparison of these results with the performance of RACER on the $\mathcal{ALCQ}$-translations of C, D and their extensions to larger n is unfair since the culprit is the exponential translation rather than RACER. However, this is the only known translation of $\mathcal{EL}^{(n)}$-concept descriptions into a DL that can be handled by RACER, and it is the one originally used in the process engineering application.

Acyclic TBoxes. In the process engineering application, acyclic TBoxes are used to introduce abbreviations for complex concept descriptions. In order to extend the polynomial-time algorithm for subsumption between restricted $\mathcal{EL}^{(n)}$-concept descriptions to subsumption w.r.t. acyclic TBoxes, we must first define what it means that an $\mathcal{EL}^{(n)}$-TBox is restricted. An acyclic $\mathcal{EL}^{(n)}$-TBox is called *restricted* iff its concept definitions are of the form

$$A \equiv P_1 \sqcap \ldots \sqcap P_n \sqcap \exists r_1.(A_{1,1}, \ldots, A_{1,\ell_1}) \sqcap \ldots \sqcap \exists r_m.(A_{m,1}, \ldots, A_{m,\ell_m}),$$

where $A, A_{1,1}, \ldots, A_{m,\ell_m}$ are defined concepts, $P_1, \ldots, P_n$ are primitive concepts, and $r_1, \ldots, r_m$ are *distinct* role names.

Given defined concepts A, B in such a TBox $\mathcal{T}$, we can decide subsumption between A and B w.r.t. $\mathcal{T}$ by first expanding A and B, i.e., replacing defined concept names by their definitions until no more defined concepts occur, and then testing the expanded concept descriptions obtained this way for subsumption. The definition of restricted $\mathcal{EL}^{(n)}$-TBoxes ensures that these expanded concept descriptions are restricted, and thus we can use the subsumption algorithm described above. However, it is well-know that the expansion process may lead to an exponential blow-up, i.e., the expanded concept descriptions can be exponential in the size of the TBox [19].

To overcome this problem, we represent restricted $\mathcal{EL}^{(n)}$-TBoxes as directed acyclic graphs (DAG), and define a notion of embedding that, (i) can be tested in time polynomial in the size of the DAG, and (ii) implies the existence of an embedding between the description trees of the expanded concept descriptions (see [5] for details).

Corollary 2. *Subsumption between defined concepts with respect to restricted acyclic $\mathcal{EL}^{(n)}$-TBoxes can be decided in polynomial time.*

Disjointness Statements. In the chemical process engineering application motivating this paper, the real-world concepts expressed by primitive concept names are often disjoint. For example, an object cannot be both an apparatus and a plant. Disjointness statements of the form $dis(P, Q)$, where P, Q are primitive concepts, allow us to express such additional knowledge. An interpretation $\mathcal{I}$ is a model of this statement iff $P^{\mathcal{I}} \cap Q^{\mathcal{I}} = \emptyset$.

For restricted $\mathcal{EL}^{(n)}$-concept descriptions, the only effect that disjointness statements have is that they can make concepts unsatisfiable. It is easy to see that the $\mathcal{EL}^{(n)}$-concept description C is unsatisfiable w.r.t. the set of disjointness statements $\mathcal{D}$ iff there is a statement $dis(P, Q)$ in $\mathcal{D}$ and a node v in T_C whose label contains P and Q. Now, assume that C, D are restricted $\mathcal{EL}^{(n)}$-concept descriptions. Then C is subsumed by D w.r.t. $\mathcal{D}$ iff (i) either C is unsatisfiable w.r.t. $\mathcal{D}$, or (ii) both are satisfiable w.r.t. $\mathcal{D}$ and C is subsumed by D without considering $\mathcal{D}$.

Corollary 3. *Subsumption between restricted $\mathcal{EL}^{(n)}$-concept descriptions w.r.t. disjointness statements can be decided in polynomial time.*

6   Unrestricted $\mathcal{EL}^{(n)}$-Concept Descriptions

In such concept descriptions, several n-ary existential restrictions for the same role r can occur in a conjunction, such as in the description

$$C_u := A \sqcap \exists r.(A, B) \sqcap \exists r.(\exists r.A \sqcap \exists r.A).$$

If we translate this unrestricted $\mathcal{EL}^{(n)}$-concept description into a description tree, then we obtain the tree on the right-hand side of Fig. 3, which is also obtained as a translation of the restricted $\mathcal{EL}^{(n)}$-concept description

$$C_r := A \sqcap \exists r.(A, B, \exists r.(A, A)).$$

To distinguish between these two descriptions, we introduce *distinctness classes*: for each node x in the tree and each role r, the r-successors of x are partitioned into such classes. For example, in the tree corresponding to C_u, the r-successors of y_1 are partitioned into the sets $\{y_2, y_3\}$, $\{y_4\}$, whereas there is only one distinctness class $\{y_2, y_3, y_4\}$ for these nodes in the tree corresponding to C_r.

The notion of an *embedding* that we will use in this section must take these distinctness classes into account. Instead of requiring that the homomorphism φ is injective, we require that for each node x in T_1 and each distinctness class $\{x_1, \ldots, x_k\}$ of r-successors of x, the nodes $\varphi(x_1), \ldots, \varphi(x_k)$ are distinct r-successors of $\varphi(x)$.

However, if we just change the notion of an embedding in this way, then Theorem 4 obviously does not hold for unrestricted $\mathcal{EL}^{(n)}$-concept descriptions.

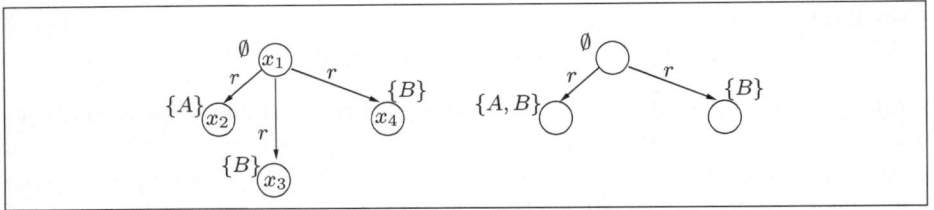

Fig. 4. Identification of $\mathcal{EL}^{(n)}$-description trees

In fact, if $\varphi(x_1), \ldots, \varphi(x_k)$ do not belong to the same distinctness class, then we cannot be sure that they really represent distinct individuals. For example, if $C = \exists r.A \sqcap \exists r.B$ and $D = \exists r.(A, B)$, then there is an embedding from T_D into T_C, but D does not subsume C.

Thus, an obvious conjecture could be that the embedding must *respect distinctness classes*, i.e., we must require $\varphi(x_1), \ldots, \varphi(x_k)$ to belong to the same distinctness class. However, the following example shows that this requirement is too strong. Let $C = \exists r.A \sqcap \exists r.(B, B)$ and $D = \exists r.(A, B)$. There is no embedding from T_D to T_C that respects distinctness classes, but it is easy to see that D subsumes C.

Before we can formulate a correct characterization of subsumption between unrestricted $\mathcal{EL}^{(n)}$-concept descriptions, we must introduce some notation. Given a description tree $T = (V, E, v_0, \ell)$ where role successors are partitioned into distinctness classes, an *identification* on T is an equivalence relation $\sim$ on V such that $v_1 \sim v_2$ implies that

- there are $u_1, u_2 \in V$ and a role r such that v_1 is an r-successor of u_1, v_2 is an r-successor of u_2, and $u_1 \sim u_2$;
- if $v_1 \neq v_2$, then v_1, v_2 do not belong to the same distinctness class.

Any identification $\sim$ on T induces a description tree $T/\sim$ whose nodes are the $\sim$-equivalence classes $[v]_\sim := \{u \in V \mid u \sim v\}$, whose root is $[v_0]_\sim$, and whose edges and node labels are defined as follows:

$$E_\sim := \{([u]_\sim, r, [v]_\sim) \mid \text{there is } u' \in [u]_\sim, v' \in [v]_\sim \text{ such that } (u', r, v') \in E\},$$
$$\ell_\sim([u]_\sim) := \bigcup_{u' \in [u]_\sim} \ell(u').$$

For example, the $\mathcal{EL}^{(n)}$-description tree T_C corresponding to $C = \exists r.A \sqcap \exists r.(B, B)$ is depicted on the left-hand side of Fig. 4, where the r-successors of x_1 are partitioned into the distinctness classes $\{x_2\}, \{x_3, x_4\}$. There are three different identifications: the identity relation, the relation where in addition $x_2 \sim x_3$, and the relation where in addition $x_2 \sim x_4$. The $\mathcal{EL}^{(n)}$-description tree induced by the identity relation is T_C itself, whereas the trees induced by the other two identifications are isomorphic to the tree depicted on the right-hand side of Fig. 4. Obviously, there is an embedding of the $\mathcal{EL}^{(n)}$-description tree T_D corresponding to $D = \exists r.(A, B)$ into each of these two trees.

Theorem 5. *Let C, D be (unrestricted) $\mathcal{EL}^{(n)}$-concept descriptions and T_C, T_D the corresponding description trees. Then $C \sqsubseteq D$ iff for every identification $\sim$ on T_C there exists an embedding from T_D into $T_C/\sim$.*

This theorem yields an NP-algorithm for testing *non*-subsumption of unrestricted $\mathcal{EL}^{(n)}$-concept descriptions: guess in non-deterministic polynomial time an identification $\sim$ of T_C, and then check in polynomial time (by a simple adaptation of the algorithm described in Section 5) whether there is an embedding from T_D into $T_C/\sim$.

Corollary 4. *The subsumption problem for (unrestricted) $\mathcal{EL}^{(n)}$-concept descriptions is in coNP.*

Disjointness Statements. The characterization of subsumption between (unrestricted) $\mathcal{EL}^{(n)}$-concept descriptions given in Theorem 5 can easily be extended to deal with disjointness statements. In fact, the only thing that must be changed is the definition of an identification: we must additionally require that $u \sim v$ implies $\{P, Q\} \not\subseteq \ell(u) \cup \ell(v)$ for all $dis(P, Q)$ in $\mathcal{D}$. With this new notion of an identification, Theorem 5 also holds w.r.t. a set of disjointness statements $\mathcal{D}$. This shows that the subsumption problem for (unrestricted) $\mathcal{EL}^{(n)}$-concept descriptions w.r.t. disjointness statements is in coNP. In the presence of disjointness statements, we can also show the matching hardness result.

Corollary 5. *The subsumption problem for (unrestricted) $\mathcal{EL}^{(n)}$-concept descriptions w.r.t. disjointness statements is coNP-complete.*

The hardness result can be shown by a reduction of *graph 3-colorability* to non-subsumption.[6] A given undirected graph $G = (V, E)$ is 3-colorable iff there is a mapping $f : V \longrightarrow \{1, 2, 3\}$ such that $\{u, v\} \in E$ implies $f(u) \neq f(v)$. It is well-known (see [12]) that the 3-colorability problem, i.e., the question whether a given graph is 3-colorable, is NP-complete.

Let $G = (V, E)$ be an undirected graph with n vertices, i.e., $V = \{v_1, \ldots, v_n\}$. Without loss of generality we assume that this graph has no loops, i.e., $\{u, v\} \in E$ implies $u \neq v$. Let $A_1, \ldots, A_n$ be concept names. The graph $G = (V, E)$ is represented by the set of disjointness statements

$$\mathcal{D}_G := \{ dis(A_i, A_j) \mid \{v_i, v_j\} \in E \}.$$

Let $C := \exists r.A_1 \sqcap \ldots \sqcap \exists r.A_n$ and $D := \exists r.(\top, \top, \top, \top)$. In [5], it is shown that C is *not* subsumed by D w.r.t. $\mathcal{D}_G$ iff G is 3-colorable.

7   Related and Future Work

Polynomiality of the subsumption problem in $\mathcal{EL}$ was shown in [4] as a by-product of the characterization of subsumption via the existence of homomorphisms between the corresponding description trees. This result can also be

---

[6] The idea underlying this reduction was suggested by an anonymous reviewer.

obtained as a consequence of the fact that the containment problem $Q_1 \subseteq Q_2$ for conjunctive queries is polynomial if Q_2 is acyclic [25,20]. Since it is easy to see that $\mathcal{EL}^{(n)}$-concept descriptions can be expressed by acyclic conjunctive queries with disequations [17], one might conjecture that polynomiality of subsumption in $\mathcal{EL}^{(n)}$ follows from the corresponding result for acyclic conjunctive queries with disequations. This is not true, however. In fact, the containment problem for conjunctive queries becomes considerably harder if disequations (i.e., atoms of the form $x \neq y$ for variables x, y) are allowed to occur in the conjunctive queries. For general conjunctive queries with disequations, the containment problem is Π_2^p-complete rather than NP-complete as in the case of conjunctive queries without disequations. Surprisingly, the problem remains Π_2^p-complete if Q_2 is restricted to being acyclic [17]. And even if both queries contain only disequations (and no database predicates), it is not hard to show by a reduction of 3-colorability to non-containment that the containment problem is coNP-hard. Thus, the polynomiality results shown in the present paper does *not* follow from known results for containment of conjunctive queries with disequations.

In [8], it was shown that subsumption in $\mathcal{EL}$ remains polynomial even in the presence of GCIs, and this result was recently extended to a DL extending $\mathcal{EL}$ by several other interesting constructors [1]. Unfortunately, the results in [1] imply that subsumption in $\mathcal{EL}^{(n)}$ becomes EXPTIME-hard in the presence of GCIs.

The most interesting topics for future research are, on the one hand, to show that the exponential translation from $\mathcal{EL}^{(n)}\mathcal{C}$ into $\mathcal{ALCQ}$ given in Section 3 is optimal, i.e., to prove that there is no polynomial translation. On the other hand, the exact complexity of subsumption between *unrestricted* $\mathcal{EL}^{(n)}$-concept descriptions is not yet known. The best complexity upper-bound that we currently have is coNP (see Corollary 4). We conjecture that the problem is coNP-hard, but have not yet found an appropriate reduction from a coNP-complete problem.

References

1. F. Baader, S. Brandt, and C. Lutz. Pushing the $\mathcal{EL}$-envelope. In *Proc. 19th Int. Joint Conf. on Artificial Intelligence*, 2005. To appear.
2. F. Baader, D. Calvanese, D. McGuinness, D. Nardi, and P. F. Patel-Schneider, editors. *The Description Logic Handbook: Theory, Implementation, and Applications*. Cambridge University Press, 2003.
3. F. Baader and P. Hanschke. Extensions of concept languages for a mechanical engineering application. In *Proc. 16th German Workshop on Artificial Intelligence*, volume 671 of *LNCS*, 1992. Springer-Verlag.
4. F. Baader, R. Küsters, and R. Molitor. Computing least common subsumers in description logics with existential restrictions. In *Proc. 16th Int. Joint Conf. on Artificial Intelligence*, 1999.
5. F. Baader, C. Lutz, E. Karabaev, and M. Theißen. A new n-ary existential quantifier in description logics. LTCS-Report 05-08, Theoretical Computer Science, TU Dresden, Germany, 2005. See http://lat.inf.tu-dresden.de/research/reports.html.

6. F. Baader, M. Milicic, C. Lutz, U. Sattler, and F. Wolter. Integrating description logics and action formalisms for reasoning about web services. LTCS-Report 05-02, Theoretical Computer Science, TU Dresden, Germany, 2005. See http://lat.inf.tu-dresden.de/research/reports.html.

7. P. Blackburn, M. de Rijke, and Y. Venema. *Modal Logic*, volume 53 of *Cambridge Tracts in Theoretical Computer Science*. Cambridge University Press, 2001.

8. S. Brandt. Polynomial time reasoning in a description logic with existential restrictions, GCI axioms, and—what else? In *Proc. 16th Eur. Conf. on Artificial Intelligence*, 2004.

9. D. Calvanese, M. Lenzerini, and D. Nardi. A unified framework for class based representation formalisms. In *Proc. 4th Int. Conf. on the Principles of Knowledge Representation and Reasoning*, 1994.

10. T. H. Cormen, C. E. Leiserson, and R. L. Rivest. *Introduction to Algorithms*. The MIT Press, 1990.

11. F. Donini. Complexity of reasoning. In *[2]*. 2003.

12. M. R. Garey and D. S. Johnson. *Computers and Intractability — A guide to NP-completeness*. W. H. Freeman and Company, San Francisco, 1979.

13. V. Haarslev and R. Möller. RACER system description. In *Proc. Int. Joint Conf. on Automated Reasoning*, 2001.

14. P. Hall. On representatives of subsets. *The Journal of the London Mathematical Society*, 10:26–30, 1935.

15. I. Horrocks. Using an expressive description logic: FaCT or fiction? In *Proc. 6th Int. Conf. on Principles of Knowledge Representation and Reasoning*, 1998.

16. I. Horrocks, U. Sattler, and S. Tobies. Practical reasoning for very expressive description logics. *J. of the Interest Group in Pure and Applied Logic*, 8(3):239–264, 2000.

17. P. G. Kolaitis, D. M. Martin, and M. N. Thakur. On the complexity of the containment problem for conjunctive queries with built-in predicates. In *Proc. 17th ACM Symp. on Principles of Database Systems*, 1998.

18. C. Lutz. Complexity of terminological reasoning revisited. In *Proc. 6th Int. Conf. on Logic for Programming and Automated Reasoning*, volume 1705 of *LNAI*. Springer-Verlag, 1999.

19. B. Nebel. Terminological reasoning is inherently intractable. *Artificial Intelligence*, 43:235–249, 1990.

20. X. Qian. Query folding. In *Proc. 12th IEEE Int. Conf. on Data Engineering*, 1996.

21. S. W. Reyner. An analysis of a good algorithm for the subtree problem. *SIAM J. on Computing*, 6(4):730–732, 1977.

22. U. Sattler. A concept language extended with different kinds of transitive roles. In *Proc. 20th German Annual Conf. on Artificial Intelligence*, volume 1137 of *LNAI*. Springer-Verlag, 1996.

23. M. Theißen and L. von Wedel. The need for an n-ary existential quantifier in description logics. In *Proc. KI-04 Workshop on Applications of Description Logics*. CEUR Electronic Workshop Proceedings, http://CEUR-WS.org/Vol-115/, 2004.

24. S. Tobies. *Complexity Results and Practical Algorithms for Logics in Knowledge Representation*. PhD thesis, Computer Science Department, RWTH Aachen, Germany, 2001.

25. M. Yannakakis. Algorithms for acyclic database schemes. In *Proc. 7th Int. Conf. on Very Large Data Bases*, 1981.

Subsumption in $\mathcal{EL}$ w.r.t. Hybrid TBoxes

Sebastian Brandt and Jörg Model

Theoretical Computer Science,
TU Dresden, D-01062 Dresden, Germany
lastname@tcs.inf.tu-dresden.de

Abstract. In the area of Description Logic (DL) based knowledge representation, two desirable features of DL systems have as yet been incompatible: firstly, the support of general TBoxes containing general concept inclusion (GCI) axioms, and secondly, non-standard inference services facilitating knowledge engineering tasks, such as build-up and maintenance of terminologies (TBoxes).

In order to make non-standard inferences available without sacrificing the convenience of GCIs, the present paper proposes *hybrid TBoxes* consisting of a pair of a general TBox $\mathcal{F}$ interpreted by descriptive semantics, and a (possibly) cyclic TBox $\mathcal{T}$ interpreted by fixpoint semantics. $\mathcal{F}$ serves as a foundation of $\mathcal{T}$ in the sense that the GCIs in $\mathcal{F}$ define relationships between concepts used as atomic concept names in the definitions in $\mathcal{T}$. Our main technical result is a polynomial time subsumption algorithm for hybrid $\mathcal{EL}$-TBoxes based on a polynomial reduction to subsumption w.r.t. cyclic $\mathcal{EL}$-TBoxes with fixpoint semantics. By virtue of this reduction, all non-standard inferences already available for cyclic $\mathcal{EL}$-TBoxes become available for hybrid ones.

1 Motivation

In the area of Description Logic (DL) based knowledge representation (KR), intensional knowledge of a given domain is represented by a terminology (TBox) that defines general properties of concepts relevant to the domain [1]. In its simplest form, a TBox comprises *definitions* of the form $A \equiv C$ by which a *concept name* A is assigned to a *concept description* C. Concept descriptions are terms built from atomic concepts by means of a set of constructors provided by the DL under consideration. In the present case, we are mostly concerned with the DL $\mathcal{EL}$ which provides conjunction ($\sqcap$) and existential restriction ($\exists r.C$).

General TBoxes additionally allow for *general concept inclusion (GCI)* axioms of the form $C \sqsubseteq D$, where both C and D may be complex concept descriptions. GCIs define implications ("D holds whenever C holds") relevant to the terminology as a whole. The utility of GCIs for practical KR applications has been examined in depth; see, e.g., [2,3,4]. Apart from constraining terminologies further without explicitly changing all its definitions, using GCIs can lead to smaller, more readable TBoxes, and can facilitate the re-use of data in applications of different levels of detail. As a consequence, GCIs are supported by most modern DL systems, such as FACT [5] and RACER [6].

U. Furbach (Ed.): KI 2005, LNAI 3698, pp. 34–48, 2005.
© Springer-Verlag Berlin Heidelberg 2005

TBoxes are interpreted w.r.t. a model-theoretic *semantics* which allows to reason over the terminology in a formally well-defined way. A model $\mathcal{I}$ satisfies a definition $A \equiv C$ iff the extensions of A and C in $\mathcal{I}$ are equal. A GCI $C \sqsubseteq D$ is satisfied by $\mathcal{I}$ iff the extension of C is a subset of the one of D. $\mathcal{I}$ is a model of a TBox $\mathcal{T}$ iff all definitions and GCIs in $\mathcal{T}$ are satisfied. This semantics for TBoxes is usually called *descriptive semantics* [7]. In contrast, *greatest fixpoint semantics* only considers models that interpret concepts as large as possible.

One of the most important reasoning services provided by DL systems is computing the subsumption hierarchy. A concept A is *subsumed* by a concept B w.r.t. a TBox $\mathcal{T}$ iff the extension of A is a subset of that of B in every model of $\mathcal{T}$. Before DL systems can be used to reason over terminologies, however, the relevant TBoxes must be built-up and maintained. In order to support these knowledge engineering tasks, additional so-called 'non-standard' inference services have been proposed, most notably *least-common subsumer (lcs)* [8,9,10], *most specific concept (msc)* [10], and *matching* [11,12,13]. It has been argued in [14] that the lcs and msc facilitate the build-up of DL terminologies in a 'bottom-up' fashion suitable for domain experts with limited KR background. 'Bottom-up' fashion means to begin by selecting a set of example instances and use them to construct a new concept description intended to represent them. Matching, on the other hand, can be used as a means of querying TBoxes for concepts of a certain structure [15]. This can be utilized to construct new concepts by retrieving and modifying structurally similar ones in the TBox.

The practical utility of GCIs and non-standard inferences motivates the question for a DL in which both can be provided. The problem encountered here relates to the appropriate choice of semantics. General TBoxes need to be interpreted w.r.t. *descriptive* semantics. For this kind of semantics, however, it has been shown in [16] that lcs and msc need not always exist, even w.r.t. cyclic $\mathcal{EL}$-TBoxes. This result carries over to general $\mathcal{EL}$-TBoxes and any extension of $\mathcal{EL}$. The same holds for matching which relies on the lcs. On the other hand, lcs and msc are available for cyclic $\mathcal{EL}$-TBoxes interpreted by *fixpoint semantics* [16].

In order to provide lcs and msc without sacrificing the convenience of GCIs, the present paper proposes *hybrid TBoxes*. A hybrid $\mathcal{EL}$-TBox is a pair $(\mathcal{F}, \mathcal{T})$ of a general TBox $\mathcal{F}$ ('foundation') and a possibly cyclic TBox $\mathcal{T}$ ('terminology') defined over the same set of atomic concepts and roles. $\mathcal{F}$ serves as a foundation of $\mathcal{T}$ in that the GCIs in $\mathcal{F}$ define relationships between concepts used as atomic concept names in the definitions in $\mathcal{T}$. Hence, $\mathcal{F}$ lays a foundation of general implications constraining $\mathcal{T}$. Models of $(\mathcal{F}, \mathcal{T})$ are greatest fixpoint models of $\mathcal{T}$ that respect all GCIs in $\mathcal{F}$. Hence, the foundation is interpreted by descriptive semantics while the terminology is interpreted by greatest fixpoint semantics. Note that hybrid $\mathcal{EL}$-TBoxes cannot be reduced to ordinary general $\mathcal{EL}$-TBoxes.

Having introduced hybrid TBoxes, the main purpose of the present paper is to show that subsumption w.r.t. hybrid $\mathcal{EL}$-TBoxes can be decided by a polynomial reduction to cyclic $\mathcal{EL}$-TBoxes for which a polynomial time decision algorithm exists [17]. This yields a polynomial time subsumption algorithm for hybrid $\mathcal{EL}$-TBoxes. An implication of this reduction is that non-standard inferences available for cyclic $\mathcal{EL}$-TBoxes can directly be utilized for hybrid $\mathcal{EL}$-TBoxes. In this sense, our initial goal of providing non-standard inferences in the presence of GCIs is met.

Another application for hybrid TBoxes might be DL systems supporting users with limited KR background. By restricting the view to the definitions in the terminology while hiding the GCIs in the foundation, the system could provide a simplified version of its knowledge base while preserving correct inferences.

The present paper is organized as follows. Basic definitions related to cyclic $\mathcal{EL}$-TBoxes are introduced in Section 2 while Section 3 formally introduces hybrid $\mathcal{EL}$-TBoxes. In Section 3.2 we show how subsumption w.r.t. hybrid TBoxes can be reduced to subsumption w.r.t. cyclic $\mathcal{EL}$-TBoxes interpreted by greatest fixpoint semantics.

2   Cyclic $\mathcal{EL}$-TBoxes

We begin by formally introducing syntax and semantics of cyclic $\mathcal{EL}$ TBoxes. In fact, we will consider two different semantics first introduced by Nebel [7], namely descriptive semantics and greatest[1] fixpoint semantics. Most of our preliminary Sections 2.2 and 2.3 recall basic definitions and results from [17].

2.1   Syntax and (Descriptive) Semantics

Concept descriptions are inductively defined with the help of a set of concept *constructors*, starting with disjoint sets N_{prim} and N_{def} of *primitive concept names* and *defined concept names*, respectively, and a set N_{role} of *role names*. In this paper, we consider the small DL $\mathcal{EL}$ which provides the concept constructors top-concept ($\top$), conjunction ($C \sqcap D$), and existential restrictions ($\exists r.C$).

As usual, the semantics of concept descriptions is defined in terms of an *interpretation* $\mathcal{I} = (\Delta^{\mathcal{I}}, \cdot^{\mathcal{I}})$. The domain $\Delta^{\mathcal{I}}$ of $\mathcal{I}$ is a non-empty set and the interpretation function $\cdot^{\mathcal{I}}$ maps each concept name $P \in N_{\mathrm{prim}} \cup N_{\mathrm{def}}$ to a subset $P^{\mathcal{I}} \subseteq \Delta^{\mathcal{I}}$ and each role name $r \in N_{\mathrm{role}}$ to a binary relation $r^{\mathcal{I}} \subseteq \Delta^{\mathcal{I}} \times \Delta^{\mathcal{I}}$. The extension of $\cdot^{\mathcal{I}}$ to arbitrary concept descriptions is defined inductively as follows.

$$\top^{\mathcal{I}} := \Delta^{\mathcal{I}}$$
$$(C \sqcap D)^{\mathcal{I}} := C^{\mathcal{I}} \cap D^{\mathcal{I}}$$
$$(\exists r.C)^{\mathcal{I}} := \{x \in \Delta^{\mathcal{I}} \mid \exists y \colon (x, y) \in r^{\mathcal{I}} \wedge y \in C^{\mathcal{I}}\}$$

Definition 1. *An $\mathcal{EL}$-terminology (called $\mathcal{EL}$-TBox) is a finite set $\mathcal{T}$ of definitions of the form $A \equiv C$, where $A \in N_{\mathrm{def}}$ is a concept name and C is an $\mathcal{EL}$-concept description over N_{prim}, N_{def}, and N_{role}. For every such definition, A is called* defined in $\mathcal{T}$ *and may occur on the left-hand side of no other definition in $\mathcal{T}$. Note that cyclic definitions are allowed, i.e., every defined concept may occur on the right-hand side of every definition.*

---

[1] It has been argued in [17] that least fixpoint semantics is not interesting for cyclic $\mathcal{EL}$-TBoxes because cycles are always interpreted by the empty set.

The size of $\mathcal{T}$ is defined as the sum of the sizes of all definitions in $\mathcal{T}$. Denote by $N_{\mathrm{prim}}^{\mathcal{T}}$, $N_{\mathrm{def}}^{\mathcal{T}}$, and $N_{\mathrm{role}}^{\mathcal{T}}$ the set of all primitive concepts, defined concept names, and role names, respectively, occurring in $\mathcal{T}$.

The semantics of a cyclic $\mathcal{EL}$-TBox can now be defined as follows.

Definition 2. *An interpretation $\mathcal{I} = (\Delta^{\mathcal{I}}, \cdot^{\mathcal{I}})$ is a model of $\mathcal{T}$ $(\mathcal{I} \models \mathcal{T})$ iff $A^{\mathcal{I}} = C^{\mathcal{I}}$ for all definitions $A \equiv C \in \mathcal{T}$.*

This semantics has been introduced as *descriptive semantics* by Nebel [7]. One of the most basic inference services provided by DL systems is computing the subsumption hierarchy. Formally, (descriptive) subsumption is defined as follows.

Definition 3. *Let $\mathcal{T}$ be an $\mathcal{EL}$-TBox and let $A, B \in N_{\mathrm{def}}^{\mathcal{T}}$. Then, A is subsumed by B w.r.t. descriptive semantics $(A \sqsubseteq_{\mathcal{T}} B)$ iff $A^{\mathcal{I}} \subseteq B^{\mathcal{I}}$ holds for all models $\mathcal{I}$ of $\mathcal{T}$.*

The other type of semantics relevant for us is introduced in the following section.

2.2 Greatest Fixpoint Semantics

In contrast to descriptive semantics, some more formal preliminaries are necessary to define greatest fixpoint (gfp) semantics. Recalling the relevant definitions from [17], we begin by introducing a normal form for cyclic $\mathcal{EL}$-TBoxes.

Definition 4. *An $\mathcal{EL}$-TBox $\mathcal{T}$ is normalized iff $A \equiv D \in \mathcal{T}$ implies that D is of the form*

$$P_1 \sqcap \cdots \sqcap P_m \sqcap \exists r_1.B_1 \sqcap \ldots \exists r_\ell.B_\ell,$$

where for $m, \ell \geq 0$, $P_1, \ldots, P_m \in N_{\mathrm{prim}}$ and $B_1, \ldots, B_\ell \in N_{\mathrm{def}}$. If $m = \ell = 0$ then $D = \top$.

In order to refer to the definition of a defined concept more conveniently, the following notation is introduced for normalized $\mathcal{EL}$-TBoxes.

Definition 5. *For a normalized $\mathcal{EL}$-TBox $\mathcal{T}$ and every $A \in N_{\mathrm{def}}^{\mathcal{T}}$, let*

$$\mathrm{def}_{\mathcal{T}}(A) := \{P_1, \ldots, P_m\} \cup \{\exists r_1.B_1, \ldots, \exists r_\ell.B_\ell\}$$

iff A is defined in $\mathcal{T}$ by $A \equiv P_1 \sqcap \cdots \sqcap P_m \sqcap \exists r_1.B_1 \sqcap \ldots \exists r_\ell.B_\ell$.

A gfp-model for a given $\mathcal{EL}$-TBox $\mathcal{T}$ is obtained in two steps. In the first step, only the primitive concepts and roles occurring in $\mathcal{T}$ are interpreted. The second step comprises an iteration by which the interpretation of the defined names in $\mathcal{T}$ is changed until a fixpoint is reached. The following definition formalizes the first step.

Definition 6. *Let $\mathcal{T}$ be an $\mathcal{EL}$-TBox over N_{prim}, N_{role}, and N_{def}. A primitive interpretation $(\Delta^{\mathcal{J}}, \cdot^{\mathcal{J}})$ of $\mathcal{T}$ interprets all primitive concepts $P \in N_{\mathrm{prim}}$ by subsets of $\Delta^{\mathcal{J}}$ and all roles $r \in N_{\mathrm{role}}$ by binary relations on $\Delta^{\mathcal{J}}$. An Interpretation $(\Delta^{\mathcal{I}}, \cdot^{\mathcal{I}})$ is based on $\mathcal{J}$ iff $\Delta^{\mathcal{J}} = \Delta^{\mathcal{I}}$ and $\cdot^{\mathcal{J}}$ and $\cdot^{\mathcal{I}}$ coincide on N_{role} and N_{prim}. The set of all initerpretations based on $\mathcal{J}$ is denoted by*

$$\mathrm{Int}(\mathcal{J}) := \{\mathcal{I} \mid \mathcal{I} \text{ is an interpretation based on } \mathcal{J}\}.$$

On $\mathrm{Int}(\mathcal{J})$, a binary relation $\preceq_{\mathcal{J}}$ is defined for all $\mathcal{I}_1, \mathcal{I}_2 \in \mathrm{Int}(\mathcal{J})$ by

$$\mathcal{I}_1 \preceq_{\mathcal{J}} \mathcal{I}_2 \quad \text{iff} \quad A^{\mathcal{I}_1} \subseteq A^{\mathcal{I}_2} \text{ for all } A \in N_{\mathrm{def}}^{\mathcal{T}}.$$

Primitive interpretations do not interpret defined concepts from N_{def}. It is easy to see that $\preceq_{\mathcal{J}}$ is a complete lattice on $\mathrm{Int}(\mathcal{J})$, so that every subset of $\mathrm{Int}(\mathcal{J})$ has a least upper bound (lub) and a greatest lower bound (glb). Hence, by Tarski's fixpoint theorem [18], every monotonic function on $\mathrm{Int}(\mathcal{J})$ has a fixpoint. In particular, this applies to the function $O_{\mathcal{T},\mathcal{J}}$ to be introduced next.

Definition 7. *Let $\mathcal{T}$ be an $\mathcal{EL}$-TBox over N_{prim}, N_{role}, and N_{def}, and $\mathcal{J}$ a primitive interpretation of N_{prim} and N_{role}. Then $O_{\mathcal{T},\mathcal{J}}$ is defined as follows.*

$$O_{\mathcal{T},\mathcal{J}} \colon \mathrm{Int}(\mathcal{J}) \to \mathrm{Int}(\mathcal{J})$$
$$\mathcal{I}_1 \mapsto \mathcal{I}_2 \text{ iff } A^{\mathcal{I}_2} = C^{\mathcal{I}_1} \text{ for all } A \equiv C \in \mathcal{T}.$$

As shown in [17], $O_{\mathcal{T},\mathcal{J}}$ is in fact monotonous and can be used as a fixpoint operator on $\mathrm{Int}(\mathcal{J})$. As a result, we obtain the following proposition.

Proposition 1. *Let $\mathcal{I}$ be an interpretation based on the primitive interpretation $\mathcal{J}$. Then $\mathcal{I}$ is a fixpoint of $O_{\mathcal{T},\mathcal{J}}$ iff $\mathcal{I}$ is a model of $\mathcal{T}$.*

With this, the general notion of fixpoint models for $\mathcal{EL}$-TBoxes can be defined as follows.

Definition 8. *Let $\mathcal{T}$ be an $\mathcal{EL}$-TBox. The model $\mathcal{I}$ of $\mathcal{T}$ is called* gfp-model *iff there is a primitive interpretation $\mathcal{J}$ such that $\mathcal{I} \in \mathrm{Int}(\mathcal{J})$ is the greatest fixpoint of $O_{\mathcal{T},\mathcal{J}}$. Greatest fixpoint semantics considers only gfp-models as admissible models.*

As $(\mathrm{Int}(\mathcal{J}), \preceq_{\mathcal{J}})$ is a complete lattice, the gfp-model is uniquely determined for a given TBox $\mathcal{T}$ and a primitive interpretation $\mathcal{J}$. This allows us to refer to the gfp-model $\mathrm{gfp}(\mathcal{T}, \mathcal{J})$ for any given $\mathcal{T}$ and $\mathcal{J}$.

In order to show how the gfp-model $\mathrm{gfp}(\mathcal{T}, \mathcal{J})$ can be obtained, we need to introduce the iteration of $O_{\mathcal{T},\mathcal{J}}$ over *ordinals*.

Definition 9. *Let $\mathcal{T}$ be an $\mathcal{EL}$-TBox over N_{prim}, N_{role}, and N_{def} and $\mathcal{J}$ a primitive interpretation of N_{prim} and N_{role}. Define $\mathcal{I}^{\mathrm{top}} \in \mathrm{Int}(J)$ by $\mathcal{I}_{\mathcal{J}}^{\mathrm{top}}(A) := \Delta^{\mathcal{J}}$ for every $A \in N_{\mathrm{def}}$. For every ordinal α, define*

- $\mathcal{I}_{\mathcal{T},\mathcal{J}}^{\downarrow\alpha} := \mathcal{I}_{\mathcal{J}}^{\text{top}}$ if $\alpha = 0$;
- $\mathcal{I}_{\mathcal{T},\mathcal{J}}^{\downarrow\alpha+1} := O_{\mathcal{T},\mathcal{J}}(\mathcal{I}^{\downarrow\alpha})$;
- $\mathcal{I}_{\mathcal{T},\mathcal{J}}^{\downarrow\alpha} := \text{glb}(\{\mathcal{I}^{\downarrow\beta} \mid \beta < \alpha\})$ if α is a limit ordinal.

The following corollary now shows that computing $\text{gfp}(\mathcal{T}, \mathcal{J})$ is equivalent to computing $\mathcal{I}_{\mathcal{T},\mathcal{J}}^{\downarrow\alpha}$, given an appropriate ordinal α.

Corollary 1. *Let $\mathcal{T}$ be an $\mathcal{EL}$-TBox over N_{prim}, N_{role}, and N_{def}. Let $\mathcal{J}$ be a primitive interpretation of N_{prim} and N_{role}. Then there exists an ordinal α such that $\text{gfp}(\mathcal{T}, \mathcal{J}) = \mathcal{I}_{\mathcal{T},\mathcal{J}}^{\downarrow\alpha}$.*

Note that if α is a limit ordinal then $\mathcal{I}_{\mathcal{T},\mathcal{J}}^{\downarrow\alpha}$ equals $\bigcap_{\beta<\alpha} \mathcal{I}_{\mathcal{T},\mathcal{J}}^{\downarrow\beta}$. With this preparation, we are ready to introduce gfp-subsumption.

Definition 10. *Let $\mathcal{T}$ be an $\mathcal{EL}$-TBox and let $A, B \in N_{\text{def}}^{\mathcal{T}}$. Then, A is subsumed by B w.r.t. gfp-semantics ($A \sqsubseteq_{\text{gfp},\mathcal{T}} B$) iff $A^{\mathcal{I}} \subseteq B^{\mathcal{I}}$ holds for all gfp-models $\mathcal{I}$ of $\mathcal{T}$.*

Note that descriptive semantics considers a superset of the set of gfp-models, implying that descriptive subsumption entails gfp-subsumption. Hence, all subsumption relations w.r.t. $\sqsubseteq_{\mathcal{T}}$ also hold w.r.t. $\sqsubseteq_{\text{gfp},\mathcal{T}}$. The question of how to *decide* subsumption w.r.t. gfp-semantics is addressed in the following section.

2.3   Deciding Subsumption w.r.t. Cyclic $\mathcal{EL}$-TBoxes with Descriptive Semantics

As in the previous section, we begin by recalling some definitions from [17] elementary for the decision procedure for gfp-subsumption.

Definition 11. *An $\mathcal{EL}$-description graph is a graph $\mathcal{G} = (V, E, L)$ where*

- *V is a set of nodes;*
- *$E \subseteq V \times N_{\text{role}} \times V$ is a set of edges labeled by role names;*
- *$L: V \to 2^{N_{\text{prim}}}$ is a function that labels nodes with sets of primitive concepts.*

Description graphs can be used to represent TBoxes and primitive interpretations. The description graph of a TBox is defined as follows.

Definition 12. *Let $\mathcal{T}$ be a normalized $\mathcal{EL}$-TBox. Then the $\mathcal{EL}$-description graph $\mathcal{G}_{\mathcal{T}} = (N_{\text{def}}^{\mathcal{T}}, E_{\mathcal{T}}, L_{\mathcal{T}})$ of $\mathcal{T}$ is defined as follows:*

- *the nodes of $\mathcal{G}_{\mathcal{T}}$ are the defined concepts of $\mathcal{T}$;*
- *if A is defined in $\mathcal{T}$ and*

$$A \equiv P_1 \sqcap \cdots \sqcap P_m \sqcap \exists r_1.B_1 \sqcap \cdots \sqcap \exists r_\ell.B_\ell$$

is its definition then $L_{\mathcal{T}}(A) := \{P_1, \ldots, P_m\}$, and A is the source of the edges $(A, r_1, B_1), \ldots, (A, r_\ell, B_\ell) \in E_{\mathcal{T}}$.

Note that for every $A \in N_{\mathrm{def}}^{\mathcal{T}}$, $L_{\mathcal{T}}(A)$ can be written as $\mathrm{def}_{\mathcal{T}}(A) \cap N_{\mathrm{prim}}^{\mathcal{T}}$. For primitive definitions, we define description graphs in the following way.

Definition 13. *Let $\mathcal{J} = (\Delta^{\mathcal{J}}, \cdot^{\mathcal{J}})$ be a primitive interpretation. Then the $\mathcal{EL}$-description graph $\mathcal{G}_{\mathcal{J}} = (\Delta^{\mathcal{J}}, E_{\mathcal{J}}, L_{\mathcal{J}})$ of $\mathcal{J}$ is defined as follows:*

- *the nodes of $\mathcal{G}_{\mathcal{J}}$ are the elements of $\Delta^{\mathcal{J}}$;*
- *$E_{\mathcal{J}} := \{(x, r, y) \mid (x, y) \in r^{\mathcal{J}}\}$;*
- *$L_{\mathcal{J}}(x) = \{P \in N_{\mathrm{prim}} \mid x \in P^{\mathcal{J}}\}$ for all $x \in \Delta^{\mathcal{J}}$.*

In preparation for the characterization of subsumption we need to introduce simulation relations on description graphs.

Definition 14. *Let $\mathcal{G}_i = (V_i, E_i, L_i)$, $i = 1, 2$, be two $\mathcal{EL}$-description graphs. The binary relation $Z \subseteq V_1 \times V_2$ is a simulation relation from $\mathcal{G}_1$ to $\mathcal{G}_2$ ($Z : \mathcal{G}_1 \rightleftharpoons \mathcal{G}_2$) iff*

(S1) $(v_1, v_2) \in Z$ implies $L_1(v_1) \subseteq L_2(v_2)$; and
(S2) if $(v_1, v_2) \in Z$ and $(v_1, r, v_1') \in E_1$ then there exists a node $v_2' \in V_2$ such that $(v_1', v_2') \in Z$ and $(v_2, r, v_2') \in E_2$.

It has been shown in [17] that simulation relations can be concatenated in the sense of the following lemma.

Lemma 1. *Let $\mathcal{G}_i := (V_i, E_i, L_i)$, $i = 1, 2, 3$, be $\mathcal{EL}$-description graphs, and let $Z_1 : \mathcal{G}_1 \rightleftharpoons \mathcal{G}_2$ and $Z_2 : \mathcal{G}_2 \rightleftharpoons \mathcal{G}_3$. Then $Z_1 \circ Z_2 : \mathcal{G}_1 \rightleftharpoons \mathcal{G}_3$, where*

$$Z_1 \circ Z_2 := \{(v, v'') \mid \exists v' \in V_2 : (v, v') \in Z_1 \wedge (v', v'') \in Z_2\}.$$

One of the main results in [17] is a characterization of gfp-subsumption by simulation relations over description graphs. The following results provide the relevant characterizations.

Proposition 2. *Let $\mathcal{T}$ be an $\mathcal{EL}$-TBox over N_{prim}, N_{role}, and N_{def} and $A \in N_{\mathrm{def}}^{\mathcal{T}}$. Let J be a primitive interpretation of N_{prim} and N_{role}. Then $x \in A^{\mathrm{gfp}(\mathcal{T}, \mathcal{J})}$ iff there is a simulation relation $Z : \mathcal{G}_{\mathcal{T}} \rightleftharpoons \mathcal{G}_{\mathcal{J}}$ such that $(A, x) \in Z$.*

Theorem 1. *Let $\mathcal{T}$ be an $\mathcal{EL}$-TBox and A, B be defined concepts in $\mathcal{T}$. Then $A \sqsubseteq_{\mathrm{gfp}, \mathcal{T}} B$ iff there is a simulation relation $Z : \mathcal{G}_{\mathcal{T}} \rightleftharpoons \mathcal{G}_{\mathcal{T}}$ such that $(B, A) \in Z$.*

Since the description graph of a TBox is of polynomial size in the size of the TBox and since the existence of simulation relations with the required properties can be tested in polynomial time, the following complexity result is obtained [17].

Corollary 2. *Subsumption w.r.t. gfp-semantics in $\mathcal{EL}$ can be decided in polynomial time.*

With this result, the prerequisites for the introduction of hybrid $\mathcal{EL}$-TBoxes are complete.

3   Hybrid $\mathcal{EL}$-TBoxes

In the present section, we start by defining syntax and semantics of hybrid $\mathcal{EL}$-TBoxes formally before showing in Section 3.2 how subsumption w.r.t. hybrid $\mathcal{EL}$-TBoxes can be decided in polynomial time.

3.1   Syntax and Semantics

The following definition introduces hybrid $\mathcal{EL}$-TBoxes, the central notion of the present paper.

Definition 15. *A general concept inclusion axiom (GCI) over N_{prim} and N_{role} is of the form $C \sqsubseteq D$, where C and D are $\mathcal{EL}$-concept descriptions over N_{prim} and N_{role}. A finite set of GCIs over N_{prim} and N_{role} is called a general $\mathcal{EL}$-TBox over N_{prim} and N_{role}. A primitive concept $P \in N_{\mathrm{prim}}$ (an existential restriction $\exists r.P$ with $r \in N_{\mathrm{role}}$) occurs in $\mathcal{T}$ iff there is a GCI $C \sqsubseteq D \in \mathcal{T}$ such that P ($\exists r.P$) is a conjunct of C or D.*

A hybrid $\mathcal{EL}$-TBox is a pair $(\mathcal{F}, \mathcal{T})$, where $\mathcal{F}$ is a general $\mathcal{EL}$-TBox over N_{prim} and N_{role} and $\mathcal{T}$ is an $\mathcal{EL}$-TBox over N_{prim}, N_{role}, and N_{def}.

Note that our general TBoxes are restricted 'over N_{prim} and N_{role}' to rule out the use of defined concepts from $\mathcal{T}$. Similar to the case of cyclic $\mathcal{EL}$-TBoxes, we introduce a normal form for hybrid $\mathcal{EL}$-TBoxes in order to simplify our solution for the respective subsumption problem.

Definition 16. *Let $(\mathcal{F}, \mathcal{T})$ be a hybrid TBox over N_{prim}, N_{role}, and N_{def}. Then, $(\mathcal{F}, \mathcal{T})$ is normalized iff*

1. *Every GCI in $\mathcal{F}$ is of one of the following forms: $A \sqsubseteq B$, $A_1 \sqcap A_2 \sqsubseteq B$, $A \sqsubseteq \exists r.B$, or $\exists r.A \sqsubseteq B$, where $r \in N_{\mathrm{role}}$ and $A, A_1, A_2, B \in N_{\mathrm{prim}} \cup \{\top\}$;*
2. *$\mathcal{T}$ is normalized in the sense of Definition 4; and*
3. *for every primitive concept P and for every existential restriction $\exists r.P$ occurring in $\mathcal{F}$, $\mathcal{T}$ contains a definition of the form $A_P \equiv P$ and $A_{\exists r.P} \equiv \exists r.A_P$, respectively.*

Note that the first two normalization conditions can be satisfied easily for any hybrid TBox $(\mathcal{F}, \mathcal{T})$, see [17]. For the third condition, a conservative extension $\mathcal{T}'$ of $\mathcal{T}$ of size at most the size of $(\mathcal{F}, \mathcal{T})$ can be found such that $(\mathcal{F}, \mathcal{T}')$ is normalized. All subsumption relations between concept names defined in $\mathcal{T}$ remain unchanged.

Example 1. In order to get an impression of how an actual hybrid TBox might look like, consider Figure 1. Shown is an extremely simplified part of a medical terminology[2] represented by a hybrid TBox $(\mathcal{F}, \mathcal{T})$. $\mathcal{T}$ is supposed to define the concepts 'disease of the connective tissue', 'bacterial infection' and 'bacterial pericarditis'. For instance, bacterial Pericarditis is defined as an inflammation

---

[2] Our example is only supposed to show the features of hybrid $\mathcal{EL}$-TBoxes and in no way claims to be adequate from a Medical KR perspective.

$\mathcal{T}$:
$$ConnTissDisease \equiv \mathsf{Disease} \sqcap \exists \mathsf{acts\_on.ConnTissue}$$
$$BactInfection \equiv \mathsf{Infection} \sqcap \exists \mathsf{causes.} BactPericarditis$$
$$BactPericarditis \equiv \mathsf{Inflammation} \sqcap \exists \mathsf{has\_loc.Pericardium}$$
$$\sqcap \exists \mathsf{caused\_by.} BactInfection$$

$\mathcal{F}$:
$$\mathsf{Disease} \sqcap \exists \mathsf{has\_loc.ConnTissue} \sqsubseteq \exists \mathsf{acts\_on.ConnTissue}$$
$$\mathsf{Inflammation} \sqsubseteq \mathsf{Disease}$$
$$\mathsf{Pericardium} \sqsubseteq \mathsf{ConnTissue}$$

Fig. 1. Example hybrid $\mathcal{EL}$-TBox

located in the Pericardium caused by a bacterial infection. Note that $\mathcal{T}$ is cyclic. For the primitive concepts in $\mathcal{T}$, the foundation $\mathcal{F}$ states, e.g., that a disease located in connective tissue acts on connective tissue.

The hybrid TBox $(\mathcal{F}, \mathcal{T})$ from Example 1 can be normalized in three steps. Firstly, the first GCI in $\mathcal{F}$ has to be normalized to, e.g.,

$$\exists \mathsf{has\_loc.ConnTissue} \sqsubseteq \mathsf{HasLocConnTissue}$$
$$\mathsf{ActsOnConnTissue} \sqsubseteq \exists \mathsf{acts\_on.ConnTissue}$$
$$\mathsf{Disease} \sqcap \mathsf{HasLocConnTissue} \sqsubseteq \mathsf{ActsOnConnTissue}.$$

Secondly, $\mathcal{T}$ has to be extended by a definition of the form $A_P \equiv P$ for the primitive concepts Disease, ConnTissue, Infection, Inflammation, Pericardium and also for HasLocConnTissue and ActsOnConnTissue. Thirdly, the primitive names ConnTissue and Pericardium occurring in $\mathcal{T}$ have to be replaced by $A_{\mathsf{ConnTissue}}$ and $A_{\mathsf{Pericardium}}$, respectively.

Normalization serves as an internal preprocessing step to classification and does not replace the original hybrid TBox from the perspective of the user of a DL system. Having introduced hybrid TBoxes formally, it remains to define an appropriate semantics for them.

Definition 17. *Let $(\mathcal{F}, \mathcal{T})$ be a hybrid TBox over N_{prim}, N_{role}, and N_{def}. A primitive interpretation $\mathcal{J}$ is a model of $\mathcal{F}$ ($\mathcal{J} \models \mathcal{F}$) iff $C^{\mathcal{J}} \subseteq D^{\mathcal{J}}$ for every GCI $C \sqsubseteq D$ in $\mathcal{F}$. A model $\mathcal{I} \in \mathrm{Int}(\mathcal{J})$ is a gfp-model of $(\mathcal{F}, \mathcal{T})$ iff $\mathcal{J} \models \mathcal{F}$ and $\mathcal{I}$ is a gfp-model of $\mathcal{T}$.*

Note that $\mathcal{F}$ ("foundation") is interpreted w.r.t. descriptive semantics while $\mathcal{T}$ ("terminology") is interpreted w.r.t. gfp-semantics. Note also that every gfp-model of $(\mathcal{F}, \mathcal{T})$ can be expressed as $\mathrm{gfp}(\mathcal{T}, \mathcal{J})$ for some primitive interpretation $\mathcal{J}$ with $\mathcal{J} \models \mathcal{F}$.

In order to complete the semantics of hybrid $\mathcal{EL}$-TBoxes, it remains to introduce an appropriate notion of subsumption.

Definition 18. *Let $(\mathcal{F}, \mathcal{T})$ be a hybrid $\mathcal{EL}$-TBox over N_{prim}, N_{role}, and N_{def}. Let A, B be defined concepts in $\mathcal{T}$. Then A is subsumed by B w.r.t. $(\mathcal{F}, \mathcal{T})$ ($A \sqsubseteq_{\mathrm{gfp}, \mathcal{F}, \mathcal{T}} B$) iff $A^{\mathcal{I}} \subseteq B^{\mathcal{I}}$ for all gfp-models $\mathcal{I}$ of $(\mathcal{F}, \mathcal{T})$.*

For Example 1, we shall see that the subsumption *BactPericarditis* $\sqsubseteq_{\mathrm{gfp},\mathcal{F},\mathcal{T}}$ *ConnTissDisease* holds, i.e., Pericarditis is classified as a disease of the connective tissue. How subsumption w.r.t. hybrid TBoxes can be decided in general is the subject of the following section.

Observe that hybrid TBoxes generalize cyclic TBoxes with gfp-semantics in the sense that every cyclic $\mathcal{EL}$-TBox $\mathcal{T}$ can be viewed as a hybrid TBox with an empty foundation. Thus, gfp-subsumption w.r.t. $\mathcal{T}$ coincides with subsumption w.r.t. the hybrid TBox $(\emptyset, \mathcal{T})$. Also note that, every general TBox $\mathcal{T}'$ can be seen as a hybrid TBox $(\mathcal{T}', \emptyset)$. In this case, a descriptive subsumption $P \sqsubseteq_{\mathcal{T}'} Q$ holds iff A_P is subsumed by A_Q w.r.t. the normalized instance of $(\mathcal{T}', \emptyset)$.

3.2 Deciding Subsumption w.r.t. Hybrid $\mathcal{EL}$-TBoxes

In this section we show that subsumption w.r.t hybrid $\mathcal{EL}$-TBoxes $(\mathcal{F}, \mathcal{T})$ can be reduced to subsumption w.r.t. cyclic $\mathcal{EL}$-TBoxes interpreted by gfp-semantics. The underlying idea is to use the *descriptive* subsumption relations induced by the GCIs in $\mathcal{F}$ to extend the definitions in $\mathcal{T}$ accordingly. To this end, we view the union of $\mathcal{F}$ and $\mathcal{T}$ as a general TBox and ask for all descriptive implications in $\mathcal{T}$ directly involving names from $\mathcal{F}$. These implications are then added to the definitions in $\mathcal{T}$. This notion is formalized as follows.

Definition 19. *Let $(\mathcal{F}, \mathcal{T})$ be a normalized hybrid $\mathcal{EL}$-TBox over N_{prim}, N_{role}, and N_{def}. For every $A \in N_{\mathrm{def}}^{\mathcal{T}}$, let*

$$f(A) := \bigsqcap_{P \in \{P' \in N_{\mathrm{prim}}^{\mathcal{F}} | A \sqsubseteq_{\mathcal{F} \cup \mathcal{T}} P'\}} P \sqcap \bigsqcap_{r \in N_{\mathrm{role}}^{\mathcal{F}}} \bigsqcap_{Q \in \{Q' \in N_{\mathrm{prim}}^{\mathcal{F}} | A \sqsubseteq_{\mathcal{F} \cup \mathcal{T}} \exists r.Q'\}} \exists r.A_Q \ .$$

The $\mathcal{F}$-completion $f(\mathcal{T})$ extends the definitions in $\mathcal{T}$ as follows.

$$f(\mathcal{T}) := \{A \equiv C \sqcap f(A) \mid A \equiv C \in \mathcal{T}\}$$

Note that $f(\mathcal{T})$ is still a normalized $\mathcal{EL}$-TBox. To preserve normalization, $f(A)$ adds $\exists r.A_Q$ instead of $\exists r.Q$ whenever A implies $\exists r.Q$.

Example 2. Consider the hybrid TBox $(\mathcal{F}, \mathcal{T})$ from Example 1 after normalization. Our goal is to compute the $\mathcal{F}$-completion of $\mathcal{T}$. To this end, for every defined concept in $\mathcal{T}$, we need to find all descriptive consequences of the form P and $\exists r.P$ implied by $\mathcal{F} \cup \mathcal{T}$, where $P \in N_{\mathrm{prim}}^{\mathcal{F}}$. Obviously, $A_{\mathsf{Inflammation}}$ implies Disease and $A_{\mathsf{Pericardium}}$ implies ConnTissue. Moreover, $A_{\mathsf{ActsOnConnTissue}}$ yields $\exists\mathsf{acts\_on}.\mathsf{ConnTissue}$. Finally, it is easy to check that *BactPericarditis* implies both Disease and HasLocConnTissue, and therefore also ActsOnConnTissue, yielding $\exists\mathsf{acts\_on}.\mathsf{ConnTissue}$.

Using these descriptive consequences, the completion $f(A)$ can be computed for every defined name A. Figure 2 shows the "interesting" part of the resulting description graph $\mathcal{G}_{f(\mathcal{T})}$ of the $\mathcal{F}$-completion $f(\mathcal{T})$, omitting some isolated vertices. Long concept names are abbreviated, i.e., the vertex A_{AOCT} stands for the concept $A_{\mathsf{ActsOnConnTissue}}$, A_{HLCT} for $A_{\mathsf{HasLocConnTissue}}$ and so on. For every vertex A, the label set $L_{f(\mathcal{T})}(A)$ is denoted above or underneath the relevant vertex. Underlined entries are descriptive consequences absent in the original TBox $\mathcal{T}$. As $L_{f(\mathcal{T})}(CTD) \subseteq L_{f(\mathcal{T})}(BP)$ and as BP has the same successor w.r.t. the edge acts_on, it is easy to check that there exists a simulation relation Z on $\mathcal{G}_{f(\mathcal{T})}$ with $(CTD, BP) \in Z$. Therefore, $BactPericarditis$ is subsumed by $ConnTissueDisease$ w.r.t. $f(\mathcal{T})$ interpreted with gfp-semantics.

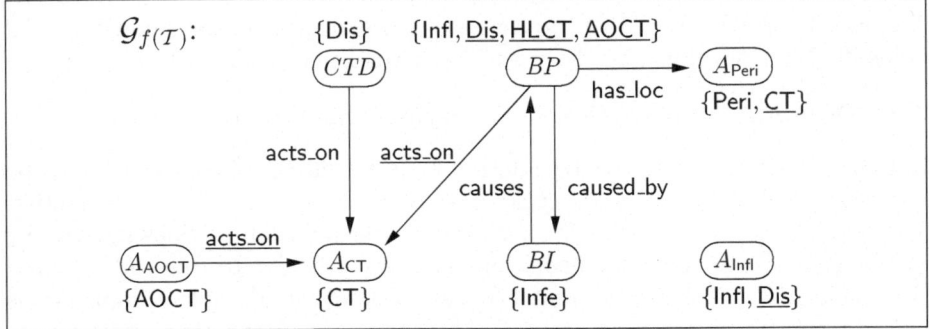

Fig. 2. Example $\mathcal{EL}$-description graph

Our goal now is to show for a given hybrid $\mathcal{EL}$-TBox $(\mathcal{F}, \mathcal{T})$ and arbitrary names A, B defined in $\mathcal{T}$ that B subsumes A w.r.t. $(\mathcal{F}, \mathcal{T})$ if and only if B subsumes A w.r.t. the $\mathcal{F}$-completion of $\mathcal{T}$ interpreted by gfp-semantics. To this end, we first show that $(\mathcal{F}, \mathcal{T})$ and the $\mathcal{F}$-completed hybrid TBox $(\mathcal{F}, f(\mathcal{T}))$ induce the same subsumption relations.

Lemma 2. *Let $(\mathcal{F}, \mathcal{T})$ be a normalized hybrid $\mathcal{EL}$-TBox over N_{prim}, N_{role}, and N_{def}. Let $A, B \in N_{\mathrm{def}}^{\mathcal{T}}$. Then, $A \sqsubseteq_{\mathrm{gfp},\mathcal{F},\mathcal{T}} B$ iff $A \sqsubseteq_{\mathrm{gfp},\mathcal{F},f(\mathcal{T})} B$.*

PROOF. We show that $\mathrm{gfp}(\mathcal{T}, \mathcal{J}) = \mathrm{gfp}(f(\mathcal{T}), \mathcal{J})$, implying for every primitive interpretation $\mathcal{J}$ with $\mathcal{J} \models \mathcal{F}$ that $A^{\mathrm{gfp}(\mathcal{T},\mathcal{J})} \subseteq B^{\mathrm{gfp}(\mathcal{T},\mathcal{J})}$ iff $A^{\mathrm{gfp}(f(\mathcal{T}),\mathcal{J})} \subseteq B^{\mathrm{gfp}(f(\mathcal{T}),\mathcal{J})}$, implying the proposition.

In order to show $\mathrm{gfp}(\mathcal{T}, \mathcal{J}) = \mathrm{gfp}(f(\mathcal{T}), \mathcal{J})$, it suffices to show that, firstly, every model $\mathcal{I} \in \mathrm{Int}(\mathcal{J})$ of $\mathcal{T}$ is also a model of $f(\mathcal{T})$, implying $\mathrm{gfp}(\mathcal{T}_f, \mathcal{J}) \succeq_{\mathcal{J}} \mathrm{gfp}(\mathcal{T}, \mathcal{J})$; and secondly, $\mathrm{gfp}(\mathcal{T}_f, \mathcal{J}) \preceq_{\mathcal{J}} \mathrm{gfp}(\mathcal{T}, \mathcal{J})$.

Consider some model $\mathcal{I} \in \mathrm{Int}(\mathcal{J})$ with $\mathcal{I} \models \mathcal{T}$ and an arbitrary $A \in N_{\mathrm{def}}^{\mathcal{T}}$. As $\mathcal{I} \models \mathcal{T}$, $A^{\mathcal{I}} = \mathrm{deft}_{\mathcal{T}}(A)^{\mathcal{I}}$. Since also $\mathcal{J} \models \mathcal{F}$, $\mathcal{I}$ respects all descriptive implications of $\mathcal{F}$. Hence, we have $A^{\mathcal{I}} \subseteq f(A)^{\mathcal{I}}$, implying $A^{\mathcal{I}} = \mathrm{deft}_{\mathcal{T}}(A)^{\mathcal{I}} \cap f(A)^{\mathcal{I}} = (\mathrm{deft}_{\mathcal{T}}(A) \sqcap f(A))^{\mathcal{I}}$. By definition of $f(\mathcal{T})$, this yields $\mathcal{I} \models f(\mathcal{T})$.

We show $\mathrm{gfp}(f(\mathcal{T}), \mathcal{J}) \preceq_{\mathcal{J}} \mathrm{gfp}(\mathcal{T}, \mathcal{J})$ by transfinite induction on the fixpoint iteration. By Corollary 1, there exists an ordinal α such that $\mathrm{gfp}(f(\mathcal{T}), \mathcal{J}) = \mathcal{I}_{f(\mathcal{T}),\mathcal{J}}^{\downarrow\alpha}$ and $\mathrm{gfp}(\mathcal{T}, \mathcal{J}) = \mathcal{I}_{\mathcal{T},\mathcal{J}}^{\downarrow\alpha}$. We distinguish the case of α being a successor or a limit ordinal.

(α successor ordinal). Induction base: if $\alpha = 0$ then $\mathcal{I}_{f(\mathcal{T}),\mathcal{J}}^{\downarrow\alpha} = \mathcal{I}^{\text{top}} = \mathcal{I}_{\mathcal{T},\mathcal{J}}^{\downarrow\alpha}$, implying $\mathcal{I}_{f(\mathcal{T}),\mathcal{J}}^{\downarrow 0} \preceq_{\mathcal{J}} \mathcal{I}_{\mathcal{T},\mathcal{J}}^{\downarrow 0}$. Induction step: for every $\beta < \alpha$, assume (IH) that $\mathcal{I}_{f(\mathcal{T}),\mathcal{J}}^{\downarrow\beta} \preceq_{\mathcal{J}} \mathcal{I}_{\mathcal{T},\mathcal{J}}^{\downarrow\beta}$. Consider an arbitrary $A \in N_{\text{def}}^{\mathcal{T}}$ defined in $\mathcal{T}$ by

$$A \equiv P_1 \sqcap \cdots \sqcap P_m \sqcap \exists r_1.B_1 \sqcap \ldots \exists r_\ell.B_\ell.$$

We have to show $A^{\mathcal{I}_{f(\mathcal{T}),\mathcal{J}}^{\downarrow\beta+1}} \subseteq A^{\mathcal{I}_{\mathcal{T},\mathcal{J}}^{\downarrow\beta+1}}$. The concept name A is interpreted by $\mathcal{I}_{f(\mathcal{T}),\mathcal{J}}^{\downarrow\beta+1}$ as

$$A^{\mathcal{I}_{f(\mathcal{T}),\mathcal{J}}^{\downarrow\beta+1}} = \bigcap_{1 \leq i \leq m} P_i^{\mathcal{J}} \cap \bigcap_{1 \leq j \leq \ell} (\exists r_j.B_j)^{\mathcal{I}_{f(\mathcal{T}),\mathcal{J}}^{\downarrow\beta}} \cap f(A)^{\mathcal{I}_{f(\mathcal{T}),\mathcal{J}}^{\downarrow\beta}}.$$

For the original TBox $\mathcal{T}$ we analogously have

$$A^{\mathcal{I}_{\mathcal{T},\mathcal{J}}^{\downarrow\beta+1}} = \bigcap_{1 \leq i \leq m} P_i^{\mathcal{J}} \cap \bigcap_{1 \leq j \leq \ell} (\exists r_j.B_j)^{\mathcal{I}_{\mathcal{T},\mathcal{J}}^{\downarrow\beta}}.$$

Hence, it suffices to show for every $r \in N_{\text{role}}^{\mathcal{T}}$ and every $B \in N_{\text{def}}^{\mathcal{T}}$ that the subset relation $(\exists r.B)^{\mathcal{I}_{f(\mathcal{T}),\mathcal{J}}^{\downarrow\beta}} \subseteq (\exists r.B)^{\mathcal{I}_{\mathcal{T},\mathcal{J}}^{\downarrow\beta}}$ holds. By (IH), $B^{\mathcal{I}_{f(\mathcal{T}),\mathcal{J}}^{\downarrow\beta}} \subseteq B^{\mathcal{I}_{\mathcal{T},\mathcal{J}}^{\downarrow\beta}}$. As $r^{\mathcal{I}_{f(\mathcal{T}),\mathcal{J}}^{\downarrow\beta}} = r^{\mathcal{J}} = r^{\mathcal{I}_{\mathcal{T},\mathcal{J}}^{\downarrow\beta}}$, the subset relation immediately carries over to the interpretations of $\exists r.B$.

(α *limit ordinal*). Assume (IH) that $\mathcal{I}_{f(\mathcal{T}),\mathcal{J}}^{\downarrow\beta} \preceq_{\mathcal{J}} \mathcal{I}_{\mathcal{T},\mathcal{J}}^{\downarrow\beta}$ for every $\beta < \alpha$. By definition, in the limit ordinal case it holds for every $B \in N_{\text{def}}^{\mathcal{T}}$ that $B^{\mathcal{I}_{f(\mathcal{T}),\mathcal{J}}^{\downarrow\alpha}}$ equals $\bigcap_{\beta < \alpha} B^{\mathcal{I}_{f(\mathcal{T}),\mathcal{J}}^{\downarrow\beta}}$ which due to (IH) is a subset of $\bigcap_{\beta < \alpha} B^{\mathcal{I}_{\mathcal{T},\mathcal{J}}^{\downarrow\beta}}$ which in turn equals $B^{\mathcal{I}_{\mathcal{T},\mathcal{J}}^{\downarrow\alpha}}$. ∎

Hence, the $\mathcal{F}$-completion $(\mathcal{F}, f(\mathcal{T}))$ of preserves the same subsumption relations as the original. The next lemma shows that, after $\mathcal{F}$-completing $\mathcal{T}$, we may 'forget' $\mathcal{F}$ and still obtain the same subsumptions.

Lemma 3. *Let $(\mathcal{F}, \mathcal{T})$ be a normalized hybrid $\mathcal{EL}$-TBox over N_{prim}, N_{role}, and N_{def}. Let $A, B \in N_{\text{def}}^{\mathcal{T}}$. Then, $A \sqsubseteq_{\text{gfp},\mathcal{F},f(\mathcal{T})} B$ iff $A \sqsubseteq_{\text{gfp},f(\mathcal{T})} B$*

PROOF. ($\Leftarrow$) trivial. ($\Rightarrow$) Assume $A \not\sqsubseteq_{\text{gfp},f(\mathcal{T})} B$. We construct a countermodel showing $A \not\sqsubseteq_{\text{gfp},\mathcal{F},f(\mathcal{T})} B$, i.e., a primitive interpretation $\mathcal{J}$ with $\mathcal{J} \models \mathcal{F}$ and $A^{\text{gfp}(f(\mathcal{T}),\mathcal{J})} \not\subseteq B^{\text{gfp}(f(\mathcal{T}),\mathcal{J})}$.

Define $\mathcal{J} =: (\Delta^{\mathcal{J}}, \cdot^{\mathcal{J}})$ as follows.

- $\Delta^{\mathcal{J}} := \{x_A \mid A \in N_{\text{def}}^{f(\mathcal{T})}\}$;
- $P^{\mathcal{J}} := \{x_A \in \Delta^{\mathcal{J}} \mid P \in \text{def}_{f(\mathcal{T})}(A)\}$ for all $P \in N_{\text{prim}}^{f(\mathcal{T})}$;
- $r^{\mathcal{J}} := \{(x_A, x_B) \in (\Delta^{\mathcal{J}})^2 \mid \exists r.B \in \text{def}_{f(\mathcal{T})}(A)\}$ for all $r \in N_{\text{role}}^{f(\mathcal{T})}$.

We first show $x_A \in A^{\text{gfp}(f(\mathcal{T}),\mathcal{J})} \setminus B^{\text{gfp}(f(\mathcal{T}),\mathcal{J})}$. By Proposition 2 it suffices to find a simulation relation $Z: \mathcal{G}_{f(\mathcal{T})} \rightleftarrows \mathcal{G}_{\mathcal{J}}$ with $(A, x_a) \in Z$. Define $Z := \{(A, x_A) \mid$

$A \in N_{\text{def}}^{f(\mathcal{T})}\}$. As obviously $(A, x_A) \in Z$, it remains to show that Z respects Definition 14. (S1) For every $(A, x_A) \in Z$, $L_{\mathcal{G}_{f(\mathcal{T})}}(A)$ equals $\text{def}_{f(\mathcal{T})}(A) \cap N_{\text{prim}}^{f(\mathcal{T})}$ which equals $\{P \in N_{\text{prim}}^{f(\mathcal{T})} \mid P \in \text{def}_{f(\mathcal{T})}(A)\} = L_{\mathcal{G}_{\mathcal{J}}}(A)$. (S2) If $(A, x_A) \in Z$ and $(A, r, B) \in E_{\mathcal{G}_{f(\mathcal{T})}}$ then $\exists r.B \in \text{def}_{f(\mathcal{T})}(A)$, implying $(x_A, x_B) \in r^{\mathcal{J}}$, implying $(x_A, r, x_B) \in E_{\mathcal{G}_{\mathcal{J}}}$. Moreover, $(B, x_B) \in Z$. Hence, by (S1) and (S2), $Z \colon \mathcal{G}_{f(\mathcal{T})} \rightleftarrows \mathcal{G}_{\mathcal{J}}$.

Observe that under (S1) we proved *equality* of $L_{\mathcal{G}_{f(\mathcal{T})}}(A)$ and $L_{\mathcal{G}_{\mathcal{J}}}(A)$. Moreover, (S2) also holds in the direction from $\mathcal{G}_{\mathcal{J}}$ to $\mathcal{G}_{f(\mathcal{T})}$: whenever $(A, x_A) \in Z$ and $(x_A, r, x_B) \in E_{\mathcal{G}_{\mathcal{J}}}$ then $(A, r, B) \in E_{\mathcal{G}_{f(\mathcal{T})}}$. Hence, $Z^{-1} \colon \mathcal{G}_{\mathcal{J}} \rightleftarrows \mathcal{G}_{f(\mathcal{T})}$.

Assume $x_A \in B^{\text{gfp}(f(\mathcal{T}), \mathcal{J})}$. Then, by Proposition 2, there is a simulation relation $Y \colon \mathcal{G}_{f(\mathcal{T})} \rightleftarrows \mathcal{G}_{\mathcal{J}}$ with $(B, x_A) \in Y$. But then, by Lemma 1, $Y \circ Z^{-1}$ is a simulation relation on $\mathcal{G}_{f(\mathcal{T})}$ with $(B, A) \in Y \circ Z^{-1}$, implying $A \sqsubseteq_{\text{gfp}, f(\mathcal{T})} B$, in contradiction to the assumption. It remains to show that $\mathcal{J} \models \mathcal{F}$. As $\mathcal{F}$ is normalized, we have four types of GCIs in $\mathcal{F}$.

1. $P \sqsubseteq Q \in \mathcal{F}$. If $x_A \in P^{\mathcal{J}}$ then $P \in \text{def}_{f(\mathcal{T})}(A)$, implying $A \sqsubseteq_{\mathcal{F} \cup \mathcal{T}} Q$ which implies $f(A) \sqsubseteq Q$. Hence, $Q \in \text{def}_{f(\mathcal{T})}(A)$, implying $x_A \in Q^{\mathcal{J}}$.
2. $P_1 \sqcap P_2 \sqsubseteq Q \in \mathcal{F}$. If $x_A \in P_1^{\mathcal{J}} \cap P_2^{\mathcal{J}}$ then $P_1, P_2 \in \text{def}_{f(\mathcal{T})}(A)$. This implies $A \sqsubseteq_{\mathcal{F} \cup \mathcal{T}} Q$ which analogously yields $x_A \in Q^{\mathcal{J}}$.
3. $P \sqsubseteq \exists r.Q \in \mathcal{F}$. If $x_A \in P^{\mathcal{J}}$ then $P \in \text{def}_{f(\mathcal{T})}(A)$, implying $A \sqsubseteq_{\mathcal{F} \cup \mathcal{T}} \exists r.Q$. Hence, $\exists r.A_Q \in \text{def}_{f(\mathcal{T})}(A)$, implying $(x_A, r, x_{A_{\exists r.Q}}) \in r^{\mathcal{J}}$. By definition, $Q \in \text{def}_{f(\mathcal{T})}(A_{\exists r.Q})$, implying $x_{A_{\exists r.Q}} \in Q^{\mathcal{J}}$.
4. $\exists r.Q \sqsubseteq P \in \mathcal{F}$. If $x_A \in (\exists r.Q)^{\mathcal{J}}$ then there exists some $x_B \in \Delta^{\mathcal{J}}$ such that $(x_A, r, x_B) \in r^{\mathcal{J}}$ and $x_B \in Q^{\mathcal{J}}$. Hence, $\exists r.B \in \text{def}_{f(\mathcal{T})}(A)$ and $Q \in \text{def}_{f(\mathcal{T})}(B)$. This implies $A \sqsubseteq_{\mathcal{F} \cup \mathcal{T}} P$, implying $P \in \text{def}_{f(\mathcal{T})}(A)$ which yields $x_A \in P^{\mathcal{J}}$. ∎

As an immediate consequence of Lemmas 2 and 3, we obtain the following theorem summarizing our reduction from hybrid $\mathcal{EL}$-TBoxes to cyclic $\mathcal{EL}$-TBoxes.

Theorem 2. *Let $(\mathcal{F}, \mathcal{T})$ be a hybrid $\mathcal{EL}$-TBox over N_{prim}, N_{role}, and N_{def}. Let $A, B \in N_{\text{def}}^{\mathcal{T}}$. Then, $A \sqsubseteq_{\text{gfp}, \mathcal{F}, \mathcal{T}} B$ iff $A \sqsubseteq_{\text{gfp}, f(\mathcal{T})} B$.*

It remains to show that subsumption w.r.t. hybrid TBoxes can be decided in polynomial time.

Corollary 3. *Subsumption w.r.t. hybrid $\mathcal{EL}$-TBoxes can be decided in polynomial time.*

PROOF. By Corollary 2, gfp-subsumption w.r.t. cyclic $\mathcal{EL}$-TBoxes can be decided in polynomial time. Hence, given $(\mathcal{F}, \mathcal{T})$, it suffices to show for every $A \in N_{\text{def}}^{\mathcal{T}}$ that $f(A)$ is of polynomial size in the size of $(\mathcal{F}, \mathcal{T})$ and can be computed in polynomial time.

By definition, every concept description $f(A)$ contains only conjuncts of the form P and $\exists r.A_P$ with $P \in N_{\text{prim}}^{\mathcal{T}}$ occurring in $\mathcal{F}$. The size of $f(A)$ is therefore linear in the size of $(\mathcal{F}, \mathcal{T})$. It has been shown in [19], that subsumption w.r.t. general $\mathcal{EL}$-TBoxes can be decided in polynomial time, implying that subsumption w.r.t. $\mathcal{F} \cup \mathcal{T}$ is polynomial. ∎

4   Conclusion

Motivated by the goal to make non-standard inference services available to DL systems supporting general TBoxes, the present paper has introduced hybrid $\mathcal{EL}$-TBoxes in which a general $\mathcal{EL}$-TBox $\mathcal{F}$ provides the foundation for a cyclic $\mathcal{EL}$-TBox $\mathcal{T}$ that uses names from $\mathcal{F}$ as primitive concepts. The reduction from Section 3.2 shows that hybrid $\mathcal{EL}$-TBoxes do not extend the expressive power of cyclic $\mathcal{EL}$-TBoxes with gfp-semantics. However, the explicit separation between definitions and implications valid for *all* definitions often leads to smaller and more readable knowledge bases.

The reduction from Section 3.2 also makes non-standard inferences accessible to hybrid TBoxes. It has been shown in [16] that, w.r.t. cyclic $\mathcal{EL}$-TBoxes interpreted by gfp-semantics, the lcs and msc can be computed in polynomial time. A DL system based on hybrid $\mathcal{EL}$-TBoxes could therefore compute the lcs or msc by first (internally) applying the above reduction to the relevant subset of the TBox and then computing the lcs or msc in the way defined in [16].

The technical motivation for choosing $\mathcal{EL}$ as the underlying DL for hybrid TBoxes is that we obtain a polynomial time subsumption problem and can utilize the non-standard inferences known for cyclic $\mathcal{EL}$-TBoxes with gfp-semantics. Our choice, however, is also motivated by applications of $\mathcal{EL}$-TBoxes in the life sciences. For instance, the widely used medical terminology SNOMED [20] corresponds to an $\mathcal{EL}$-Tbox [21]. Similarly, the Gene Ontology [22] can be represented by an $\mathcal{EL}$-TBox with transitive roles, and large parts of the medical knowledge base GALEN [23] can be expressed by a general $\mathcal{EL}$-TBox with transitive roles.

The above applications give rise to the question whether the polynomiality result for subsumption also holds for hybrid TBoxes defined over extensions of $\mathcal{EL}$. An interesting construct to add might be restricted role value maps (RVMs) of the form $r \circ s \sqsubseteq t$ by which, e.g., transitive roles can be defined. Due to positive results for cyclic $\mathcal{EL}$-TBoxes with gfp-semantics [16] and general $\mathcal{EL}$-TBoxes [24], we strongly conjecture that hybrid $\mathcal{EL}$-TBoxes with restricted RVMs can also be classified in polynomial time. For this extension, however, lcs and msc are not yet available. Extending $\mathcal{EL}$ by the bottom concept ($\bot$) would allow to express disjointness constraints of the form $P \sqcap Q \sqsubseteq \bot$ defining P and Q as mutually exclusive concepts.

References

1. Nardi, D., Brachmann, R.: An introduction to description logics. In: The Description Logic Handbook: Theory, Implementation, and Applications. Cambridge University Press (2003) 1–40
2. Rector, A., Nowlan, W., Glowinski, A.: Goals for concept representation in the GALEN project. In: Proc. of SCAMC, Washington (1993) 414–418
3. Rector, A.: Medical informatics. In: The Description Logic Handbook: Theory, Implementation, and Applications. Cambridge University Press (2003) 406–426
4. Horrocks, I., Rector, A.L., Goble, C.A.: A description logic based schema for the classification of medical data. In: Proc. of KRDB'96, CEUR-WS (1996)

5. Horrocks, I.R.: Using an expressive description logic: FaCT or fiction? In: Proc. of KR'98. Morgan Kaufmann Publishers (1998) 636–645
6. Haarslev, V., Möller, R.: RACER system description. Lecture Notes in Computer Science **2083** (2001) 701–712
7. Nebel, B.: Terminological cycles: Semantics and computational properties. In: Principles of Semantic Networks: Explorations in the Representation of Knowledge. Morgan Kaufmann Publishers, San Mateo (CA), USA (1991) 331–361
8. Cohen, W.W., Hirsh, H.: Learning the classic description logic: Theoretical and experimental results. In: Proc. of KR'94, Morgan Kaufmann Publishers (1994) 121–133
9. Frazier, M., Pitt, L.: CLASSIC learning. Machine Learning **25** (1996) 151–193
10. Baader, F., Küsters, R.: Computing the least common subsumer and the most specific concept in the presence of cyclic $\mathcal{ALN}$-concept descriptions. In: Proc. of KI-98. Volume 1504 of Lecture Notes in Computer Science, Bremen, Germany, Springer–Verlag (1998) 129–140
11. McGuinness, D.: Explaining Reasoning in Description Logics. Ph.D. dissertation, Department of Computer Science, Rutgers University, New Brunswick, New Jersey (1996)
12. Borgida, A., McGuinness, D.L.: Asking queries about frames. In: Proc. of KR'96 Morgan Kaufmann Publishers, San Francisco, California (1996) 340–349
13. Baader, F., Küsters, R., Borgida, A., McGuinness, D.: Matching in description logics. Journal of Logic and Computation **9** (1999) 411–447
14. Baader, F., Küsters, R., Molitor, R.: Computing least common subsumers in description logics with existential restrictions. In: Proc. of IJCAI-99, Morgan Kaufmann Publishers (1999) 96–101
15. Brandt, S., Turhan, A.-Y.: Using non-standard inferences in description logics— what does it buy me? In: Proc. of KI-2001 Workshop on Applications of Description Logics (KIDLWS'01), Number 44 in CEUR-WS (2001)
16. Baader, F.: Least common subsumers and most specific concepts in a description logic with existential restrictions and terminological cycles. In Proc. of IJCAI-03, Morgan Kaufmann Publishers (2003) 319–324
17. Baader, F.: Terminological cycles in a description logic with existential restrictions. In: Proc. of IJCAI-03, Morgan Kaufmann Publishers (2003) 325–330
18. Tarski, A.: A lattice-theoretic fixpoint theorem and its applications. Pacific Journal of Mathematics **5** (1955) 285–309
19. Brandt, S.: Polynomial time reasoning in a description logic with existential restrictions, GCI axioms, and—what else? In Proc. of ECAI-2004, IOS Press (2004) 298–302
20. Cote, R., Rothwell, D., Palotay, J., Beckett, R., Brochu, L.: The systematized nomenclature of human and veterinary medicine. Technical report, SNOMED International, Northfield, IL (1993)
21. Spackman, K.: Normal forms for description logic expressions of clinical concepts in SNOMED RT. Journal of the American Medical Informatics Association (2001)
22. Consortium, T.G.O.: Gene Ontology: Tool for the unification of biology. Nature Genetics **25** (2000) 25–29
23. Rector, A., Bechhofer, S., Goble, C.A., Horrocks, I., Nowlan, W.A., Solomon, W.D.: The GRAIL concept modelling language for medical terminology. Artificial Intelligence in Medicine **9** (1997) 139–171
24. Baader, F., Brandt, S., Lutz, C.: Pushing the $\mathcal{EL}$ envelope. In: Proc. of IJCAI-05, Edinburgh, UK, Morgan Kaufmann Publishers (2005). To appear.

Dependency Calculus:
Reasoning in a General Point Relation Algebra

Marco Ragni and Alexander Scivos

Institut für Informatik, Albert-Ludwigs-Universität Freiburg,
Georges-Köhler-Allee 52, 79110 Freiburg, Germany
{ragni, scivos}@informatik.uni-freiburg.de

Abstract. Reasoning about complex dependencies between events is a crucial task. However, qualitative reasoning has so far concentrated on spatial and temporal issues. In contrast, we present a new dependency calculus (DC) that is created for specific questions of reasoning about causal relations and consequences. Applications in the field of spatial representation and reasoning are, for instance, modeling traffic networks, ecological systems, medical diagnostics, and Bayesian Networks. Several extensions of the fundamental linear point algebra have been investigated, for instance on trees or on nonlinear structures. DC is an improved generalization that meets all requirements to describe dependencies on networks. We investigate this structure with respect to satisfiability problems, construction problems, tractable subclassses, and embeddings into other relation algebras. Finally, we analyze the associated interval algebra on network structures.

1 Introduction

Reasoning about complex dependencies between events is a crucial task in many applications when decisions need to be made. Whenever the required answer is a decision or classification, Qualitative Reasoning (QR) is best-suited: It abstracts from metrical details of the physical world and enables computers to make predictions about relations, even when precise quantitative information is not available or irrelevant [5]. QR is an abstraction that summarizes similar quantitative states into one qualitative characterization. From the cognitive perspective, the qualitative method *categorizes* features within the object domain rather than by *measuring* them in terms of some external scale [7]. This is the reason why qualitative descriptions are quite natural for humans.

The two main directions in QR so far are spatial and temporal reasoning. In terms of spatial reasoning, topological reasoning about regions [11], positional reasoning about point configurations [7], and reasoning about directions can be distinguished. For temporal reasoning, either points [14] or intervals [1] are used as basic entities.

In contrast, we present a calculus that is created for a new direction in QR: specific questions of reasoning about causal relations and consequences. Our approach is purely qualitative, hence in comparison with Bayesian networks, no time for computing the probabilities is needed. Nevertheless, dependencies can reliably be derived. We show how known relation algebras can be refined to achieve this goal.

The linear point algebra PA_{lin} introduced by Vilain [14] is one of the most prominent and fundamental formalisms in the domain of qualitative spatial and temporal rea-

U. Furbach (Ed.): KI 2005, LNAI 3698, pp. 49–63, 2005.

soning. Its basic relations are $\{\prec, =, \succ\}$. Many widely used algebras, like Allen's interval algebra [1] or the Cardinal Directions calculus, can be constructed by PA_{lin} relations. The general satisfiability problem for PA_{lin} is in P [14], but for the associated interval algebra IA_{lin}, it is NP-complete. NP-hard problems usually have interesting fragments - so called tractable subclasses. A subclass of a relational algebra is a subset of relations closed under composition and converse. A subclass is called tractable if satisfiability can be decided in polynomial time. Normally, for these classes the path-consistency method decides satisfiability. Nebel et al. [10] identified all maximal tractable subclasses for IA_{lin}.

However, real-world problems have not necessarily linear structures as underlying space. Starting in the nineties, there have been extensions of the linear case into treelike and nonlinear structures [2], [3] to address such problems. The point algebra PA_{br} for treelike, often called branching structures, consists of the basic relations $\{\prec, =, \succ, \|\}$, whereby $\|$ states that two points are on different branches. Contrary to the point algebra of linear time, satisfiability testing for branching time is NP-hard. Moreover, the associated satisfiability problem for PA_{br} is NP-complete. Broxvall [3] identified five maximal tractable subsets of PA_{br}. The most general known point algebra, up to now, consists of the basic relations $\{\prec, =, \succ, \|\}$ which is the point algebra on nonlinear structures PA_{po}. Broxvall showed that the satisfiability problem is in the same class as for branching structures and identified three maximal tractable fragments.

But, these approaches are still too coarse for some applications. For instance, to identify dependencies, it is not only important to state that two points are unrelated, but also to qualify if two points or states have a common ancestor or if there is no such 'decision' point. For this reason, we define a relation algebra qualifying decision points and get a proper generalization of Broxvall's nonlinear relation algebra [3]. This new algebra, called dependency calculus (DC), meets all requirements to describe dependencies in networks. This is demonstrated by applications in the field of spatial representation and reasoning. There are two aspects: dependencies of points are described by the point algebra PA_{dc}, and based on this we define the associated interval algebra IA_{dc}.

Analyzing the composition tables shows that the dependency calculus for point structures PA_{dc} is a proper generalization of PA_{po}, which leads to a more precise reasoning. For that reason, it would be also useful to use the new algebra PA_{dc} instead of using PA_{po} in nonlinear structures, if PA_{dc} is still in the same complexity class. This shows the need to analyze the complexity of the depencency calculus.

One main concept is that it is not only important to analyze questions concerning satisfiability problems, but also to show the correspondence between DC and other relation algebras. To be more precise: Is there a relation algebra which is so similar to this dependency calculus that there is an isomorphism between these two algebras? An isomorphism with the property to preserve the tractability of subclasses? The method of identifying isomorphisms, or at least homomorphisms between calculi seems to be a promising idea to structure the field of relation algebras by the mathematical concept of homomorphism. Such mappings transfer algebraic aspects and complexity results from one algebra to another.

This paper is organized as follows: In Section 2, we present application areas and pose some important questions. In Section 3, we review the theory of partial orders and sketch known results concerning the point algebra on different structures. We present the set PA_{dc} of extensions of basic relations in our general framework. Then, the concept of correspondence of relation algebras is outlined. In a next step, we investigate the computational complexity of constraint satisfaction problems of PA_{dc}. In particular, we show that the satisfiability problem is **NP**-complete and all maximal tractable subclasses of PA_{dc} are presented by an algebraic correspondence to **RCC-5**. In Section 4, we introduce the associated interval algebra IA_{dc} on directed acyclic graphs, which allows to distinguish more constellations than classical interval algebras (even on nonlinear structures). We identify tractable subclasses and prove **NP**-completeness of the satisfiability problem for IA_{dc}. Finally, Section 5 summarizes our results and raises questions that are left open.

2   Application Areas

Various tasks in management and science require a good understanding of complex dependencies, for instance monitoring the distribution of pollution in ecological networks, identifying delivery bottlenecks in a supply chain network, inhibiting the spreading of deseases, or minimizing delays in a railway system.

For instance, if we observe pollution in an ecosystem of flowing water, we can draw conclusions about pollution at other points. Regard the flow network shown in Fig. 1. In

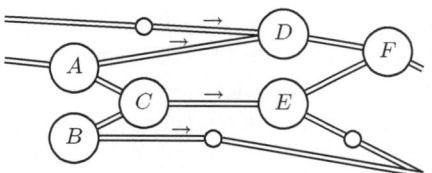

Fig. 1. A pipe network. Flow occurs along the "pipes" from left to right. Is there a difference between the pairs (A, B) and (D, E)?

this example, if pollution is found at point D, point F will be polluted as well. It might be caused from a source at point A. Points B, C, and E could not cause the pollution at point D. Nevertheless, there is a connection between C and D: Knowing that C is polluted, it is likely that D is polluted by the same substance (and vice versa). Does the same hold true for B and D? It does not because they have no common point upstream. We present a calculus that directly represents such differences. Therefore, it is vital to have a new relation $\curlywedge$ which we call "fork" for pairs like (C, D) or (D, E) that share a common ancestor. The set of basic relations "fork" ($\curlywedge$), "before" ($\prec$), "equal" ($=$), and "after" ($\succ$), we call "dependent". The only other case, like (A, B), we call "independent" ($\asymp$). If not all dependencies are known, some relations are not completely specified. Such imprecise knowledge is described by unions of these basic relations.

Dependencies of probabilities (when observations depend on each other, like in the water flow example) are often described by a Bayesian network. The answer whether two random variables are independent is based on the structure of the network. Can we

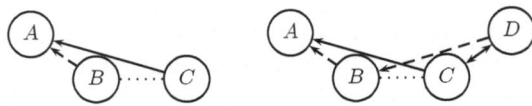

Fig. 2. A virus transmittance scheme. Arrows indicate assured, dashed ones unassured donorship, dotted lines mean that both persons carry the same virus. No lines means that we do not have prior knowledge. The situation on the left must be incomplete. The existence of a fourth person D accounts for this. PA_{dc} concludes that indeed there was indirect transmittance from D to A.

automatically deduce from a given set of direct dependencies, which pairs are "independent" and which are "dependent"? For instance, in Fig. 1, observations at A and B are independent, but E and D are not.

Another important type of task is network design, that is finding a feasible partial order for a given system of nodes under certain order contraints. For example, in a supply network, constraints could be that a warehouse has (indirectly) to be delivered from a specific production site, and that two warehouses do not have a common "bottleneck" site delivering both. Other examples are train scheduling (which connecting trains should wait for a delayed train, which not, to avoid critical dependencies?), automated planning, project organization and program verification. The task is to find a cycle-free order of the procedures such that one procedure delivers its result before the next routine is started. How time-consuming is the task of finding such an order?

In all these cases, the order may remain partly unspecified between elements for which the order does not matter. In other cases, the specific order (dependency) is vital, and it is necessary to deduce the hidden truth. For example, when tracking the spreading of a contagious disease, it is not always clear who was the donor, or if there was contact at all. Often, reasoning based on the known dependencies restricts the possibilities in cases of uncertainty or helps to detect formerly unknown causes (as in the example given in Fig. 2). These examples show that reasoning about dependencies, even with uncertainty, is highly important. In all these problems it matters if two nodes in the dependency graph have a common ancestor or not. In order to do automated reasoning, a calculus is needed that expresses this relation between a pair of nodes. This helps to distinguish between possibly affected spots and those that are unaffected. In which cases do efficient algorithms exist to detect causes and implications, to discover dependencies, and to find an order compliant with a given specification?

We present and investigate a calculus that suits to represent and deduce knowledge about dependencies by extending the language of partial ordering. This calculus is useful in various applications dealing with reasoning about spatial, temporal, spatio-temporal, topological, competitive, or causal relations.

3   The Dependency Calculus and Partial Orders

3.1   Formal Definition of PA_{dc}

To begin with the formalization of these concepts, let us recall the definition of a partial order: A partial order is a relation $\leq \subset A \times A$ that satisfies the following three properties for any $a, b, c \in A$:

i. Reflexivity: $a \leq a$.
ii. Antisymmetry: If $a \leq b$ and $b \leq a$, then $a = b$.
iii. Transitivity: If $a \leq b$ and $b \leq c$, then $a \leq c$.

A total order is a partial order that satisfies a fourth property:

iv. Comparability: For any $a, b \in A$, either $a \leq b$ or $b \leq a$.

PA_{dc} is based on the notion of relations between pairs of variables interpreted as elements of a partial order. We consider five basic relations, which we denote by $\prec, =, \succ, \curlywedge, \times$. If x, y are points in a partial order $\langle T, \leq \rangle$, then we define these relations in terms of the partial order as follows:

$$x \prec y \text{ iff } x \leq y \text{ and not } y \leq x.$$
$$x = y \text{ iff } x \leq y \text{ and } y \leq x.$$
$$x \succ y \text{ iff } y \leq x \text{ and not } x \leq y.$$
$$x \curlywedge y \text{ iff } \exists z\, z \leq y \wedge z \leq x \text{ and neither } x \leq y \text{ nor } y \leq x.$$
$$x \times y \text{ iff neither } \exists z\, z \leq y \wedge z \leq x \text{ nor } x \leq y \text{ nor } y \leq x.$$

Semantically, a constraint $v_1 R v_2$ holds in a partial order $(T, \leq)$ that complies with this definition. All relations between nodes in Fig. 1 can be described by these five basic relations. For describing trees, the four relations $\{\prec, =, \succ, \curlywedge\}$ are sufficient. Therefore, whenever the relation $\times$ occurs, the graph cannot be a tree.

A *boolean algebra* contains all unions $\cup$, intersections $\cap$, and complements $^-$ of a set of basic relations. If a boolean algebra is closed under additional operations composition $\circ$ and inverse $^{-1}$, it is called *relation algebra*, whereby

$$x \quad R^{-1} \quad y \quad \text{holds iff} \quad y\, R\, x,$$
$$x\, (R_1 \circ R_2)\, y \quad \text{holds iff} \quad \exists z \quad x\, R_1\, z \text{ and } z\, R_2\, y.$$

PA_{dc} is the concrete binary finite relation algebra [8] generated by $\{\prec, =, \succ, \curlywedge, \times\}$. Furthermore, given an atomic relation algebra A with finite atom set $B(A)$ (i.e. $B(A)$ is the set of all basic relations), each relation $r \in A$ can be written in a unique manner as a union of basic relations $b_1, \ldots, b_n$. Algebraic functions such as composition, converse, intersection, union, and complement, can be computed from basic relations:

$$(b_1 \cup \cdots \cup b_t) \circ (b_1' \cup \cdots \cup b_l') = \bigcup_{1 \leq i \leq t, 1 \leq j \leq l} (b_i \circ b_j') \tag{1}$$

$$(b_1 \cup \cdots \cup b_l)^{-1} = (b_1^{-1} \cup \cdots \cup b_l^{-1}) \tag{2}$$

Hence, it is sufficient to analyze the basic relations. For $B(PA_{dc})$, the composition operator is defined in Table 1. Composition tables are important for constraint based reasoning. For instance, the path-consistency algorithm [9], which can be used to identify some inconsistent networks, uses the composition table. PA_{dc} is the relation algebra generated by the 5 basic relations $\prec, =, \succ, \curlywedge, \times$ and the associated composition as shown by Table 1. The universal relation $\{\prec, =, \succ, \times, \curlywedge\}$ is denoted by $\top$.

Table 1. The composition table of PA_{dc}

∘	≺	=	≻	≍	λ
≺	≺	≺	⊤	≍	≺, λ, ≍
=	≺	=	≻	≍	λ
≻	≺, =, ≻, λ	≻	≻	≻, ≍, λ	≻, λ
≍	≺, ≍, λ	≍	≍	⊤	≺, ≍, λ
λ	≺, λ	λ	≻, ≍, λ	≻, ≍, λ	⊤

3.2 Computational Complexity

Assume that a set of constraints between some points is given. One question might be whether this set is consistent: Is it possible to construct a network in which all the constraints are satisfied? And, what is the computational effort for constructing it?

Definition 1. *Let $\mathcal{R} \subseteq PA_{dc}$ be a set of point relations and $\mathcal{C}$ a class of partial orders. A constraint system is a directed multigraph $\Pi = (V, E)$, where the nodes in V are point variables and $E \subseteq V \times \mathcal{R} \times V$ denotes the constraints imposed on the variables. A tuple $(f, (T, \leq))$ where $f : V \rightarrow T$ is a total function and $(T, \leq) \in \mathcal{C}$ is called an interpretation of Π. It is satisfiable iff there exists an interpretation $\mathcal{M} = (f, (T, \leq))$ such that $f(u) \, R \, f(v)$ holds for every $(u, R, v) \in E$. $\mathcal{M}$ is called a model of Π. There are two types of problems for such a given constraint system:*

   i. The satisfiability problem $SAT_{\mathcal{C}}(\mathcal{R})$: Is Π satisfiable?
   ii. The network design problem $NDP_{\mathcal{C}}(\mathcal{R})$: Find a model of Π.

The size of a problem instance (V, E) is $|V| + |E|$.

In an interpretation, a partial order is chosen such that each variable V is assigned to an element of the partial order. But this assignment is not required to be surjective: Not all elements of the partial order must correspond to a variable.

This definition of a satisfiability problem is a generalization of the classical definition of a *Constraint Satisfaction Problem* (**CSP**): Given a description consisting of

  – a set V of n variables $\{v_1, ..., v_n\}$,
  – the possible values D_i of variables v_i,
  – constraints (sets of relations) over subsets of variables,

Is it possible to find an assignment $v_i \mapsto D_i$ satisfying all constraints?

In our case, the possible values D_i are not fixed. We only know that the possible values D_i are part of a member in the class of partial orders $\mathcal{P}$. A CSP is a satisfiability problem with a class that consists just of one given partial order $(T, \leq)$.

Lemma 1. *For a class $\mathcal{C}$ of partial orders and set of relations $\mathcal{R}$,*
a. if $SAT_{\mathcal{C}}(\mathcal{R})$ is in **P**, *the same holds for $NDP_{\mathcal{C}}(\mathcal{R})$, and vice versa.*
b. if $SAT_{\mathcal{C}}(\mathcal{R})$ is **NP**-*complete, $NDP_{\mathcal{C}}(\mathcal{R})$ is* **NP**-*complete, too.*

Proof. Obviously, $NDP_{\mathcal{C}}(\mathcal{R})$ is at least as difficult as $SAT_{\mathcal{C}}(\mathcal{R})$.
For a., assume that $SAT_{\mathcal{C}}(\mathcal{R})$ is in **P**. A **PTIME** algorithm for $NDP_{\mathcal{C}}(\mathcal{R})$ is given by:

```
for  c = (x₁, R, x₂) ∈ E:                         \\ choose a  constraint
    set  possible [c] ← false;
    for  b ∈ R:                                    \\ choose a basic  relation
        set  c ← (x₁, {b}, x₂);                    \\ fix  basic  relation  b
        if  satisfiable (new constraint system):   \\ test  is  in  PTIME, by assumption
            set  possible [c] ← true; exit inner loop;
    endfor;                                         \\ try next basic  relation
    if  possible [c] = false:
        return NOT SATISFIABLE;                     \\ no feasible  basic  relation found
endfor;                                             \\ take  next  constraint
return SATISFIABLE.                                 \\ as  all   constraints  are  possible
```

For b., a scenario can be guessed, i.e. a jointly satisfiable set of basic relations. Therefore, by assumption, a model can be guessed and verified in polynomial time.

It is easy to transform a directed acyclic graph into a partial order by taking the transitive and reflexive closure. Let $\mathcal{DAG}$ be the class of partial orders obtained by such transformations. What can we expect in terms of complexity for $SAT_{\mathcal{DAG}}(PA_{dc})$? We call this the "general" case and write $SAT(PA_{dc})$.

An interesting question is whether the relation $\asymp$ in PA_{dc} provides additional complexity in comparison with PA_{po}. Or is the general satisfiability problem still NP-complete (cf. [14])? Assume that we have two relation algebras given. One of these relation algebras has been perfectly analyzed in terms of complexity questions and tractable fragments. Can we find a mapping from one algebra to the other, preserving all relevant properties, to transfer all results? By the following definitions, we introduce the concept of homomorphisms and isomorphisms between relation algebras, and we will extend this to a certain kind of reduction - a reduction which preserves the tractability.

Definition 2. *For relation algebras Γ, Γ' a homomorphism is a function γ from Γ to Γ' such that γ preserves all operations of the boolean algebra and for relations R, S:*

1. $\gamma(R^{-1}) = \gamma(R)^{-1}$
2. $\gamma(R \circ S) = \gamma(R) \circ \gamma(S)$

A homomorphism $\gamma : \Gamma \to \Gamma'$ is called a monomorphism if $\gamma(R) = \gamma(S)$ implies $R = S$ for all relations R, S and an isomorphism if $\gamma(R) = \gamma(S) \iff R = S$.

Definition 3. *For two relation algebras Γ, Γ' a tractability-preserving-homomorphism (tph) is a homomorphism γ from Γ to Γ' such that each subset $\beta \subseteq \Gamma$ is tractable iff $\gamma(\beta) \subseteq \Gamma'$ is tractable. An isomorphic tph is called tpi.*

Of course, tph and tpi are reflexive and transitive. In the literature there can be implicitly found a tph on PA_{po} [4] and another one for directed intervals [13].

Definition 4. *A coarsening of a relation algebra is a monomorphism on its set of relations which is not an isomorphism.*

An example of a coarsening of RCC-8 is RCC-5 (cf. Fig. 3). The tph and tpi are important for problems like the following: Given are two relation algebras Γ and Γ'. If we have a tph $\gamma : \Gamma \xrightarrow{\gamma} \Gamma'$ and know that $\beta \subseteq \Gamma$ is (in)tractable, then we know that

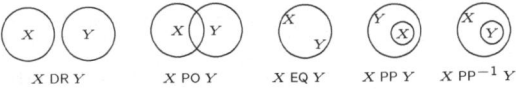

X DR Y X PO Y X EQ Y X PP Y X PP^{-1} Y

Fig. 3. The RCC-5 Relations

the image $\gamma(\beta)$ is (in)tractable in Γ'. But, how do we know if a subset β' of Γ' is the image of a tractable subset β of Γ ? The advantage of a tpi compared to a tph is that for a tpi there exists an inverse function, so that it is possible to transfer the tractable classes in both directions. If we know a maximal tractable subclass $\beta' \subseteq \Gamma'$, the tph does not allow to say anything about the maximal tractable subclasses of Γ. But with a tpi, we could identify the maximal tractable subclasses of Γ.

In order to identify maximal tractable subclasses of PA_{dc}, the tph given in [4] is here not sufficient. Instead, we use the following tpi γ:

Lemma 2. *A tpi γ from PA_{dc} to* **RCC-5** *(cf. Fig. 3) is given by:*

$$\gamma : \prec \mapsto \mathsf{PP} \qquad = \; \mapsto \mathsf{EQ} \qquad \succ \mapsto \mathsf{PP}^{-1}$$
$$\curlywedge \mapsto \mathsf{PO} \qquad \asymp \; \mapsto \mathsf{DR}$$

Proof. The claim that γ is a homomorphism follows from a comparision of the composition table of PA_{dc} with **RCC-5** and the observation that for the only non-symmetric relation $\prec$ holds: $\gamma(\prec)^{-1} = (\mathsf{PP})^{-1} = \mathsf{PP}^{-1} = \gamma(\succ)$. Therefore, it is sufficient to prove that γ is tractability-preserving and isomorphic. The reduction is an one-to-one identification of the basic relations in linear time. As any relation of a relation algebra consists of a union of basic relations, we can identify each relation of PA_{dc} with a relation of **RCC-5** in linear time. Therefore, the reduction is isomorphic and tractability-preserving if it preserves satisfiability.

In order to show that for each model of PA_{dc} there is a model in **RCC-5**, let a set of constraints of PA_{dc} with model M be given. For each element y in M, we introduce a point P_y. We define for each element x in M a set $S_x = \{P_y \mid y \prec x\}$ so that:

$$
\begin{aligned}
S_x \subsetneqq S_y & &\iff& \quad x \prec y \\
S_x \supsetneqq S_y & &\iff& \quad x \succ y \\
S_x = S_y & &\iff& \quad x = y \\
S_x \cap S_y = \emptyset &\iff \neg \exists a : a \in S_x \cup S_y &\iff& \quad x \asymp y \\
S_x \cup S_y \notin \{S_x, S_y\} & & & \\
\wedge S_x \cap S_y \neq \emptyset &\iff \exists a : a \in S_x \cap S_y &\iff& \quad x \curlywedge y.
\end{aligned}
$$

So, the reduction from PA_{dc} to **RCC-5** preserves the satisfiability. Finally, it has to be shown that a model of **RCC-5** is a model of PA_{dc}. We interpret the PA_{dc} relations as derived from the partial order $\subseteq = \{\mathsf{PP}, \mathsf{EQ}\}$.

$$
\begin{aligned}
X\,\mathsf{PP}\ \ Y &\iff X\,\{\mathsf{PP}, \mathsf{EQ}\}\,Y \quad \wedge \neg(X\,\mathsf{EQ}\,Y) \implies X \prec Y \\
X\,\mathsf{EQ}\ \ Y &\iff & X = Y \\
X\,\mathsf{PP}^{-1}\,Y &\iff X\,\{\mathsf{PP}, \mathsf{EQ}\}^{-1}\,Y \wedge \neg(X\,\mathsf{EQ}\,Y) \implies X \succ Y
\end{aligned}
$$

X PO Y implies $\exists Z \ \ Z \subseteq X \wedge Z \subseteq Y$. From this, it follows that there is a Z with $Z \prec X$ and $Z \prec Y$, and this is the definition of $\curlywedge$. Similarly, from X DR Y follows $\neg \exists Z \ \ Z \subseteq X \wedge Z \subseteq Y$, hence $X \asymp Y$.

The tpi is not only an isomorphism between RCC-5 and PA_{dc}, but also a polynomial reduction in both directions. For that reason, the following theorem is an easy implication:

Theorem 1. *SAT(PA_{dc}) is* NP-*complete.*

With Lemma 1, we get the next consequence.

Corollary 1. *NDP(PA_{dc}) is* NP-*complete.*

The tpi provides us not only with the results about the NP-completeness of the general satisfiability problem, but also allows to identify all maximal tractable subclasses by a reduction to the classes presented in [12]. The results are depicted in Table 2.

Table 2. The tractable subclasses of PA_{dc}

	τ_{28}	τ_{20}	τ_{17}	τ_{14}		τ_{28}	τ_{20}	τ_{17}	τ_{14}
$\bot$	•	•	•	•	$\{=\}$	•	•	•	•
$\{\asymp\}$	•	•			$\{=,\asymp\}$	•	•	•	
$\{\lambda\}$	•	•			$\{=,\lambda\}$	•	•	•	
$\{\asymp,\lambda\}$	•	•			$\{=,\lambda,\asymp\}$	•	•	•	
$\{\prec\}$	•			•	$\{=,\prec\}$	•		•	•
$\{\asymp,\prec\}$	•	•			$\{=,\prec,\asymp\}$	•	•	•	
$\{\lambda,\prec\}$	•				$\{=,\prec,\lambda\}$	•		•	
$\{\asymp,\lambda,\prec\}$	•	•			$\{=,\prec,\lambda,\asymp\}$	•	•	•	
$\{\succ\}$	•			•	$\{=,\succ\}$	•		•	•
$\{\asymp,\succ\}$	•	•			$\{=,\succ,\asymp\}$	•	•	•	
$\{\lambda,\succ\}$	•				$\{=,\succ,\lambda\}$	•		•	
$\{\asymp,\lambda,\succ\}$	•	•			$\{=,\succ,\lambda,\asymp\}$	•	•	•	
$\{\prec,\succ\}$				•	$\{=,\prec,\succ\}$			•	•
$\{\asymp,\prec,\succ\}$		•		•	$\{=,\prec,\succ,\asymp\}$		•	•	•
$\{\lambda,\prec,\succ\}$	•			•	$\{=,\prec,\succ,\lambda\}$	•		•	•
$\{\asymp,\lambda,\prec,\succ\}$	•	•		•	$\top$	•	•	•	•

In fact, the relations including $\{=\}$ are contained in τ_{28}, τ_{20}, τ_{14} if and only if $R \setminus \{=\}$ is contained in τ_{28} (τ_{20},τ_{14}). τ_{17} contains all relations including $\{=\}$ and the empty relation $\bot$.

Theorem 2. *The four classes* $\tau_{28}, \tau_{20}, \tau_{17}, \tau_{14}$ *(cf. Tab. 2) are the only maximal tractable subclasses of* PA_{dc}.

Such correspondence functions have clear advantages: Not only do we have the answers for all questions concerning the complexity investigations, and we know that RCC-5 and PA_{dc} are satisfiability equivalent up to isomorphism, but we have also found a connection between the field of reasoning on directed graphs and on topological constraint problems. Since there are many techniques known for handling graphs, some benefits for the RCC-5 algebra by such a translation should be possible.

4 The Associated Interval Algebra

4.1 Definition of the Interval Algebra IA_{dc}

There are applications in which it is not sufficient to compare single points in a network. For instance, pollution in a pipe network is not restricted to single points but extends to whole sections, and automated planning and project management deal with tasks that span over time intervals. This shows the need for a calculus with intervals as its basic elements. The basic relations are the relations between intervals that are definable by PA_{dc} relations of its endpoints (cf. Fig. 4).

Definition 5. *An interval* $I = [s_I, e_I]$ *is a pair of points satisfying* $s_I \prec e_I$. *The interval algebra* IA_{dc} *is the relation algebra generated by quadruples of relations as basic relations*

$$\mathcal{B} = \left\{ \begin{pmatrix} R_{ss} & R_{se} \\ R_{es} & R_{ee} \end{pmatrix} \mid R_{ss}, R_{se}, R_{es}, R_{ee} \in \{\prec, =, \succ, \asymp, \curlywedge\} \right\}$$

closed under $\cap, \cup, ^-, \circ, ^{-1}$. *For* $I = [s, e]$ *and* $I' = [s', e']$, *being in relation* $I \, R \, I'$ *means* $s \, R_{ss} \, s'$, $s \, R_{se} \, e'$, $e \, R_{es} \, s'$, *and* $e \, R_{ee} \, e'$.

In an identical way, the associated interval algebra IA_{po} for partial order PA_{po} can be introduced based on the set $\{\prec, =, \succ, \|\}$. The concepts and results for IA_{dc} are also applicable to IA_{po} as for IA_{po}, the two relations $\{\curlywedge, \asymp\}$ collapse into one relation $\{\|\}$. Some examples for IA_{dc} relations are given in Fig. 5. In IA_{po}, the three relations in the upper row collapse to the relation $fa = \left(\begin{smallmatrix} \| & \| \\ \succ & \succ \end{smallmatrix} \right)$ and in the lower row to fa^{-1}.

In the following, we write (R) for $\left(\begin{smallmatrix} R & R \\ R & R \end{smallmatrix} \right)$ $(R \in PA_{dc})$ and eq for $\left(\begin{smallmatrix} = & \prec \\ \succ & = \end{smallmatrix} \right)$. Further, $\left(\begin{smallmatrix} \{R_1, R_2\} & R_{se} \\ R_{es} & R_{ee} \end{smallmatrix} \right)$ abbreviates $\left(\begin{smallmatrix} R_1 & R_{se} \\ R_{es} & R_{ee} \end{smallmatrix} \right) \cup \left(\begin{smallmatrix} R_2 & R_{se} \\ R_{es} & R_{ee} \end{smallmatrix} \right)$, etc.

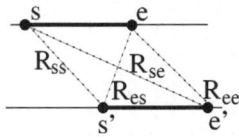

Fig. 4. A basic IA_{dc} relation is a combination of the basic PA_{dc} relations $R_{ss}, R_{se}, R_{es}, R_{ee}$

Fig. 5 and Fig. 6 show advantages of the new, finer interval calculus: More situations can be distinguished and more conclusions are possible. Fig. 6 shows a situation in a railway network. An obstruction B blocks the way from A to C. Can A reach its aim C? The situation is formally described by: $A \, (\prec) \, A_1$, $A_1 \, (\prec) \, C$, $A_2 \, (\prec) \, C$, $B \, (\prec) \, C$,

$$B \left(\begin{smallmatrix} \{\prec, =, \succ\} & \prec \\ \{\prec, =, \succ\} & \prec \end{smallmatrix} \right) A_1, \quad B \left(\begin{smallmatrix} \{\asymp, \curlywedge\} & \prec \\ \{\asymp, \curlywedge\} & \prec \end{smallmatrix} \right) A_2, \quad A_1 \left(\begin{smallmatrix} \{\asymp, \curlywedge\} & \prec \\ \succ & = \end{smallmatrix} \right) A_2$$

By specifying the latter relation, e.g. to $A_1 \left(\begin{smallmatrix} \asymp & \prec \\ \succ & = \end{smallmatrix} \right) A_2$, new conclusions can be drawn, in this case, $A \, (\prec) \, A_2$ becomes impossible. This means, if B is an obstacle on the path from A to C via A_1, then there is no alternative route via A_2.

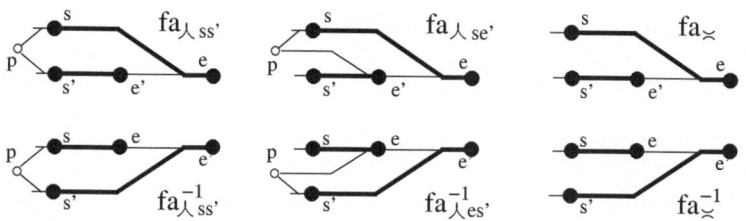

Fig. 5. Six examples for IA_{dc} relations. Refinements of the "finally after" relation and its inverse.

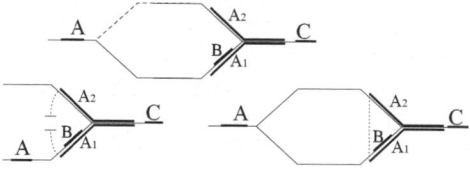

Fig. 6. Reasoning with uncertainty. The dashed line indicates that it is unknown if there is a path from A to A_2. In contrast to IA_{po}, IA_{dc} concludes such knowledge: Depending on the IA_{dc} relation between A_1 and A_2, the interval B can or cannot be bypassed using direction A_2.

Remark 1. IA_{dc} is a finite relation algebra consisting of 45 basic relations.

For each $\parallel$ occurring as a point relation in an IA_{po} relation, specifiying it to either $\asymp$ or $\curlywedge$, leads to different IA_{dc} relations. Omitting impossible cases like $\left(\begin{smallmatrix} \curlywedge & \asymp \\ * & * \end{smallmatrix}\right), \left(\begin{smallmatrix} \curlywedge & * \\ \asymp & * \end{smallmatrix}\right), \left(\begin{smallmatrix} \curlywedge & * \\ * & \asymp \end{smallmatrix}\right),$ $\left(\begin{smallmatrix} * & \curlywedge \\ * & \asymp \end{smallmatrix}\right), \left(\begin{smallmatrix} * & * \\ \curlywedge & \asymp \end{smallmatrix}\right)$ leaves 45 possible cases, where $*$ can be any basic relation.

For IA_{dc}, the set $\mathcal{U}$ of unions of basic relations with $\cup, \cap$, and $^-$ forms a boolean algebra. For basic relations, the operators $\circ$ and $^{-1}$ lead to the following relations of $\mathcal{U}$:

$$\begin{pmatrix} R_{ss} & R_{se} \\ R_{es} & R_{ee} \end{pmatrix}^{-1} = \begin{pmatrix} R_{ss}^{-1} & R_{es}^{-1} \\ R_{se}^{-1} & R_{ee}^{-1} \end{pmatrix} \tag{3}$$

$$R \circ R' = \begin{pmatrix} R_{ss} \circ R'_{ss} \wedge R_{se} \circ R'_{es} & R_{ss} \circ R'_{se} \wedge R_{se} \circ R'_{ee} \\ R_{es} \circ R'_{ss} \wedge R_{ee} \circ R'_{es} & R_{es} \circ R'_{se} \wedge R_{ee} \circ R'_{ee} \end{pmatrix} \tag{4}$$

Because of equations (1) and (2), $\mathcal{U}$ contains the result of operations $\circ, ^{-1}$ on arbitrary relations. Fig. 7 gives examples for such compositions.

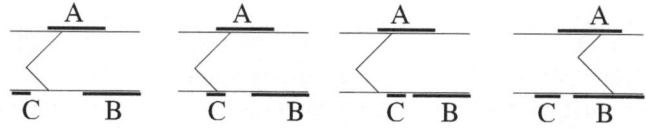

Fig. 7. In contrast to IA_{po}, IA_{dc} discerns these specifications of $A(\{\asymp, \curlywedge\})B \circ B(\succ)C$

4.2 The Complexity of the Interval Algebra IA_{dc}

For a given set of IA_{dc} constraints, how hard is it to decide the satisfiability?

Theorem 3. *The satisfiability problem of IA_{dc} is NP-hard.*

Proof. (Sketch) We give a tpi from PA_{dc} to a subset of IA_{dc}. We set $\gamma(=) = eq$ and $\gamma(R) = \left(\begin{smallmatrix} R & R \\ R & R \end{smallmatrix}\right)$ for all other basic relations, and extend it by $\gamma(R \cup R') = \gamma(R) \cup \gamma(R')$. By applying the rules (3) and (4) above, it is possible to show that $\circ, ^{-1}$ are preserved. For any solution of the interval case, a solution of the point case is found by picking an arbitrary point from the interval. Vice versa, for every solution of the point algebra, a solution of the interval case can be constructed by "inflating" the points, i.e. for each point x, we introduce an interval $\bar{x} = [x^-, x^+]$. As satisfiability is preserved and the translation can be done in linear time, the reduction is tractability-preserving.

Thus, for specific subclasses of IA_{dc} we know if they are tractable, namely for the ones containing only unions of relations of the type (R) or eq. But, are there larger tractable subclasses? We define special subclasses of unions of basic relations: *pointizable* relations $\mathcal{P}$ and *gadgetable* relations $\mathcal{G}$. The class of pointizable relations has been for the first time introduced for IA_{lin}, and shown to be a tractable subclass [10]. However, we will show that the general class is not tractable for IA_{dc} and neither for IA_{po}.

Definition 6. *For a subset $\mathcal{S} \subseteq PA_{dc}$, a relation R is called $\mathcal{S}$-pointizable if it belongs to the class*

$$\mathcal{P}_{\mathcal{S}} = \left\{ \left(\begin{smallmatrix} R_{ss} & R_{se} \\ R_{es} & R_{ee} \end{smallmatrix}\right) | R_{es}, R_{se}, R_{es}, R_{ee} \in \mathcal{S} \right\}$$

$\mathcal{P}_{\mathcal{S}}$ consists of those IA_{dc} relations that can exactly be expressed by a set $\mathcal{S}$ of PA_{dc} relations between its endpoints. What if we generalize this concepts to IA_{dc} relations that can exactly be expressed by a set of PA_{dc} relations between a larger set of points? A superset of the endpoints with such relations is called a gadget.

Definition 7. *A gadget (V_m, E_m) for an IA_{dc} relation R is a set of point variables $V_m = \{p_1, \ldots, p_m\}$ ($m \geq 4$) with PA_{dc} relations between them so that:*

1. *In each satisfying assignment of (V_m, E_m), $[p_1, p_2]R[p_3, p_4]$ is satisfied.*
2. *For each assignment of $p_1, \ldots, p_4$ holds: It can be extended to $p_1, \ldots, p_m$ in a way satisfying all relations of (V_m, E_m) iff $[p_1, p_2]R[p_3, p_4]$ is satisfied.*

For $\mathcal{S} \subseteq PA_{dc}$, an IA_{dc} relation R is called $\mathcal{S}$-gadgetable ($R \in \mathcal{G}_{\mathcal{S}}$) if all the PA_{dc} relations of a gadget for R are relations of $\mathcal{S}$. We write $\mathcal{G}$ for $\mathcal{G}_{PA_{dc}}$ and $\mathcal{P}$ for $\mathcal{P}_{PA_{dc}}$.

Example. With $I = [p_1, p_2]$ and $I' = [p_3, p_4]$, the points $p_1, .., p_6$ with the relations indicated in Figure 8 form a gadget for the IA_{dc} relation $I \left(\begin{smallmatrix} \asymp & \prec \\ \succ & \{\prec, \curlywedge\} \end{smallmatrix}\right) I'$.

Consider the first property: $p_1 \asymp p_3$ is directly a relation of the gadget. The point p_5 with its relations enforces that $p_1 \prec p_4$, $p_2 \succ p_3$, and $p_2 \not\curlywedge p_4$. As p_6 with its relations enforces that $p_2 \neq p_4$, $p_2 \not\curlywedge p_4$, only $p_2\{\prec, \curlywedge\}p_4$ remains satisfiable. The second property holds since any model of $I \left(\begin{smallmatrix} \asymp & \prec \\ \succ & \{\prec, \curlywedge\} \end{smallmatrix}\right) I'$ can be extended to include p_5, p_6 in a way satisfying all these PA_{dc} relations.

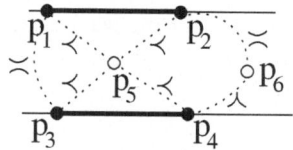

Fig. 8. A gadget with 6 points (with PA_{dc} relations as indicated at the dotted edges)

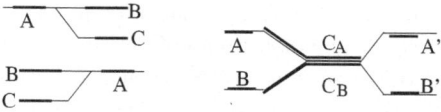

Fig. 9. The left part shows the two only cases satisfying the gadget for $A\ (\prec)\cup(\succ)\ B$, and the right part shows the two possibilities (A, B) and (A', B') satisfying the gadget for $(\curlywedge)\cup(\asymp)$

All basic IA_{dc} relations and relations (R) are pointizable, and pointizability is closed under $\cap$ and $^{-1}$, but not under $\cup$. For instance, the relation $(\prec)\cup(\succ)$ [1] is not pointizable. But all relations are subsets of a pointizable relation because

$$\begin{pmatrix} R_{ss} & R_{se} \\ R_{es} & R_{ee} \end{pmatrix} \cup \begin{pmatrix} R'_{ss} & R'_{se} \\ R'_{es} & R'_{ee} \end{pmatrix} \subseteq \begin{pmatrix} R_{ss} \cup R'_{ss} & R_{se} \cup R'_{se} \\ R_{es} \cup R'_{es} & R_{ee} \cup R'_{ee} \end{pmatrix}.$$

Theorem 4. $\mathcal{G}, \mathcal{P}$ are intractable subclasses of IA_{dc}.

Proof. By definition, $\mathcal{P}_S \subseteq \mathcal{G}_S$. Consider the set $\mathcal{R} := \{(\{\asymp, \curlywedge\}), (\{\prec, \succ\}), (\begin{smallmatrix} \asymp & \prec \\ \succ & = \end{smallmatrix})\}$. We show that $\mathcal{R} \subsetneq \mathcal{P}$ is intractable. Due to Table 2, $\{\{\asymp, \curlywedge\}, \{\prec, \succ\}\}$ is not contained in a tractable subclass of PA_{dc}. Applying γ from Theorem 3 shows that the satisfiability problem over $\mathcal{I} := \{(\asymp)\cup(\curlywedge), (\prec)\cup(\succ)\} \subsetneq IA_{dc}$ is intractable. A satisfiability problem over $\mathcal{I}$ can be reduced to $\mathcal{R}$ because $A(\prec)\cup(\succ)B$ is satisfiable iff the corresponding gadget $A(\{\prec, \succ\})B$, $A(\{\prec, \succ\})C$, $B(\{\asymp, \curlywedge\})C$ is satisfiable, and $A(\asymp)\cup(\curlywedge)B$ is satisfiable iff $A(\{\asymp, \curlywedge\})B$, $A(\prec)\cup(\succ)C_A$, $B(\prec)\cup(\succ)C_B$, $C_A(\begin{smallmatrix} \asymp & \prec \\ \succ & = \end{smallmatrix})C_B$ is satisfiable (cf. Fig. 9). Hence, intractability is inherited from $\mathcal{I}$ via $\mathcal{R}$ to $\mathcal{P}$ and $\mathcal{G}$.

Not all gadgetable relations are pointizable. For instance, $(\prec)\cup(\succ)$ is gadgetable, but not pointizable: Hence $\mathcal{B} \subsetneq \mathcal{P} \subsetneq \mathcal{G} \subsetneq IA_{dc}$.

If $\mathcal{S}$ is tractable in PA_{dc}, $\mathcal{P}_S$ is tractable in IA_{dc}. How about the larger class $\mathcal{G}_S$?

Theorem 5. *If $\mathcal{R}$ is a class of $\mathcal{S}$-gadgetable IA_{dc} relations and the satisfiability problem over $\mathcal{S}$ is tractable, then the satisfiability problem over $\mathcal{R}$ is tractable.*

Proof. For a given set of constraints for n intervals, we need to find a polynomial algorithm that decides if it is satisfiable. The idea is to translate the $\mathcal{S}$-gadgetable IA_{dc} constraints into corresponding PA_{dc} constraints over $\mathcal{S}$. Since $\mathcal{S}$ is a tractable subclass of PA_{dc}, there is a polynomial procedure to decide consistency. We have to show that only a polynomial number of PA_{dc} relations is needed, and that the translation is possible in polynomial time. For each of the finitely many interval relations, a gadget is fixed. The

---

[1] This relation may describe that two projects need to be done by the same person in any order.

largest one defines an upper bound M for the used gadget's size (number of points in it). Hence, replacing the $\mathcal{S}$-gadgetable IA_{dc} constraints by the associated $\mathcal{S}$ constraints can be done in $O((M \cdot n^2)^2) = O(n^4)$.

Corollary 2. $\mathcal{B}, \mathcal{G}_{\tau_{28}}, \mathcal{G}_{\tau_{20}}, \mathcal{G}_{\tau_{17}}, \mathcal{G}_{\tau_{14}}$ *are tractable subclasses of* IA_{dc}.

$\mathcal{B}$ is tractable since $\mathcal{B} \subseteq \mathcal{G}_{\{\{\prec\},\{=\},\{\succ\},\{\asymp\},\{\curlywedge\}\}} \subseteq \mathcal{G}_{\tau_{28}}$.

From this, the NP-completeness of IA_{dc} can be concluded. The NP-hardness follows directly from Theorem 3. The membership in NP follows from the tractability of constraints of basic relations. For an arbitrary set of constraints of IA_{dc}, we guess a scenario (i.e. all constraints are basic relations), and because of Corollary 2, we can test if it is satisfiable. With Lemma 1, we get the complexity class of the IA_{dc} problems.

Corollary 3. *SAT(IA_{dc}) and NDP(IA_{dc}) are* NP-complete.

5   Outlook and Summary

Starting from the question how a relation algebra useful for reasoning about dependencies in general networks should be designed, we have identified a calculus that qualifies points in such a network. This calculus consists of five basic relations.

Traditional reasoning for points in networks uses only four relations: before, equal, after, and unrelated. This language has been used so far for modelling networks usually thought of as models for relational systems in space, time or space-time. We showed that an algebra for reasoning about dependencies needs a new 'fork' relation. This relation, which states that two points have a common ancestor, proved useful in various application areas from ecological systems, over transportation networks, to planning and medical diagnosis. Also dependencies in a Bayesian network can be expressed.

In a formal analysis of this calculus, we proved that the complexity of the general satisfiability problem is NP-complete as well as the satisfiability problem of the corresponding interval algebra IA_{dc}. This means, in general, the problem of finding out if there is a network satisfying certain conditions, is rather difficult. But in many cases, there is a polynomial algorithm. We have identified all tractable subclasses of PA_{dc}, via a tractability preserving isomorphism into the RCC-5 calculus. Classes of the new type of $\mathcal{S}$-gadgetable relations were identified as tractable subclasses of IA_{dc}. We have shown how these results for IA_{dc} can also be applied to the proper subclass IA_{po}.

This work opens the field in many directions: By the tph and tpi techniques presented here, the satisfiability equivalence of different relation algebras can be shown, and the expressibility and complexity of these algebras can be transfered and compared. Further investigations should reveal that the class of $\mathcal{S}$-gadgetables are maximal tractable subclasses for IA_{po} and IA_{dc}. Variations of the presented calculi are possible. Making the similar distinction for common consequences like for common causes could easily be expressed by our calculus with a direction-reversed interpretation. The combination of both provides a framework to satisfy the suggested extension of the linear directed intervals on networks [13]. Another promising idea is the temporalization of the dependency calculus for modeling dependencies that vary over time.

Acknowledgments

This work was partially supported by the Deutsche Forschungsgemeinschaft (DFG) as part of the Transregional Collaborative Research Center SFB/TR 8 Spatial Cognition. We like to thank Bernhard Nebel for various helpful discussions.

References

1. J. F. Allen. Maintaining knowledge about temporal intervals. *Comm. ACM*, 26(11): 832–843, 1983.
2. F. Anger, P. Ladkin, and R. Rodriguez. Atomic temporal interval relations in branching time: Calculation and application. In *Actes 9th SPIE Conference on Applications of AI*, Orlando, FL, USA, 1991.
3. M. Broxvall and P. Jonsson. Towards a complete classification of tractability in point algebras for nonlinear time. In *Proc. of CP-99*: 129–143, 1999.
4. M. Broxvall, P. Jonsson, and J. Renz. Refinements and Independence: A Simple Method for Identifying Tractable Disjunctive Constraints. In *CP*: 114–127, 2000.
5. A.G. Cohn. Qualitative spatial representation and reasoning techniques. In KI-97: Advances in AI, Brewka, G. and Habel, C. and Nebel, B (eds), LNAI, 1–30, 1997.
6. T. Drakengren and P. Jonsson. A complete classification of tractability in Allen's algebra relative to subsets of basic relations. *Artificial Intelligence*, 106(2): 205–219, 1998.
7. C. Freksa. Using Orientation Information for Qualitative Spatial Reasoning. In Theories and Methods of Spatial-Temporal in Geographic Space. Reasoning. Frank, A. U. and Campari, I. and Formentini, U. (eds.), 162–178, 1992.
8. P. B. Ladkin and R. D. Maddux. On binary constraint problems. *J. ACM*, 1994.
9. U. Montanari. Networks of constraints: Fundamental properties and applications to picture processing. *Inform. Sci.*, 7:95-132,1974.
10. B. Nebel and H.-J. Bürckert. Reasoning about temporal relations: A maximal tractable subclass of Allen's interval algebra.*J.ACM*, 42(1):43–66, 1995.
11. Randell, D. and Cui, Z. and Cohn, A. A Spatial Logic Based on Regions and Connection. Proceedings KR-92, 165–176, 1992.
12. J. Renz and B. Nebel. On the complexity of qualitative spatial reasoning: A maximal tractable fragment of the Region Connection Calculus. *AIJ*, 108(1-2):69–123, 1999.
13. J. Renz. A Spatial Odyssey of the Interval Algebra: 1. Directed Intervals. In *Proc. of IJ-CAI'01*, 2001.
14. M. B. Vilain, H. A. Kautz, and P. G. van Beek. Contraint propagation algorithms for temporal reasoning: A revised report. *Reasoning about Physical Systems*: 373–381, 1989.

Temporalizing Spatial Calculi:
On Generalized Neighborhood Graphs

Marco Ragni and Stefan Wölfl

Institut für Informatik, Albert-Ludwigs-Universität Freiburg,
Georges-Köhler-Allee, 79110 Freiburg, Germany
{ragni, woelfl}@informatik.uni-freiburg.de

Abstract. To reason about geographical objects, it is not only necessary to have more or less complete information about where these objects are located in space, but also how they can change their position, shape, and size over time. In this paper we investigate how calculi discussed in the field of qualitative spatial reasoning (QSR) can be temporalized in order to gain reasoning formalisms that can be used to express spatial configurations and their dynamics. In a first step, we briefly discuss temporalized spatial constraint languages. In particular, we investigate how the notion of continuous change can be expressed in such languages and how continuous change is represented in the so-called conceptual neighborhood graph of the spatial calculus at hand. In a second step, we focus on a special reasoning problem, which occurs quite naturally in the context of temporalized spatial calculi: Given an initial spatial scenario of some physical objects, which scenarios are accessible if the set of all possible paths of these objects is constrained by some further conditions? We show that for many spatial calculi this general problem cannot be dealt with by using the information encoded in the classical neighborhood graphs, as usually discussed in the literature. Rather, we introduce a generalized concept of neighborhood graph, which allows for reasoning about objects in such dynamic settings.

1   Introduction

To reason about geographical objects, it is not only important to have information about where these objects are located in space, but also how they can change their position, shape, and size over time. Some physical objects such as chairs, towers, and stars are usually assumed to be rather robust to changes in shape and size (at least from the point of what we can experience without using scientific instruments). Other objects such as hurricanes, clouds, and balloons may vary their size and shape quite rapidly. Obviously, how physical objects can change such spatial properties depends on the physical quality structure of the respective object and its environment. A crucial notion in this context is the notion of continuous change since it seems common-sense that many property changes occur continuously. Topic of the paper will be to discuss how continuity concepts can be integrated into the formal calculi discussed in the qualitative spatial reasoning literature.

Under the heading of *qualitative spatial reasoning*, many formalisms for representing, and reasoning with, spatial configurations have been discussed in the past two

U. Furbach (Ed.): KI 2005, LNAI 3698, pp. 64–78, 2005.

decades. In recent years also the issue of how to temporalize such spatial formalisms has gained more attention in the literature. Obviously, temporalizations of spatial calculi can be developed by exploiting different research strategies. First, they can be embedded into rich first-order theories by integrating mereotopological and temporal concepts. For example, Muller [16] has proposed a first-order theory of spatio-temporal entities, which is based on the first-order theory of the *region connection calculus* [19]. Second, temporalizations may be discussed in the framework of temporal logics. The combination of RCC8 and linear time temporal logic, for example, has been investigated by Wolter and Zakharyaschev [21] and Gabelaia et al. [10]. Third, spatio-temporal representation languages can be obtained via temporalizing a spatial constraint language (e. g., RCC8) by a suitable temporal constraint language such as Allen's interval calculus. Bennett et al. [3] proposed such a reasoning formalism, which was further investigated by Gerevini and Nebel [13]. From a more philosophical perspective, Galton [11] discussed various facets of continuous change, in particular, how such changes can be consistently described at different levels of granularity and how their qualitative and quantitative descriptions are related to each other.

In this paper we focus on the third research strategy outlined previously. In more detail, we discuss temporal constraint languages, which are enriched by formulae expressing time-dependent spatial constraints. In these languages it is possible to express temporally annotated spatial information as well as their temporal relationships. Then two kinds of reasoning tasks may be distinguished: The *static reasoning problem* is to determine whether such a spatio-temporal description is consistent, i. e., whether there exists a temporal model satisfying each temporal constraint as well as each temporally annotated spatial constraint. The *dynamic reasoning problem* is to determine whether, given a set of *transformation constraints*, there exists a continuous transformation between two spatial configurations such that none of the transformation constraints is violated. For example, the well-known *Towers of Hanoi* puzzle may be cast as such a problem.

A central notion in the context of temporalized spatial constraint networks is that of a *neighborhood graph*. For many qualitative spatial reasoning formalisms researchers have intensively discussed the so-called conceptual neighborhood graph, a concept introduced first by Freksa [9]. The neighborhood graph is usually understood as to describe which relation transitions be possible if the objects are subject to minimal changes. This interpretation of neighborhood graphs is clearly *temporal*, that means, it aims at describing the dynamics of the relations at hand.[1] Interestingly, the neighborhood graph often is not uniquely determined by the underlying background theory of the respective calculus [e. g., 18]. For instance, different neighborhood graphs can be found depending on whether objects are allowed to change their size or whether one allows two objects to be changed at the same moment. This means that, in principle, neighborhood graphs are a suitable means for encoding spatial information about the kind of objects that are described by the qualitative spatial calculus at hand. But in our opinion, the traditional concept of neighborhood graph is too restricted to be really useful for reasoning with spatio-temporal constraints. For this reason we present a gen-

---

[1] In the literature there is a further research stream, in which neighborhood graphs are discussed in the context of *conceptual* neighborhoodness of relations Knauff [14].

eralized concept of neighborhood graph, which may be considered a first step towards developing a more general theory on dynamic reasoning problems.

The paper is organized as follows: In section 2 we briefly introduce the calculi that will be of interest in the following sections. In section 3 we discuss spatio-temporalized constraint languages and their models. Moreover, we present a precise notion of continuous change that enables us to analytically prove the correctness of neighborhood graphs. In section 4 we explain how generalized neighborhood graphs can be applied in order to solve dynamic reasoning problems. In more detail, we present the generalized neighborhood graphs for the point algebra and for RCC5. Finally, section 5 gives a short summary of our results and a brief outlook on interesting future work.

2   Preliminaries

Let us start by briefly sketching the qualitative calculi that will be of interest in the following sections. Readers familiar with these calculi may wish to skip this section.

Constraint Satisfaction Problems. Qualitative reasoning problems are usually cast as constraint satisfaction problems (CSP), i. e., as a problem to determine whether a constraint network (a finite set of constraints) is satisfiable or entailed by another constraint network. Typically, a qualitative constraint network is a finite set of constraints of the form xRy where x and y are variables taking values in a given domain D, and a binary relation R defined on D. For modelling imprecise knowledge, one usually considers sets of relations that are closed with respect to unions. In more detail, given a specific level of granularity chosen describing the domain at hand, one starts by identifing a (finite) set jointly exhaustive and pairwise disjoint sets of *base relations* on the domain. A *composition table* gives information about which constraints xRy are possible if one has complete knowledge about how x and y are related (via base relations) to a third object z. Speaking more algebraically, from a set of base relations (containing the identity relation) and a composition table (satisfying some requirements), one can build up a *relation algebra*, i. e., a set of relations that contains the identy relation and is closed with respect to unions, intersections, converse formation, and composition of relations.

To put these notions in a more precise context, we introduce the following terminology: A qualitative constraint satisfaction problem is defined by a constraint language $\mathscr{L}$ and a class of *(intended) models*. The constraint language usually consists of an infinite set of variables and a finite set of (binary) base relation symbols. A *constraint* is a formula of the form $x\{R_1,\ldots,R_n\}y$ (meaning $xR_1y \vee \cdots \vee xR_ny$), where x and y are variables and each R_i is a base relation symbol. Finite sets of constraints are referred to as *constraint networks*. A *model* is a first- or higher-order structure $\mathcal{M} = \langle \ldots, D, \ldots \rangle$ assigning an interpretation $R^{\mathcal{M}} \subseteq D^2$ to each base relation symbol R. A *(variable) assignment* in $\mathcal{M}$ is a function a that assigns an element $a(x) \in D$ to each variable x. Given an assignment a and an element $d \in D$, the function $a(x/d)$ is defined as the function that coincided with a in all variables distinct from x and assigns object d to variable x. A constraint $x\{R_1,\ldots,R_n\}y$ is said to be *satisfied* in $\mathcal{M}$ by a (denoted by $\mathcal{M}, a \models x\{R_1,\ldots,R_n\}y$) if $(a(x),a(y)) \in R_i^{\mathcal{M}}$ for some $1 \leq i \leq n$. A *constraint network* C is said to be *satisfiable* in a class of models if there exists a model $\mathcal{M}$ in this class as well as an assignment a in $\mathcal{M}$ such that all constraints in C are satisfied. Furthermore, *a composition table* is a map assigning to each pair of base relations R_i and R_j a set of base relations $R_i \circ R_j = \{R_{k_1}, \ldots, R_{k_m}\}$. A composition table is said to be *extensionally correct* for $\mathcal{M}$ if for each pair of base relations R_i and R_j and each assignment a,

$$\mathcal{M}, a \models x(R_i \circ R_j)y \iff \exists d \in D \text{ s.t. } \mathcal{M}, a(z/d) \models xR_i z \text{ and } \mathcal{M}, a(z/d) \models zR_j y.$$

If the base relations defined by $\mathcal{M}$ are (a) jointly exhaustive (i.e., $\bigcup_{1 \leq i \leq n} R_i^{\mathcal{M}} = D^2$), (b) pairwise disjoint (i.e., $R_i^{\mathcal{M}} \cap R_j^{\mathcal{M}} = \emptyset$ for $i \neq j$), (c) closed with respect to converses (i.e., $R_i^{\mathcal{M}} = (R_j^{\mathcal{M}})^{\smile} := \{(x,y) : (y,x) \in R_j^{\mathcal{M}}\}$), and (d) have an extensionally correct composition table, then there exists a uniquely determined algebra of binary relations on the domain set D.

Point Algebra. The point algebra (for linear time) may be considered the most simple qualitative calculus. The point algebra (PA) describes the relations between instants of linear flows of time. Hence, this algebra considers the three base relations $<$ ("earlier"), $=$ ("equal"), and $>$ ("later"), as well as unions of them. Point algebras can also be defined for much weaker relational structures such as branching flows of time, partial orders, etc.

Interval Algebra. Given a linear flow of time, an *Allen interval* is a pair of instants $\langle t_1, t_2 \rangle$ with $t_1 < t_2$. By comparing the relative positions of start and endpoints of two intervals, one can identify thirteen jointly exhaustive and pairwise disjoint base relations between intervals, which are known in the literature as the *Allen 13 relations* (cf. Table 1).

RCC5 and RCC8. The most prominent calculi in the domain of spatial qualitative reasoning are the region connection calculi RCC5 and RCC8. These calculi allow for

Table 1. The 13 base relations of Allen's interval algebra

Relation	Converse	Pictorial Representation
I b J	J bi I	
I m J	J mi I	
I o J	J oi I	
I d J	J di I	
I s J	J si I	
I f J	J fi I	
I e J	J e I	

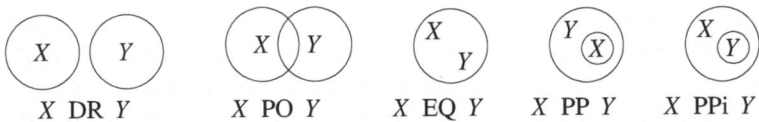

Fig. 1. The RCC5 relations

expressing relations between *regions*, which often are represented as non-void, connected, and regular closed (or regular open) subsets of some topological space. The set of RCC5 base relations consists of the relations DR ("discrete"), PO ("partially overlap"), PP ("proper part"), PPi ("proper part inverse"), and EQ ("equal") (cf. Fig. 1). RCC8 refines these relations by splitting DR into the relations DC ("disconnected") and EC ("externally connected") and by splitting PP (analogously PPi) into the relations TPP ("tangential proper part") and NTPP ("non-tangential proper part"). From the semantical point of view, a *topological model* is a tuple $\mathcal{M} = \langle S, O, \text{Reg} \rangle$, where $\langle S, O \rangle$ is a topological space (O being its set of open sets) and Reg is a non-void set of regular closed subsets of S. Topological models induce RCC5 and RCC8 models in a natural manner: For example, for regions X and Y, the pair (X, Y) is in $\text{DR}^{\mathcal{M}}$ if and only if $X \cap Y = \emptyset$ and $(X, Y) \in \text{NTPP}^{\mathcal{M}}$ if and only if there exists an $U \in O$ such that $X \subsetneq U \subseteq Y$.

The composition table of RCC5 (cf. Table 2) is known to be correct if the relations are interpreted on closed discs in the Euclidean plane. In this case the RCC5 relations coincide with the relations definable in terms of the subset relation. For this reason, RCC5 is sometimes also referred to as *containment algebra*.

Table 2. The composition table of RCC5

	EQ	DR	PO	PP	PPi
EQ	EQ	DR	PO	PP	PPi
DR	DR	EQ, DR, PO, PP, PPi	DR, PO, PP	DR, PO, PP	DR
PO	PO	DR, PO, PPi	EQ, DR, PO, PP, PPi	PO, PP	DR, PO, PPi
PP	PP	DR	DR, PO, PP	PP	EQ, DR, PO, PP, PPi
PPi	PPi	DR, PO, PPi	PO, PPi	EQ, PO, PP, PPi	PPi

3   Temporalizing Spatial Calculi

The general method for temporalizing the language of a given spatial calculus is the following: Let $\langle V_T, \mathcal{R}_T \rangle$ be the language of a temporal calculus $\mathfrak{T}$ and $\langle V_S, \mathcal{R}_S \rangle$ be the language of a spatial calculus $\mathfrak{S}$, that is, V_T and V_S are disjoint sets of variables and $\mathcal{R}_S$

and $\mathcal{R}_T$ are the sets of base relation symbols of the respective calculi. In general, the temporal calculus will be the point algebra or the interval algebra for linear time (but, of course, the method is not restricted to these calculi). In what follows, constraints of the temporal calculus, i. e., formulae of the form

$$i\{R_1,\ldots,R_n\}j \qquad (i,j \in V_T,\ R_1,\ldots,R_n \in \mathcal{R}_T)$$

are referred to as *temporal constraints*. We now enrich the language of $\mathfrak{T}$ by *temporalized spatial constraints*, namely formulae of the form:

$$i:x\{S_1,\ldots,S_n\}y \qquad (i \in V_T,\ x,y \in V_S,\ S_1,\ldots,S_n \in \mathcal{R}_S).$$

In the sequel, the combined calculus will be referred to as $\mathfrak{T}:\mathfrak{S}$.

How can we define models for this language in terms of respective models for the temporal and spatial languages? To illustrate this, let us start by defining the models of a spatial calculus chosen from the RCC-family (denoted by RCCx), which is temporalized by PA. The key concept for defining such models is that of a temporalized topological model (note that the concept used here presents a modified version of the concept introduced by Wolter and Zakharyaschev [21]):

Definition 1. *A temporalized topological model (abbr. by tt-model) is a tuple* $\mathcal{M} = \langle T,<,S,O,\mathrm{Reg},\Pi \rangle$, *where* $\langle T,< \rangle$ *is a linear flow of time,* $\langle S,O \rangle$ *is a topological space,* Reg *is a set of* regions *(i. e., non-void, connected, and regular closed subsets of S), and* Π *is a non-void set of* (object) paths $\pi: T \longrightarrow \mathrm{Reg}$.

The idea on which the definition is based is the following: We assume that at each instant, an object occupies a specific region in a fixed topological space. Since we are only interested in the path of an object, i. e., in the sequence of regions occupied by the object, it is reasonable to represent objects as functions assigning regions to instants.

Obviously, each temporalized topological model $\mathcal{M}$ induces a (temporal) model for PA (denoted by $\mathcal{M}_T$) and a (spatial) model for RCCx (denoted by $\mathcal{M}_S$). A *PA:RCCx (variable) assignment* in a tt-model is a pair $a = \langle a_T, a_S \rangle$, where $a_T: V_T \longrightarrow T$ is a function assigning instants to temporal variables and $a_S: V_S \times T \longrightarrow \mathrm{Reg}$ is a function assigning a region to each spatial variable at each instant t such that $a_{S,x}(t) := a_S(t,x)$ defines an object path of $\mathcal{M}$. Note that for each instant t, a_S also defines an RCC5 assignment by $a_{S,t}(x) := a_S(t,x)$. The model relation is then introduced as follows: For temporal PA:RCCx constraints we set

$$\mathcal{M},a \models i\{R_1,\ldots,R_n\}j \iff \mathcal{M}_T,a_T \models i\{R_1,\ldots,R_n\}j,$$

and for temporalized spatial constraints we define

$$\mathcal{M},a \models i:x\{S_1,\ldots,S_n\}y \iff \mathcal{M}_S,a_{S,t} \models x\{S_1,\ldots,S_n\}y,$$

where $t = a_T(i)$.

In the case that a spatial calculus is temporalized with respect to the interval algebra, we need to modify this semantics as follows: An *IA:RCC5 assignment* in a tt-model is a pair $a = \langle a_T, a_S \rangle$, where a_T assigns an ordered pair $(a_T(i)^-, a_T(i)^+) \in T^2$ with

$a_T(i)^- < a_T(i)^+$ to each interval variable and a_S is defined as above. Here the model relation is defined as follows:

$$\mathcal{M}, a \models i : x\{S_1, \ldots, S_n\}y \iff \mathcal{M}_S, a_{S,t} \models x\{S_1, \ldots, S_n\}y, \text{ for each}$$
$$t \in T \text{ with } a_T(i)^- < t < a_T(i)^+.$$

Note that we only require that the spatial constraints hold in the interior of the interval. This is necessary since if these spatial constraints need to hold at starting and endpoints of the interval as well, then it would not be possible that a base relation holding between objects X and Y in interval I changes to a different base relation between these objects in any interval met by I. Hence, it would follow that a base relation holding between two object would remain the same all the time, which is apparently unacceptable.

Let us illustrate these notions by some examples: If we temporalize the region connection calculus RCC5 with respect to the point algebra, we can express that two objects X and Y are disjoint at some instant t, but overlap at some later instant t' by the following constraints:

$$t : X \text{ DR } Y, \, t < t', \, t' : X \text{ PO } Y.$$

If we temporalize RCC8 with respect to the interval algebra, we obtain the calculus STCC introduced by Gerevini and Nebel [13]. Here we can state constraints such as

$$I\{m,b\}J, \; I : X \text{ DC } Y, \; I : Y \text{ DC } Z, \; J : X\{\text{NTPP}, \text{TPP}\}Y, \; J : Y \text{ PO } Z,$$

which express that interval I (weakly) precedes interval J, that during I region X is disconnected from region Y and Y is disconnected from region Z, that during J region X is proper part of region Y, and so on.

The semantical definitions presented so far do not impose any restrictions on how objects can change their size, shape, or position. But how can we introduce such conditions on the semantic level? To explain this, let us focus on the condition that objects need to change their position, size, and shape continuously. For the sake of simplicity, we will assume that each region in a tt-model is a homeomorphic image of the n-dimensional closed unit circle E_n (for $n = 2, 3$) — these circles provide typical examples of connected, regular closed subsets. This means that for each region $X \in \text{Reg}$, we have a continuous function $\varepsilon_X : E_n \longrightarrow S$ induced by a fixed homeomorphism between E_n and X. We will be refer to such models as *simple tt-models*.

Definition 2. *Let $\mathcal{M}$ be a simple tt-model. A path $\pi : T \longrightarrow \text{Reg}$ of $\mathcal{M}$ is said to be* continuous *if the function*

$$\tau : E_n \times T \longrightarrow S, \quad \tau(p,t) := \varepsilon_{\pi(t)}(p)$$

is continuous (in both variables, think of T as equipped with the order topology). A simple tt-model is said to be continuous *if each of its object paths is continuous, and it is* strictly continuous *if Π consists exactly of all continuous object paths possible for the regions of $\mathcal{M}$.*

Apparently, this concept of continuous object paths is closely related to the topological notion of homotopic functions, i. e., continuous transformations between continuous functions. In fact, a continuous object path π defines homotopies between arbitrary regions $\pi(t)$ and $\pi(t')$. The important point is not that $\pi(t)$ and $\pi(t')$ are homotopic (which is trivial since both are homeomorphic to E_n), but that the object path itself defines such a homotopy. For example, let π be an object path from $\mathbb{R}$ into a suitable set of all subsets of $\mathbb{R}^n$ assigning the unit circle to each $t \neq 0$ and the unit cube at $t = 0$. Then obviously this object path cannot be continuous.

Prepared with these notions, we can define a precise notion of the neighborhood graph of a spatial calculus. For this let $\mathcal{M}$ be a simple tt-model. We define the *RCCx neighborhood graph* associated to $\mathcal{M}$ as follows: Let S be an RCCx base relation. The set of $\mathcal{M}$-neighbors of S is defined as the smallest set of base relations, $N(S)$, satisfying the following two conditions:

- $S \notin N(S)$;
- For each pair of object paths π, π' and each pair of instants $t_0 < t_1$ of $\mathcal{M}$ with $\pi(t_0) S^{\mathcal{M}} \pi'(t_0)$ and not $\pi(t_1) S^{\mathcal{M}} \pi'(t_1)$, there exists a relation $S' \in N(S)$ and an instant $t_0 < t \leq t_1$ such that $\pi(t) S'^{\mathcal{M}} \pi'(t)$.

Thus the RCCx neighborhood graph w. r. t. $\mathcal{M}$ is defined as the directed graph $G_{\text{RCC}x,\mathcal{M}}$ that has the RCCx base relations as vertices and for each base relation an edge to each of its $\mathcal{M}$-neighbors. A graph G with vertex set $R_{\text{RCC}x}$ is said to be *correct* for *a class of models* if G is the neighborhood graphs w. r. t. all models of that class.

Lemma 3. *The neighborhood graph of RCC5 (cf. Fig. 2) is correct for each class of strictly continuous tt-models that instantiate all RCC5 base relations.*

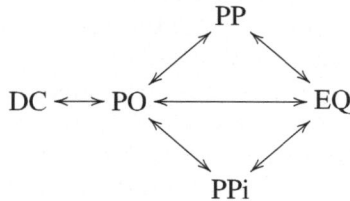

Fig. 2. The neighborhood graph of RCC5

Given a neighborhood graph G, we define the *neighborhood distance* between spatial relations as follows: For base relations B and B', $\Delta_G(B, B')$ is defined as the length of the shortest path in G between B and B'. For arbitrary relations S and S' we set

$$\Delta_G(S, S') = \min_{B \in S, B' \in S'} \Delta_G(B, B').$$

Obviously, $\Delta_G(S, S') = 0$ if and only if S intersects with S', and $\Delta_G(S, S') = 1$ if and only if S and S' are disjoint, but contain base relations B and B' respectively such that B' is a neighbor of B in G.

Finally, let us turn to the question whether continuity is expressible in the temporalized spatial constraint language presented here. The quick answer is that continuity is not expressible by formulae, but is expressible via *rules* in the language IA:RCCx. To see this, suppose that we have a constraint set, which contains the temporalized spatial constraint $I : X\{DC,PP,EQ\}Y$. This constraint is satisfied by an assignment in a *continuous* tt-model if and only if either $I : X$ DC Y or $I : X\{PP,EQ\}Y$ is satisfied by that assignment. To show this, let t_0 be an instant such that $I^- < t_0 < I^+$ and X DC Y is false at t_0, and let t_1 be an arbitrary instant in the interior of I. Then we obtain that at t_1 $X\{DC,PP,EQ\}Y$ is true. Without restriction we may assume that $t_1 < t_0$. Now if $X\{DC\}Y$ holds at t_1, there must be an instant $t_1 < t \le t_0$ such that $X\{PO\}Y$ is true at t. But this cannot be since at t one of the constraints X DC Y, X PP Y, or X EQ Y must be true.

In fact, continuity rules could be applied in tableau algorithms as well as in natural deduction systems. But this goes beyond the scope of this paper.

4   Generalized Neighborhood Graphs

In the previous section we presented a precise notion of continuous change in spatial settings, which can be described in terms of the RCC relations. In this section we will deal with the question whether there exists a continuous transformation from an initial spatial scenario to a final spatial scenario, even when the set of all possible transformations is restricted by some further constraints (so-called *transformation constraints*).

In the following, we will argue within concrete models (i. e., within the reals as flow of time and a fixed topological model such as as the Euclidean plane or the three-dimensional Euclidean space). Then the reasoning problems we are now concerned with have the following form: Let σ_s and σ_f be two spatial scenarios, each describing the same set of objects $X_1, \ldots X_n$, i. e., σ_s and σ_f are sets of "interpreted" constraints where between each pair of objects a spatial base relation holds. Furthermore, let Σ be a set of constraints in which at most variables for $X_1, \ldots, X_n$ occur. Now the question is whether these objects can be continuously transformed from the first into the second scenario so that none of the constraints in Σ is violated. In terms of Definition 1, we may reformulate this as follows: Are there continuous paths for the objects $X_1, \ldots, X_n$ so that the constraints of σ_s and σ_f hold at the starting and the endpoint, respectively, and the constraints of Σ hold everywhere in the *closed* interval defined by starting and endpoint.

It is clear that this problem can also be expressed in terms of the temporalized spatial constraint language presented in the previous section since a problem instance $\langle \Sigma, \sigma_s, \sigma_f \rangle$ is satisfiable in a fixed topological model if and only if the constraint set

$$I_s \text{ m } J, \quad J \text{ m } I_f, \quad I_s : \sigma_s \cup \Sigma, \quad J : \Sigma, \quad I_f : \sigma_f \cup \Sigma$$

is satisfiable in a suitably chosen strictly continuous tt-model based on the reals and the topological model at hand (here I_s, I_f, and J denote intervals: I_s is an interval in which the start scenario holds, I_f an interval in which the final scenario is true, and J is the interval in which the transformation occurs).

Transformation problems $\langle \Sigma, \sigma_s, \sigma_f \rangle$ can easily be solved by using the information encoded in the classical neighborhood graph of the spatial calculus at hand (e. g., Fig. 2) *if* transformations of at most two objects are considered. But, in general, this method already fails for more than two objects. As an examples consider the scenario $\{X \text{ EQ } Y, Y \text{ EQ } Z, X \text{ EQ } Z\}$. By only applying the information encoded in the classical neighborhood graph, we cannot conclude that every change of the first constraint $X \text{ EQ } Y$ results in a change of at least one of the other two constraints [cf. 13].

The main idea to solve transformation problems is the following: Try to find a partitioning of the transformation interval into subintervals such that in each of these subintervals only a minimal number of objects has to be transformed, but the sequence of subintervals describes for all objects a continuous transformation from the start into the final scenario. This means that a transformation problem is satisfiable if there exists a sequence of satisfiable transformation problems $\langle \Sigma, \sigma_s^1, \sigma_f^1 \rangle, \ldots, \langle \Sigma, \sigma_s^m, \sigma_f^m \rangle$ with $\sigma_s^1 = \sigma_s$, $\sigma_f^m = \sigma_f$, and $\sigma_f^k = \sigma_s^{k+1}$. This means, there is a chance to solve large and complicated transformation problems by solving transformation problems for a restricted number of objects.

For such restricted problems it is interesting to precompute a generalized neighborhood graph, which encodes possible transformations for a fixed number of objects. For this we represent possible scenarios for n objects $X_1, \ldots, X_n$ as $\binom{n}{2}$-tuples of spatial base relations $(S_{ij})_{1 \leq i < j \leq n}$ where S_{ij} is the spatial base relation that holds between X_i and X_j in that scenario. Note that not each such tuple is a consistent representation of a spatial configuration, but when it does, we will refer to it as an n-*scenario*. Note that, for a spatial calculus with k relations, usually only a subset of all $k^{\binom{n}{2}}$ many tuples represents a scenario. For RCC5, for example, only 54 from $5^3 = 125$ possible triples are scenarios of three objects.

Definition 4. *The (n,l)-neighborhood graph (for a fixed spatial model of some spatial calculus) is the directed graph $G = (V, E)$ defined by the following data: V is the set of all n-scenarios and E contains a directed edge from an n-scenario s to a distinct n-scenario s' if and only if s can be continuously transformed to s' by changing at most l objects.*

In the case of RCC5, both the $(2,1)$- and the $(2,2)$-neighborhood graph coincide with the classical neighborhood graph presented in Fig. 2.[2] But as previously argued, a necessary condition for solving transformation problems is to solve for each triple of objects X_i, X_j, X_k, the transformation problem restricted to these three objects. Hence in what follows we focus on the $(3,1)$-neighborhood graph. In this graph an edge from one vertex to another edge represents that exactly one object is subject to a continuous transformation, while all others are considered fixed (i. e., their object paths are (locally) constant functions). If, for instance, the first object changes its position with respect to the second object, then a scenario (r_1, r_2, r_3) can be connected to (r_1', r_2', r_3') only if (in the classical neighborhood graph) $\Delta(r_1, r_1') = 1$, $\Delta(r_2, r_2') = 0$, and $\Delta(r_3, r_3') \leq 1$.

The $(3,1)$-neighborhood has the nice feature that it can be computed easily by extensively using the information encoded in the classical neighborhood graph and in the composition table of the spatial calculus at hand. More precisely, the algorithm *GNG* presented in Fig. 3 takes as input a list of base relations (denoted by *rel_list[]*) and an array representing the composition table (*compTable[i][j]* refers to the set of base re-

---

[2] Note that the $(2,1)$- and the $(2,2)$-neighborhood are not necessarily identical [cf. 18].

lations obtained by composing relations i and j). The function $neighbor(i)$ assigns to each base relation its set of neighbors w. r. t. the classical neighborhood graph.

The algorithm GNG works as follows: First, it generates a possible scenario (i, j, k), checks if it is consistent. If so, it calls a function that generates a list of all continuous successors of the scenario (i, j, k). Finally, this list is returned. In more detail, the Boolean function $isConsistent()$ checks for a triple (i, j, k), whether k is contained in $compTable[i][j]$, in other words, whether relation k can hold between objects X and Y if there exists an object Z such that $X i Z$ and $Z j Y$ are true. The function $Succ()$ generates for a triple (i, j, k) all consistent successors into which the scenario can be transformed. This is done in the following way: For the input (i, j, k) the algorithm successively generates a neighbor relation for each relation of the triple. If, for instance, l is a neighbor relation of i, the algorithm checks if (l, j, k) is consistent. If so, the relation is added to the list of possible successors. This models the qualitative change of object X in relation to some fixed objects Y and Z. By changing object X, its qualitative relation to Z can also be affected. Since in this case object Y and Z are considered fixed, the relation j cannot change. The same is analogously done for the second and third relation of the triple.

Proposition 5. *The $(3,1)$-neighborhood graph computed by the algorithm GNG is (semantically) correct, if the algorithm is applied to a correct $(2,1)$-neighborhood graph and a correct composition table.*

By applying this proposition to Lemma 2 we obtain that for RCC5 the $(3,1)$-graph computed by GNG is correct for each class of strictly continuous tt-models that instantiate all base relations. This graph (a subgraph of it is depicted in Fig. 5) has 54 vertices and 291 edges.[3] We also applied this algorithm for computing the $(3,1)$-neighborhood graphs of the point algebra (PA) thought of as a spatial calculus. In this case the graph consists of 13 vertices and 24 edges (cf. Fig. 4).

To put things a little bit further, we can define a refined consistency concept for transformation problems.

Definition 6. *A transformation problem $\langle \Sigma, \sigma_s, \sigma_f \rangle$ is said to be (n, l)-consistent if for each subscenario of σ_s consisting of n objects $X_1, \ldots, X_n$, there exists a path in the (n, l)-neighborhood graph to the corresponding subscenario of σ_f (the subscenario for $X_1, \ldots, X_n$) such that no constraint of Σ is violated.*

This consistency concept can be useful, when impossible transformations are to be identified. Since a problem instance with m objects is satisfiable only if it is (n, l)-consistent for all $n, l \le m$, we can apply the (n, l)-neighborhood graph in order to find impossible transformations. To illustrate this, let us discuss the following example for the point algebra: Consider the PA-scenarios $\sigma_s = \{a < b, b < c, a < c\}$ and $\sigma_f = \{a < b, b > c, a < c\}$. Can σ_s be transformed into σ_f if we forbid that $b = c$, i. e., $b\{<, >\}c \in \Sigma$? Certainly not, because a lookup in the $(3,1)$-neighborhood graph shows that there is no path between the corresponding vertices. This can of course also be used

---

[3] A representation of the full $(3,1)$-neighborhood graph for RCC5 is available to the public at ftp://ftp.informatik.uni-freiburg.de/documents/papers/ki/ragni-woelfl-nghood.pdf.

```
def GNG ( rel_list [], compTable[][])
for  i, j, k in rel_list[]:
    if  isConsistent (i, j, k) :
       Succ(i, j, k)
    else :
       output "scenario (i, j, k) is inconsistent";
    output "all successors of (i, j, k)": Succ(i, j, k);


def function  isConsistent (i, j, k)
  if  k ∈ compTable[i][j]:
     return true ;
  else return  false ;


def function  Succ(i, j, k)
  succArray[];
  for   l ∈ {i, j, k}:
   if  l = i :
      for  m ∈ neighbor(i):
         if  isConsistent (m, j, k) && (m, j, k) ∉ succArray[]:
            succArray[] = succArray[] ∪ (m, j, k);
         for  n ∈ neighbor(k):
            if  isConsistent (m, j, n) && (m, j, n) ∉ succArray[]:
               succArray[] = succArray[] ∪ (m, j, n);
   if  l = j :
      for  m ∈ neighbor(j):
         if  isConsistent (i, m, k) && (i, m, k) ∉ succArray[]:
            succArray[] = succArray[] ∪ (i, m, k);
         for  n ∈ neighbor(i):
            if  isConsistent (n, m, k) && (n, m, k) ∉ succArray[]:
               succArray[] = succArray[] ∪ (n, m, k);
   if  l = k:
      for  m ∈ neighbor(k):
         if  isConsistent (i, j, m) && (i, j, m) ∉ succArray[]:
            succArray[] = succArray[] ∪ (i, j, m);
         for  n ∈ neighbor(j):
            if  isConsistent (n, j, m) && (n, j, m) ∉ succArray[]:
               succArray[] = succArray[] ∪ (n, j, k);
  output succArray[]                    .
```

Fig. 3. The algorithm GNG computes the $(3,1)$-neighborhood graph from a set of base relations, a composition table, and a classical neighborhood graph

in more complex formal calculi. For a given transformation problem, check first if for each pair of objects X and Y with $X r_1 Y \in \sigma_s$ and $X r_2 Y \in \sigma_f$, there is a path from r_1 to r_2 in the classical neighborhood graph. Then for each triple of objects X, Y, and Z with $X r_1 Y, Y r_2 Z, X r_3 Z \in \sigma_s$ and $X r_1' Y, Y r_2' Z, X r_3' Z \in \sigma_f$, check if there is a path in the $(3,1)$-, $(3,2)$-, or $(3,3)$-neighborhood graph from (r_1, r_2, r_3) to (r_1', r_2', r_3') so that during that none of the constraints in Σ is violated, and so on.

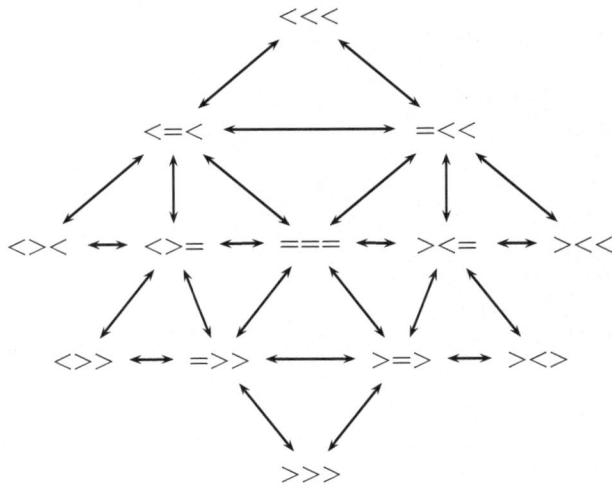

Fig. 4. The generalized neighborhood graph of the point algebra. The relation triple $r_1 r_2 r_3$ in the node encode a scenario for the constraints $X r_1 Y$, $Y r_2 Z$, and $X r_3 Z$.

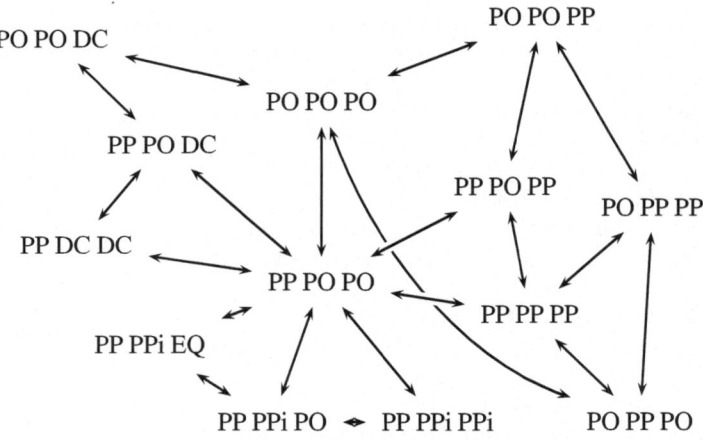

Fig. 5. A subgraph of the $(3, 1)$-neighborhood graph for RCC5

5   Summary and Outlook

We started from the question how spatial constraint calculi can be temporalized. In this context, we presented a precise notion of continuous change that seems to be conceptually adequate for temporalized topological calculi. Such a precise concept is necessary for a well-founded semantics of temporalized calculi dealing with continuous transformations. Moreover, it can be used to analytically prove the correctness of neighborhood graphs.

In a second step we considered so-called transformation problems, i. e., problems of the kind whether some spatial configuration can be continuously transformed into another configuration, even if these transformations are constrained by further conditions. Solving such problems may be especially interesting, for instance, if we want to plan how objects have to be moved in space in order to reach a specific goal state.

The classical neighborhood graphs discussed in the literature only represent possible continuous transformations of at most two objects. We proposed a concept that eliminates this limitation. These generalized neighborhood graphs may also be considered an appropriate tool for solving transformation problems, because they mirror the notion of k-consistency known from static reasoning problems. This idea is reflected in our definition of (n, l)-consistency.

Future work will be concerned with the following questions: What is the exact relationship between (n, l)-consistency and satisfiability of transformation problems? Are there tractable classes of these problems? How can the notion of (n, l)-consistency be used to identify such classes? Finally, how can temporalized QSR be used to solve spatial planning scenarios?

Acknowledgments

This work was partially supported by the Deutsche Forschungsgemeinschaft (DFG) as part of the Transregional Collaborative Research Center SFB/TR 8 Spatial Cognition. We would like to thank Bernhard Nebel for helpful discussions. We also gratefully acknowledge the suggestions of three anonymous reviewers, who helped improving the paper.

References

[1] J. F. Allen. Maintaining knowledge about temporal intervals. *Communications of the ACM*, 26(11):832–843, 1983.

[2] B. Bennett. Space, time, matter and things. In *FOIS*, pages 105–116, 2001.

[3] B. Bennett, A. G. Cohn, F. Wolter, and M. Zakharyaschev. Multi-dimensional modal logic as a framework for spatio-temporal reasoning. *Applied Intelligence*, 17(3):239–251, 2002.

[4] B. Bennett, A. Isli, and A. G. Cohn. When does a composition table provide a complete and tractable proof procedure for a relational constraint language? In *Proceedings of the IJCAI97 Workshop on Spatial and Temporal Reasoning*, Nagoya, Japan, 1997.

[5] A. G. Cohn. Qualitative spatial representation and reasoning techniques. In G. Brewka, C. Habel, and B. Nebel, editors, *KI-97: Advances in Artificial Intelligence*. Springer, 1997.

[6] E. Davis. Continuous shape transformation and metrics on regions. *Fundamenta Informaticae*, 46(1-2):31–54, 2001.

[7] M. J. Egenhofer and K. K. Al-Taha. Reasoning about gradual changes of topological relationships. In A. U. Frank, I. Campari, and U. Formentini, editors, *Spatio-Temporal Reasoning*, Lecture Notes in Computer Science 639, pages 196–219. Springer, 1992.

[8] M. Erwig and M. Schneider. Spatio-temporal predicates. *IEEE Transactions on Knowledge and Data Engineering*, 14(4):881–901, 2002.

[9] C. Freksa. Conceptual neighborhood and its role in temporal and spatial reasoning. In *Decision Support Systems and Qualitative Reasoning*, pages 181–187. North-Holland, 1991.

[10] D. Gabelaia, R. Kontchakov, A. Kurucz, F. Wolter, and M. Zakharyaschev. Combining spatial and temporal logics: Expressiveness vs. complexity. To appear in Journal of Artificial Intelligence Research, 2005.

[11] A. Galton. *Qualitative Spatial Change*. Oxford University Press, 2000.

[12] A. Galton. A generalized topological view of motion in discrete space. *Theoretical Compututer Science*, 305(1-3):111–134, 2003.

[13] A. Gerevini and B. Nebel. Qualitative spatio-temporal reasoning with RCC-8 and Allen's interval calculus: Computational complexity. In *Proceedings of the 15th European Conference on Artificial Intelligence (ECAI-02)*, pages 312–316. IOS Press, 2002.

[14] M. Knauff. The cognitive adequacy of allen's interval calculus for qualitative spatial representation and reasoning. *Spatial Cognition and Computation*, 1:261–290, 1999.

[15] P. Muller. A qualitative theory of motion based on spatio-temporal primitives. In A. G. Cohn, L. K. Schubert, and S. C. Shapiro, editors, *Proceedings of the Sixth International Conference on Principles of Knowledge Representation and Reasoning (KR'98), Trento, Italy, June 2-5, 1998*, pages 131–143. Morgan Kaufmann, 1998.

[16] P. Muller. Topological spatio-temporal reasoning and representation. *Computational Intelligence*, 18(3):420–450, 2002.

[17] B. Nebel and H.-J. Bürckert. Reasoning about temporal relations: A maximal tractable subclass of Allen's interval algebra. Technical Report RR-93-11, Deutsches Forschungszentrum für Künstliche Intelligenz GmbH, Kaiserslautern, Germany, 1993.

[18] M. Ragni and S. Wölfl. Branching Allen: Reasoning with intervals in branching time. In C. Freksa, M. Knauff, B. Krieg-Brückner, B. Nebel, and T. Barkowsky, editors, *Spatial Cognition*, Lecture Notes in Computer Science 3343, pages 323–343. Springer, 2004.

[19] D. A. Randell, Z. Cui, and A. G. Cohn. A spatial logic based on regions and connection. In B. Nebel, W. Swartout, and C. Rich, editors, *Principles of Knowledge Representation and Reasoning: Proceedings of the 3rd International Conference (KR-92)*, pages 165–176. Morgan Kaufmann, 1992.

[20] M. B. Vilain, H. A. Kautz, and P. G. van Beek. Contraint propagation algorithms for temporal reasoning: A revised report. In D. S. Weld and J. de Kleer, editors, *Readings in Qualitative Reasoning about Physical Systems*, pages 373–381. Morgan Kaufmann, 1989.

[21] F. Wolter and M. Zakharyaschev. Spatio-temporal representation and reasoning based on RCC-8. In A. Cohn, F. Giunchiglia, and B. Selman, editors, *Principles of Knowledge Representation and Reasoning: Proceedings of the 7th International Conference (KR2000)*. Morgan Kaufmann, 2000.

Design of Geologic Structure Models with Case Based Reasoning

Mirjam Minor and Sandro Köppen

Humboldt-Universität zu Berlin, Institut für Informatik,
Unter den Linden 6, D-10099 Berlin
minor@informatik.hu-berlin.de

Abstract. This paper describes a new approach of case based reasoning for the design of geologic structure models. It supports geologists in analysing building ground by retrieving and adapting old projects. Projects are divided in several complex cases consisting of pairs of drilling cores with their outcrop and the spatial context. A three-step retrieval algorithm (1) selects structurally similar candidate cases, (2) estimates the quality of a transfer of the layer structures to the query, and (3) retrieves from the remaining set of promising cases the most geologically and geometrically similar cases. The case based system is able to adapt a similar case to the query situation and to export the result to a 3D visualisation tool. A first evaluation has shown that the case based system works well and saves time.

1 Introduction

Designing geologic structure models is a time-consuming process that requires a lot of geological knowledge. Legal regulations require the careful analysis of drilling data resulting in a 3D model of building ground. CAD (computer aided design) programs support the geologists by drawing tools when they determine the stone layers between drilling cores.

In this paper, we present a case based approach to support the design process by the reuse of experience from former geologic projects. Section 2 describes the design procedure for geologic structure models. Section 3 deals with the division of geologic projects into reusable units in form of cases. Section 4.3 deals with the extended similarity computation and the adaptation. A sample application with first results is given in Section 5. Section 6 discusses related work, and, finally, Section 7 draws a conclusion.

2 Construction of Geological Structure Models

When a real estate is developed, a geological structure model has to be constructed according to law. Therefore, a geologist uses information from geologic outcrops (drillings) and geologic basic knowledge about the area. For instance, if the land lies in the 'Nordische Tiefebene' in North Germany, certain types of glacial stone like sand and silt are expected.

U. Furbach (Ed.): KI 2005, LNAI 3698, pp. 79–91, 2005.

Usually, a structure model is a network of geologic outcrops in their spatial context which are connected via cross sections. A cross section describes the distribution of stone layers between two geological outcrops. Fig. 1 shows an example of three outcrops with two cross sections. The third cross section is analogical to the second one. The letters A, B, and C stand for different layers of stone.

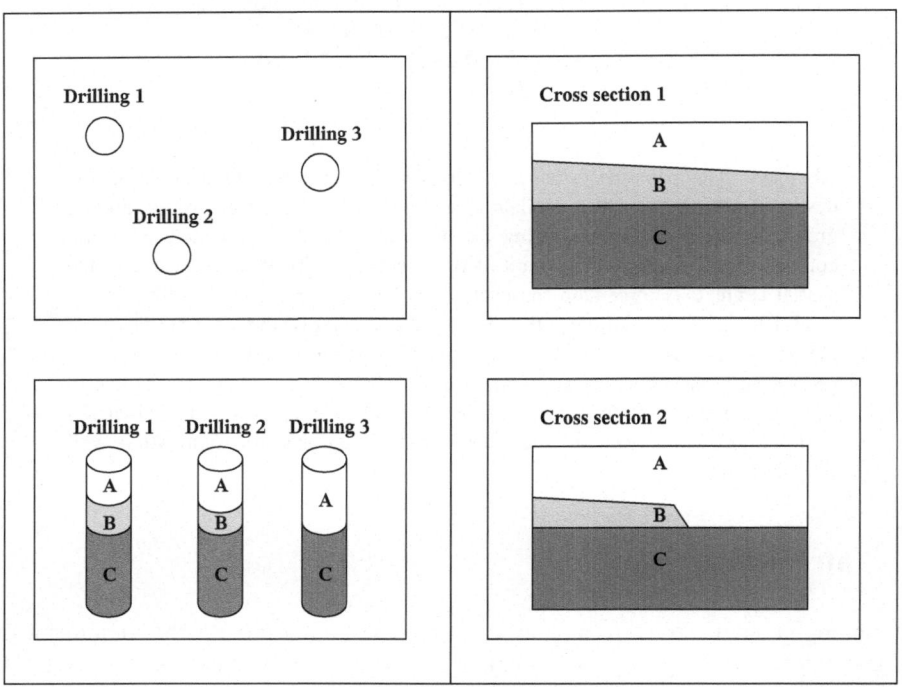

Fig. 1. Example of outcrops and cross sections between two drillings each

A **layer** is characterised by many properties:

- two depth parameters for the beginning and the ending of the layer,
- the strike and fall (vertical and horizontal direction) of the layer, and
- geologic parameters:
  - stratigraphy (ages, e.g., Jurassic, Cretaceous),
  - petrography (major and minor stone components),
  - genesis,
  - colour, and
  - additional information, e.g., bulged or canted.

The construction of the model, especially the specification of the particular cross sections, is performed in a two-dimensional representation as connected, planar graph. The nodes are the outcrops and the edges are the cross sections. To avoid conflicts at intersecting points, the topology of the graph consists of non-intersecting triangles that are computed by a standard triangulation algorithm.

The complete structural model is visualised in 3D. An example for the three out-crops from Fig. 1 with the triangle of three cross sections is shown in Fig. 2. A CAD (computer aided design) program should support both. The main focus of our work is on the 2D construction of a cross section graph.

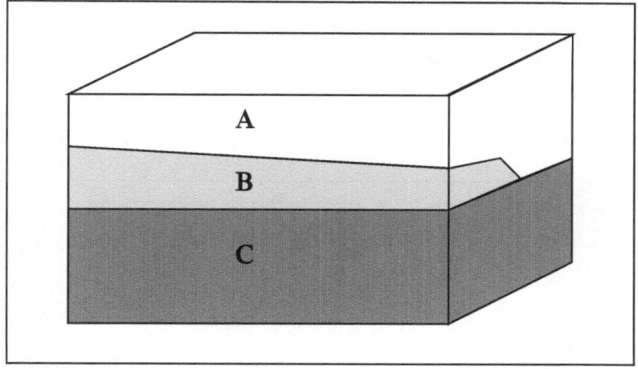

Fig. 2. Examples of a 3D structural model

3   Cases in a Cross Section Graph

The number of outcrops per project varies from only a few for the building land of a one-family house up to some hundred drillings for a rail route. So, we decided not to use projects as cases, they are hardly comparable. Instead, a **case** in our approach is a single cross section with its spatial context:

– The **problem part of a case** are the two outcrops at the border of the cross section and one or two additional outcrops;
– the **solution** is the distribution of layers in the cross section.

We distinguish two topological settings. Either the wanted cross section is at the border or within the convex closure of the triangulation graph (see Fig. 3). In case the cross section is within the convex closure of the graph, we need two additional outcrops (D3, D4) to describe the context, only one (D3), otherwise.

An **outcrop** is represented as sequence of layers beginning with the top layer and its location. Each layer is described by the parameters explained above. The location consists of x and y coordinates concerning the dimensions of the whole project plus the overall height of the outcrop. In fact, the 'height' means the depth of the layer, but the geologic term is height.

A **cross section** is represented as a graph (see Fig. 4). A node stands for the oc-currence of a layer in an outcrop. The vertical edges model the sequential order of the layers within an outcrop; the horizontal edges are for continuous layers from different outcrops. To represent the ending of a layer, we use virtual outcrops. In Fig. 4, the two virtual outcrops in the middle have non-bold nodes.

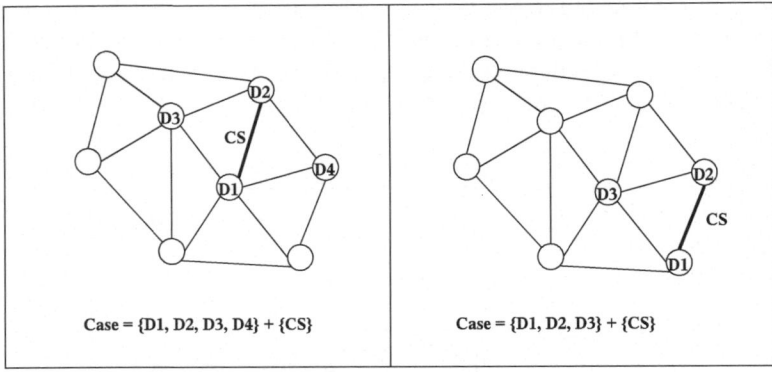

Fig. 3. Two different topological settings for cases

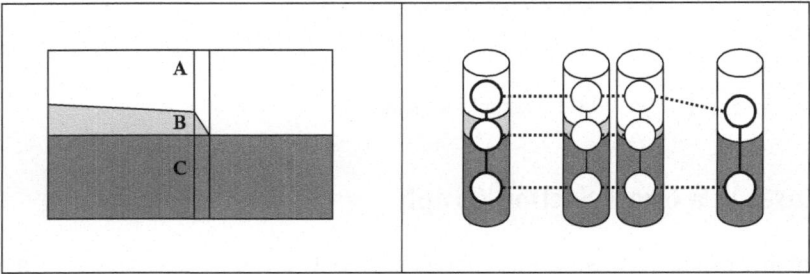

Fig. 4. Sample representation of a cross section

In this way, a project with 20 outcrops is divided into up to 54 cases (one per edge of the triangulation graph). The generated cases belong to the complex cases due to their structure.

4   Determining Useful Cases

To find useful cases to a query, we designed a three-step retrieval mechanism including a quality estimation in the middle:

1. Retrieve candidate cases that are structurally similar to the query,
2. Retrieve promising candidate cases that are structurally transferable to the query with a promising result:
    - reuse the structure of each candidate case for the query,
    - estimate the quality of this transfer, and
    - reduce the set of candidates by means of a threshold,
3. Retrieve the best matching cases by means of a detailed, composite similarity computation.

We will discuss the three steps and the adaptation in the following subsections after two general remarks on the mechanism. (1) We integrated step two afterwards what improved our retrieval results significantly. (2) In step one and three, we compared the according outcrops as illustrated in Fig. 5. And step two employs them analogically. Partner outcrops can be assigned directly or in mirrored form. To avoid effort at retrieval time, the mirrored cases could be stored in the case base, too.

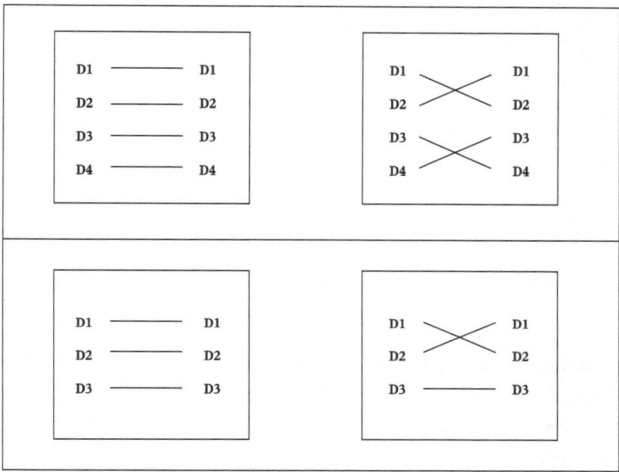

Fig. 5. Direct and mirrored comparison of outcrops for two topological settings

4.1 Structural Similarity

The first step of the retrieval mechanism is to find candidate cases that are structurally similar to the query. Therefore, we count the number of layers in the two main outcrops. A case must have an equal or higher number of layers on both sides than the query. The idea behind this is, that cases covering more than the query might be useful if there are not any cases available that have exactly the same numbers of layers. The lower layers of the case need to be cut off, as the upper layers are more important than the deeper ones. We have structured the case storage by means of kd-trees [7] (see Fig. 6) where the first $\geq$ means the number of layers on the bigger side and the second $\geq$ the number of layers on the smaller side.

Note, that the queries are asked automatically for both, the direct and the mirrored form. The tree contains many redundant case pointers what simplifies the retrieval step one. It needs only to identify the matching leaf. Is this leaf not available as the cases are all too small we take the rightmost node to handle at least the upper part of the query. The result of the retrieval step one is the set of all cases connected with the selected leaf node.

4.2 Structural Transferability

The second step of the retrieval mechanism reduces the set of candidate cases to those that are structurally transferable with a promising quality estimation. Each candiate

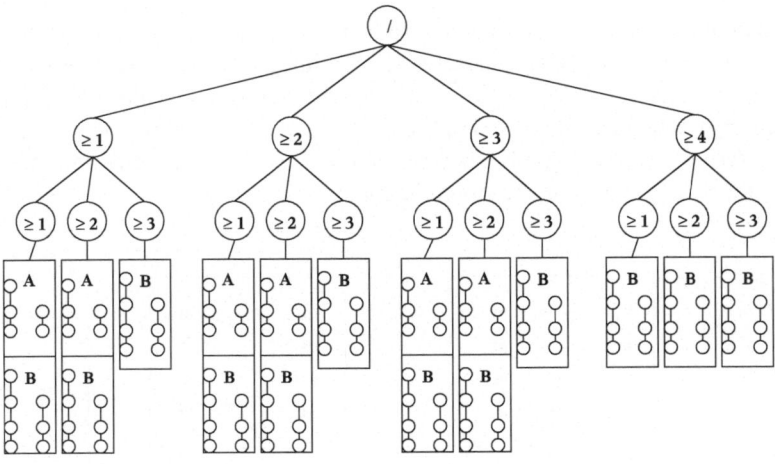

Fig. 6. Organisation of the case storage with a kd-tree

case from step one is applied to the query as follows: Its structure is transferred to the query outcrops to build hypothetical connections of layers in the query. The connections are tranferred top down. The overall quality of the structural transfer is estimated by composing two values:

1. similarity of the proportion of layers,
2. quality of the hypothetical connection of layers.

The similarity of the proportion of layers is computed by $sim_{pol} : CASE \times CASE \rightarrow [0,1]$:

$$sim_{pol}(case_1, case_2) = \begin{cases} \frac{\alpha_1}{\alpha_2} & , \quad \alpha_1 < \alpha_2 \\ \frac{\alpha_2}{\alpha_1} & , \quad \text{else,} \end{cases}$$

where α_i is the ratio of the numbers of layers in $case_i$:

$$\alpha_i = \frac{number\ of\ layers\ of\ the\ first\ outcrop}{number\ of\ layers\ of\ the\ second\ outcrop}.$$

For instance, if $case_1$ has two and three layers and $case_2$ has three and three layers then is $\alpha_1 = 2/3$, $\alpha_2 = 1$, and $sim_{pol}(case_1, case_2) = \frac{2/3}{1} = 2/3$.

The quality of the hypothetical connection of layers is estimated layer by layer l_i from both outcrops of the hypothetical case:

$$quality_{hypo\_sim}(case, hypothetical\_case) =$$

$$\frac{1}{N+M}(\textstyle\sum_{i=1}^{N} quality(l_i^{outcrop\_1}, case, hypothetical\_case)+$$

$$\textstyle\sum_{i=1}^{M} quality(l_i^{outcrop\_2}, case, hypothetical\_case)).$$

N and M are the numbers of layers in outcrop_1 and outcrop_2 of the hypothetical case. The quality function for single layers $quality(l^{outcrop\_i}, case, hypothetical\_case)$ uses

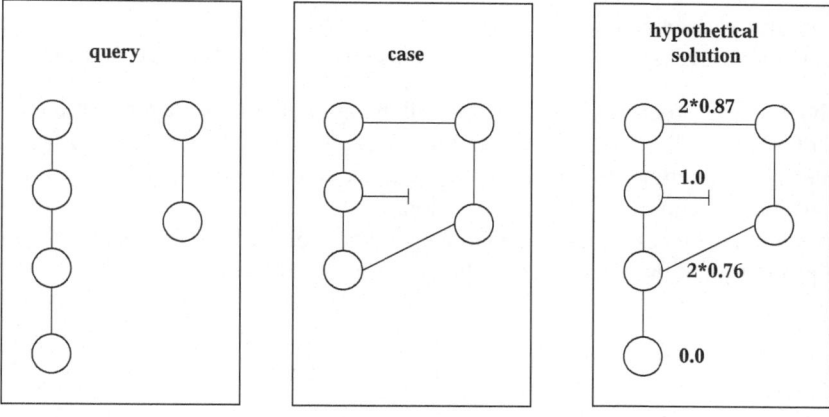

Fig. 7. The quality of a hypothetical solution reusing the structure of a case

the geologic similarity function for layers sim_l (see Section 4.3) to compute the similarity between a layer and its hypothetical partner section in the connected outcrop:

$$quality(l^{o-i}, case, h\_case) = \begin{cases} 0, & l^{o-i} \notin case \\ 1, & l^{o-i} \in case \text{ not connected} \\ sim_l(l^{o-i}, l^{o-i+-1}), & \text{else.} \end{cases}$$

An example of $quality_{hypo\_sim}$ is illustrated in Fig. 7.

The composite value for the quality estimation is $1/2 * (sim_{pol} + quality_{hypo\_sim})$. For the sample in Fig. 7, it is $1/2 * [\frac{3/2}{4/2} + \frac{1}{4+2}((0.87 + 1 + 0.76 + 0) + (0.87 + 0.76))]$ $= 1/2 * [0.75 + 0.71] = 0.73$.

The result of retrieval step two is the set of candidate cases that reached a quality estimation value above a threshold.

4.3   Final Similarity Computation

After determining promising candidate cases in step one and two of the retrieval mechanism, in step three the most similar cases to the query are computed by a 'usual' retrieval process.

The detailed and final similarity computation for a query and a case uses a composite similarity function $sim(case_1, case_2)$ that aggregates geologic and geometric similarity values:

$$sim(case_1, case_2) = 1/2(sim_{geo}(case_1, case_2) + sim_{geometric}(case_1, case_2)).$$

Geologic Similarity. The geologic similarity relates to properties of the stone. It needs four levels:

1. The *first level similarity* compares single attributes of a layer.
2. The *second level similarity* aggregates first level similarity values to compare one layer with another.

3. The *third level similarity* aggregates similarity values of layers to compare two outcrops with each other, i.e. sequences of layers.
4. The *fourth level similarity* regards the constellation of all outcrops of a case.

The *first level similarity* considers the following attributes of a layer (compare properties of a level in Section 2): upper and lower depth of the layer, strike and fall, stone parameters for stratigraphy, petrography, genesis, colour, and additional information. Each attribute has an own similarity function.

The depth values are transformed to the thickness of the layer. The similarity sim_Δ is the proportion of thicknesses Δ_1 of layer l_1 and Δ_2 of layer l_2:

$$sim_\Delta(l_1, l_2) = \begin{cases} \frac{\Delta_1}{\Delta_2} & , \quad \Delta_1 < \Delta_2 \\ \frac{\Delta_2}{\Delta_1} & , \quad \text{else.} \end{cases}$$

The strike and fall of a layer is compared by means of angles α and β. α describes the difference between the strike direction of the layer and the cross section. β is the fall of the layer itself as the relation to the cross section is already described by the thicknesses within the outcrops. A sample illustration is shown in Fig. 8.

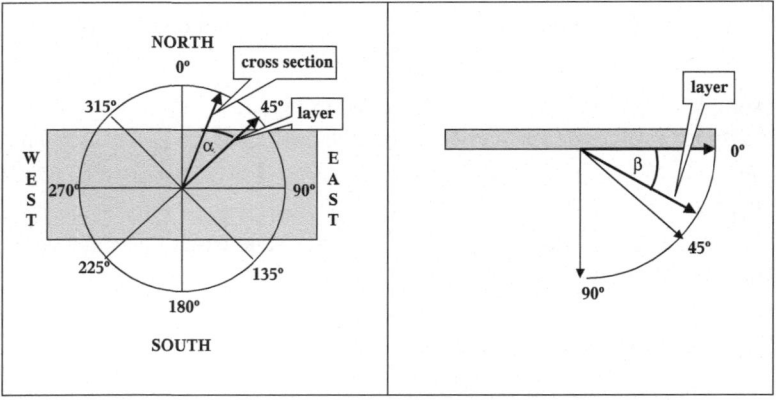

Fig. 8. Strike and fall of a layer within a cross section

The similarity $sim_{s\&f}$ composes the strike and fall in equal shares:
$sim_{s\&f}(l_1, l_2) = 1/2(\frac{180-(\alpha_1-\alpha_2)}{180} + \frac{90-(\beta_1-\beta_2)}{90})$.
The similarity values for stratigraphy, petrography, genesis, colour, and additional information are stored in similarity tables of possible values. We used manyfold types of geological knowledge to specify the values for the tables, e.g., on the hierarchy of ages or on the grain size of sedimentary earth. Due to space limits, we present just a section of the table for petrographic similarity, namely for magma stones in Table 4.3.

The *second level similarity* computes the similarity of two layers $sim_l(l_1, l_2)$ by composing the first level similarity values with the following weights:

Table 1. Similarity values sim_{pet} for magma stones

	+G	+Sy	+DR	+Gb	+R	+Dz	+B	+Lt	+Tr
+G	1	0.6	0.75	0.4	0.3	0.2	0.1	0.1	0.1
+Sy	0.6	1	0.5	0.4	0.1	0.1	0.1	0.2	0.3
+Dr	0.75	0.5	1	0.75	0.1	0.1	0.3	0.1	0.1
+Gb	0.4	0.4	0.75	1	0.1	0.1	0.3	0.1	0.1
+R	0.3	0.1	0.1	0.1	1	0.75	0.4	0.5	0.5
+Dz	0.2	0.1	0.1	0.1	0.75	1	0.75	0.4	0.3
+B	0.1	0.1	0.3	0.3	0.4	0.75	1	0.7	0.4
Lt+	0.1	0.2	0.1	0.1	0.5	0.4	0.7	1	0.7
Tr+	0.1	0.3	0.1	0.1	0.5	0.3	0.4	0.7	1

$$sim_l(l_1, l_2) = \frac{2sim_\triangle + 2sim_{s\&f} + 2sim_{str} + 9sim_{pet}^m + 3sim_{pet}^s + 7sim_g + sim_c + sim_a}{28}.$$

$sim_\triangle$ compares the thickness of l_1 and l_2, $sim_{s\&f}$ the strike and fall, sim_{str} the stratography, sim_{pet}^m the main mixture, sim_{pet}^s the secondary mixture, sim_g the genesis, sim_c the colour, and sim_a the additional information.

The *third level similarity* computes the similarity of two outcrops $sim_o(o_1, o_2)$ by composing the second level similarity values as normalized sum:

$sim_o(o_1, o_2) = (1/N \sum_{i=1}^{N} sim_l(l_i^{o_1}, l_i^{o_2}))$.

N is the minimum number of layers of o_1 and o_2.

The *forth level similarity* computes the similarity of two constellations of outcrops $sim_{geo}(case_1, case_2)$ by composing the third level similarity values with a double weight for the main outcrops o_1 and o_2:

$sim_{geo}(case_1, case_2) = 1/6(2sim_o(o_1^{case_1}, o_1^{case_2}) + 2sim_o(o_2^{case_1}, o_2^{case_2}) + sim_o(o_3^{case_1}, o_3^{case_2}) + sim_o(o_4^{case_1}, o_4^{case_2}))$,

where $sim_o(o_4^{case_1}, o_4^{case_2})$ is substituted by 0 if one of the outcrops $o_4^{case_1}$, $o_4^{case_2}$ is not existing, and 1 if both are not existing.

Geometric Similarity. The geometric similarity relates to the location of outcrops. The location of outcrops consists of the height values of the outcrops as well as their position to each other. Both components have to be computed from the absolute values stored in the outcrop representations.

The height values of the two main outcrops provide the angle of the slope of the land. If we have, for instance, two pairs of drillings with ascending slopes of 17 and 5 degree. Then the straight-forward similarity is the difference of the angles (12 in the example) divided through 180 as the maximum difference is 180 degrees. The similarity of the two angles in our example is about 0.93. To reach more realistic values, we think about taking the square of this value.

The absolute positions of the three or four outcrops in a case can be used to estimate how 'spacious' the case is. We compute the sides of the triangle or tetragon of outcrops and compare them with those of the query by $1/3(a/a' + b/b' + c/c')$ for triangles and $1/4(a/a' + b/b' + c/c' + d/d')$ for tetragons. An alternative is to compare the areas, e.g., computed by taking the formula of Heron for triangles and Green's theorem for tetragons.

The results of the difference of angles and the size estimation are combined to
$sim_{geometric}(case_1, case_2) = 1/2(sim_{angle}(case_1, case_2) + sim_{size}(case_1, case_2))$.

All values and functions used in this section can be modified independently as the similarity function is a composite one - despite of the manyfold ways to determine the partial values. The overall result of retrieval step three is an ordered set of the best matching cases from the set of promising candidate cases that has been the output of step two.

4.4   Adaptation of a Cross Section Case

The user can select a case from the retrieval result and ask the system for adaptation. Then, the user modifies or cancels the adapted case. The automatic adaptation concerns

- the coordinates of outcrops and layers,
- the layer names, and
- the thickness of layers in the virtual outcrops.

The coordinates of the cross section have to be transferred: The original section is rotated and its length has to be scaled to the new outcrop coordinates.The relative position of a virtual outcrop must be preserved. The height of the virtual outcrop has to be modified.

The names of the layers can be copied top down to the layers of the query case. Deeper layers of the original case might be cut off for a query case.

The layers of the virtual outcrops have to be adapted to the thickness of the layers in the main outcrops (see Fig. 9). A linear interpolation is not realistic as more than one virtual outcrops may lie between two real outcrops. Instead, we go top down for each layer of the virtual outcrop and choose the nearest connected layer in a real outcrop. The proportion of the thickness of both layer parts of the original case is transferred to the dimensions of the new case. Has the virtual layer in the original half the size of the real layer, the virtual layer in the new case gets also half the size of the real layer in the new case.

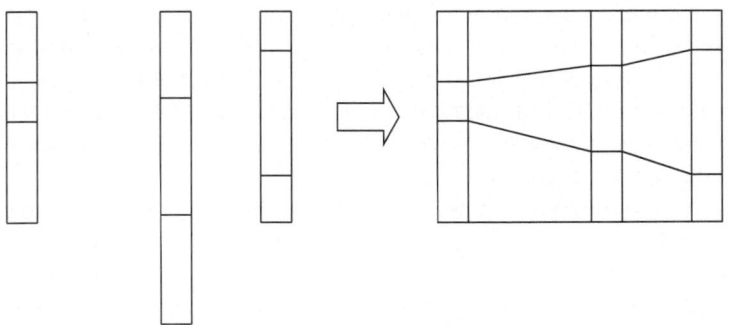

Fig. 9. Scaling virtual layers

5   Application and Evaluation

A sample application is prototypically implemented in Visual Basic. It has an interface to the CAD program Autocad for the 3D visualisation according to the DIN4023 norm. The sample case base contains 37 cases of six solved projects from the 'Norddeutsche Tiefebene', a glacial part of North Germany. The most projects cover an area of about 100 x 100 meters with a depth of 10 meters.

The test case base consists of cases with horizontal bedding and cases with ending layers. A frequent geologic structure in projects is a lens. An example of a lens structure is shown in Fig. 10. The project of this example has four outcrops at the corners and one outcrop in the middle of the lens. It is described by eight cases.

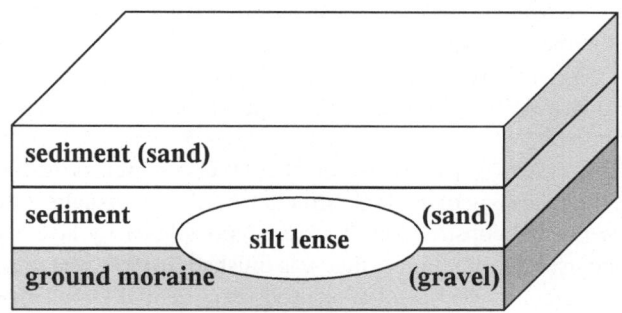

Fig. 10. 3D visualisation of a project with a lens structure

To fill the sample case base, we solved only five cases with different structures by hand. The remaining 32 cases have been generated incrementally by the case based system. 9 of the 32 cases needed to be modified manually, 23 cases have been solved successfully and required only slight or none adaptation. Table 5 shows a manual evaluation of the test cases grouped by overlapping geologic phenomena.

For further tests, the five completely manually solved cases have been asked as queries to the overall case base. The case itself was found in all cases and reached always one of the highest similarity values. The time savings during the editing of the

Table 2. Six test projects with evaluation (from 0 = bad to 10 = perfect

description	out-crops	cases	new	;-)	;-(	evalu-ation
horizontal	4	5	1	0	4	9
ending layer	4	5	1	1	3	8
lens structure	5	8	0	1	7	9
strong slope	4	5	0	3	2	7
shallow slope	4	5	0	2	3	8
alluvial channel	6	9	3	2	4	6

32 cases were about 15 minutes to half an hour per case compared to the construction of a cross section with Autocad only. This lets us hope that the system will be very useful in real scenarios.

Reducing the system to the third step of the retrieval led to unacceptable results. We think about presenting also the result of the second retrieval step to allow the user to select one of those cases in case the third retrieval step has not produced sufficient cases.

6 Related Work

Conventional software to support the design of geologic structure models are CAD programs. They use fix rules to construct cross sections or provide only graphical support. Surpac Minex Groups' [6] Drillking has been developed for the collection of drilling data for mines. Drillking provides tables to entry the data and generates 3D models with the tool Surpac-Vision. The design task has to be solved manually. Drillking is not conform to the German DIN norm.

Geographix' [3] GESXplorer has been developed for the oil production and works similar to Drillking. Additionally, it provides support for project management. It is not DIN conform, too.

The PROFILE system [5] is the result of a DFG research project on developing methods for the digital design of geologic structure models. It visualises the drilling data and supports the manual construction of cross sections layer by layer or drill core by drill core. A conclusion of the project that was finished in 1997 was that the automatic design is not possible at the moment as it needs experiential knowledge. We confirm this statement and have shown that case based reasoning technique helps to go one step further than PROFILE in supporting human geologists.

The FABEL project [1] employs case based reasoning for design tasks in architecture. It uses conceptual clustering and conceptual analogy to find useful cases from the past. The case memory is clustered according to hierarchically organised concepts, e.g. the numbers of pipes and connections in a pipe system. The cases are represented as graphs. The geologic design task in our approach is easier to solve than the architectural design: The projects can be divided into independent cases of a less complex structure. So, we can use a composite similarity function instead of graph matching in FABEL.

O'Sullivan et al. ([4]) use case based reasoning for the analysis of geo-spatial images. They use an annotation-based retrieval on textual annotations to series of geospatial images. In [2], they presented an algorithm to perform shape matching with grids of pixels. The structural similarity in our approach is comparable to the combination of topological, orientation, and distance properties in [2]. Fortunately, we don't have to deal with noisy data like images.

Our work is not related to spatial reasoning as it does not use an explicit or general model of space. It just compares particular domain-specific spatial parameters with a specific similarity function, e.g. the depth and orientation of two outcrops.

7 Conclusion and Future Work

In this paper, we presented a case based approach to support geologists in designing structure models. We divided projects concerning a whole building ground into several cases to reduce the complexity and increase the comparability of the stored pieces of experience. Before a composite similarity function performs a classical retrieval process, the structurally best matching cases are preselected by means of an estimation how successfull their adaptation to the query might be. Many types of geologic knowledge have been integrated to define the composite similarity function for geologic and geometric properties. We implemented a prototype and modelled the knowledge for the 'Norddeutsche Tiefebene'. The knowledge is reusable for projects with similar glacial imprinting. The evaluation with a small set of cases was successfull and saved about one man-day of geological work.

We would like to acknowledge the geo software company Fell-Kernbach for their support of this work!

References

1. K. Börner. CBR for Design. In M. Lenz, H.-D. Burkhard, B. Bartsch-Spörl, and S. Weß, editors, *Case-Based Reasoning Technology — From Foundations to Applications*, LNAI 1400, Berlin, 1998. Springer Verlag.
2. J. D. Carswell, D. C. Wilson, and M. Bertolotto. Digital Image Similarity for Geo-spatial Knowledge Management. In S. Craw and A. Preece, editors, *Advances in Case-Based Reasoning: 6th European Conference, ECCBR-2002*, LNAI 2416, Berlin, 2002. Springer Verlag.
3. http://www.geographix.com/, 2005.
4. D. O'Sullivan, E. McLoughlin, M. Bertolotto, and D. C. Wilson. A case-based approach to managing geo-spatial imagery tasks. In P. Funk and P. A. González Calero, editors, *Advances in Case-Based Reasoning: 7th European Conference, ECCBR-2004*, number 3155 in Lecture Notes in Artificial Intelligence, pages 702–716, Berlin, 2004. Springer.
5. H. Preuß, M. Homann, and U. Dennert. Methodenentwicklung zur digitalen Erstellung von geologischen Profilschnitten als Vorbereitung zur 3D-Modellierung. Final report PR 236/3-1, DFG, Germany, 1997.
6. http://www.surpac.com/, 2005.
7. S. Weß, K.-D. Althoff, and G. Derwand. Using *k*d-trees to improve the retrieval step in case-based reasoning. In S. Wess, K.-D. Althoff, and M. M. Richter, editors, *Topics in Case-Based Reasoning: First European Workshop, EWCBR-9 3, selected papers*, number 837 in Lecture Notes in Artificial Intelligence, pages 167–181, Berlin, 1994. Springer.

Applying Constrained Linear Regression Models to Predict Interval-Valued Data

Eufrasio de A. Lima Neto, Francisco de A.T. de Carvalho,
and Eduarda S. Freire

Centro de Informatica - CIn / UFPE, Av. Prof. Luiz Freire,
s/n Cidade Universitaria, CEP: 50740-540 - Recife - PE - Brasil
{ealn, fatc}@cin.ufpe.br

Abstract. Billard and Diday [2] were the first to present a regression method for interval-value data. De Carvalho et al [5] presented a new approach that incorporated the information contained in the ranges of the intervals and that presented a better performance when compared with the Billard and Diday method. However, both methods do not guarantee that the predicted values of the lower bounds ($\hat{y}_{Li}$) will be lower than the predicted values of the upper bounds ($\hat{y}_{Ui}$). This paper presents two approaches based on regression models with inequality constraints that guarantee the mathematical coherence between the predicted values $\hat{y}_{Li}$ and $\hat{y}_{Ui}$. The performance of these approaches, in relation with the methods proposed by Billard and Diday [2] and De Carvalho et al [5], will be evaluated in framework of Monte Carlo experiments.

1 Introduction

The classical model of regression for usual data is used to predict the behaviour of a dependent variable Y as a function of other independent variables that are responsible for the variability of variable Y. However, to fit this model to the data, the estimation is necessary of a vector $\boldsymbol{\beta}$ of parameters from the data vector $\mathbf{Y}$ and the model matrix $\mathbf{X}$, supposed with complete rank p. The estimation using the *method of least square* does not require any probabilistic hypothesis on the variable Y. This method consists of minimizing the sum of the square of residuals. A detailed study on linear regression models for classical data can be found in [6], [9], [10] among others.

This paper presents two new methods for numerical prediction for interval-valued data sets based on the constrained linear regression model theory. The probabilistic assumptions that usually are behind this model for usual data will not be considered in the case of symbolic data (interval variables), since this is still an open research topic. Thus, the problem will be investigated as an optimization problem, in which we desire to fit the best hyper plan that minimizes a predefined criterion.

In the framework of *Symbolic Data Analysis*, Billard and Diday [2] presented for the first time an approach to fitting a linear regression model to an interval-valued data-set. Their approach consists of fitting a linear regression model to

U. Furbach (Ed.): KI 2005, LNAI 3698, pp. 92–106, 2005.

the mid-point of the interval values assumed by the variables in the learning set and applies this model to the lower and upper bounds of the interval values of the independent variables to predict, respectively, the lower and upper bounds of the interval value of the dependent variable.

De Carvalho et al [5] presented a new approach based on two linear regression models, the first regression model over the mid-points of the intervals and the second one over the ranges, which reconstruct the bounds of the interval-values of the dependent variable in a more efficient way when compared with the Billard and Diday method.

However, both methods do not guarantee that the predicted values of the lower bounds ($\hat{y}_{Li}$) will be lower than the predicted values of the upper bounds ($\hat{y}_{Ui}$). Judge and Takayama [7] consider the use of restrictions in regression models for usual data to guarantee the positiveness of the dependent variable. This paper suggests the use of constraints linear regression models in the approaches proposed by [2] and [5] to guarantee the mathematical coherence between the predicted values $\hat{y}_{Li}$ and $\hat{y}_{Ui}$.

In the first approach we consider the regression linear model proposed by [2] with inequality constrained in the vector of parameters $\boldsymbol{\beta}$ which guarantee that the values of $\hat{y}_{Li}$ will be lower than the values of $\hat{y}_{Ui}$. The lower and upper bounds of the dependent variable are predicted, respectively, from the the lower and upper bounds of the interval values of the dependent variables.

The second approach considers the regression linear model proposed by [5] that consist in two linear regression models, respectively, fitted over the mid-point and range of the interval values assumed by the variables on the learning set. However, the regression model fitted over the ranges of the intervals will take in consideration inequality constraints to guarantee that $\hat{y}_{Li}$ will be lower than $\hat{y}_{Ui}$.

In order to show the usefulness of these approaches will be predicted according with the proposed methods in independent data sets, the lower and upper bound of the interval values of a variable which is linearly related to a set of independent interval-valued variables. The evaluation of the proposed prediction methods will be based on the estimation of the average behaviour of the *root mean squared error* and of the *squared of the correlation coefficient* in the framework of a Monte Carlo experiment.

Section 2 presents the new approaches that fits a linear regression model with inequality constraints for interval-valued data. Section 3 describes the framework of the Monte Carlo simulations and presents experiments with artificial and real interval-valued data sets. Finally, the section 4 gives the concluding remarks.

2   Linear Regression Models for Interval-Valued Data

2.1   The Constrained Center Method

The first approach consider the regression linear model proposed by [2] with inequality constrained in the vector of parameters $\boldsymbol{\beta}$ to guarantee that the values of $\hat{y}_{Li}$ will be lower than the values of $\hat{y}_{Ui}$.

Let $E = \{e_1, \ldots, e_n\}$ be a set of the examples which are described by $p + 1$ intervals quantitative variables $Y, X_1, \ldots, X_p$. Each example $e_i \in E$ ($i = 1, \ldots, n$) is represented as an interval quantitative feature vector $\mathbf{z}_i = (\mathbf{x}_i, y_i)$, $\mathbf{x}_i = (x_{i1}, \ldots, x_{ip})$, where $x_{ij} = [a_{ij}, b_{ij}] \in \Im = \{[a, b] : a, b \in \Re, a \leq b\}$ ($j = 1, \ldots, p$) and $y_i = [y_{Li}, y_{Ui}] \in \Im$ are, respectively, the observed values of X_j and Y.

Let us consider Y (a dependent variable) related to $X_1, \ldots, X_p$ (the independent predictor variables), according to a linear regression relationship:

$$y_{Li} = \beta_0 + \beta_1 a_{i1} + \ldots + \beta_p a_{ip} + \epsilon_{Li}$$
$$y_{Ui} = \beta_0 + \beta_1 b_{i1} + \ldots + \beta_p b_{ip} + \epsilon_{Ui}, \tag{1}$$
$$\text{constrained to } \beta_i \geq 0, i = 0, \ldots, p.$$

From the equations (1), we will denote the *sum of squares of deviations* in the first approach by

$$S_1 = \sum_{i=1}^{n} (\epsilon_{Li} + \epsilon_{Ui})^2 = \sum_{i=1}^{n} (y_{Li} - \beta_0 - \beta_1 a_{i1} - \ldots - \beta_p a_{ip} +$$
$$+ y_{Ui} - \beta_0 - \beta_1 b_{i1} - \ldots - \beta_p b_{ip})^2, \tag{2}$$
$$\text{constrained to } \beta_i \geq 0, i = 0, \ldots, p,$$

that represents the sum of the square of the lower and upper bounds errors.

Additionally, let us consider the same set of examples $E = \{e_1, \ldots, e_n\}$ but now described by $p+1$ continuous quantitative variables Y^{md} and $X_1^{md}, \ldots, X_p^{md}$, which assume as value, respectively, the mid-point of the interval assumed by the interval-valued variables Y and $X_1, \ldots, X_p$. This means that each example $e_i \in E$ ($i = 1, \ldots, n$) is represented as a continuous quantitative feature vector $\mathbf{w}_i = (\mathbf{x}_i^{md}, y_i^{md})$, $\mathbf{x}_i^{md} = (x_{i1}^{md}, \ldots, x_{ip}^{md})$, where $x_{ij}^{md} = (a_{ij} + b_{ij})/2$ ($j = 1, \ldots, p$) and $y_i^{md} = (y_{Li} + y_{Ui})/2$ are, respectively, the observed values of X_j^{md} and Y^{md}.

The first approach proposed in this paper is equivalent to fit a linear regression model with inequality constraints in the vector of parameters $\boldsymbol{\beta}$ over the mid-point of the interval values assumed by the variables on the learning set, and applies this model on the lower and upper bounds of the interval values of the independent variables to predict, respectively, the lower and upper bounds of the interval value of the dependent variable. This method guarantee that the values of $\hat{y}_{Li}$ will be lower than the values of $\hat{y}_{Ui}$.

In matrix notation, this first approach, called here *constrained center method* (CCM), can be rewritten as

$$\mathbf{y}^{md} = \mathbf{X}^{md} \boldsymbol{\beta}^{md} + \boldsymbol{\epsilon}^{md}, \text{ with constraints } \beta_i \geq 0, i = 0, \ldots, p \tag{3}$$

where $\mathbf{y}^{md} = (y_1^{md}, \ldots, y_n^{md})^T$, $\mathbf{X}^{md} = ((\mathbf{x}_1^{md})^T, \ldots, (\mathbf{x}_n^{md})^T)^T$, $(\mathbf{x}_i^{md})^T = (1, x_{i1}^{md}, \ldots, x_{ip}^{md})$ ($i = 1, \ldots, n$), $\boldsymbol{\beta}^{md} = (\beta_0^{md}, \ldots, \beta_p^{md})^T$ with constraints $\beta_i \geq 0$, $i = 0, \ldots, p$ and $\boldsymbol{\epsilon}^{md} = (\epsilon_1^{md}, \ldots, \epsilon_n^{md})^T$.

Notice that the model proposed in (3) will be fitted according to the inequality constraints $\beta_i \geq 0$, $i = 0, 1, \ldots, p$. Judge and Takayama [7] presented a solution to this problem which uses iterative techniques from quadratic programming. Lawson and Hanson [8] presented an algorithm that guarantee the positiveness of the least squares estimates of the vector of parameters $\boldsymbol{\beta}$, if $\mathbf{X}^{md}$ has full rank $p+1 \leq n$. The basic idea of the Lawson and Hanson [8] algorithm is identify the values that present incompatibility with the restriction, and change them to positive values through a re-weighting process. The algorithm has two main steps: initially, the algorithm find the classical least squares estimates to the vector of parameters $\boldsymbol{\beta}$ and identify the negative values in the vector of parameters. In the second step, the values which are incompatible with the restriction are re-weighted until they become positive. The authors gives a formal proof of the convergence of the algorithm. The approach proposed in this section uses the algorithm developed by Lawson and Hanson in the expression (3) to obtain the vector of parameters $\boldsymbol{\beta}$.

Given a new example e, described by $\mathbf{z} = (\mathbf{x}, y)$, $\mathbf{x} = (x_1, \ldots, x_p)$, where $x_j = [a_j, b_j]$ $(j = 1, \ldots, p)$, the value $y = [y_L, y_U]$ of Y will be predicted by $\hat{y} = [\hat{y}_L, \hat{y}_U]$ as follow:

$$\hat{y}_L = (\mathbf{x}_L)^T \hat{\boldsymbol{\beta}} \text{ and } \hat{y}_U = (\mathbf{x}_U)^T \hat{\boldsymbol{\beta}}, \tag{4}$$

where, $(\mathbf{x}_L)^T = (1, a_1, \ldots, a_p)$, $(\mathbf{x}_U)^T = (1, b_1, \ldots, b_p)$ and $\hat{\boldsymbol{\beta}} = (\hat{\beta}_0, \hat{\beta}_1, \ldots, \hat{\beta}_p)^T$ with $\hat{\beta}_i \geq 0$, $i = 0, \ldots, p$.

2.2   The Constrained Center and Range Method

The second approach consider the regression linear model proposed by [5]. However, the regression model fitted over the ranges of the intervals will take in consideration inequality constraints to guarantee that $\hat{y}_{Li}$ will be lower than $\hat{y}_{Ui}$.

Let $E = \{e_1, \ldots, e_n\}$ be the same set of examples described by the interval quantitative variables $Y, X_1, \ldots, X_p$. Let Y^{md} and X_j^{md} $(j = 1, 2, \ldots, p)$ be, respectively, quantitative variables that represent the midpoint of the interval variables Y and X_j $(j = 1, 2, \ldots, p)$ and let Y^r and X_j^r $(j = 1, 2, \ldots, p)$ be, respectively, quantitative variables that represent the ranges of these interval variables.

This means that each example $e_i \in E$ $(i = 1, \ldots, n)$ is represented by two vectors $\mathbf{w}_i = (\mathbf{x}_i^{md}, y_i^{md})$, $\mathbf{x}_i^{md} = (x_{i1}^{md}, \ldots, x_{ip}^{md})$ and $\mathbf{r}_i = (\mathbf{x}_i^r, y_i^r)$, $\mathbf{x}_i^r = (x_{i1}^r, \ldots, x_{ip}^r)$, where $x_{ij}^{md} = (a_{ij} + b_{ij})/2$, $x_{ij}^r = (b_{ij} - a_{ij})$, $y_i^{md} = (y_{Li} + y_{Ui})/2$ and $y_i^r = (y_{Ui} - y_{Li})$ are, respectively, the observed values of X_j^{md}, X_j^r, Y^{md} and Y^r.

Let us consider Y^{md} and Y^r being dependent variables and X_j^{md} and X_j^r $(j = 1, 2, \ldots, p)$ being independent predictor variables, which related according to the following linear regression relationship:

$$y_i^{md} = \beta_0^{md} + \beta_1^{md} x_{i1}^{md} + \ldots + \beta_p^{md} x_{ip}^{md} + \epsilon_i^{md}$$
$$y_i^r = \beta_0^r + \beta_1^r x_{i1}^r + \ldots + \beta_p^r x_{ip}^r + \epsilon_i^r \tag{5}$$
$$\text{with constraints } \beta_i^r \geq 0, i = 0, \ldots, p.$$

From equation (5), we will express the *sum of squares of deviations* in the second approach as

$$S_2 = \sum_{i=1}^{n}(y_i^{md} - \beta_0^{md} - \beta_1^{md}x_{i1}^{md} - \cdots - \beta_p^{md}x_{i1}^{md})^2 +$$

$$+ \sum_{i=1}^{n}(y_i^r - \beta_0^r - \beta_1^r x_{i1}^r - \cdots - \beta_p^r x_{ip}^r)^2 \qquad (6)$$

$$\text{constrained to } \beta_i^r \geq 0, i = 0, \ldots, p,$$

that represent the sum of the midpoint square error plus the sum of the range square error considering independent vectors of parameters to predict the midpoint and the range of the intervals.

This second approach consider the regression linear model proposed by [5] that fits two independent linear regression models, respectively, over the midpoint and range of the interval values assumed by the variables on the learning set. However, the regression model fitted over the ranges of the intervals take in consideration inequality constraints to guarantee that $\hat{y}_{Li}$ is lower than $\hat{y}_{Ui}$. The prediction of the lower and upper bound of an interval value of the dependent variable is accomplished from its mid-point and range which are estimated from the fitted linear regression models applied to the mid-point and range of each interval value of the independent variables.

In matrix notation, this second approach, called here *constrained center and range method* (CCRM) can be rewritten as

$$\mathbf{y}^{md} = \mathbf{X}^{md}\boldsymbol{\beta}^{md} + \epsilon^{md},$$

$$\mathbf{y}^r = \mathbf{X}^r \boldsymbol{\beta}^r + \epsilon^r, \text{ with constraints } \beta_i^r \geq 0, i = 0, \ldots, p, \qquad (7)$$

where, $\mathbf{X}^{md}$ and $\mathbf{X}^r$ has full rank $p + 1 \leq n$, $\mathbf{y}^{md} = (y_1^{md}, \ldots, y_n^{md})^T$, $\mathbf{X}^{md} = ((\mathbf{x}_1^{md})^T, \ldots, (\mathbf{x}_n^{md})^T)^T$, $(\mathbf{x}_i^{md})^T = (1, x_{i1}^{md}, \ldots, x_{ip}^{md})$, $\boldsymbol{\beta}^{md} = (\beta_0^{md}, \ldots, \beta_p^{md})$, $\mathbf{y}_r = (y_1^r, \ldots, y_n^r)^T$, $\mathbf{X}^r = ((\mathbf{x}_1^r)^T, \ldots, (\mathbf{x}_n^r)^T)^T$, $(\mathbf{x}_i^r)^T = (1, x_{i1}^r, \ldots, x_{ip}^r)$, $\boldsymbol{\beta}^r = (\beta_0^r, \ldots, \beta_p^r)$, $x_{ij}^{md} = (a_{ij} + b_{ij})/2$, $x_{ij}^r = (b_{ij} - a_{ij})$, $y_i^{md} = (y_{Li} + y_{Ui})/2$ and $y_i^r = (y_{Ui} - y_{Li})$.

Notice that the *least square estimates* of the vector of parameters $\boldsymbol{\beta}^{md}$ in the first equation of the expression (7) is given by

$$\hat{\boldsymbol{\beta}}^{md} = ((\mathbf{X}^{md})^T\mathbf{X}^{md})^{-1}(\mathbf{X}^{md})^T\mathbf{y}^{md}. \qquad (8)$$

However, the second equation in expression (7) must be fitted according to the inequality constraints $\beta_i^r \geq 0, i = 0, 1, \ldots, p$. To accomplish this task we will use the algorithm proposed by Lawson and Hanson [8]. In this way we will obtain the vector of parameters $\boldsymbol{\beta}^r$ according to these inequality constraints.

Given a new example e, described by $\mathbf{z} = (\mathbf{x}, y)$ with $\mathbf{x} = (x_1, \ldots, x_p)$, $\mathbf{w} = (\mathbf{x}^{md}, y^{md})$ with $\mathbf{x}^{md} = (x_1^{md}, \ldots, x_p^{md})$ and $\mathbf{r} = (\mathbf{x}^r, y^r)$ with $\mathbf{x}^r = (x_1^r, \ldots, x_p^r)$, where $x_j = [a_j, b_j]$, $x_j^{md} = (a_j + b_j)/2$ and $x_j^r = (b_j - a_j)$ $(j = 1, \ldots, p)$, the value $y = [y_L, y_U]$ of Y will be predicted from the predicted values $\hat{y}^{md}$ of Y^{md} and $\hat{y}^r$ of Y^r as follow:

$$\hat{y}_L = \hat{y}^{md} - (1/2)\hat{y}^r \text{ and } \hat{y}_U = \hat{y}^{md} + (1/2)\hat{y}^r \tag{9}$$

where $\hat{y}^{md} = (\tilde{\mathbf{x}}^{md})^T \hat{\boldsymbol{\beta}}^{md}$, $\hat{y}^r = (\tilde{\mathbf{x}}^r)^T \hat{\boldsymbol{\beta}}^r$, $(\tilde{\mathbf{x}}^{md})^T = (1, x_1^{md}, \ldots, x_p^{md})$, $(\tilde{\mathbf{x}}^r)^T = (1, x_1^r, \ldots, x_p^r)$, $\hat{\boldsymbol{\beta}}^{md} = (\hat{\beta}_0^{md}, \hat{\beta}_1^{md}, \ldots, \hat{\beta}_p^{md})^T$ and $\hat{\boldsymbol{\beta}}^r = (\hat{\beta}_0^r, \hat{\beta}_1^r, \ldots, \hat{\beta}_p^r)^T$ with $\hat{\beta}_i^r \geq 0, i = 0, \ldots, p$.

3 The Monte Carlo Experiences

To show the usefulness of the approaches proposed in this paper in comparison with the methods proposed by [2] (here called *Center Method - CM*) and [5] (here called *Center and Range Method - CRM*), experiments with simulated interval-valued data sets with different degrees of difficulty to fit a linear regression model are considered in this section, along with a cardiological data set (see [2]).

3.1 Simulated Interval-Valued Data Sets

We consider standard quantitative data sets in $\Re^2$ and $\Re^4$. Each data set (in $\Re^2$ or $\Re^4$) has 375 points partitioned in a learning set (250 points) and a test set (125 points). Each data point belonging to the standard data set is a seed for an interval data (a rectangle in $\Re^2$ or a hypercube in $\Re^4$). In this way, the interval data sets are obtained from these standard data sets.

The construction of the standard data sets and the corresponding interval data sets is accomplished in the following steps:

s_1) Let us suppose that each random variables X_j^{md} is uniformly distributed in the interval $[a, b]$; at each iteration 375 values of each variable X_j^{md} are randomly selected;

s_2) The random variable Y^{md} is supposed to be related to variables X_j^{md} according to $Y^{md} = (\mathbf{X}^{md})^T \boldsymbol{\beta} + \epsilon$, where $(\mathbf{X}^{md})^T = (1, X_1^{md})$ and $\boldsymbol{\beta} = (\beta_0 = U[c, d], \beta_1 = U[c, d])^T$ in $\Re^2$, or $(\mathbf{X}^{md})^T = (1, X_1^{md}, X_2^{md}, X_3^{md})$ and $\boldsymbol{\beta} = (\beta_0 = U[c, d], \beta_1 = U[c, d], \beta_2 = U[c, d], \beta_3 = U[c, d])^T$ in $\Re^4$ and $\epsilon = U[e, f]$; the mid-points of these 375 intervals are calculated according this linear relation;

s_3) Once the mid-points of the intervals are obtained, let us consider the range of each interval. Let us suppose that each random variable Y^r, X_j^r $(j = 1, 2, 3)$ is uniformly distributed, respectively, in the intervals $[g, h]$ and $[i, j]$; at each iteration 375 values of each variable Y^r, X_j^r are randomly selected, which are the range of these intervals;

s_4) The interval-valued data set is partitioned in learning (250 observations) and test (125 observations) data sets.

Table 1 shows nine different configurations for the interval data sets which are used to measure de performance of the **CCM** and **CCRM** methods in comparison with the approaches proposed by [2] and [5].

These configurations take into account the combination of two factors (range and error on the mid-points) with three degrees of variability (low, medium and

Table 1. Data set configurations

C_1	$X_j^{md} \sim U[20, 40]$	$X_j^r \sim U[20, 40]$	$Y^r \sim U[20, 40]$	$\epsilon \sim U[-20, 20]$
C_2	$X_j^{md} \sim U[20, 40]$	$X_j^r \sim U[20, 40]$	$Y^r \sim U[20, 40]$	$\epsilon \sim U[-10, 10]$
C_3	$X_j^{md} \sim U[20, 40]$	$X_j^r \sim U[20, 40]$	$Y^r \sim U[20, 40]$	$\epsilon \sim U[-5, 5]$
C_4	$X_j^{md} \sim U[20, 40]$	$X_j^r \sim U[10, 20]$	$Y^r \sim U[10, 20]$	$\epsilon \sim U[-20, 20]$
C_5	$X_j^{md} \sim U[20, 40]$	$X_j^r \sim U[10, 20]$	$Y^r \sim U[10, 20]$	$\epsilon \sim U[-10, 10]$
C_6	$X_j^{md} \sim U[20, 40]$	$X_j^r \sim U[10, 20]$	$Y^r \sim U[10, 20]$	$\epsilon \sim U[-5, 5]$
C_7	$X_j^{md} \sim U[20, 40]$	$X_j^r \sim U[1, 5]$	$Y^r \sim U[1, 5]$	$\epsilon \sim U[-20, 20]$
C_8	$X_j^{md} \sim U[20, 40]$	$X_j^r \sim U[1, 5]$	$Y^r \sim U[1, 5]$	$\epsilon \sim U[-10, 10]$
C_9	$X_j^{md} \sim U[20, 40]$	$X_j^r \sim U[1, 5]$	$Y^r \sim U[1, 5]$	$\epsilon \sim U[-5, 5]$

high): low variability range ($U[1, 5]$), medium variability range ($U[10, 20]$), high variability range ($U[20, 40]$, low variability error ($U[-5, 5]$), medium variability error ($U[-10, 10]$) and high variability error ($U[-20, 20]$).

The configuration C_1, for example, represents observations with a high variability range and with a poor linear relationship between Y and the dependent variables due the high variability error on the mid-points. Figure 1 shows the configuration C_1 when the data is in $\Re^2$.

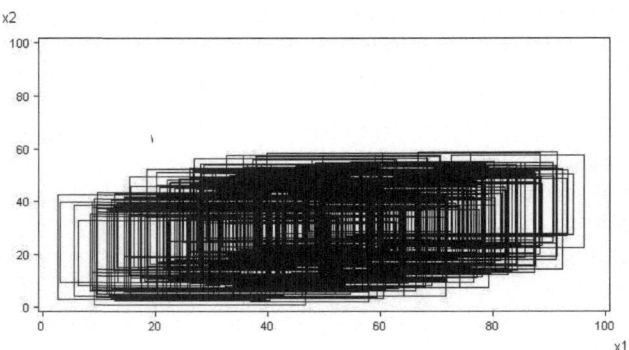

Fig. 1. Configuration C_1 showing a poor linear relationship between Y and X_1

On the other hand, the configuration C_9 represents observations with a low variability range and with a rich linear relationship between Y and the dependent variables due the low variability error on the mid-points. Figure 2 shows the configuration C_9 when the data is in $\Re^2$.

3.2 Experimental Evaluation

The performance assessment of these linear regression models is based on the following measures: the *lower bound root mean-square error* ($RMSE_L$), the *upper bound root mean-square error* ($RMSE_U$), the *square of the lower bound*

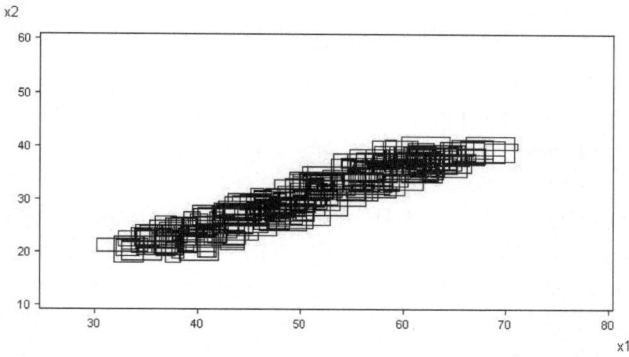

Fig. 2. Configuration C_9 showing a rich linear relationship between Y and X_1

correlation coefficient (r_L^2) and the *square of the upper bound correlation coefficient* (r_U^2). These measures are obtained from the observed values $y_i = [y_{Li}, y_{Ui}]$ $(i = 1, \ldots, n)$ of Y and from their corresponding predicted values $\hat{y}_i = [\hat{y}_{Li}, \hat{y}_{Ui}]$, denoted as:

$$RMSE_L = \sqrt{\frac{\sum_{i=1}^{n}(y_{Li} - \hat{y}_{Li})^2}{n}} \text{ and } RMSE_U = \sqrt{\frac{\sum_{i=1}^{n}(y_{Ui} - \hat{y}_{Ui})^2}{n}} \quad (10)$$

$$r_L^2 = \frac{Cov(\mathbf{y}_L, \hat{\mathbf{y}}_L)}{S_{\mathbf{y}_L} S_{\hat{\mathbf{y}}_L}} \text{ and } r_U^2 = \frac{Cov(\mathbf{y}_U, \hat{\mathbf{y}}_U)}{S_{\mathbf{y}_U} S_{\hat{\mathbf{y}}_U}} \quad (11)$$

where: $\mathbf{y}_L = (y_{L1}, \ldots, y_{Ln})^T$, $\hat{\mathbf{y}}_L = (\hat{y}_{L1}, \ldots, \hat{y}_{Ln})^T$, $\mathbf{y}_U = (y_{U1}, \ldots, y_{Un})^T$, $\hat{\mathbf{y}}_U = (\hat{y}_{U1}, \ldots, \hat{y}_{Un})^T$, $Cov(\mathbf{y}_\bullet, \hat{\mathbf{y}}_\bullet)$ is the covariance between $\mathbf{y}_L$ and $\hat{\mathbf{y}}_L$ or between $\mathbf{y}_U$ and $\hat{\mathbf{y}}_U$, $S_{\mathbf{y}_\bullet}$ is the standard deviation of $\mathbf{y}_L$ or $\mathbf{y}_U$ and $S_{\hat{\mathbf{y}}_\bullet}$ is the standard deviation of $\hat{\mathbf{y}}_L$ or $\hat{\mathbf{y}}_U$.

These measures are estimated for the **CM**, **CCM**, **CRM** and **CCRM** methods in the framework of a Monte Carlo simulation with 100 replications for each independent test interval data set, for each of the nine fixed configurations, as well as for different numbers of independent variables in the model matrix **X**. At each replication, we fitted a linear regression model to the training interval data set using the **CM**, **CCM**, **CRM** and **CCRM** methods. Thus, the fitted regression models are used to predict the interval values of the dependent variable Y in the test interval data set and these measures are calculated. For each measure, the average and standard deviation over the 100 Monte Carlo simulations is calculated and then a statistical t-test is applied to compare the performance of the approaches. Additionally, this procedure is repeated considering 50 different values for the parameter vector $\boldsymbol{\beta}$ selected randomly in the interval $[-10, 10]$.

The comparison between these approaches is achieved by a statistical t-test for independent samples at a significance level of 1%, applied to each measure considered. For each configuration presented in the Table 1, number of independent variables and methods compared is calculated the ratio of times that the hypothesis H_0 is rejected for each measure presented in the equations (10) and (11), considering the 50 different vector of parameters $\boldsymbol{\beta}$.

For any two methods compared (A and B, in this order), concerning $RMSE_L$ and $RMSE_U$ measures (the higher the measures, the worse the method), the null and alternative hypotheses are structured as:

H_0 : **Method A** $\geq$ **Method B** versus H_1: **Method A** $<$ **Method B**;

Concerning r_L^2 and r_U^2 measures (the higher the measures, the better the method), the hypotheses are structured as:

H_0 : **Method A** $\leq$ **Method B** versus H_1: **Method A** $>$ **Method B**;

In the Table 2, we illustrate for the **CRM** and **CCRM** methods (in this order) the results of the Monte Carlo experience for a specific vector of parameters β. This table shows that, for any considered situation and for all the measures considered, the **CRM** and **CCRM** methods presented an average performance very similar.

The comparison between the **CRM** and **CCRM** methods (in this order) in relation to 50 different values for the parameter vector β showed that the null hypothesis was never rejected at a significance level of 1%, regardless the number of variables, the performance measure considered or the configuration. This means that the average performance of the **CRM** method is not better than the **CCRM** method. The same conclusion was again found in the comparisons of the **CCRM** method against **CRM** method (in this order). This corroborates the hypothesis that the approaches present, in average, the same performance. Then, concerning the model presented by [5] we conclude that the use of inequality constraints to guarantee that $\hat{y}_{Li}$ will be lower than $\hat{y}_{Li}$ does not penalize, statistically, its prediction performance. Then, the **CCRM** method can be preferred when compared with the **CRM**

In the Table 3 we present the results of the comparison between **CM** and **CCM** methods (in this order). In contrast with the results presented above, we conclude that the use of inequality constraints in the method proposed by [2] penalizes its prediction performance. We can see that the growing of the number of independent variables produced an increase in the percentage of rejection of the null hypothesis H_0. This means that the performance difference between the **CM** and **CCM** methods increases when the number of independent variables in the model increases. Comparing the configurations C_1, C_4 and C_7, that differ with respect to the variability on the range of the intervals, notice that when the interval's range decreases, the percentage of rejection of the null hypothesis H_0 increases. We have a similar conclusion comparing the configurations (C_2, C_5, C_8) and (C_3, C_6, C_9). However, it's important remember that the **CM** method does not guarantee that the values of $\hat{y}_{Li}$ will be lower than the values of $\hat{y}_{Ui}$.

Finally, the results presented in Table 4 show the superiority of the **CCRM** method when compared with the **CM** method (in this order). We can see that the percentage of rejection of the null hypothesis at a significance level of 1%, regardless the number of variables or the performance measure considered, is always higher than 80%. This indicate that the **CCRM** method has a better performance than the **CM** approach. When the number of variables is 3, in all configurations except C_7, the null hypothesis H_0 was rejected in 100% of the cases.

Table 2. Comparision between **CRM** and **CCRM** methods - Average and standard deviation of each measure calculated from the 100 replications in the framework of a Monte Carlo experience for a specific vector of parameters β

Config.	p	Stat.	$RMSE_L$		$RMSE_U$		R_L^2 (%)		R_U^2 (%)	
			CRM	**CCRM**	**CRM**	**CCRM**	**CRM**	**CCRM**	**CRM**	**CCRM**
C_1	1	$\bar{x}$	19.874	19.874	19.718	19.718	93.85	93.86	94.07	94.07
		S	1.180	1.179	1.121	1.123	0.78	0.78	0.76	0.76
	3	$\bar{x}$	19.870	19.866	19.697	19.696	97.28	97.29	97.25	97.25
		S	1.099	1.096	1.070	1.069	0.38	0.38	0.42	0.42
C_2	1	$\bar{x}$	16.919	16.919	17.076	17.076	94.76	94.76	95.01	95.01
		S	0.751	0.750	0.763	0.764	0.73	0.74	0.63	0.63
	3	$\bar{x}$	19.778	19.770	19.648	19.644	91.98	91.99	91.83	91.83
		S	1.254	1.251	1.082	1.083	1.02	1.02	1.10	1.10
C_3	1	$\bar{x}$	16.310	16.310	16.337	16.337	97.32	97.32	97.33	97.34
		S	0.643	0.644	0.599	0.599	0.35	0.34	0.36	0.36
	3	$\bar{x}$	16.342	16.340	16.316	16.315	98.28	98.28	98.27	98.27
		S	0.616	0.617	0.642	0.643	0.22	0.22	0.24	0.25
C_4	1	$\bar{x}$	14.060	14.059	14.122	14.120	72.52	72.52	72.41	72.41
		S	0.800	0.800	0.880	0.879	3.31	3.31	3.46	3.46
	3	$\bar{x}$	14.082	14.081	13.959	13.957	94.64	94.64	94.62	94.62
		S	0.825	0.825	0.771	0.769	0.72	0.72	0.76	0.76
C_5	1	$\bar{x}$	9.915	9.915	9.872	9.871	96.91	96.61	96.63	96.64
		S	0.612	0.612	0.583	0.584	0.43	0.43	0.36	0.36
	3	$\bar{x}$	9.955	9.956	9.905	9.904	97.91	97.91	97.93	97.93
		S	0.493	0.494	0.498	0.498	0.25	0.25	0.28	0.28
C_6	1	$\bar{x}$	8.490	8.490	8.545	8.545	99.02	99.02	99.00	99.00
		S	0.370	0.370	0.370	0.370	0.14	0.13	0.13	0.13
	3	$\bar{x}$	8.530	8.529	8.577	8.577	99.53	99.53	99.53	99.53
		S	0.391	0.390	0.394	0.395	0.07	0.07	0.08	0.08
C_7	1	$\bar{x}$	11.863	11.863	11.851	11.851	27.14	27.14	27.20	27.20
		S	0.540	0.540	0.515	0.515	6.81	6.80	6.90	6.91
	3	$\bar{x}$	11.700	11.701	11.753	11.752	94.48	94.48	94.50	94.50
		S	0.524	0.524	0.431	0.431	0.63	0.63	0.60	0.60
C_8	1	$\bar{x}$	6.121	6.121	6.103	6.103	98.85	98.85	96.86	96.86
		S	0.285	0.285	0.296	0.295	0.13	0.13	0.14	0.14
	3	$\bar{x}$	6.145	6.145	6.125	6.124	98.85	98.85	98.86	98.86
		S	0.298	0.296	0.328	0.329	0.16	0.16	0.15	0.15
C_9	1	$\bar{x}$	3.476	3.475	3.455	3.456	81.56	81.57	81.85	81.85
		S	0.206	0.205	0.187	0.187	2.51	2.51	2.29	2.29
	3	$\bar{x}$	3.455	3.454	3.489	3.489	99.52	99.52	99.51	99.51
		S	0.181	0.180	0.202	0.202	0.06	0.06	0.07	0.07

Additionally, in all configurations, the growing of the number of independent variables produced an increase in the percentage of rejection of the null hypothesis H_0. This means that the performance difference between the **CCRM** and **CM** methods increases when the number of independent variables in the model increases.

Table 3. Comparison between **CM** and **CCM** methods - Percentage of rejection of the null hypothesis H_0 (T-test for independent samples following a Student's t distribution with 198 degrees of freedom)

Configuration	p	$RMSE_L$	$RMSE_U$	r_L^2 (%)	r_U^2 (%)
C_1	1	22%	1%	0%	0%
	3	26%	20%	34%	32%
C_2	1	22%	2%	0%	0%
	3	24%	24%	36%	36%
C_3	1	18%	2%	0%	0%
	3	26%	16%	28%	28%
C_4	1	52%	44%	0%	0%
	3	74%	80%	22%	22%
C_5	1	62%	48%	0%	0%
	3	80%	88%	38%	38%
C_6	1	46%	36%	0%	0%
	3	78%	88%	38%	38%
C_7	1	60%	62%	0%	0%
	3	86%	88%	42%	42%
C_8	1	58%	56%	0%	0%
	3	80%	84%	42%	42%
C_9	1	60%	60%	0%	0%
	3	90%	90%	36%	36%

Now, comparing the configurations C_1, C_2, and C_3, that differ with respect to the error on the mid-points of the intervals, notice that when the variability of the error decreases, the percentage of rejection of the null hypothesis H_0 increases. This shows that, as much linear is the relationship between the variables as higher is the statistical significance of the difference between the **CCRM** and **CM** methods. We arrive to the same conclusion comparing the configurations (C_4, C_5, C_6) and (C_7, C_8, C_9).

On the other hand, comparing the configurations C_1, C_4 and C_7, that differ with respect to the variability on the range of the intervals, notice that when the interval's range decreases, the percentage of rejection of the null hypothesis H_0 decreases too. We have a similar conclusion comparing the configurations (C_2, C_5, C_8) and (C_3, C_6, C_9).

3.3 Cardiological Interval Data Set

This data set (Table 5) concerns the record of the pulse rate Y, systolic blood pressure X_1 and diastolic blood pressure X_2 for each of eleven patients (see [2]). The aim is to predict the interval values y of Y (the dependent variable) from x_j $(j = 1, 2)$ through a linear regression model.

The fitted linear regression models to the **CM**, **CCM**, **CRM** and **CCRM** methods in the cardiological interval data set are presented below:

Table 4. Comparison between **CCRM** and **CM** methods - Percentage of rejection of the null hypothesis H_0 (T-test for independent samples following a Student's t distribution with 198 degrees of freedom)

Configuration	p	$RMSE_L$	$RMSE_U$	r_L^2 (%)	r_U^2 (%)
C_1	1	100%	100%	97%	97%
	3	100%	100%	100%	100%
C_2	1	100%	100%	96%	96%
	3	100%	100%	100%	100%
C_3	1	100%	100%	98%	98%
	3	100%	100%	100%	100%
C_4	1	100%	100%	94%	94%
	3	100%	100%	100%	100%
C_5	1	100%	100%	96%	94%
	3	100%	100%	100%	100%
C_6	1	100%	100%	100%	100%
	3	100%	100%	100%	100%
C_7	1	94%	94%	84%	80%
	3	98%	98%	98%	98%
C_8	1	100%	100%	94%	94%
	3	100%	100%	100%	100%
C_9	1	100%	100%	100%	100%
	3	100%	100%	100%	100%

Table 5. Cardiological interval data set

u	Pulse rate	Systolic blood pressure	Diastolic blood pressure
1	[44-68]	[90-100]	[50-70]
2	[60-72]	[90-130]	[70-90]
3	[56-90]	[140-180]	[90-100]
4	[70-112]	[110-142]	[80-108]
5	[54-72]	[90-100]	[50-70]
6	[70-100]	[130-160]	[80-110]
7	[72-100]	[130-160]	[76-90]
8	[76-98]	[110-190]	[70-110]
9	[86-96]	[138-180]	[90-110]
10	[86-100]	[110-150]	[78-100]
11	[63-75]	[60-100]	[140-150]

CM: $\hat{y}_L = 21.17 + 0.33a_1 + 0.17a_2$ and $\hat{y}_U = 21.17 + 0.33b_1 + 0.17b_2$;
CCM: $\hat{y}_L = 21.17 + 0.33a_1 + 0.17a_2$ and $\hat{y}_U = 21.17 + 0.33b_1 + 0.17b_2$;
CRM: $\hat{y}^{md} = 21.17 + 0.33x_1^{md} + 0.17x_2^{md}$ and $\hat{y}^r = 20.13 - 0.15x_1^r + 0.35x_2^r$.
CCRM: $\hat{y}^{md} = 21.17 + 0.33x_1^{md} + 0.17x_2^{md}$ and $\hat{y}^r = 17.96 + 0.20x_2^r$.

From the fitted linear regression models for interval-valued data listed above we present, in the Table 6, the predicted interval values of the dependent variable *pulse rate*, in accordance with **CM**, **CCM**, **CRM** and **CCRM** models.

Table 6. Predicted interval values of the dependent variable *pulse rate*, according to **CM**, **CCM**, **CRM** and **CCRM** models

CM	CCM	CRM	CCRM
[59-66]	[59-66]	[50-75]	[52-74]
[63-79]	[63-79]	[60-82]	[60-82]
[83-97]	[83-97]	[81-99]	[80-100]
[71-86]	[71-86]	[66-99]	[67-90]
[59-66]	[59-66]	[50-75]	[52-74]
[78-92]	[78-92]	[72-98]	[73-97]
[77-89]	[77-89]	[73-93]	[73-93]
[69-102]	[69-102]	[75-97]	[73-99]
[82-99]	[82-99]	[80-101]	[79-101]
[71-87]	[71-87]	[68-90]	[68-90]
[65-80]	[65-80]	[63-81]	[62-82]

Table 7. Performance of the methods in the cardiological interval data set

Method	$RMSE_L$	$RMSE_L$	r_L^2 (%)	r_U^2 (%)
CM	11.09	10.41	30.29	53.47
CCM	11.09	10.41	30.29	53.47
CRM	9.81	8.94	41.54	63.34
CCRM	9.73	9.18	41.04	59.41

The performance of the **CM**, **CCM**, **CRM** and **CCRM** methods is evaluated through the calculation of $RMSE_L$, $RMSE_U$, r_L^2 and r_U^2 measures on this cardiological interval data set. From Table 7, we can see that the performance of the methods in this cardiological data set were similar to what was demonstrated in the Monte Carlo simulations. We conclude that the **CRM** and **CCRM** methods outperforms the **CM** and **CCM** methods, that present the same performance in this particular case, and that the **CRM** and **CCRM** methods presented a very close performance.

4   Concluding Remarks

In this paper we presented two new methods to fit a regression model on interval-valued data which considers inequality constraints to guarantee that the predicted values of the $\hat{y}_{Li}$ will be lower than the predicted values of the upper bounds $\hat{y}_{Ui}$. The first approach consider the regression linear model proposed by [2] and add inequality constraints in the vector of parameters β. The second approach consider the regression linear model proposed by [5] and add, in the regression model fitted over the ranges of the intervals, inequality constraints to guarantee that $\hat{y}_{Li}$ will be lower than $\hat{y}_{Ui}$.

The assessment of the proposed prediction methods was based on the average behaviour of the *root mean square error* and the *square of the correlation coefficient* in the framework of a Monte Carlo simulation. The performance of the proposed approaches was also measured in a cardiological interval data set

The comparison between the **CRM** and **CCRM** showed that the use of inequality constraints to guarantee that $\hat{y}_{Li}$ would be be lower than $\hat{y}_{Li}$ did not represent, statistically, penalization in the model prediction performance. Then, the **CCRM** method was preferred when compared with the **CRM**.

On the other hand, we conclude that the use of inequality constraints in the method proposed by [2] penalizes its prediction performance. We observed that the performance difference between the **CM** and **CCM** methods increases when the number of independent variables in the model increases.

The results showed the superiority of the **CCRM** method when compared with the **CM** method. The percentage of rejection of the null hypothesis at a significance level of 1%, regardless the number of variables or the performance measure considered, is always higher than 80%. The performance difference between the **CCRM** and **CM** methods increases when the number of independent variables in the model increases. Moreover, as much linear is the relationship between the variables as higher is the statistical significance of the difference between the **CCRM** and **CM** methods.

Finally, the performance of the methods in the cardiological data set was similar to that showed in the Monte Carlo simulations. We concluded that the **CRM** and **CCRM** methods outperforms the **CM** and **CCM** methods and that the **CRM** and **CCRM** methods presented a very similar performance.

Acknowledgments. The authors would like to thank CNPq and CAPES (Brazilian Agencies) for its financial support.

References

1. Bock, H.H. and Diday, E.: Analysis of Symbolic Data: Exploratory Methods for Extracting Statistical Information from Complex Data. Springer, Berlin Heidelberg (2000)
2. Billard, L., Diday, E.: Regression Analysis for Interval-Valued Data. Data Analysis, Classification and Related Methods: Proceedings of the Seventh Conference of the International Federation of Classification Societies, IFCS-2000, Namur (Belgium), Kiers, H.A.L. et al. Eds, Vol. 1 (2000), Springer, 369–374
3. Billard, L., Diday, E.: Symbolic Regression Analysis. Classification, Clustering and Data Analysis: Proceedings of the Eighenth Conference of the International Federation of Classification Societies, IFCS-2002, Crakow (Poland), Jajuga, K. et al. Eds, Vol. 1 (2002), Springer, 281–288
4. Billard, L., Diday, E.: From the Statistics of Data to the Statistics of Knowledge: Symbolic Data Analysis. Journal of the American Statistical Association, Vol. 98 (2003) 470-487
5. De Carvalho, F.A.T, Lima Neto, E.A., Tenorio, C.P.: A New Method to Fit a Linear Regression Model for Interval-Valued Data. Advances in Artificial Intelligence: Proceedings of the Twenty Seventh Germany Conference on Artificial the International Intelligence, KI-04, Ulm (Germany), Biundo, S. et al. Eds, Vol. 1 (2004), Springer, 295-306.

6. Draper, N.R., Smith, H.: Applied Regression Analysis. John Wiley, New York (1981)
7. Judge, G.G., Takayama, T.: Inequality Restrictions in Regression Analysis. Journal of the American Statistical Association, Vol. 61 (1966) 166-181
8. Lawson, C.L., Hanson, R.J.: Solving Least Squares Problem. Prentice-Hall, New York (1974)
9. Montgomery, D.C., Peck, E.A.: Introduction to Linear Regression Analysis. John Wiley, New York (1982)
10. Scheffé, H.: The Analysis of Variance. John Wiley, New York, (1959)

On Utilizing Stochastic Learning Weak Estimators for Training and Classification of Patterns with Non-stationary Distributions

B. John Oommen[1,*] and Luis Rueda[2,**]

[1] School of Computer Science, Carleton University,
1125 Colonel By Dr., Ottawa, ON, K1S 5B6, Canada
oommen@scs.carleton.ca
[2] School of Computer Science, University of Windsor,
401 Sunset Avenue, Windsor, ON, N9B 3P4, Canada
lrueda@cs.uwindsor.ca

Abstract. Pattern recognition essentially deals with the training and classification of patterns, where the distribution of the features is assumed unknown. However, in almost all the reported results, a fundamental assumption made is that this distribution, although unknown, is *stationary*. In this paper, we shall relax this assumption and assume that the class-conditional distribution is non-stationary. To now render the training and classification feasible, we present a novel estimation strategy, involving the so-called Stochastic Learning Weak Estimator (SLWE). The latter, whose convergence is *weak*, is used to estimate the parameters of a binomial distribution using the principles of stochastic learning. Even though our method includes a learning coefficient, λ, it turns out that the mean of the final estimate is independent of λ, the variance of the final distribution decreases with λ, and the speed decreases with λ. Similar results are true for the multinomial case. To demonstrate the power of these estimates in data which is truly "non-stationary", we have used them in *two* pattern recognition exercises, the first of which involves artificial data, and the second which involves the recognition of the *types* of data that are present in news reports of the Canadian Broadcasting Corporation (CBC). The superiority of the SLWE in both these cases is demonstrated.

1 Introduction

Every Pattern Recognition (PR) problem essentially deals with two issues : the training and then the classification of the patterns. Generally speaking, the fundamental assumption made is that the class-conditional distribution, although unknown, is *stationary*. It is interesting to note that this tacit assumption is so all-pervading - all the data sets included in the acclaimed UCI machine learning repository [2] represent data in which the class-conditional distribution is assumed to be stationary. Our intention here is to

---

* *Fellow of the IEEE*. Phone: +1 (613) 520-2600 Ext. 4358, Fax: +1 (613) 520-4334.
** Member of the IEEE.

U. Furbach (Ed.): KI 2005, LNAI 3698, pp. 107–120, 2005.
© Springer-Verlag Berlin Heidelberg 2005

propose training (learning) and classification strategies when the class-conditional distribution is non-stationary. Indeed, we shall approach this problem by studying the fundamental statistical issue, namely that of estimation.

Since the problem involves random variables, the training/classification decisions are intricately dependent on the system obtaining reliable estimates on the parameters that characterize the underlying random variable. These estimates are computed from the observations of the random variable itself. Thus, if a problem can be modeled using a random variable which is normally (Gaussian) distributed, the underlying statistical problem involves estimating the mean and the variance. The theory of estimation has been studied for hundreds of years [1,3,8,24,26,27]. It is also easy to see that the learning (training) phase of a statistical Pattern Recognition (PR) system is, indeed, based on estimation theory [4,5,6,32]. Estimation methods generally fall into various categories, including the Maximum Likelihood Estimates (MLE) and the Bayesian family of estimates [1,3,4,5].

Although the MLEs and Bayesian estimates have good computational and statistical properties, the fundamental premise for establishing the quality of estimates is based on the assumption that the parameter being estimated does not change with time. In other words, the distribution is assumed to be *stationary*. Thus, it is generally assumed that there is an underlying parameter θ, *which does not change with time*, and as the number of samples increases, we would like the estimate $\hat{\theta}$ to converge to θ with probability one, or in a mean square sense.

There are numerous problems which we have recently encountered, where these strong estimators pose a real-life concern. One scenario occurs in pattern classification involving moving scenes. We also encounter the same situation when we want to perform adaptive compression of files which are interspersed with text, formulae, images and tables. Similarly, if we are dealing with adaptive data structures, the structure changes with the estimate of the underlying data distribution, and if the estimator used is "strong" (i.e., w. p. 1), it is hard for the learned data structure to emerge from a structure that it has converged to. Indeed, we can conclusively demonstrate that it is sub-optimal to work with strong estimators in such application domains, i.e., when the data is truly *non-stationary*.

In this paper, we shall present one such "weak" estimator, referred to as the Stochastic Learning Weak Estimator (SLWE), and which is developed by using the principles of stochastic learning. In essence, the estimate is updated at each instant based on the value of the current sample. However, this updating is not achieved using an *additive* updating rule, but rather by a *multiplicative* rule, akin to the family of linear action-probability updating schemes [13,14]. The formal results that we have obtained for the binomial distribution are quite fascinating. Similar results have been achieved for the multinomial case, except that the variance (covariance) has not been currently derived.

The entire field of obtaining weak estimators is novel. In that sense, we believe that this paper is of a pioneering sort – we are not aware of any estimators which are computed as explained here. We also hope that this is the start of a whole new body of research, namely those involving weak estimators for various distributions, and their applications in various domains. In this regard, we are currently in the process of deriving weak estimators for various distributions. To demonstrate the power of the SLWE in real-life problems, we conclude the paper with results applicable to news classification problems, in which the underlying distribution is non-stationary.

To devise a new estimation method, we have utilized the principles of learning as achieved by the families of Variable Structure Stochastic Automata (VSSA) [10,11,13]. In particular, the learning we have achieved is obtained as a consequence of invoking algorithms related to families of linear schemes, such as the Linear Reward-Inaction (L_{RI}) scheme. The analysis is also akin to the analysis used for *these* learning automata (LA). This involves first determining the updating equations, and taking the conditional expectation of the quantity analyzed. The condition disappears when the expectation operator is invoked a second time, leading to a difference equation for the specified quantity, which equation is later explicitly solved. We have opted to use these families of VSSA in the design of our SLWE, because it turns out that the analysis is considerably simplified and (in our opinion) fairly elegant.

The traditional strategy to deal with non-stationary environments has been one of using a *sliding window* [7]. The problem with such a "sliding window" scheme for estimation is that the width of the window is crucial. If the width is too "small", the estimates are necessarily poor. On the other hand, if the window is "large", the estimates *prior* to the change of the parameters influence the new estimates significantly. Also, the observations during the entire sliding window must be retained and updated during the entire process. A comparison of our new estimation strategy with some of the other "non-sliding-window" approaches is currently under way. However, the main and fundamental difference that our method has over all the other reported schemes is the fact that our updating scheme is not additive, but multiplicative.

The question of deriving weak estimators based on the concepts of *discretized* LA [16,18] and *estimator-based* LA [12,17,19,21,23,29] remains open. We believe, however, that even if such estimator-based SLWE are designed, the analysis of their properties will not be trivial.

2   Weak Estimators of Binomial Distributions

Through out this paper we assume that we are estimating the parameters of a binomial/multinomial distribution. The binomial distribution is characterized by two parameters, namely, the *number* of Bernoulli trials, and the parameter characterizing *each* Bernoulli trial. In this regard, we assume that the number of observations is the number of trials. Thus, all we have to do is to estimate the *Bernoulli* parameter for each trial. This is what we endeavor to do using stochastic learning methods.

Let X be a binomially distributed random variable, which takes on the value of either '1' or '2'[1]. We assume that X obeys the distribution S, where $S = [s_1, s_2]^T$. In other words,

$X = $ '1' with probability s_1

   $ = $ '2' with probability s_2 ,

where, $s_1 + s_2 = 1$.

---

[1] We depart from the traditional notation of the random variable taking values of '0' and '1', so that the notation is consistent when we consider the multinomial case.

Let $x(n)$ be a concrete realization of X at time 'n'. The intention of the exercise is to estimate S, i.e., s_i for $i = 1, 2$. We achieve this by maintaining a running estimate $P(n) = [p_1(n), p_2(n)]^T$ of S, where $p_i(n)$ is the estimate of s_i at time 'n', for $i = 1, 2$. Then, the value of $p_i(n)$ is updated as per the following simple rule :

$$p_1(n + 1) \leftarrow \lambda p_1(n) \qquad \text{if } x(n) = 2 \qquad (1)$$
$$\leftarrow 1 - \lambda p_2(n) \quad \text{if } x(n) = 1. \qquad (2)$$

where λ is a user-defined parameter, $0 < \lambda < 1$.

In order to simplify the notation, the vector $P(n) = [p_1(n), p_2(n)]^T$ refers to the estimates of the probabilities of '1' and '2' occurring at time 'n', namely $p_1(n)$ and $p_2(n)$ respectively. In the interest of simplicity, we omit the index n, whenever there is no confusion, and thus, P is an abbreviation $P(n)$.

The first theorem, whose proof can be found in [15], concerns the distribution of the vector P which estimates S as per Equations (1) and (2). We shall state that P converges in distribution. The mean of P is shown to converge exactly to the mean of S. The proof, which is a little involved, follows the types of proofs used in the area of stochastic learning [15].

Theorem 1. *Let X be a binomially distributed random variable, and $P(n)$ be the estimate of S at time 'n'. Then, $E[P(\infty)] = S$.* $\qquad\square$

The next results that we shall prove indicates that $E[P(n + 1)]$ is related to $E[P(n)]$ by means of a stochastic matrix. We derive the explicit dependence, and allude to the resultant properties by virtue of the stochastic properties of this matrix. This leads us to two results, namely that of the limiting distribution of the chain, and that which concerns the rate of convergence of the chain. It turns out that while the former is independent of the learning parameter, λ, the latter is determined *only* by λ.

The reader will observe that the results we have derived are asymptotic results. In other words, the mean of $P(n)$ is shown to converge exactly to the mean of S. The implications of the "asymptotic" nature of the results will be clarified presently. The proofs of Theorems 2, 3 and 4 can be found in [15].

Theorem 2. *If the components of $P(n + 1)$ are obtained from the components of $P(n)$ as per Equations (1) and (2), $E[P(n + 1)] = M^T E[P(n)]$, where M is a stochastic matrix. Thus, the limiting value of the expectation of $P(.)$ converges to S, and the rate of convergence of P to S is fully determined by λ.* $\qquad\square$

Theorem 3. *Let X be a binomially distributed random variable governed by the distribution S, and $P(n)$ be the estimate of S at time 'n' obtained by (1) and (2). Then, the rate of convergence of P to S is fully determined by λ.* $\qquad\square$

We now derive the explicit expression for the asymptotic variance of the SLWE. A small value of λ leads to fast convergence and a large variance and vice versa.

Theorem 4. *Let X be a binomially distributed random variable governed by the distribution S, and $P(n)$ be the estimate of S at time 'n' obtained by (1) and (2). Then, the algebraic expression for the variance of $P(\infty)$ is fully determined by λ.* $\qquad\square$

When $\lambda \to 1$, it implies that the variance tends to zero, implying mean square convergence.

Our result seems to be contradictory to our initial goal. When we motivated our problem, we were working with the notion that the environment was non-stationary. However, the results we have derived are asymptotic, and thus, are valid only as $n \to \infty$. While this could prove to be a handicap, realistically, and for all practical purposes, the convergence takes place after a relatively small values of n. Thus, if λ is even as "small" as 0.9, after 50 iterations, the variation from the asymptotic value will be of the order of 10^{-50}, because λ also determines the rate of convergence, and this occurs in a geometric manner [13]. In other words, even if the environment switches its Bernoulli parameter after 50 steps, the SLWE will be able to track this change. Observe too that we do not need to introduce or consider the use of a "sliding window".

We conclude this Section by presenting the updating rule given in (1) and (2) in the context of some schemes already reported in the literature. If we assume that X can take values of '0' and '1', then the probability of '1' at time $n + 1$ can be estimated as follows:

$$p_1(n+1) \leftarrow \frac{n}{n+1}p_1 + \frac{1}{n+1}x(n+1) \tag{3}$$

This expression can be seen to be a particular case of the rule (1) and (2), where the parameter $\lambda = 1 - \frac{1}{n+1}$. This kind of rule is typically used in stochastic approximation [9], and in some reinforcement learning schemes, such as the Q-learning [31]. What we have done is to show that when such a rule is used in estimation, the mean converges to the true mean (independent of the learning parameter λ), and that the variance and rate of convergence are determined only by λ. Furthermore, we have derived the explicit relationships for these dependencies.

3   Weak Estimators of Multinomial Distributions

In this section, we shall consider the problem of estimating the parameters of a multinomial distribution. The multinomial distribution is characterized by two parameters, namely, the *number* of trials, and a probability vector which determines the probability of a specific event (from a pre-specified set of events) occurring. In this regard, the number of observations is the number of trials. Thus, we are to estimate the latter probability *vector* associated with the set of possible outcomes or trials. Specifically, let X be a multinomially distributed random variable, which takes on the values from the set $\{'1', \ldots, 'r'\}$. We assume that X is governed by the distribution $S = [s_1, \ldots, s_r]^T$ as:
$X = 'i'$ with probability s_i, where $\sum_{i=1}^{r} s_i = 1$.

Also, let $x(n)$ be a concrete realization of X at time 'n'. The intention of the exercise is to estimate S, i.e., s_i for $i = 1, \ldots, r$. We achieve this by maintaining a running estimate $P(n) = [p_1(n), \ldots, p_r(n)]^T$ of S, where $p_i(n)$ is the estimate of s_i at time 'n', for $i = 1, \ldots, r$. Then, the value of $p_1(n)$ is updated as per the following simple rule (the rules for other values of $p_j(n)$ are similar):

$$p_1(n+1) \leftarrow p_1 + (1 - \lambda) \sum_{j \neq 1} p_j \quad \text{when } x(n) = 1 \tag{4}$$

$$\leftarrow \lambda p_1 \quad \text{when } x(n) \neq 1 \tag{5}$$

As in the binomial case, the vector $P(n) = [p_1(n), p_2(n), \ldots, p_r(n)]^T$ refers to the estimate of $S = [s_1, s_2, \ldots, s_r]^T$ at time 'n', and we will omit the reference to time 'n' in $P(n)$ whenever there is no confusion.

The results that we present now are as in the binomial case, i.e. the distribution of $E[P(n+1)]$ follows $E[P(n)]$ by means of a stochastic matrix. This leads us to two results, namely to that of the limiting distribution of the vector, and that which concerns the rate of convergence of the recursive equation. It turns out that both of these, in the very worst case, could only be dependent on the learning parameter λ. However, to our advantage, while the former is *independent* of the learning parameter, λ, the latter is *only* determined by it (and not a function of it). The complete proofs of Theorems 5, 6 and 7 are found in [15].

Theorem 5. *Let X be a multinomially distributed random variable governed by the distribution S, and $P(n)$ be the estimate of S at time 'n' obtained by (4) and (5). Then,* $E[P(\infty)] = S$. $\qquad\square$

We now derive the explicit dependence of $E[P(n+1)]$ on $E[P(n)]$ and the consequences.

Theorem 6. *Let the parameter S of the multinomial distribution be estimated at time 'n' by $P(n)$ obtained by (4) and (5). Then, $E[P(n+1)] = \mathbf{M}^T E[P(n)]$, in which every off-diagonal term of the stochatic matrix, $\mathbf{M}$, has the* same *multiplicative factor, $(1-\lambda)$. Furthermore, the final solution of this vector difference equation is independent of λ.*

$\qquad\square$

The convergence and eigenvalue properties of $\mathbf{M}$ follow.

Theorem 7. *Let the parameter S of the multinomial distribution be estimated at time 'n' by $P(n)$ obtained by (4) and (5). Then, all the non-unity eigenvalues of $\mathbf{M}$ are exactly λ, and thus the rate of convergence of P is fully determined by λ.* $\qquad\square$

A small value of λ leads to fast convergence and a large variance, and vice versa. Again, although the results we have derived are asymptotic, if λ is even as "small" as 0.9, after 50 iterations, the variation from the asymptotic value will be of the order of 10^{-50}. In other words, after 50 steps, the SLWE will be able to track this change. Our experimental results demonstrate this fast convergence.

4   Experiments on Pattern Classification

The power of the SLWE has been earlier demonstrated in the "vanilla" estimation problem [20]. We now argue that these principles also have significance in the PR of non-stationary data, by demonstrating in two pattern recognition experiments. The classification in these cases was achieved on non-stationary data respectively derived synthetically, and from real-life data files obtained from the Canadian Broadcasting Corporation's (CBC's) news files. We now describe both these experiments and the comparison of the SLWE with the well-known MLE that uses a sliding window (MLEW).

4.1 Classification of Synthetic Data

In the case of synthetic data, the classification problem was the following. Given a stream of bits, which are drawn from two different sources (or classes), say, S_1 and S_2, the aim of the classification exercise was to *identify* the source which each symbol belonged to. To train the classifier, we first generated blocks of bits using two binomial distributions (20 blocks for each source), where the probability of 0 for each distribution was s_{11} and s_{12} respectively. For the results which we report, (other settings yielding similar results are not reported here) the specific values of s_{11} and s_{12} were randomly set to be $s_{11} = 0.18$ and $s_{12} = 0.76$, which were assumed unknown to the classifier. These data blocks were then utilized to compute MLE estimates of s_{11} and s_{12}, which we shall call s'_{11} and s'_{12} respectively. In the testing phase, 40 blocks of bits were generated randomly from either of the sources, and the order in which they were generated was kept secret to the classifier. Furthermore, the size of the blocks was also unknown to the estimators, MLEW and SLWE. In the testing phase, each bit read from the testing set was classified using two classifiers, which were designed using the probability of the symbol '0' estimated by the MLEW and SLWE, which we refer to as $p_{M1}(n)$ and $p_{S1}(n)$ respectively. The classification rule was based on a linear classifier, and the bit was assigned to the class that minimized the Euclidean distance between the estimated probability of 0, and the "learned" probability of 0 (for the two sources) obtained during the training. Thus, based on the MLWE classifier, the nth bit read from the test set was assigned to class S_1, if (ties were resolved arbitrarily) :

$$(p_{1M} - s'_{11}) < (p_{1M} - s'_{12}) \, ,$$

and assigned to class S_2 otherwise. The classification rule based on the SLWE was analogous, except that it used p_{1S} instead of p_{1M}.

The two classifiers were tested for different scenarios. For the SLWE, the value of λ was arbitrarily set to 0.9. In the first set of experiments, the classifiers were tested for blocks of different sizes, $b = 50, 100, \ldots, 500$. For each value of b, an ensemble of 100 experiments was performed. The MLEW used a window of size w, which was centered around b, and was computed as the nearest integer of a randomly generated value obtained from a uniform distribution $U[b/2, 3b/2]$. In effect, we thus gave the MLEW the advantage of having some *a priori* knowledge of the block size. The results obtained are shown in Table 1, from which we see that classification using the SLWE was uniformly superior to classification using the MLEW, and in many cases yielded more than 20% accuracy than the latter. The accuracy of the classifier increased with the block size as is clear from Figure 1.

In the second set of experiments, we tested both classifiers, the MLEW and SLWE, when the block size was fixed to be $b = 50$. The window size of the MLEW, w was again randomly generated from $U[b/2, 3b/2]$, thus giving the MLE the advantage of having some *a priori* knowledge of the block size. The SLWE again used a conservative value of $\lambda = 0.9$. In this case, we conducted an ensemble of 100 simulations, and in each case, we computed the classification accuracy of the first k bits in each block. The results are shown in Table 2. Again, the uniform superiority of the SLWE over the MLEW can be observed. For example, when $k = 45$, the MLEW yielded an accuracy

Table 1. The results obtained from testing linear classifiers which used the MLEW and SLWE respectively. In this case, the blocks were generated of sizes $b = 50, 100, \ldots, 500$, and the window size w used for the MLEW was randomly generated from $U[b/2, 3b/2]$. The SLWE used a value of $\lambda = 0.9$. The results were obtained from an ensemble of 100 simulations, and the accuracy was computed using *all* the bits in the block.

b	MLEW	SLWE
50	76.37	93.20
100	75.32	96.36
150	73.99	97.39
200	74.63	97.95
250	74.91	98.38
300	74.50	98.56
350	75.05	98.76
400	74.86	98.88
450	75.68	98.97
500	75.55	99.02

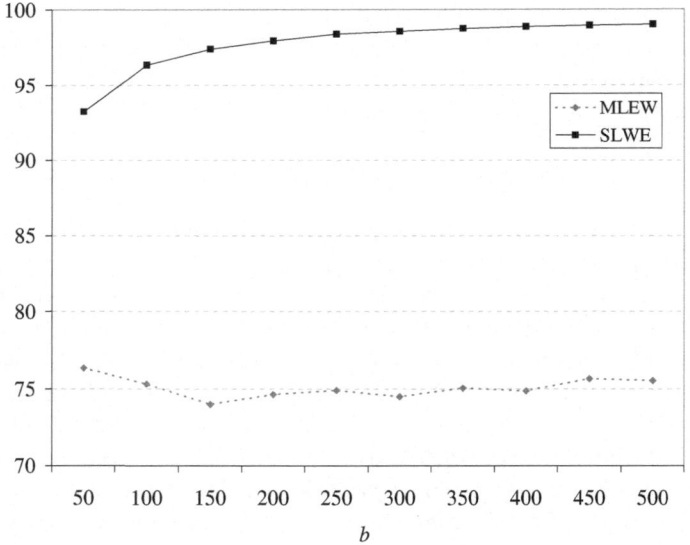

Fig. 1. Plot of the accuracies of the MLEW and SLWE classifiers for different block sizes. The settings of the experiments are as described in Table 1.

of only 73.77%, but the corresponding accuracy of the SLWE was 92.48%. From the plot shown in Figure 2, we again observe the increase in accuracy with k.

In the final set of experiments, we tested both classifiers, the MLEW and SLWE, when the *block size* was made random (as opposed to being fixed as in the previous cases). This was, in one sense, the most difficult setting, because there was no fixed "periodicity" for the switching from one "environment" to the second, and vice versa.

Table 2. The results obtained from testing linear classifiers which used the MLEW and SLWE respectively. In this case, the blocks were generated of size $b = 50$, and the window size w used for the MLEW was randomly generated from $U[b/2, 3b/2]$. The SLWE used a value of $\lambda = 0.9$. The results were obtained from an ensemble of 100 simulations, and the accuracy was computed using the first k bits in each block.

k	MLEW	SLWE
5	50.76	56.46
10	51.31	70.00
15	53.15	78.65
20	56.03	83.66
25	59.47	86.78
30	63.19	88.90
35	67.03	90.42
40	70.65	91.57
45	73.77	92.48
50	76.37	93.20

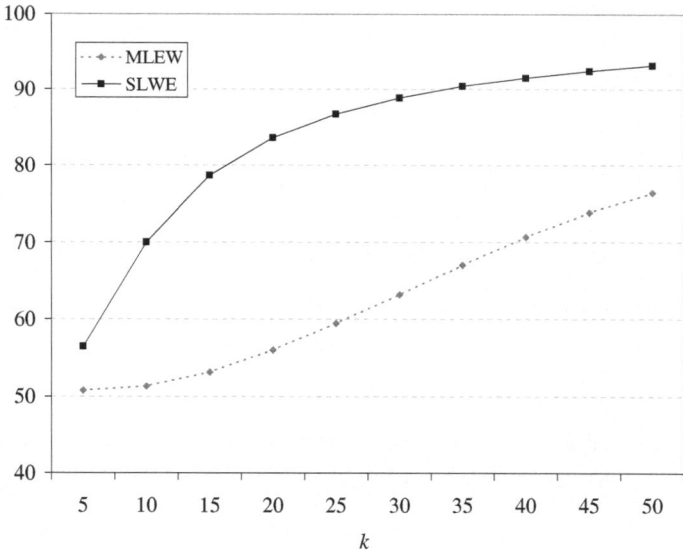

Fig. 2. Plot of the accuracies of the MLEW and SLWE classifiers for different values of k. The settings of the experiments are as described in Table 2.

This *block size* was thus randomly generated from $U[w/2, 3w/2]$, where w was the width used by the MLEW. Observe again that this implicitly assumed that the MLE had some (albeit, more marginal than in the previous cases) advantage of having some *a priori* knowledge of the system's behaviour, but the SLWE used the same conservative value of $\lambda = 0.9$. Again, we conducted an ensemble of 100 simulations, and in each case, we computed the classification accuracy of the first k bits in each block. The

Table 3. The results obtained from testing linear classifiers which used the MLEW and SLWE respectively. In this case, the blocks were generated of a random width b uniform in $U[w/2, 3w/2]$, where w was the window size used by the MLEW. The SLWE used a value of $\lambda = 0.9$. The results were obtained from an ensemble of 100 simulations, and the accuracy was computed using the first k bits in each block.

b	MLEW	SLWE
5	51.53	56.33
10	51.79	70.05
15	52.06	78.79
20	53.09	83.79
25	56.17	86.88

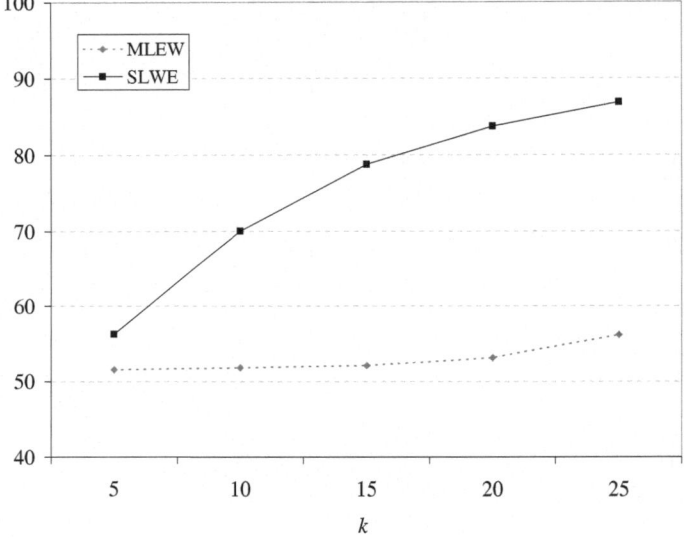

Fig. 3. Plot of the accuracies of the MLEW and SLWE classifiers for different values of k. The numerical results of the experiments are shown in Table 3.

results are shown in Table 3. As in the above two cases, the SLWE was uniformly superior to the MLEW. For example, when $b = 20$, the MLEW yielded a recognition accuracy of only 53.09%, but the corresponding accuracy of the SLWE was 83.79%. From the plot shown in Figure 3, we again observe the increase in accuracy with k for the case of the SLWE, but the increase in the case of the MLEW is far less marked.

4.2   Result on Real-Life Classification

In this section, we present the results of classification obtained from real-life data. The problem can be described as follows. We are given a stream of news items from different sources, and the aim of the PR exercise is to detect the source that the news come

Table 4. Empirical results obtained by classification of CBC files into *Sports* (S) and *Business* (B) items using the MLEW and SLWE

File name	Class	Size	MLEW		SLWE	
			Hits	Accuracy	Hits	Accuracy
s3.txt	S	1162	721	62.05	909	78.23
b11.txt	B	1214	1045	86.08	1154	95.06
b17.txt	B	1290	1115	86.43	1261	97.75
s16.txt	S	1044	610	58.43	656	62.84
s9.txt	S	2161	1542	71.36	1789	82.79
b7.txt	B	1021	877	85.90	892	87.37
b5.txt	B	1630	932	57.18	1087	66.69
b12.txt	B	1840	1289	70.05	1502	81.63
s20.txt	S	2516	1582	62.88	1919	76.27
b1.txt	B	1374	826	60.12	882	64.19
b2.txt	B	1586	940	59.27	1124	70.87
s4.txt	S	1720	1051	61.10	1169	67.97
b8.txt	B	1280	801	62.58	917	71.64
b15.txt	B	1594	831	52.13	928	58.22
s10.txt	S	2094	1202	57.40	1486	70.96
b20.txt	B	2135	1361	63.75	1593	74.61
s14.txt	S	2988	1749	58.53	2163	72.39
b10.txt	B	3507	2086	59.48	2291	65.33
b9.txt	B	2302	1425	61.90	1694	73.59
s7.txt	S	2209	928	42.01	1116	50.52
s5.txt	S	2151	1181	54.90	1411	65.60
s12.txt	S	3197	1636	51.17	2071	64.78
s8.txt	S	1935	1149	59.38	1398	72.25
s1.txt	S	2334	1079	46.23	1248	53.47
b3.txt	B	1298	563	43.37	640	49.31
s2.txt	S	2147	917	42.71	1218	56.73
b19.txt	B	1943	991	51.00	1210	62.27
s13.txt	S	2534	1380	54.46	1568	61.88

from. For example, consider the scenario when a TV channel broadcasts live video, or a channel that releases news in the form of text, for example, on the internet. The problem of detecting the source of the news item from a live channel has been studied by various researchers (in the interest of brevity, we merely cite one reference [25]) with the ultimate aim being that of extracting *shots* for the videos and using them to classify the video into any of the pre-defined classes. Processing images is extremely time-consuming, which makes real time processing almost infeasible. Thus, to render the problem tractable, we consider the scenario in which the news items arrive in the form of text blocks, which are extracted from the closed-captioning text embedded in the video streams. A classification problem of this type is similar to that of classifying files posted, for example, on the internet.

To demonstrate the power of our SLWE over the MLE when used in a classification scheme for real-life problems, we used them both as in the case of the artificial data described above, followed by a simple linear classifier. We first downloaded news files from the Canadian Broadcasting Corporation (CBC), which were from either of two different sources (the classes): *business* or *sports*[2]. The data was then divided into two subsets, one for training and the other for testing purposes, each subset having 20 files of each class. With regard to the parameters, for the SLWE we randomly selected a value of λ uniformly from the range $[0.99, 0.999]$, and for the case of the MLEW, the window size was uniformly distributed in $[50, 150]$. In our particular experiments, the value for λ was 0.9927, and the value for the size of the window was 79, where the latter was obtained by rounding the resulting random value to the nearest integer. The results of the classification are shown in Table 4, where we list the name of the files that were selected (randomly) from the *testing* set. The column labeled "hits" contains the number of characters in the file, which were correctly classified, and "accuracy" represents the percentage of those characters. The results reported here are only for the cases when the classification problem is meaningful. In a *few* cases, the bit pattern did not even yield a 50% (random choice) recognition accuracy. We did not report these results because, clearly, classification using these "features" and utilizing such a classifier is meaningless. But we have reported all the cases when the MLEW gave an accuracy which was less than 50%, but for which the SLWE yielded an accuracy which was greater than 50%. The superiority of the SLWE over the MLEW is clearly seen from the table - for example in the case of the file *s9.txt*, the MLEW yielded a classification accuracy of 71.36%, while the corresponding accuracy for the SLWE was 82.79%. Indeed, *in every single case*, the SLWE yielded a classification accuracy better than the corresponding accuracy obtained by the MLEW !

We conclude this section by observing that a much better accuracy can be obtained using similar features if a more powerful classifier (for example, a higher order classifier) is utilized instead of the simple linear classifier that we have invoked.

5   PR Using SLWE Obtained with Other LA Schemes

Throughout this paper, we have achieved PR by deriving training schemes obtained by utilizing the principles of learning as achieved by the families of Variable Structure Stochastic Automata (VSSA) [10,11,13,28]. In particular, the learning we have achieved is obtained using an learning algorithm of the linear scheme family, namely the L_{RI} scheme, and provided the corresponding analysis.

We would like to mention that the estimation principles introduced here leads to various "open problems". A lot of work has been done in the last decades which involve the so-called families of Discretized LA, and the families of Pursuit and Estimator algorithms. Discretized LA were pioneered by Thathachar and Oommen [28] and since then, all the families of continuous VSSA have found their counterparts in the corresponding discretized versions [16,18]. The design of SLWE using discretized VSSA is open. Similarly, the Pursuit and Estimator algorithms were first pioneered by Thathachar, Sastry and their colleagues [23,29,30]. These involve specifically utilizing running estimates of the penalty probabilities of the actions in enhancing the stochastic

---

[2] The files used in our experiments were obtained from the CBC web-site, www.cbc.ca, between September 24 and October 2, 2004, and can be made available upon request.

learning. These automata were later discretized [12,19] and currently, extremely fast continuous and discretized pursuit and estimator LA have been designed [17,21,22]. The question of designing SLWE using Pursuit and Estimator updating schemes is also open[3]. We believe that "weak" estimates derivable using any of these principles will lead to interesting PR and classification methods that could also be applicable for non-stationary data.

6   Conclusions and Future Work

In this paper we have considered the pattern recognition (i.e., the training and classification) problem in which we assume class-conditional distribution of the features is assumed unknown, but non-stationary. To achieve the training and classification feasible within this model, we presented a novel estimation strategy, involving the so-called Stochastic Learning Weak Estimator (SLWE), which is based on the principles of stochastic learning. The latter, whose convergence is *weak*, has been used to estimate the parameters of a binomial/multinomial distribution. Even though our method includes a learning coefficient, λ, it turns out that the mean of the final estimate is independent of λ, the variance of the final distribution decreases with λ, and the speed decreases with λ. To demonstrate the power of these estimates in data which is truly "non-stationary", we have used them in *two* PR exercises, the first of which involves artificial data, and the second which involves the recognition of the *types* of data that are present in news reports of the Canadian Broadcasting Corporation (CBC). The superiority of the SLWE in both these cases is demonstrated.

Acknowledgements. This research work has been partially supported by NSERC, the Natural Sciences and Engineering Research Council of Canada, CFI, the Canadian Foundation for Innovation, and OIT, the Ontario Innovation Trust.

References

1. P. Bickel and K. Doksum. *Mathematical Statistics: Basic Ideas and Selected Topics*, volume I. Prentice Hall, second edition, 2000.
2. C. Blake and C. Merz. UCI repository of machine learning databases, 1998.
3. G. Casella and R. Berger. *Statistical Inference*. Brooks/Cole Pub. Co., second edition, 2001.
4. R. Duda, P. Hart, and D. Stork. *Pattern Classification*. John Wiley and Sons, Inc., New York, NY, 2nd edition, 2000.
5. K. Fukunaga. *Introduction to Statistical Pattern Recognition*. Academic Press, 1990.
6. R. Herbrich. *Learning Kernel Classifiers: Theory and Algorithms*. MIT Press, Cambridge, Massachusetts, 2001.
7. Y. M. Jang. Estimation and Prediction-Based Connection Admission Control in Broadband Satellite Systems. *ETRI Journal*, 22(4):40–50, 2000.
8. B. Jones, P. Garthwaite, and Ian Jolliffe. *Statistical Inference*. Oxford University Press, second edition, 2002.

---

[3] We believe, however, that even if such estimator-based SLWEs are designed, the analysis of their properties will not be trivial.

9. H. Kushner and G. Yin. *Stochastic Approximation and Recursive Algorithms and Applications*. Springer, 2nd. edition, 2003.
10. S. Lakshmivarahan. *Learning Algorithms Theory and Applications*. Springer-Verlag, New York, 1981.
11. S. Lakshmivarahan and M. A. L. Thathachar. Absolutely Expedient Algorithms for Stochastic Automata. *IEEE Trans. on System, Man and Cybernetics*, SMC-3:281–286, 1973.
12. J. K. Lanctôt and B. J. Oommen. Discretized Estimator Learning Automata. *IEEE Trans. on Systems, Man and Cybernetics*, 22(6):1473–1483, 1992.
13. K. Narendra and M. Thathachar. *Learning Automata. An Introduction*. Prentice Hall, 1989.
14. J. Norris. *Markov Chains*. Springer Verlag, 1999.
15. B. J. Oommen and L. Rueda. Stochastic Learning-based Weak Estimation of Multinomial Random Variables and Its Applications to Non-stationary Environments. 2004. Submitted for publication.
16. B.J. Oommen. Absorbing and Ergodic Discretized Two-Action Learning Automata. *IEEE Trans. on System, Man and Cybernetics*, SMC-16:282–296, 1986.
17. B.J. Oommen and M. Agache. Continuous and Discretized Pursuit Learning Schemes: Various Algorithms and Their Comparison. *IEEE Trans. on Systems, Man and Cybernetics*, SMC-31(B):277–287, June 2001.
18. B.J. Oommen and J. R. P. Christensen. Epsilon-Optimal Discretized Reward-Penalty Learning Automata. *IEEE Trans. on System, Man and Cybernetics*, SMC-18:451–458, 1988.
19. B.J. Oommen and J. K. Lanctôt. Discretized Pursuit Learning Automata. *IEEE Trans. on System, Man and Cybernetics*, 20(4):931–938, 1990.
20. B.J. Oommen and L. Rueda. A New Family of Weak Estimators for Training in Nonstationary Distributions. In *Proceedings of the Joint IAPR International Workshops SSPR 2004 and SPR 2004*, pages 644–652, Lisbon, Portugal, 2004.
21. G. I. Papadimitriou. A New Approach to the Design of Reinforcement Schemes for Learning Automata: Stochastic Estimator Learning Algorithms. *IEEE Trans. on Knowledge and Data Engineering*, 6:649–654, 1994.
22. G. I. Papadimitriou. Hierarchical Discretized Pursuit Nonlinear Learning Automata with Rapid Convergence and High Accuracy. *IEEE Trans. on Knowledge and Data Engineering*, 6:654–659, 1994.
23. K. Rajaraman and P. S. Sastry. Finite Time Analysis of the Pursuit Algorithm for Learning Automata. *IEEE Trans. on Systems, Man and Cybernetics*, 26(4):590–598, 1996.
24. S. Ross. *Introduction to Probability Models*. Academic Press, second edition, 2002.
25. M. De Santo, G. Percannella, C. Sansone, and M. Vento. A multi-expert approach for shot classification in news videos. In *Image Analysis and Recogniton, LNCS*, volume 3211, pages 564–571, Amsterdam, 2004. Elsevier.
26. J. Shao. *Mathematical Statistics*. Springer Verlag, second edition, 2003.
27. R. Sprinthall. *Basic Statistical Analysis*. Allyn and Bacon, second edition, 2002.
28. M. A. L. Thathachar and B. J. Oommen. Discretized Reward-Inaction Learning Automata. *Journal of Cybernetics and Information Sceinces*, pages 24–29, Spring 1979.
29. M. A. L. Thathachar and P.S. Sastry. A Class of Rapidly Converging Algorithms for Learning Automata. *IEEE Trans. on Systems, Man and Cybernetics*, SMC-15:168–175, 1985.
30. M. A. L. Thathachar and P.S. Sastry. Estimator Algorithms for Learning Automata. In *Proc. of the Platinum Jubilee Conference on Systems and Signal Processing*, Bangalore, India, December 1986.
31. C. Watkins. *Learning from Delayed Rewards*. PhD thesis, University of Cambridge, England, 1989.
32. A. Webb. *Statistical Pattern Recognition*. John Wiley & Sons, N.York, second edition, 2002.

Noise Robustness by Using Inverse Mutations

Ralf Salomon

Faculty of Computer Science and Electrical Engineering,
University of Rostock, 18051 Rostock, Germany
ralf.salomon@etechnik.uni-rostock.de

Abstract. Recent advances in the theory of evolutionary algorithms have indicated that a hybrid method known as the evolutionary-gradient-search procedure yields superior performance in comparison to contemporary evolution strategies. But the theoretical analysis also indicates a noticeable performance loss in the presence of noise (i.e., noisy fitness evaluations). This paper aims at understanding the reasons for this observable performance loss. It also proposes some modifications, called *inverse mutations*, to make the process of estimating the gradient direction more noise robust.

1   Introduction

The literature on evolutionary computation discusses the question whether or not evolutionary algorithms are gradient methods very controversially. Some [10] believe that by generating trial points, called offspring, evolutionary algorithms stochastically operate along the gradient, whereas others [4] believe that evolutionary algorithms might deviate from gradient methods too much because of their high explorative nature.

The goal of an optimization algorithm is to locate the optimum $x^{(\mathrm{opt})}$ (minimum or maximum in a particular application) of an N-dimensional objective (fitness) function $f(x_1, \ldots, x_N) = f(x)$ with x_i denoting the N independent variables. Particular points x or y are also known as search points. In essence, most procedures differ in how they derive future test points $x^{(t+1)}$ from knowledge gained in the past. Some selected examples are gradient descent, Newton's method, and evolutionary algorithms. For an overview , the interested reader is referred to the pertinent literature [3,6,7,8,9,14].

Recent advances in the theory of evolutionary algorithms [1] have also considered hybrid algorithms, such as the evolutionary-gradient-search (EGS) procedure [12]. This algorithm fuses principles from both gradient and evolutionary algorithms in the following manner:

- it operates along an explicitly estimated gradient direction,
- it gains information about the gradient direction by randomly generating trial points (offspring), and
- it periodically reduces the entire population to a single search point.

U. Furbach (Ed.): KI 2005, LNAI 3698, pp. 121–133, 2005.

A description of this algorithm is presented in Section 2, which also briefly reviews some recent theoretical analyses. Arnold [1] has shown that the EGS procedure performs superior, i.e., sequential efficiency, in comparison to contemporary evolution strategies.

But Arnold's analysis [1] also indicates that the performance of EGS progressively degrades in the presence of noise (noisy fitness evaluations). Due to the high importance of noise robustness in real-world applications, Section 3 briefly reviews the relevant literature and results.

Section 4 is analyzing the reason for the observable performance loss. A first result of this analysis is that the procedure might be extended by a second, independently working step size in order to de-couple the gradient estimation from performing the actual progress step.

A main reason for EGS suffering from a performance loss in the presence of noise is that large mutation steps cannot be used, because they lead to quite imprecise gradient estimates. Section 5 proposes the usage of inverse mutations, which use each mutation vector z twice, originally and in the inverse direction $-z$. Section 5 shows that inverse mutations are able to compensate for noise due to a mechanism, which is called genetic repair in other algorithms. Section 6 concludes with a brief discussion and an outlook for future work.

2   Background: Algorithms and Convergence

This section summarizes some background material as far as necessary for the understanding of this paper. This includes a brief description of the procedure under consideration as well as some performance properties on quadratic functions.

2.1   Algorithms

As mentioned in the introduction, the evolutionary-gradient-search (EGS) procedure [12] is a hybrid method that fuses some principles of both gradient and evolutionary algorithms. It periodically collapses the entire population to a single search point, and explicitly estimates the gradient direction in which it tries to advance to the optimum. The procedure estimates the gradient direction by randomly generating trial points (offspring) and processing *all* trial points in calculating a weighted average. In its simplest form, EGS works as follows[1]:

1. Generate $i=1,\ldots,\lambda$ offspring $\boldsymbol{y}_t^{(i)}$ (trial points)

$$\boldsymbol{y}_t^{(i)} = \boldsymbol{x}_t + \sigma\boldsymbol{z}_t^{(i)} \tag{1}$$

---

[1] Even though further extensions, such as using a momentum term or generating non-isotropic mutations, significantly accelerate EGS [13], they are not considered in this paper, because they are not appropriately covered by currently available theories. Furthermore, these extensions are not rotationally invariant, which limits their utility.

from the current point $\boldsymbol{x}_t$ (at time step t), with $\sigma > 0$ denoting a step size and $\boldsymbol{z}^{(i)}$ denoting a mutation vector consisting of N independent, normally distributed components.

2. Estimate the gradient direction $\tilde{\boldsymbol{g}}_t$:

$$\tilde{\boldsymbol{g}}_t = \sum_{i=1}^{\lambda} \left(f(\boldsymbol{y}_t^{(i)}) - f(\boldsymbol{x}_t) \right) \left(\boldsymbol{y}_t^{(i)} - \boldsymbol{x}_t \right) . \tag{2}$$

3. Perform a step

$$\boldsymbol{x}_{t+1} = \boldsymbol{x}_t + \sqrt{N} \frac{\tilde{\boldsymbol{g}}_t}{\|\tilde{\boldsymbol{g}}_t\|} = \boldsymbol{x}_t + \boldsymbol{z}_t^{(\text{prog})} , \tag{3}$$

with $\boldsymbol{z}^{(\text{prog})}$ denoting the progress vector.

It can be seen that Eq. (2) estimates an approximated gradient direction by calculating a weighted sum of *all* offspring regardless of their fitness values; the procedure simply assumes that for offspring with negative progress, positive progress can be achieved in the opposite direction and sets those weights to negative values. This usage of *all* offspring is in way contrast to most other evolutionary algorithms, which only use ranking information of some selected individuals of some sort. Further details can be found in the literature [1,5,12].

EGS also features some dynamic adaptation of the step size σ. This, however, is beyond the scope of this paper, and the interested reader is referred to the literature [12].

This paper compares EGS with the $(\mu/\mu, \lambda)$-evolution strategy [5,10], since the latter yields very good performance and is mathematically well analyzed. The $(\mu/\mu, \lambda)$-evolution strategy maintains μ parents, applies global-intermediate recombination on all parents, and applies normally distributed random numbers to generates λ offspring, from which the μ best ones are selected as parents for the next generation. In addition, this strategy also features some step size adaptation mechanism [1].

2.2   Progress on Quadratic Functions

Most theoretical performance analyses have yet been done on the quadratic function $f(x_1, \ldots, x_n) = \sum_i x_i^2$, also known as the sphere model. This choice has the following two main reasons: First, other functions are currently too difficult to analyze, and second, the sphere model approximates the optimum's vicinity of many (real-world) applications reasonably well. Thus, the remainder of this paper also adopts this choice.

As a first performance measure, the rate of progress is defined as

$$\varphi = f(\boldsymbol{x}_t) - f(\boldsymbol{x}_{t+1}) \tag{4}$$

in terms of the best population members' objective function values in two subsequent time steps t and $t+1$. For standard (μ, λ)-evolution strategies operating on the sphere model, Beyer [5] has derived the following rate of progress φ:

$$\varphi \approx 2Rc_{\mu,\lambda}\sigma - N\sigma^2 , \tag{5}$$

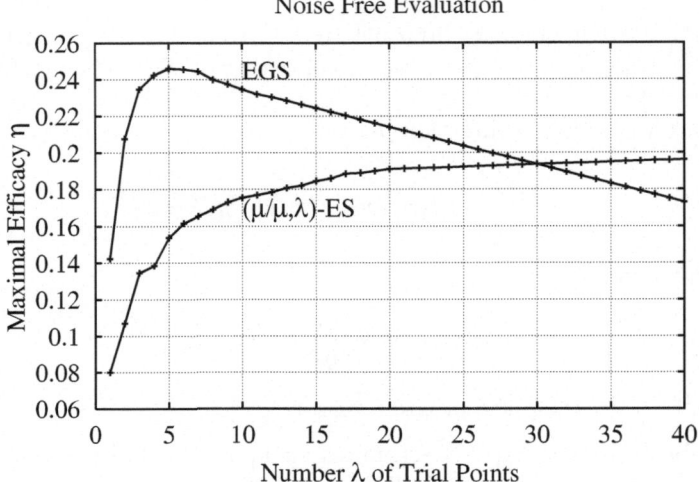

Fig. 1. Rate of progress for EGS and $(\mu/\mu, \lambda)$-evolution strategies according to Eqs. (7) and (8). Further details can be found in [1].

with $R = \|x_g\|$ denoting the distance of the best population member to the optimum and $c_{\mu,\lambda}$ denoting a constant that subsumes all influences of the population configuration as well as the chosen selection scheme. Typical values are: $c_{1,6}=1.27$, $c_{1,10}=1.54$, $c_{1,100}=2.51$, and $c_{1,1000}=3.24$. To be independent from the current distance to the optimum, normalized quantities are normally considered (i.e., relative performance):

$$\varphi^* = \sigma^* c_{\mu,\lambda} - 0.5(\sigma^*)^2 \text{ with}$$
$$\varphi^* = \varphi \frac{N}{2R^2}, \sigma^* = \sigma \frac{N}{R} \quad . \tag{6}$$

Similarily, the literature [1,5,10] provides the following rate of progress formulas for EGS and the $(\mu/\mu, \lambda)$-evolution strategies:

$$\varphi^*_{EGS} \approx \sigma^* \sqrt{\frac{\lambda}{1 + \sigma^{*2}/4}} - \frac{\sigma^{*2}}{2}, \tag{7}$$

$$\varphi^*_{\mu/\mu, \lambda} \approx \sigma^* c_{\mu/\mu, \lambda} - \frac{\sigma^*}{2\mu} \tag{8}$$

The rate of progress formula is useful to gain insights about the influences of various parameters on the performance. However, it does not consider the (computational) costs required for the evaluation of all λ offspring. For the common assumption that all offspring be evaluated sequentially, the literature often uses a second performance measure, called the efficiency $\eta = \varphi^*/\lambda$. In other words, the efficiency η expresses the sequential run time of an algorithm.

Figure 1 shows the efficiency of both algorithms according to Eqs. (7) and (8). It can be seen that for small numbers of offspring (i.e., $\lambda \approx 5$), EGS is most efficient (in terms of sequential run time) and superior to the $(\mu/\mu, \lambda)$-evolution strategy.

3   Problem Description: Noise and Performance

Subsection 2.2 has reviewed the obtainable progress rates for the undisturbed case. The situation changes, however, when considering noise, i.e., noisy fitness evaluations. Noise is present in many (if not all) real-world applications. For the viability of practially relevant optimization procedures, noise robustness is thus of high importance.

Most commonly, noise is modeled by additive $N(0, \sigma_\epsilon)$ normally distributed random numbers with standard deviation σ_ϵ. For noisy fitness evaluations, Arnold [1] has derived the following rate of progress

$$\varphi^*_{EGS} \approx \sigma^* \sqrt{\frac{\lambda}{1 + \sigma^{*2}/4 + \sigma_\epsilon^{*2}/\sigma^{*2}} - \frac{\sigma^{*2}}{2}} , \qquad (9)$$

with $\sigma_\epsilon^* = \sigma_\epsilon N/(2R^2)$ denoting the normalized noise strength.

Figure 2 demonstrates how the presence of noise σ_ϵ^* reduces the rate of progress of the EGS procedure. Eq. (9) reveals that positive progress can be achieved only if $\sigma_\epsilon^* \leq \sqrt{4\lambda + 1}$ holds. In other words, the required number of offspring (trial points) to estimate the gradient direction grows quadratically.

Another point to note is that the condition $\sigma_\epsilon^* \leq \sqrt{4\lambda + 1}$ incorporates the *normalized* noise strength. Thus, if the procedure approaches the optimum, the

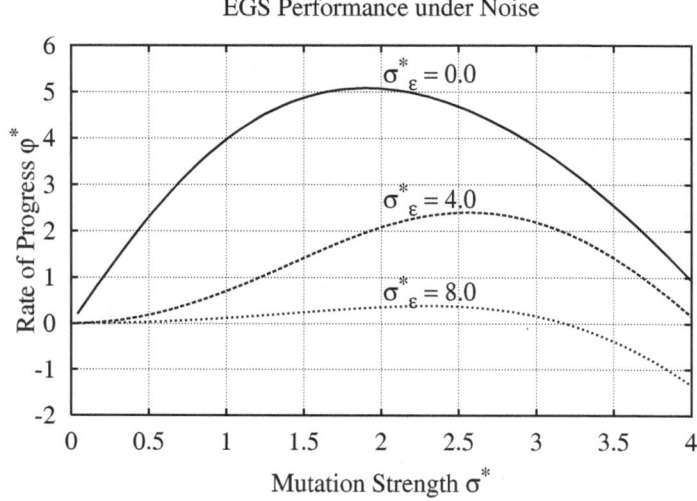

Fig. 2. The rate of progress φ^* of EGS progressively degrades under the presence of noise σ_ϵ^*. The example has used $\lambda=25$ offspring. Further details can be found in [1].

distance R decreases and the noise strength increases. Consequently, the procedure exhibits an increasing performance loss as it advances towards the optimum.

By contrast, the $(\mu/\mu, \lambda)$-evolution strategy benefits from an effect called genetic repair induced by the global-intermediate recombination, and is thus able to operate with larger mutation step sizes σ^*. For the $(\mu/\mu, \lambda)$-evolution strategy, the literature [2] suggest that only a linear growth in the number of offspring λ is required.

4   Analysis: Imprecise Gradient Approximation

This section analysis the reasons for the performance loss described in Section 3. To this end, Figure 3 illustrates how the EGS procedure estimates the gradient direction according to Eq. (2): $\tilde{g}_t = \sum_{i=1}^{\lambda}(f(y_t^{(i)}) - f(x_t))(y_t^{(i)} - x_t)$. The following two points might be mentioned here:

1. The EGS procedure uses the same step size σ for generating trial points and performing a step from x_t to x_{t+1}.
2. For small step sizes σ, the probability for an offspring to be better or worse than its parents is half chance. For larger step sizes, though, the chances for better offspring are steadily decreasing, and are approaching zero for $\sigma \geq 2R$.

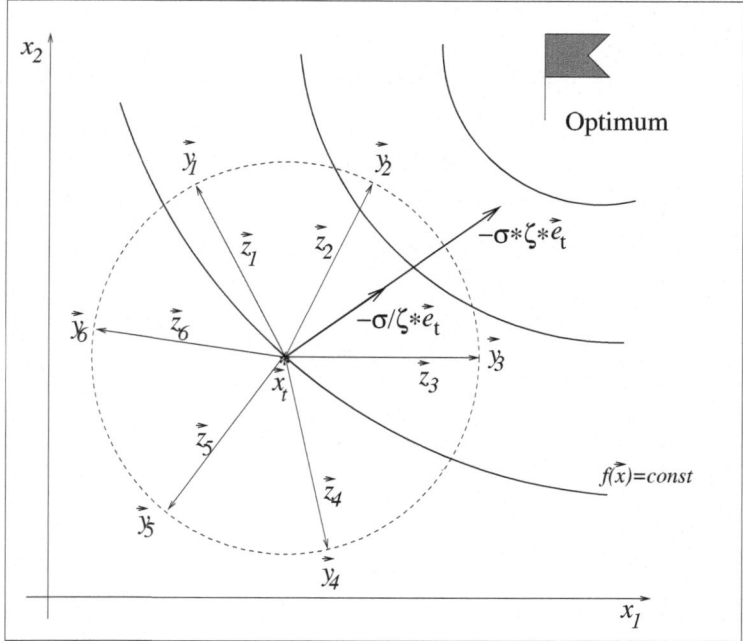

Fig. 3. This figure shows how the EGS procedure estimates the gradient direction $\tilde{g}$ by averaging weighted test points y

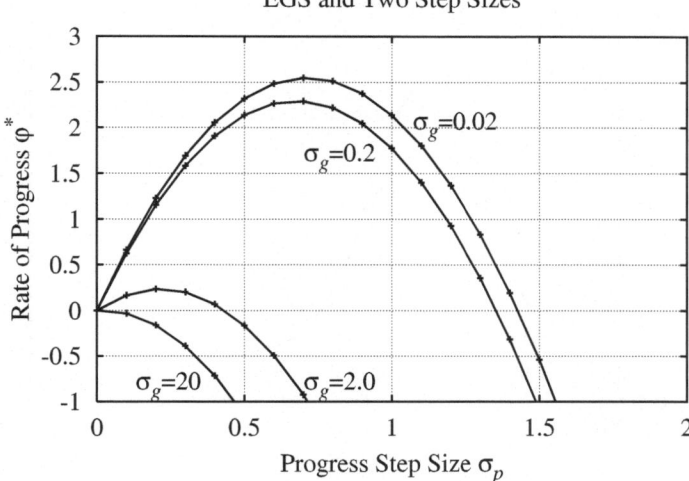

Fig. 4. Normalized rate of progress φ^* when using different step sizes σ_g and σ_p for the example $N = \lambda = 10$

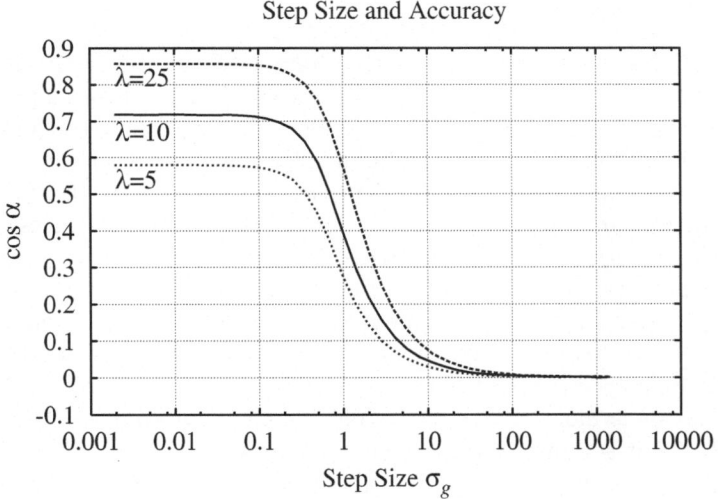

Fig. 5. $\cos\alpha$ between the true gradient g and its estimate $\tilde{g}$ as a function of the number of trial points λ

It is obvious that for small λ, unequal chances have a negative effect on the gradient approximation accuracy.

Both points can be further elaborated as follows: A modified version of the EGS procedure uses two independent step sizes σ_g and σ_p for generating test points (offspring) $y_t^{(i)} = x_t + \sigma_g z_t^{(i)}$ and performing the actual progress step $x_{t+1} = x_t + \sigma_p \sqrt{N} \tilde{g}_t / \|\tilde{g}_t\|$ according to Eqs. (1) and (3), respectively. Figure 4 illustrates

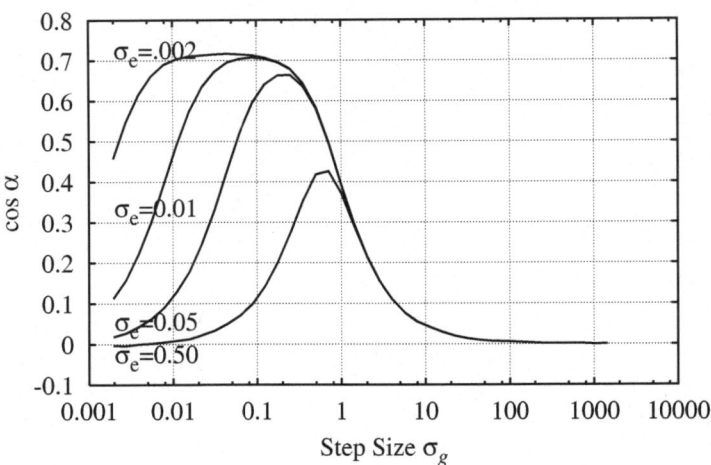

Fig. 6. $\cos \alpha$ between the true gradient g and its estimate $\tilde{g}$ as a function of the noise strength $\sigma_e \in \{0.002, 0.01, 0.05, 0.5\}$. In this example, $N = \lambda = 10$ and $R = 1$ were used.

the effect of these two step sizes for the example $R = 1$, $N = 10$ dimensions and $\lambda = 10$ trial points. It can be clearly seen that the rate of progress φ^* drastically degrades for large step sizes σ_g. Since Figure 4 plots the performance for 'all possible' step sizes σ_p, it can be directly concluded that the accuracy of the gradient estimation significantly degrades for step size σ_g being too large; if the estimation $\tilde{g} = g$ was precise, the attainable rate of progress would be $\varphi^* = (f(\boldsymbol{x}_0) - f(\boldsymbol{0}))N/(2R^2) = (1-0)10/2 = 5$.

This hypothesis is supported by Fig. 5, which shows the angle $\cos \alpha = g\tilde{g}/(\|g\|\|\tilde{g}\|)$ between the true gradient g and its estimate $\tilde{g}$. It can be clearly seen that despite being dependent on the number of trial points λ, the gradient estimate's accuracy significantly depends on the step size σ_g. The qualitative behavior is more or less equivalent for all three number of trial points $\lambda \in \{5, 10, 25\}$: it is quite good for small step sizes, starts degrading at $\sigma_g \approx R$, and quickly approaches zero for $\sigma_g > 2R$. In addition, Fig. 6 shows how the situation changes when noise is present. It can be seen that below the noise level, the accuracy of the gradient estimate degrades.

In summary, this section has show that it is generally advantageous to employ two step sizes σ_g and σ_p and that the performance of the EGS procedure degrades when the step size σ_g is either below the noise strength or above the distance to the optimum.

5   Inverse Mutations

The previous section has identified two regimes, $\sigma_g < \sigma_e$ and $\sigma_g > R$ that yield poor gradient estimates. This problem could be tackled by either resampling already generated trial points or by increasing the number of different trial points. Both options, however, have high computational costs, since the performance gain would grow at most with $\sqrt{\lambda}$ (i.e., reduced standard deviation or Eq. (9)). Since the performance gain of the $(\mu/\mu, \lambda)$-evolution strategy grows linearly in λ (due to genetic repair [1]), these options are not further considered here.

For the second problematic regime $\sigma_g > R$, this paper proposes *inverse mutations*. Here, the procedure still generates λ trial points (offspring). However, half of them are mirrored with respect to the parent, i.e., they are pointing towards the opposite direction. Inverse mutations can be formally defined as:

$$\boldsymbol{y}_t^{(i)} = \boldsymbol{x}_t + \sigma_g \boldsymbol{z}_t^{(i)} \text{ for } i = 1 \ldots \lceil \lambda/2 \rceil \tag{10}$$
$$\boldsymbol{y}_t^{(i)} = \boldsymbol{x}_t - \sigma_g \boldsymbol{z}_t^{(i - \lceil \lambda/2 \rceil)} \text{ for } i = \lceil \lambda/2 \rceil + 1, \ldots \lambda$$

In other words, each mutation vector $\boldsymbol{z}^{(i)}$ is used twice, once in its original form and once as $-\boldsymbol{z}^{(i)}$.

Figure 7 illustrates the effect of introducing inverse mutations. The performance gain is obvious: The observable accuracy of the gradient estimate is constant over the entire range of σ_g values. This is in sharp contrast to the regular case, which exhibits the performance loss discussed above. The figure, however, indicates a slight disadvantage in that the number of trial points should be twice as much in order to gain the same performance as in the regular case. In addition, the figure also shows the accuracy for $\lambda = 6$, which is smaller than the number of search space dimensions; the accuracy is with $\cos \alpha \approx 0.47$ still reasonably good.

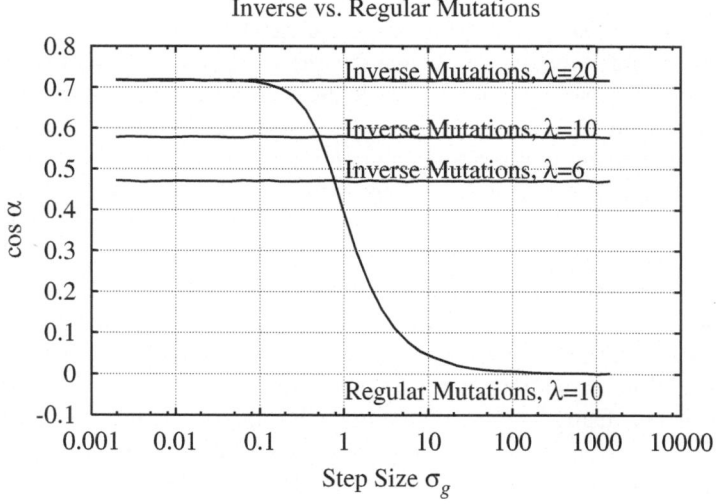

Fig. 7. $\cos \alpha$ between the true gradient $\boldsymbol{g}$ and its estimate $\tilde{\boldsymbol{g}}$ for various numbers of trial points λ. For comparison purposes, also the regular case with $\lambda = 10$ is shown. In this example, $N = 10$ and $R = 1$ were used.

Figure 8 illustrates the performance inverse mutations yield in the presence of noise. For comparison purposes, the figure also shows the regular case for $\lambda = 10$ and $\sigma_e \in \{0.05, 0.5\}$ (dashed lines). Again, the performance gain is obvious: the accuracy degrades for $\sigma_g < \sigma_e$ but sustains for $\sigma_g > R$, which is in way contrast to the regular case.

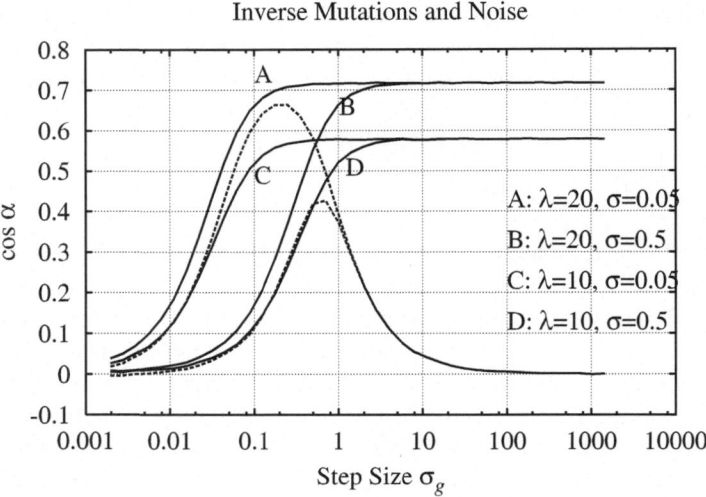

Fig. 8. $\cos\alpha$ between the true gradient g and its estimate $\tilde{g}$ for $\lambda \in \{10, 20\}$ and $\sigma_e \in \{0.05, 0.5\}$. For comparison purposes, also the regular case with $\lambda = 10$ and $\sigma_e \in \{0.05, 0.5\}$ (dashed lines). In this example, $N = 10$ and $R = 1$ were used.

Figure 9 shows the rate of progress φ^* of the EGS procedure with two step sizes σ_g and σ_p and inverse mutations for the example $N = 40$ and $\lambda = 24$ and a normalized noise strength of $\sigma_\epsilon^* = 8$. When comparing the figure with Fig. 2, it can be seen that the rate of progress is almost that of the undisturbed case, i.e., $\sigma_\epsilon^* = 0$. It can be furthermore seen that the performance starts degrading only for too small a step size σ_g. It should be mentioned here that for the case of $\sigma_\epsilon^* = 0$, the graphs are virtually identical with the cases $\sigma_\epsilon^* = 8$ and $\sigma_g = 32$, and are thus not shown.

Figure 10 demonstrates that the proposed method is not restricted to the sphere model; also the general quadratic case $f(\boldsymbol{x}) = \sum_i \xi^i x_i^2$ benefits from using inverse mutations. The performance gain is quite obvious when comparing graph "C", i.e., regular mutations, with graphs "A" and "B", i.e., inverse mutations. When using regular mutations, the angle between the true and estiamted gradient is rather eratic, whereas inverse mutations yield virtually the same accuracy as compared to the sphere model. It might be worthwhile to note that regular mutations are able to yield some reasonable accurace only for small excentricities, e.g., $\xi = 2$ in graph "D".

Even though inverse mutations are able to accurately estimate the gradient *direction*, that do not resort to second order derivatives – that in the case of excentric quadratic fitness functions, the gradient does not point towards the minimum. Please note that as all gradient-based optimization procedures that do not utilize second order derivatives suffer from the same problem.

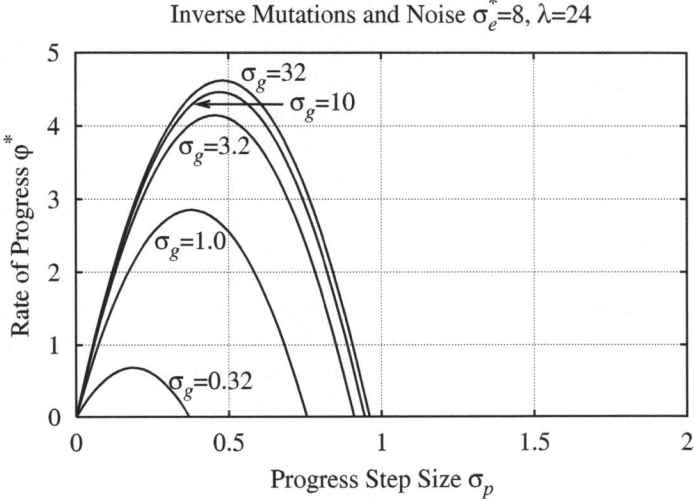

Fig. 9. The rate of progress φ^* of EGS with two step sizes and inverse mutations under the presence of noise $\sigma_\epsilon^* = 8$. The example has used $N = 40$ and $\lambda = 24$. In comparison with Fig. 2, the performance is not affected by the noise.

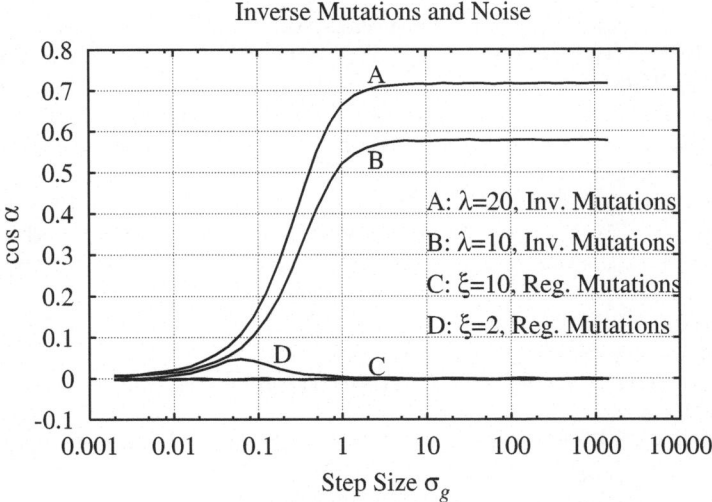

Fig. 10. $\cos\alpha$ between the true gradient g and its estimate $\tilde{g}$ for $\lambda \in \{10, 20\}$ and $\sigma_e = 0.5$ for the test function $f(x) = \sum_i \xi^i x_i^2$ when using inverse mutations (labels "A" and "B") as well as regular mutations (labels "C" and "D"). For graphs "A"-"C", $\xi = 10$, and for graph "D", $\xi = 2$ was used. In all examples, $N = 10$ and $R = 1$ were used.

6   Conclusions

This paper has briefly reviewed the contemporary $(\mu/\mu, \lambda)$-evolution strategy as well as a hybrid one known as the evolutionary-gradient-search procedure. The review of recent analyses has emphasized that EGS yields superior efficiency (sequential run time), but that it significantly degrades in the presence of noise. The paper has also analyzed the reasons of this observable performance loss. It has then proposed two modifications: inverse mutations and using two independent step sizes σ_g and σ_p, which are used to generate the λ trial points and to perform the actual progress step, respectively. With these two modifications, the EGS procedure yields a performance, i.e., the normalized rate of progress φ^*, which is very noise robust; the performance is almost not effected as long as the step size σ_g for generating trial points is sufficiently large.

The achievements reported in this paper have been motivated by both analyses and experimental evidence, and have be validated by further experiments. Future work will be devoted to conducting a mathematical analysis.

References

1. D. Arnold, An Analysis of Evolutionary Gradient Search, *Proceedings of the 2004 Congress on Evolutionary Computation.* IEEE, 47-54, 2004.
2. D. Arnold and H.-G. Beyer, Local Performance of the $(\mu/\mu, \lambda)$-Evolution Strategy in a Noisy Environment, in W. N. Martin and W. M. Spears, eds., *Proceeding of Foundation of Genetic Algorithms 6 (FOGA),* Morgan Kaufmann, San Francisco, 127-141, 2001.
3. T. Bäck, U. Hammel, and H.-P. Schwefel, Evolutionary Computation: Comments on the History and Current State. *IEEE Transactions on Evolutionary Computation,* 1(1), 1997, 3-17.
4. H.-G. Beyer. On the Explorative Power of ES/EP-like Algorithms, in V.W. Porto, N. Saravanan, D. Waagen, and A.E. Eiben, (eds), *Evolutionary Programming VII: Proceedings of the Seventh Annual Conference on Evolutionary Programming (EP'98),* Springer-Verlag, Berlin, Heidelberg, Germany, 323-334, 1998.
5. H.-G. Beyer. *The Theory of Evolution Strategies.* Springer-Verlag Berlin, 2001.
6. D.B. Fogel. *Evolutionary Computation: Toward a New Philosophy of Machine Learning Intelligence..* IEEE Press, NJ, 1995.
7. D.E. Goldberg. *Genetic Algorithms in Search, Optimization and Machine Learning.* Addison-Wesley, Reading, MA, 1989.
8. D.G. Luenberger. *Linear and Nonlinear Programming.* Addison-Wesley, Menlo Park, CA, 1984.
9. W.H. Press, B.P. Flannery, S.A. Teukolsky, and W.T. Vetterling. *Numerical Recipes.* Cambridge University Press, 1987.
10. I. Rechenberg, *Evolutionsstrategie* (Frommann-Holzboog, Stuttgart, 1994).
11. R. Salomon and L. Lichtensteiger, Exploring different Coding Schemes for the Evolution of an Artificial Insect Eye. *Proceedings of The First IEEE Symposium on Combinations of Evolutionary Computation and Neural Networks,* 2000, 10-16.

12. R. Salomon, Evolutionary Algorithms and Gradient Search: Similarities and Differences, *IEEE Transactionson Evolutionary Computation,* **2**(2): 45-55, 1998.

13. R. Salomon, Accelerating the Evolutionary-Gradient-Search Procedure: Individual Step Sizes, in T. Bäck, A. E. Eiben, M. Schoenauer, and H.-P. Schwefel, (eds.), *Parallel Problem Solving from Nature (PPSN V)*, Springer-Verlag, Berlin, Heidelberg, Germany, 408-417, 1998.

14. H.-P. Schwefel. *Evolution and Optimum Seeking.* John Wiley and Sons, NY. 1995.

Development of Flexible and Adaptable Fault Detection and Diagnosis Algorithm for Induction Motors Based on Self-organization of Feature Extraction

Hyeon Bae[1], Sungshin Kim[1], Jang-Mok Kim[1], and Kwang-Baek Kim[2]

[1] School of Electrical and Computer Engineering, Pusan National University,
Jangjeon-dong, Geumjeong-gu, 609-735 Busan, Korea
{baehyeon, sskim, jmok}@pusan.ac.kr
http://icsl.ee.pusan.ac.kr
[2] Dept. of Computer Engineering, Silla University,
Gwaebop-dong, Sasang-gu, 617-736 Busan, Korea
gbkim@silla.ac.kr

Abstract. In this study, the datamining application was achieved for fault detection and diagnosis of induction motors based on wavelet transform and classification models with current signals. Energy values were calculated from transformed signals by wavelet and distribution of the energy values for each detail was used in comparing similarity. The appropriate details could be selected by the fuzzy similarity measure. Through the similarity measure, features of faults could be extracted for fault detection and diagnosis. For fault diagnosis, neural network models were applied, because in this study, it was considered which details are suitable for fault detection and diagnosis.

1 Introduction

The most popular way of converting electrical energy to mechanical energy is an induction motor. This motor plays an important role in modern industrial plants. The risk of motor failure can be remarkably reduced if normal service conditions can be arranged in advance. In other words, one may avoid very costly expensive downtime by replacing or repairing motors if warning signs of impending failure can be headed. In recent years, fault diagnosis has become a challenging topic for many electric machine researchers.

The diagnostic methods to identify the faults listed above may involve several different types of fields of science and technology [1]. K. Abbaszadeh et al. presented a novel approach for the detection of broken rotor bars in induction motors based on the wavelet transformation [2]. Multiresolution signal decomposition based on wavelet transform or wavelet packet provides a set of decomposed signals in independent frequency bands [3]. Z. Ye et al. presented a novel approach to induction motor current signature analysis based on wavelet packet decomposition (WPD) of the stator current [4]. L. Eren et al. analyzed the stator current via wavelet packet decomposition to detect bearing defects [5]. The proposed method enables the analysis of

U. Furbach (Ed.): KI 2005, LNAI 3698, pp. 134–147, 2005.
© Springer-Verlag Berlin Heidelberg 2005

frequency bands that can accommodate the rotational speed dependence of the bearing-defect frequencies.

Fault detection and diagnosis algorithms have been developed according to the target motors, so it cannot guarantee the adaptability and flexibility under changeable field conditions. In this study, similarity measure was performed to find out the good features from signals that improve compatibility of fault detection and diagnosis algorithm.

2 Application Data from Induction Motor

2.1 Signal Acquisition from Induction Motor

As it is obvious, sometimes, different faults produce nearly the same frequency components or behave like healthy machine, which make the diagnosis impossible. This is the reason that new techniques must also be considered to reach a unique policy for distinguishing among faults. In this study, current signals were used for fault diagnosis of induction motors. Current signals can be changed by the inner resistivity, so it can indicate the status of the motor intuitionally. Figure 1 shows the measuring method of induction motors. In this study, we used this concept to detect faults of the motor.

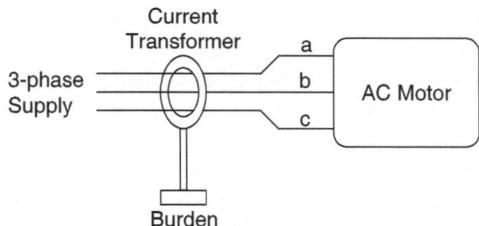

Fig. 1. Measuring the current signal using instruments

2.2 Fault Types

Many types of signals have been studied for the fault detection of induction motors. However, each technique has advantages and disadvantages with respect to the various types of faults. In this study, the proposed fault detection system dealt with bowed rotor, broken rotor bar, and healthy cases.

2.2.1 Bearing Faults
Though almost 40~50% of all motor failures are bearing related, very little has been reported in the literature regarding bearing related fault detection techniques. Bearing faults might manifest themselves as rotor asymmetry faults from the category of eccentricity related faults [6], [7]. Artificial intelligence or neural networks have been researched to detect bearing related faults on line. In addition, adaptive, statistical time frequency method is studying to find bearing faults.

2.2.2 Rotor Faults

Rotor failures now account for 5-10% of total induction motor failures. Frequency domain analysis and parameter estimation techniques have been widely used to detect this type of faults. In practice, current side bands may exist even when the machine is healthy. Also rotor asymmetry, resulting from rotor ellipticity, misalignment of the shaft with the cage, magnetic anisotropy, etc. shows up at the same frequency components as the broken bars.

2.2.3 Eccentricity Related Faults

This fault is the condition of unequal air-gap between the stator and rotor. It is called static air-gap eccentricity when the position of the minimal radial air-gap length is fixed in the space. In case of dynamic eccentricity, the center of rotor is not at the center of rotation, so the position of minimum air-gap rotates with the rotor. This maybe caused by a bent rotor shaft, bearing wear or misalignment, mechanical resonance at critical speed, etc. In practice, an air-gap eccentricity of up to 10% is permissible.

3   Feature Extraction from Current Signals

3.1   Wavelet Analysis

Time series data are of growing importance in many new database applications, such as data warehousing and datamining. A time series is a sequence of real numbers, each number representing a value at a time point. In the last decade, there has been an explosion of interest in mining time series data. Literally hundreds of papers have introduced new algorithms to index, classify, cluster and segment time series.

For fault detection and diagnosis of motors, Fourier transform has been employed. Discrete Fourier Transform (DFT) has been one of the general techniques used commonly for analysis of discrete data. One problem with DFT is that it misses the important feature o time localization. Discrete Wavelet Transform (DWT) has been found to be effective in replacing DFT in many applications in computer graphic, image, speech, and signal processing. In this paper, this technique was applied in time series for feature extraction from current signals of induction motors. The advantage of using DWT is multi-resolution representation of signals. Thus, DWT is able to give locations in both time and frequency. Therefore, wavelet representations of signals bear more information than that of DFT [8]. In the past studies, details of wavelet decomposition were used as features of the motor faults. Following sections will introduce the principal algorithm of the past researches.

3.2   Problem in Feature Extraction

In the time series datamining, the following claim is made. Much of the work in the literature suffers from two types of experimental flaws, implementation bias and data bias. Because of these flaws, much of the work has very little generalizability to real world problems. In particular, many of the contributions made (speed in the case of indexing, accuracy in the case of classification and clustering, model accuracy in the case of segmentation) offer an amount of "improvement" that would have been completely dwarfed by the variance that would have been observed by testing on many real world data sets, or the variance that would have been observed by changing minor (unstated) implementation details [9].

To solve the problem, fuzzy similarity measure was employed to analyze detail values and find good features for fault detection. The features were defined by energy values of each detail level of wavelet transform. Bed adaptability and flexibility that have been considered as important ability in fault detection and diagnosis of induction motors could be improved by the proposed algorithm.

3.2.1 Which Wavelet Function Is Proper?

Many transform methods have been used for feature extraction from fault signals. Typically, Fourier transform has been broadly employed in data analysis on frequency domain. In this study, instead of Fourier transform, wavelet transform was applied, because the wavelet transform can keep information of time domain and can extract features from time series data.

On the other hand, types of wavelet functions and detail levels of decomposition were chosen by user's decision. Therefore, it could be difficult to extract suitable features for fault detection based on optimal algorithm. Especially, when the developed algorithm in changed new environments or motors, fault detection and diagnosis could not achieved properly. Algorithms are sometimes very sensitive to environments, so the system can make incorrect results. Therefore, in this study, the new algorithm is proposed to increase adaptability of fault detection and diagnosis.

3.2.2 How Many Details Are Best?

In past studies, fault detection and diagnosis algorithms were developed and applied to specific motors, that is, the algorithms had no flexibility for general application. Therefore, under changed environments, the system cannot detect faults correctly. In actual, when the developed algorithm in the past researches was applied on the used data here, the faults could not be detected and diagnosed.

In this study, the advanced algorithm is proposed to compensate the problems of the traditional approaches in fault detection and diagnosis. The best features for fault detection were automatically extracted by fuzzy similarity measure. The purpose of the proposed algorithm is to compare features and determine function types and detail levels of wavelet transform. The algorithm was performed by fuzzy similarity measure and improved detection performance. If similarity is low, the classification can be well accomplished. Through the proposed method, the best mother wavelet function and detail level can be determined according to target signals.

3.3 Proposed Solutions

In this chapter, the solution is proposed to settle the problem in feature extraction that is mentioned before. When wavelet analysis is employed to find features from signals, two important things must be considered before starting transform. One is to select a wavelet mother function and the other is to determine a detail level for composition.

In this study, fuzzy similarity measure was applied to compare the performance of 5 functions and 12 detail levels and determine the best one for the function and the detail level. The proposed algorithm can automatically find out the best condition in feature extraction, so it can guarantee adaptability and flexibility in industrial processes. Figure 2 shows the flowchart for fault detection and diagnosis of induction motors.

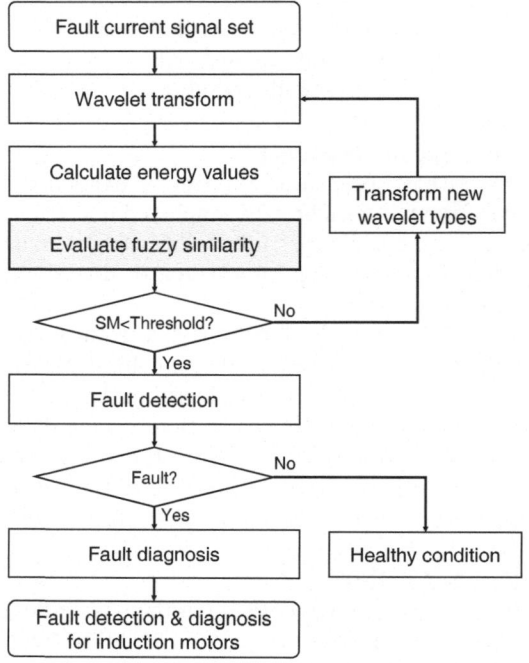

Fig. 2. The proposed algorithm with fuzzy similarity measure

3.3.1 Calculation of Energy Value from Wavelet Transforms

In this study, wavelet transform was used for feature extraction from motor signals. The energy of each detail was applied for features of faults, that is, energy index. Because the energy index has unique characteristics according to each fault on specific frequency, faults can be detected by calculated energy values. Especially, the energy index shows considerable differences between normal and abnormal, thus, it can powerfully used in fault detection.

3.3.2 Determination of Function Type and Detail Level

Similarity measure was applied to evaluate similarity among cases with energy values. For the similarity measure, fuzzy similarity measure was used because considering the distribution of the faults was good for flexible diagnosis.

The similarity between one normal case and two fault cases was achieved and the best function and level was selected that had the least similarity value, because well-separated distribution is better for fault diagnosis. Through the proposed algorithm, the function type and detail level can be automatically selected that can perfectly classify the fault type. This ability is self-organization and it plays a very important role in fault diagnosis under changeable environmental conditions. After developing the fault diagnosis systems, when the system is implemented new sites, the proposed algorithm evaluates energy values and finds the function type and detail level for better performance. After finishing feature extraction, the extracted features can be classified by logics or models. In this study, fault diagnosis was achieved by neural network models with fault detection.

3.4 Applied Methods for Feature Extraction

3.4.1 Wavelet Transform

Wavelet transform is a method for time varying or non-stationary signal analysis, and uses a new description of spectral decomposition via the scaling concept. Wavelet theory provides a unified framework for a number of techniques, which have been developed for various signal-processing applications. One of its feature is multi-resolution signal analysis with a vigorous function of both time and frequency localization. Mallat's pyramidal algorithm based on convolutions with quadratic mirror filters is a fast method similar to FFT for signal decomposition of the original signal in an orthonormal wavelet basis or as a decomposition of the signal in a set of independent frequency bands. The independence is due to the orthogonality of the wavelet function.

The DWT computes the wavelet coefficients $c_{j,k}$ and $d_{j,k}$ ($j=1,\ldots,J$) given by

$$c_{j,k} = \sum_n x[n]h_j[n-2^j k] \tag{1}$$

$$d_{j,k} = \sum_n x[n]g_j[n-2^j k] \tag{2}$$

where $x[n]$ is a discrete-time series, $h_j[n-2^j k]$ is called the discrete wavelets, is equivalent to $2^{j}/2\,\psi(2^{j}(t-2^j k))$. The term $g_j[n-2^j k]$ is called the scaling sequence. At the jth resolution, $c_{j,k}$ and $d_{j,k}$ present the approximation and the detail signal $f(t)$. At each resolution $j>0$, the scaling coefficients and the wavelet coefficients are

$$c_{j+1,k} = \sum_n g[n-2k]d_{j,k} \tag{3}$$

$$d_{j+1,k} = \sum_n h[n-2k]d_{j,k} \tag{4}$$

3.4.2 Fuzzy Similarity Measure

A recent comparative analysis of *similarity measures* between fuzzy sets has shed a great deal of light on our understanding of these measures. Different kinds of measures were investigated from both the geometric and the set-theoretic points of view. In this paper, a similarity measure was used as a measure transformed from a distance measure by using $SM=1/(1+DM)$. In this sense, Zwick *et al.* *similarity measures* are actually *distance measures* used in the derivation of our *SM*.

In a behavioral experiment, different *distance measures* were generalized to fuzzy sets. A correlation analysis was conducted to assess the goodness of these measures. More specifically, the distance measured by a particular *distance measure* was correlated with the *true* distance. It was assumed that it would be desirable for a measure to have high mean and median correlation, to have a small dispersion among its correlations, and to be free of extremely how correlations.

In this study, the other similarity of fuzzy values was employed that was proposed by Pappis and Karacapilidis [10]. The similarity measure is defined as follows:

$$M_{A,B} = \frac{|A \cap B|}{|A \cup B|} = \frac{\sum_{i=1}^{n}(a_i \wedge b_i)}{\sum_{i=1}^{n}(a_i \vee b_i)} \qquad (5)$$

4 Fault Detection of Induction Motor

4.1 Current Signals and Data Preprocessing

The motor ratings applied in this paper depend on electrical conditions. The rated voltage, speed, and horsepower are 220 [V], 3450 [RPM], and 0.5 [HP], respectively. The specifications for used motors are 34 slots, 4 poles, and 24 rotor bars. Figure 3 shows the experimental equipment. The current signals are measured under fixed conditions that consider the sensitivity of the measuring signals that is, the output of the current probes. The sensitivity of each channel is 10 mV/A, 100 mV/A, and 100mV/A, respectively. The specification of the measured input current signal under this condition consists of 16,384 sampling numbers, 3 kHz maximum frequency, and 2.1333 measuring time. Therefore, the sampling time is 2.1333 over 16,384. Fault types used in this study are broken rotor, faulty bearing, bowed rotor, unbalance, and static and dynamic eccentricity.

When wavelet decomposition is used to detect faults in induction motors, the un-synchronized current phase problem should have great influence on the detection results. Figure 4 shows original current signals that are not synchronized with the origin point. If the target signals are not synchronized with each other, unexpected results will appear in wavelet decomposition. Therefore the signals have to be re-sampled with the starting origin to a 0 (zero) phase. In this study, 64 of 128 cycles of the signals had to be re-sampled. The average value divided by one cycle signal is used to reduce the noise of the original signals.

Fig. 3. Fault detection and data acquisition equipment

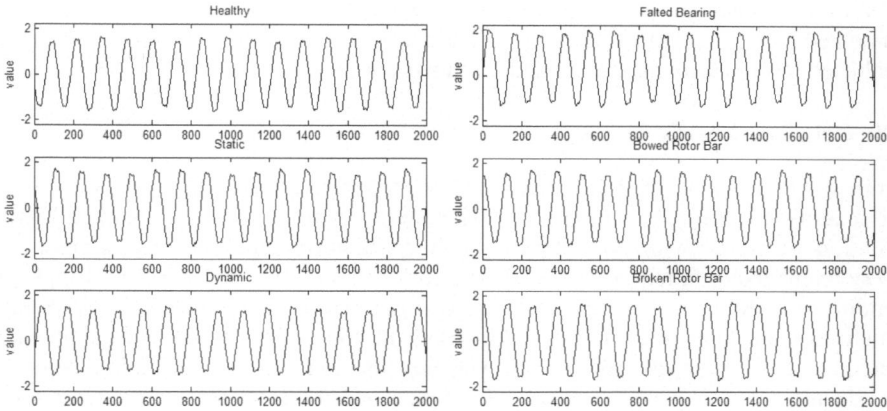

Fig. 4. Current signals measured by equipped motors

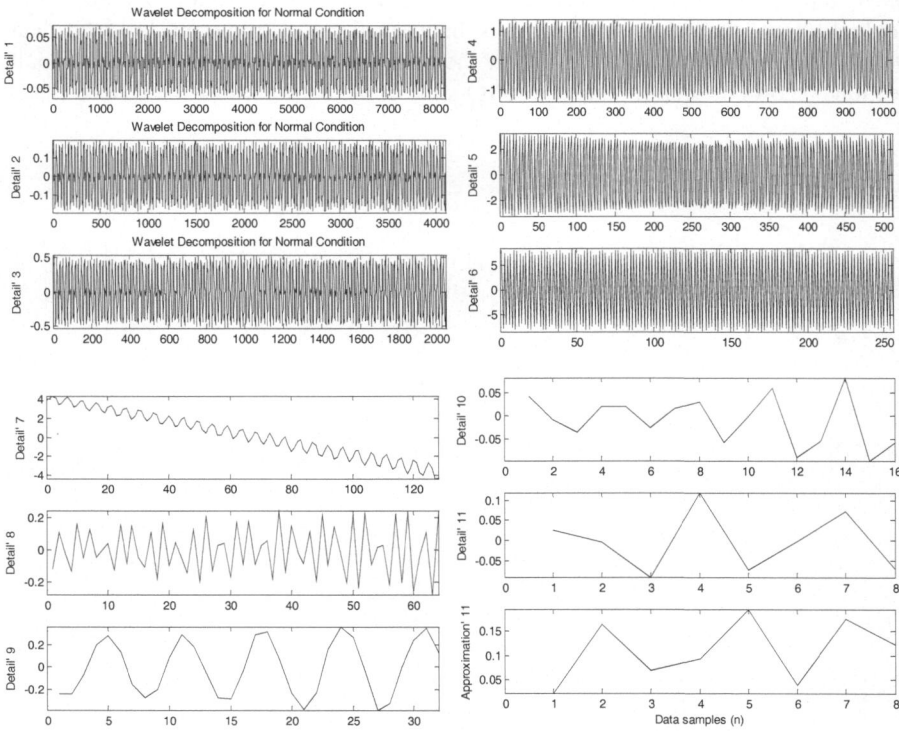

Fig. 5. Wavelet transformation for one current signal

4.2   Wavelet Transform for Current Signals

As mentioned above, wavelet transform was applied to extract features. A signal was transformed to 12 details that depend on sampling numbers. An energy value was calculated to one value, so one set consists of 12 values for one case. The energy value was computed by each detail value and the distribution of the calculated energy values was evaluated by similarity measure. Figure 5 shows one result of wavelet transform.

4.3 Calculate Energy Values

From wavelet transformed 12 details, the energy values were calculated that shows the features according to motor status. The energy values could be illustrated by distribution functions with 20 samples. Each distribution function was expressed as fuzzy membership function types and applied for fuzzy similarity measure. Employed conditions included one normal case and two fault cases, so constructed membership functions were three according to each detail. In conclusion, 12 sets of membership functions can be constructed for one wavelet function. After determining functions and details, features were extracted from signals and the extracted features were classified by neural network models, that is, fault diagnosis.

Figure 6 shows the calculated energy values that include 12 results according to each detail. And Fig. 7 illustrates the distribution of the energy values for 20 samples. One detail has three distributions for normal, broken rotor bar, and bowed rotor fault. Table 1 shows the calculated similarity between three distributions for each detail. Detail 1 and 2 of the first function show the best result for fault classification. Finally, Section 5 shows the classification results using similarity values to diagnose faults.

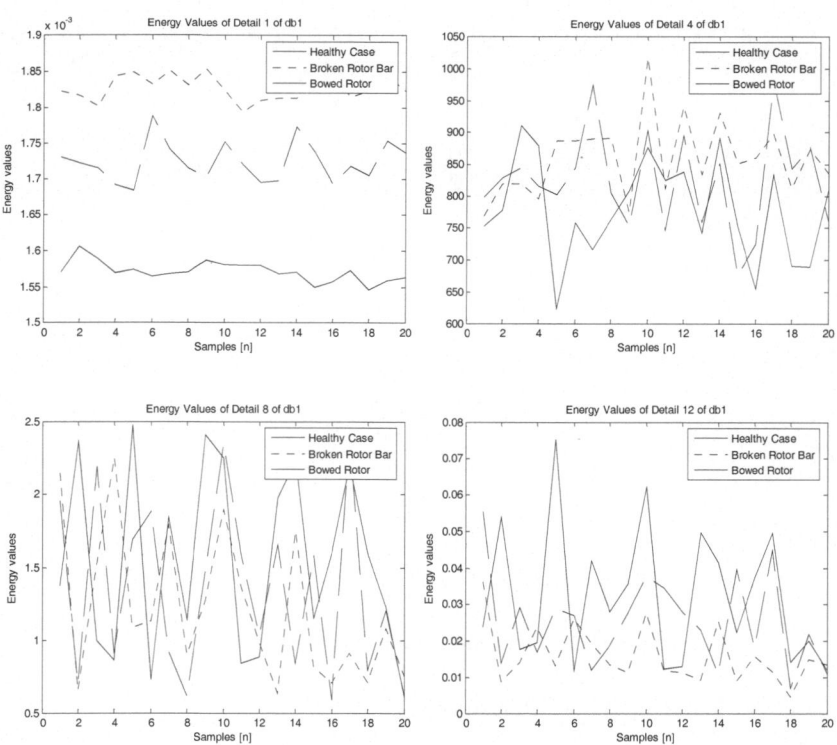

Fig. 6. Energy values of wavelet transform with Daubechies wavelet function

4.4   Distribution of Energy Values of Details

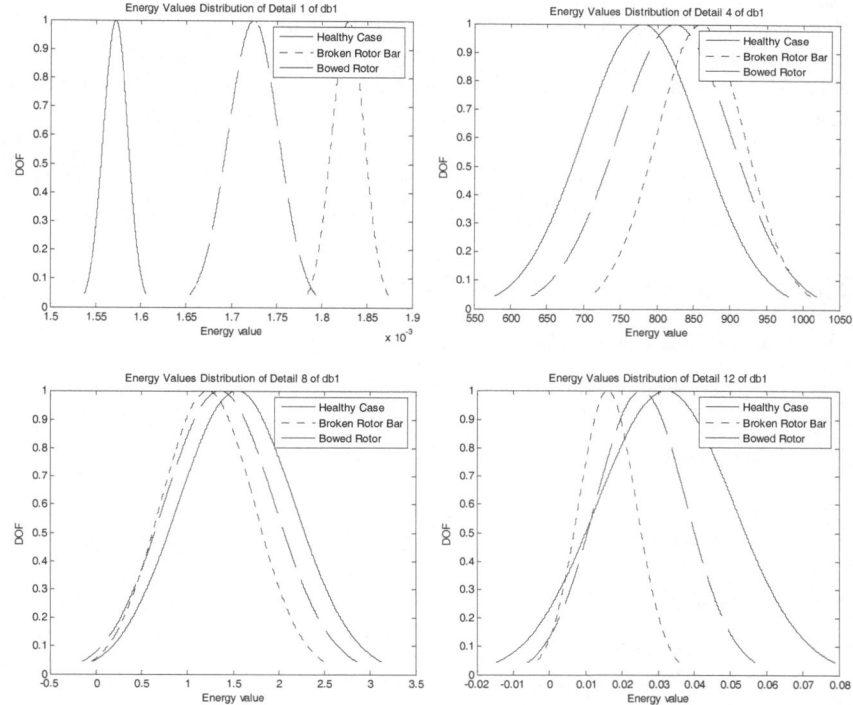

Fig. 7. Distribution for energy values of wavelet transform with Daubechies wavelet function

4.5   Results of Similarity Measure

In this section, the result of fuzzy similarity measure is shown. Using similarity measure, the proper a wavelet function and detail levels were selected that can classify

Table 1. Results of fuzzy similarity measures

Detail	Daub1	Daub2	Sym	Coif	Mey	Bior
Detail 1	0.0081	0.0000	0.0000	0.0000	1.4286	0.0081
Detail 2	0.0084	0.0000	0.0000	0.0000	0.0000	0.0084
Detail 3	0.0094	2.0664	2.0664	1.2986	0.9402	0.0094
Detail 4	1.8507	1.9684	1.9684	1.9502	2.0714	1.8507
Detail 5	0.9273	2.3579	2.3579	2.7438	2.3912	0.9273
Detail 6	2.4349	2.3521	2.3521	2.6290	2.3532	2.4349
Detail 7	2.5762	2.3616	2.3616	2.3854	2.3671	2.5762
Detail 8	2.3295	2.5403	2.5403	2.2591	2.4443	2.3295
Detail 9	1.9273	2.6427	2.6427	2.3220	2.4702	1.9273
Detail 10	1.3469	2.6122	2.6122	2.3463	2.4958	1.3469
Detail 11	2.5999	2.6327	2.6327	2.3444	2.3301	2.5999
Detail 12	1.7005	2.6177	2.6177	2.3442	2.6441	1.7005

faults correctly. For the similarity measure, fuzzy similarity measure was employed to measure similarity with distribution of 20 samples for one case. Namely, one distribution was constructed by 20 sample sets for one fault or normal case. On one detail domain, three distribution functions were illustrated for similarity measure.

After measuring fuzzy similarity, low valued details were selected for fault diagnosis, because dissimilar distribution valued details can classify the cases well. As shown in Table 1, the Daubechies wavelet function has low similarity value in general and Detail 1 to Detail 3 has low similarity generally. Therefore, Detail 1 to Detail 3 of Daubechies wavelet function is suitable for fault detection and diagnosis in this study.

5 Fault Diagnosis of Induction Motor

5.1 Fault Diagnosis Using Whole Details

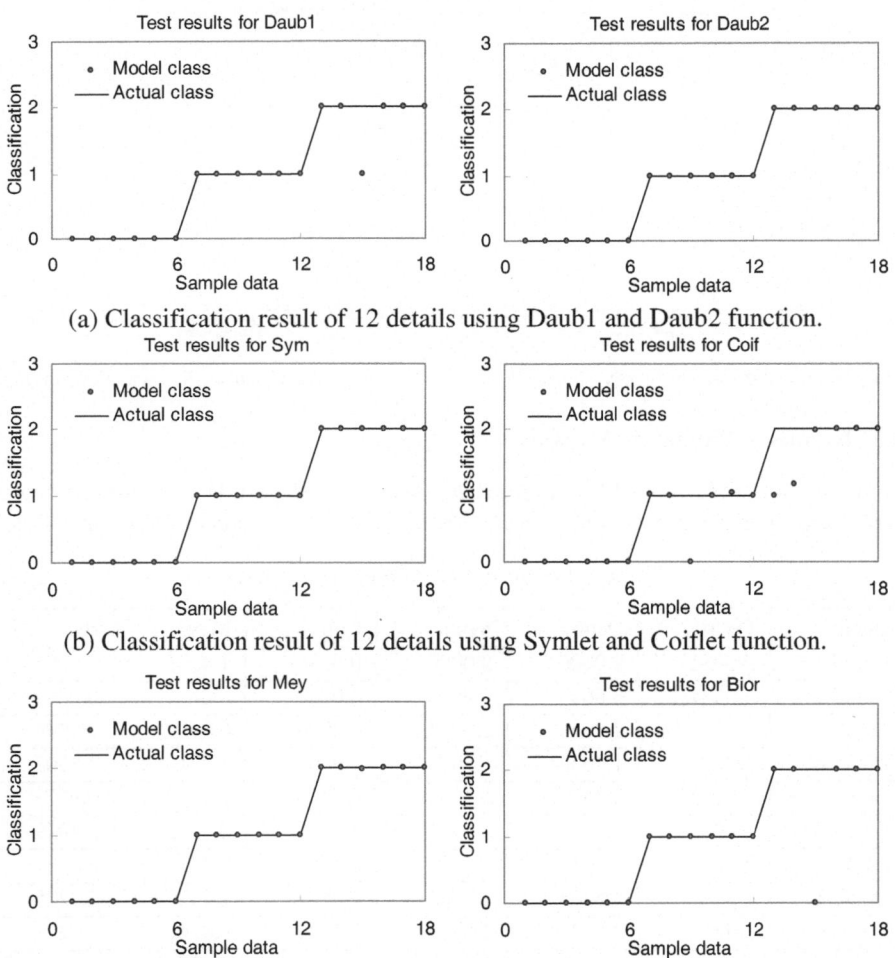

(a) Classification result of 12 details using Daub1 and Daub2 function.

(b) Classification result of 12 details using Symlet and Coiflet function.

(c) Classification result of 12 details using Meyer and Biorthogonal function

Fig. 8. Fault diagnosis result using 12 details of Wavelet transform

5.2   Fault Diagnosis Using Selected Details

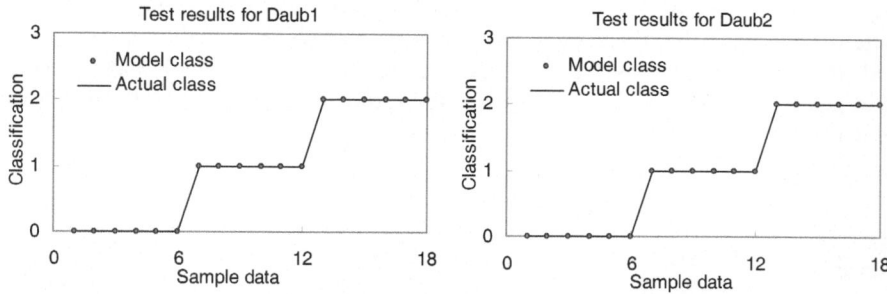

(a) Classification result of selected details using Daub1 and Daub2 function

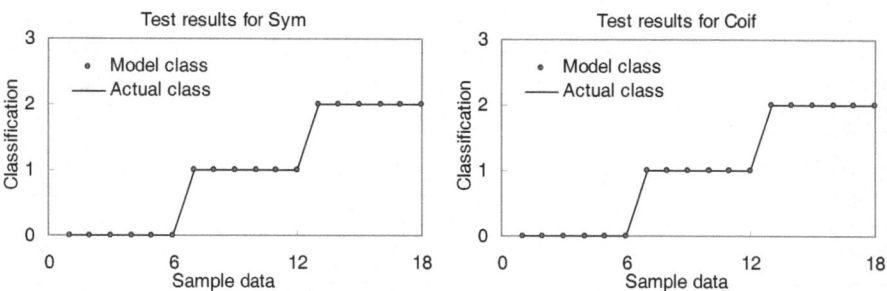

(b) Classification result of selected details using Symlet and Coiflet function

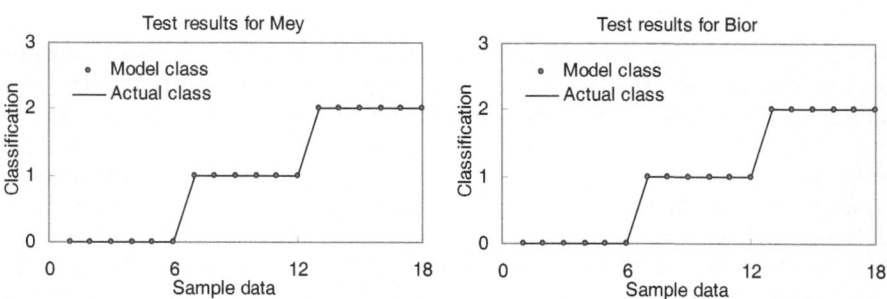

(c) Classification result of selected details using Meyer and Biorthogonal function

Fig. 9. Fault diagnosis result using selected details of Wavelet transform

5.3   Comparison of the Three Detail Sets

In this section, the result of fault diagnosis using neural network is shown. As shown in figures, fault diagnosis was achieved using three data sets. First, whole details were applied, second, selected details by fuzzy similarity measure were used, and finally, unselected details in the second case were employed for fault diagnosis. The purpose of evaluation was to check the effect of the selected details. If the selected details do not affect the performance of diagnosis, three cases will show good performance, and vice verse.

As shown in Table 2, the performance shows the best when the selected details are only used in diagnosis. The performance of diagnosis with whole details is also adequate for diagnosis, but there were some error points. When the selected details were used, the performance was the best. This means the selected details have principal information for fault diagnosis. However, the result using unselected details showed the worst performance.

Table 2. Comparison of the three data sets

Function	Whole details		Selected details		Unselected details	
	Train: 42	Test: 18	Train: 42	Test: 18	Train: 42	Test: 18
Daub1	42	17	42	18	42	16
	100%	94.44%	100%	100%	100%	88.88%
Daub2	42	18	42	18	42	12
	100%	100%	100%	100%	100%	66.66%
Symlet	42	18	42	18	42	16
	100%	100%	100%	100%	100%	88.88%
Coiflet	42	15	42	18	42	11
	100%	83.33%	100%	100%	100%	61.11%
Meyer	42	18	42	18	42	15
	100%	100%	100%	100%	100%	83.33%
Bior	42	17	42	18	42	16
	100%	94.44%	100%	100%	100%	88.88%

6   Conclusions

The motor is one of the most important equipment in industrial processes. Most production lines used the motor for power supplying in the industrial fields. The motor control algorithms have been studied for a long time, but the motor management systems have not been researched much. The management is one of popular fields that are broadly performed recently. However, it is difficult to diagnose the motor and guarantee the performance of the diagnosis, because the motors are applied in various fields with different specification.

In this study, the datamining application was achieved for fault detection and diagnosis of induction motors based on wavelet transform and classification models with current signals. Energy values were calculated from transformed signals by wavelet and distribution of the energy values for each detail was used in comparing similarity. The appropriate details could be selected by the fuzzy similarity measure. Through the similarity measure, features of faults could be extracted for fault detection and diagnosis. For fault diagnosis, neural network models were applied, because in this study, it was considered which details are suitable for fault detection and diagnosis.

Through the proposed algorithm, the fault detection and diagnosis system for induction motors can be employed in different motors and changed environments. Therefore, the importantly dealt problem related to the adaptability and flexibility in the past can be solved to improve the performance of motors.

Acknowledgement

This work has been supported by KESRI (R-2004-B-129), which is funded by the Ministry of Commerce, Industry and Energy, in Korea.

References

1. P. Vas: Parameter Estimation, Condition Monitoring, and Diagnosis of Electrical Machines, Clarendron Press, Oxford (1993)
2. G. B. Kliman and J. Stein: Induction motor fault detection via passive current monitoring. International Conference in Electrical Machines, Cambridge, MA (1990) 13-17
3. K. Abbaszadeh, J. Milimonfared, M. Haji, and H. A. Toliyat: Broken Bar Detection in Induction Motor via Wavelet Transformation. The 27th Annual Conference of the IEEE Industrial Electronics Society (IECON '01), Vol. 1 (2001) 95-99
4. Z. Ye, B. Wu, and A. Sadeghian: Current signature analysis of induction motor mechanical faults by wavelet packet decomposition. IEEE Transactions on Industrial Electronics, Vol. 50, No. 6 (2003) 1217-1228
5. L. Eren and M. J. Devaney: Bearing Damage Detection via Wavelet Packet Decomposition of the Stator Current. IEEE Transactions on Instrumentation and Measurement, Vol. 53, No. 2 (2004) 431-436
6. Y. E. Zhongming and W. U. Bin: A Review on Induction Motor Online Fault Diagnosis. The Third International Power Electronics and Motion Control Conference (PIEMC 2000), Vol. 3 (2000) 1353-1358.
7. Masoud Haji and Hamid A. Toliyat: Patern Recognition-A Technique for Induction Machines Rotor Fault Detection Eccentricity and Broken Bar Fault. Conference Record of the 2001 IEEE Industry Applications Conference, Vol. 3 (2001) 1572-1578
8. K. P. Chan and A. W. C. Fu: Efficient time series matching by wavelets. 15th International Conference on Data Engineering (1999) 126-133
9. E. Keogh and S. Kasetty: On the Need for Time Series Datamining Benchmarks: A Survey and Empirical Demonstration. In the 8th ACM SIGKDD International Conference on Knowledge Discovery and Datamining (2002) 102-111
10. C. P. Pappis and N. I. Karacapilidis: A comparative assessment of measures of similarity of fuzzy values. Fuzzy Sets and Systems 56 (1993) 171-174

Computing the Optimal Action Sequence by Niche Genetic Algorithm[*]

Chen Lin, Huang Jie, and Gong Zheng-Hu

School of Computer Science, National University of Defense Technology,
Changsha, 410073, China
chenlin@nudt.edu.cn, agnes_nudt@hotmail.com

Abstract. Diagnosis aims to identify faults that explain symptoms by a sequence of actions. Computing minimum cost of diagnosis is a NP-Hard problem if diagnosis actions are dependent. Some algorithms about dependent actions are proposed but their problem descriptions are not accurate enough and the computation cost is high. A precise problem model and an algorithm named NGAOAS (Niche Genetic Algorithm for the Optimal Action Sequence) are proposed in this paper. NGAOAS can avoid running out of memory by forecasting computing space, and obtain better implicit parallelism than normal genetic algorithm, which makes the algorithm be easily computed under grid environments. When NGAOAS is executed, population is able to hold diversity and avoid premature problem. At the same time, NGAOAS is aware of adaptive degree of each niche that can reduce the computation load. Compared with updated P/C algorithm, NGAOAS can achieve lower diagnosis cost and obtain better action sequence.

Keywords: ECD (Expected Cost of Diagnosis); Fault Diagnosis; Niche Genetic Algorithm.

1 Introduction

The purpose of diagnosis is to identify faults by a sequence of actions. It is when faults can be identified in the least time and with the lowest cost that the diagnostic result does make sense. Otherwise, any sequences of the actions can identify faults at a potential tremendous cost. The efficiency of an action is defined in papers [1-4]. The ratio $eff(a_i) = P_i/c_{a_i}$ is called the efficiency of the action a_i, where c_{a_i} is the cost of the action a_i, $P_i = p(a_i \ succeeds \,|\, e)$ is the conditional probability that the action a_i identifies faults if the evidence e is present. Corresponding P/C algorithm is presented in [1-4], which calculates efficiencies of all the actions, performs the actions in degressive order of their efficiencies and finally acquires an action sequence with minimal expected cost. The algorithm can get the optimal action sequence in the case of single fault, independent actions and cost. In [5] an updated

---

[*] This work is supported by the National Grand Fundamental Research 973 Program of China under Grant No. 2003CB314802.

U. Furbach (Ed.): KI 2005, LNAI 3698, pp. 148–160, 2005.

P/C algorithm is presented to solve dependent actions problem. The algorithm selects the action with the highest efficiency $\mathit{eff}(a_i) = p(a_i \ \mathit{succeeds}|e)/c_{a_i}$ and performs it under the current evidence e. If the action cannot successfully identify faults, it will update the evidence e with $\{a_i \ \mathit{fails}\} \cup e$, and erase the action a_i from the list of actions. Repeating the above procedure, we can obtain an action sequence with minimal expected cost. We will describe the algorithm by a simple example. There are three actions, their possibilities of successful diagnosis are $P_1 = 0.45, P_2 = 0.65$ and $P_3 = 0.55$, and their costs are $c_{a_1} = c_{a_2} = c_{a_3} = 1$. We can get $\mathit{eff}(a_2) > \mathit{eff}(a_3) > \mathit{eff}(a_1)$. First of all, we select the action a_2 with the highest efficiency and perform it. If it cannot identify any faults, we update the probabilities and the new efficiencies are turned into $\mathit{eff}(a_1) = p(a_1 \ \mathit{succeeds}|a_2 \ \mathit{fails})/c_{a_1} = 0.5714$, $\mathit{eff}(a_3) = p(a_3 \ \mathit{succeeds}|a_2 \ \mathit{fails})/c_{a_3} = 0.4286$ respectively. Since the action a_1 has higher efficiency now, we perform a_1 next. If the process of fault identification fails again, we perform the last remaining action, i.e. a_3. Finally, the sequence $< a_2, a_1, a_3 >$ gets the expected diagnosis cost, i.e. $ECD(a_2, a_1, a_3) = c_{a_2} + (1 - p(a_2 \ \mathit{succeeds}))c_{a_1}$ $+(1 - p(a_1 \ \mathit{succeeds}|a_2 \ \mathit{fails}))(1 - p(a_2 \ \mathit{succeeds}))c_{a_3} \approx 1.50$, which is lower than the expected cost acquired by the sequence $< a_2, a_3, a_1 >$ in P/C algorithm.

Suppose $\mathit{eff}(a_1) = p(a_1 \ \mathit{succeeds}|a_3 \ \mathit{fails})/c_{a_1} = 0.99$ and $\mathit{eff}(a_2) = p(a_2 \ \mathit{succeeds}|a_3 \ \mathit{fails})/c_{a_2} = 0.10$, the sequence $< a_3, a_1, a_2 >$ will have less expected diagnosis cost $ECD(a_3, a_1, a_2) \approx 1.455$ in the condition of the dependent actions. Therefore, we can conclude that the approximate approach of updated P/C algorithm is not always optimal.

In this paper, we first give detailed descriptions of the fault diagnosis problem, and then present an algorithm named NGAOAS. NGAOAS deals with the background knowledge of the problem, partitions the candidate action sequences into several niches each of which corresponds to a permutation of the candidate fault sets. Each individual will compete with others in the same niche and NGAOAS will estimate whether a niche should continue evolving and when to do the evolution at the same time, which can reduce the calculation load as much as possible and get the result as soon as possible. At last, we prove through experiments that NGAOAS has several outstanding features: 1) The space used in NGAOAS can be predicted. 2) Compared with normal genetic algorithm, NGAOAS has better implicit parallelism. 3) When NGAOAS is on execution, population is able to hold diversity and avoid premature problem. 4) NGAOAS introduces the niche adaptive value to reduce calculation load, and thus can speed up convergence rate and avoid premature problem. Compared with updated P/C algorithm, NGAOAS can achieve lower diagnosis cost and obtain better actions sequence.

2   Description of the Fault Diagnosis Problem in Network

This section describes the fault diagnosis problem in network and related definitions.

The fault evidence set $E = \{e_1, e_2, ..., e_m\}$ denotes all possible fault evidences in network. $E_c \subseteq E$ is the current fault evidence set.

The candidate fault set $F = \{f_1, f_2, ..., f_n\}$ denotes all possible faults. $f_c \in F$ is the current fault. The probability P_{f_i} of the candidate fault f_i in the evidence set E_c can be expressed as $P_{f_i} = p(f_i \; happens \,|\, E_c)$. The probability set of candidate faults in the evidence set E_c is $Prob_{E_c} = \{p(f_1 \; happens \,|\, E_c), ..., p(f_n \; happens \,|\, E_c)\}$.

The action set $A = \{a_1, a_2, ..., a_l\}$ is formed by all the possible actions which can identify the current fault f_i according to E_c. The cost of the action a_i is c_{a_i}. The action subset on f_i is an action set selected to diagnose candidate fault f_i, which is represented as A_{f_i}, $A_{f_i} \subseteq A$. The action subset on F is represented as multi-set $A_F = \bigcup_{f_i \in F} A_{f_i}$. A_F is a multi-set because the repeated actions are possibly needed while diagnosing different candidate faults.

For an arbitrary action subset A_{f_i}, we can obtain $|A_{f_i}|!$ permutations of all the elements in it. The sequence set is denoted as $S_{A_{f_i}} = \{s_{1A_{f_i}}, s_{2A_{f_i}}, ..., s_{(|A_{f_i}|)!A_{f_i}}\}$, where $s_{jA_{f_i}}$ is a permutation of A_{f_i}.

Definition 1. (Expected Cost of Diagnosis, ECD) The expected cost of diagnosis refers to the cost spent by the process that an action sequence $s \in S_{A_F}$ completes the diagnosis task. Supposing there is an action sequence, $s = (a_1, a_2, ..., a_k)$, which is a permutation of the multi-set A_F, expected cost of diagnosis is defined as:

$$ECD(s\,|\,E_c) = \sum_{j=1}^{k} p(\bigcap_{i=1}^{j-1} \{a_i \; fails\}\,|\,E_c) c_{a_j} \qquad (1)$$

$p(a_i \; fails\,|\,E_c)$ is the probability of the diagnosis failure when a_i is performed in the evidence set E_c.

Definition 2. For an arbitrary $s' \in S_{A_F}$, we can get $ECD(s\,|\,E_c) \leq ECD(s'\,|\,E_c)$, if $s \in S_{A_F}$ is the optimal action sequence when s is to identify f_c in the evidence set E_c.

Once the evidence set E_c is known, the optimal action sequence can be found by searching $s \in S_{A_F}$ and computing the least $ECD(s\,|\,E_c)$. This combinatorial optimization problem is an NP-hard problem. Considering the space limit, we omit proof of the NP-hard problem.

To solve the decision-making problem about diagnosis action, we propose the following assumptions.

Assumption 1. (Single fault assumption) Only one fault is present.

Assumption 2. (Independent cost) The cost of each action is independent with the results of former actions, which means that each action cost is a constant.

Assumption 3. (Independent diagnosis process) We can identify faults one by one according to an action sequence until a candidate fault issued by the current symptoms is proven.

Although we have made assumptive limits for the problem, it is still an NP-hard problem and it is hard to find the optimal action sequence for dependence actions. However, the assumptions can help to solve the problem, particularly assumption 3, which is originated from the actual diagnosis process.

According to assumption 3, finding the optimal action sequence is a process of calculating $s_1 \triangleright s_2 \triangleright ... \triangleright s_n$, during which $ECD(s_1 \triangleright s_2 \triangleright ... \triangleright s_n | E_c)$ remains the least, where $\triangleright$ is the append operator.. For any i, we have $s_i \in \bigcup_{j=1}^{n} S_{A_{fj}}$. If $i \neq j, s_i \in S_{A_{fk}}, s_j \in S_{A_{fi}}$, we have $S_{A_{fk}} \neq S_{A_{fi}}$.

According to the former assumptions, for a sequence $s_1 \triangleright s_2 \triangleright ... \triangleright s_n$, we can suppose $s_i \in S_{A_{fi}}$, $s_1 = (a_{11}, a_{12}, ..., a_{1\alpha})$, $s_2 = (a_{21}, a_{22}, ..., a_{2\beta})$, $s_3 = (a_{31}, a_{32}, ..., a_{3\gamma})$, ...

The value of ECD can be calculated based on the assumptions and definition of action subset on F as follow.

$$ECD(s_1 \triangleright s_2 \triangleright ... \triangleright s_n | E_c) = p(f_1 | E_c) \sum_{j=1}^{\alpha} p(\bigcap_{i=1}^{j} \{a_{1i} \quad fails\} | f_1)c_{a_j}$$

$$+ p(f_2 | E_c) \sum_{j=1}^{\beta} p((\bigcap_{i=1}^{\alpha} \{a_{1i} \quad fails\}) \cap (\bigcap_{i=1}^{j-1} \{a_{2i} \quad fails\}) | f_2)c_{a_j} \qquad (2)$$

$$+ p(f_3 | E_c) \sum_{j=1}^{\gamma} p((\bigcap_{i=1}^{\alpha} \{a_{1i} \quad fails\}) \cap (\bigcap_{i=1}^{\beta} \{a_{2i} \quad fails\}) \cap (\bigcap_{i=1}^{j-1} \{a_{3i} \quad fails\}) | f_3)c_{a_j}$$

$+ ...$

In the formula (2), $p(f_i | E_c)$ can be calculated by probability of fault f_i in evidence set E_c. According to the assumptions, c_{a_i} is a constant. $p(f_1 | E_c) \sum_{j=1}^{\alpha} p(\bigcap_{i=1}^{j} \{a_{1i} \quad fails\} | f_1)c_{a_j}$ is the expected cost spent when we verify whether the fault f_1 occurs at the first time. $p(f_2 | E_c) \sum_{j=1}^{\beta} p((\bigcap_{i=1}^{\alpha} \{a_{1i} \quad fails\}) \cap (\bigcap_{i=1}^{j-1} \{a_{2i} \quad fails\}) | f_2)c_{a_j}$ is the expected cost spent when we verify whether fault f_2 occurs if f_1 does not occur.

We have known the conditional probability that a single action can not find out the fault f_i, that is $p(a_j \quad fails | f_i)$, and the conditional probability that any two of the actions can not find out the fault f_i, that is :

$$p(a_j \quad fails, a_k \quad fails | f_i) \qquad (3)$$

In the formula (2), the possibilities $p(a_1 \quad fails, a_2 \quad fails, a_3 \quad fails, ... | f_i)$ can be calculated by the following iterative procedures:

$p(a_1 \quad fails, a_2 \quad fails, a_3 \quad fails | f_i)$

$= p(a_1 \quad fails, a_2 \quad fails | f_i) p(a_3 \quad fails | a_1 \quad fails, a_2 \quad fails, f_i)$

$\approx p(a_1 \quad fails, a_2 \quad fails | f_i) p(a_3 \quad fails | a_1 \quad fails, f_i) \qquad (4)$

$p(a_3 \quad fails | a_2 \quad fails | f_i) = p(a_1 \quad fails, a_2 \quad fails | f_i) p(a_3 \quad fails, a_1 \quad fails | f_i)$

$p(a_3 \quad fails, a_2 \quad fails | f_i) / (p(a_1 \quad fails | f_i) p(a_2 \quad fails | f_i))$

$$p(a_1 \ fails, a_2 \ fails, a_3 \ fails, a_4 \ fails \mid f_i)$$
$$= p(a_1 \ fails, a_2 \ fails, a_3 \ fails \mid f_i) p(a_4 \ fails \mid a_1 \ fails, a_2 \ fails, a_3 \ fails, f_i)$$
$$\approx p(a_1 \ fails, a_2 \ fails, a_3 \ fails \mid f_i) p(a_4 \ fails \mid a_1 \ fails, f_i)$$
$$p(a_4 \ fails \mid a_2 \ fails, f_i) p(a_4 \ fails \mid a_3 \ fails, f_i) \tag{5}$$
$$= p(a_1 \ fails, a_2 \ fails, a_3 \ fails \mid f_i) p(a_4 \ fails, a_1 \ fails \mid f_i)$$
$$p(a_4 \ fails, a_2 \ fails \mid f_i) p(a_4 \ fails, a_3 \ fails \mid f_i)$$
$$/ (p(a_1 \ fails \mid f_i) p(a_2 \ fails \mid f_i) p(a_3 \ fails \mid f_i))$$

In order to calculate the united possibilities of the multi-actions' failure, we only need the possibilities of a single action failure and the united possibilities of any two actions failure. We can calculate approximation of $ECD(s_1 \triangleright s_2 \triangleright ... \triangleright s_n \mid E_c)$ according to the formula (2) and iterative method in the formulae (3), (4), (5) according to the previous assumptions.

Based on the description of problem and simplification of assumptions, the process to calculate the optimal action sequence is the one to make $ECD(s_1 \triangleright s_2 \triangleright ... \triangleright s_n \mid E_c)$ the minimum.

3 Computing the Optimal Action Sequence by Niche Genetic Algorithm

In order to solve the problem, we should firstly find an optimal permutation of the candidate fault set F, and then find an optimal action sequence for each f_i. Because the diagnosis actions are dependent, the diagnosis costs are different for different permutations of F and the same action sequences of f_i. In the case of specified fault sequence (the permutation of F is specified), we can get a local optimal action sequence by adjusting action sequence of each f_i.

Based on the former analysis, we partition the solution into two steps. First, we should get a local optimal action sequence for each permutation of F. Then, a global optimal action sequence is selected by comparing the cost of all the local optimal action sequences.

In order to obtain the least diagnosis cost with less calculation load, we present the niche genetic algorithm.

From the formula (2), we can see that the optimal action sequences are closely related to the sequences of candidate faults. Once a fault sequence is given, we can get a local optimal action sequence by adjusting the order of each fault's action subset. From the local optimal action sequence of each candidate fault permutation, we can obtain a global optimal result. We should hold better results from different fault sequences as possible as we can during the process of the algorithm. We adopt niche algorithm of the cooperating multi-group genetic algorithms, and partition all action sequences into multi-niches, each of which corresponds to a permutation of fault set. Firstly an individual will compete with others in the same niche firstly. At the same time, the algorithm uses adaptability function to estimate whether a niche should continue evolving and when to do evolution, which can reduce the calculation load as much as possible and get the result as soon as possible.

First, we can define the code of the niche genetic algorithm and its operators.

The coding manner. [6][7]Each action sequence of f_i adopts the symbolic coding $(a_{ix}, a_{iy}, ..., a_{iz})$. Each action sequence of F adopts multi-symbolic coding, which is appended by a action sequence of f_i as :

$(a_{ix}, a_{iy}, ..., a_{iz}, a_{jx'}, a_{jy'}, ..., a_{jz'}, ..., a_{kx''}, a_{ky''}, ..., a_{kz''})$.

Crossover operator. For chromosomes

$(a_{ix}, a_{iy}, ..., a_{iz}, ..., a_{jx'}, a_{jy'}, ..., a_{jz'}, ..., a_{kx''}, a_{ky''}, ..., a_{kz''})$ and

$(a_{lx'''}, a_{ly'''}, ..., a_{lz'''}, ..., a_{jx'''}, a_{jy'''}, ..., a_{jz'''}, ..., a_{nx''''}, a_{ny''''}, ..., a_{nz''''})$, their offspring generated by

performing the crossing operation are $(a_{ix}, a_{iy}, ..., a_{iz}, a_{jx'''}, a_{jy'''}, ..., a_{jz'''}, ..., a_{kx''}, a_{ky''}, ..., a_{kz''})$

and $(a_{lx'''}, a_{ly'''}, ..., a_{lz'''}, ..., a_{jx'}, a_{jy'}, ..., a_{jz'}, ..., a_{nx''''}, a_{ny''''}, ..., a_{nz''''})$ for any integer j .

Mutation operator. For a chromosome $(..., a_{jx}, ..., a_{jy}, ...)$ and three integers j, x, y , the chromosome generated by performing the mutation operation is $(..., a_{jy}, ..., a_{jx}, ...)$.

Reversal operator. For a chromosome $(..., a_{iz'}, a_{jx}, a_{jy}, ..., a_{jz}, a_{kx'}, ...)$, and any integer j , the chromosome generated by performing the reversal operation is $(..., a_{iz'}, a_{jz}, ..., a_{jy}, a_{jx}, a_{kx'}, ...)$.

Adaptability function. For Adaptability function

$fitness(s | E_c) = length(s) \cdot \max\{C_{a_i}\} / ECD(s | E_c)$, where $ECD(s | E_c)$ can be calculated according to formula (2), $s = (a_{ix}, a_{iy}, ..., a_{iz}, a_{jx'}, a_{jy'}, ..., a_{jz'}, ..., a_{kx''}, a_{ky''}, ..., a_{kz''})$ is a chromosome and $length(s) \cdot \max\{C_{a_i}\}$ in numerator is used to keep the adaptability function value within the ability of a computer.

In the niche genetic algorithm, in order to make individuals in niche compete with each other, we adopt the fitness-sharing model. The method adjusts an individual's adaptive value by defining the individual's sharing scale in population, and thus the population holds multiple high-level modes. The sharing model is a kind of measurement transformation of a special nonlinear adaptive value, which is based on the similarity between individuals in the population. The mechanism limits the limitless increase of a special species in the population, which ensures that different species can develop themselves in their own niches.

In order to define the sharing function, we firstly define the coding distance between any two individuals.

Definition 3. $d(s_i, s_j)$ represents the distance between coding s_i and s_j as :

$d(s_i, s_j) = \sum |order_i(a_{kl}) - order_j(a_{kl})|$, where $order_i(a_{kl})$ and $order_j(a_{kl})$ are the sequence numbers of a_{kl} in s_i and s_j respectively.

Definition 4. Given a set which has $|F|!$ niches $NICHES = \{niche_1, niche_2, ..., niche_{|F|!}\}$, the radius of a niche is defined as : $\sigma_{niche_\alpha} = \max\{d(s_i, s_j)\}$, where $s_i, s_j \in niche_\alpha$.

Definition 5. The sharing function between individuals s_i and s_j is defined as

$$sh(d(s_i, s_j)) = \begin{cases} 1 - \left(\dfrac{d(s_i, s_j)}{\sigma_{niche_\alpha}}\right)^\gamma & ,if \ s_i, s_j \in niche_\alpha \\ 0 & ,otherwise \end{cases}$$,

γ is used to adjust the shape of the sharing function. Beasley's advice is to set $\gamma = 2$ [7]. The sharing value represents similar degree between two individuals. We can see that individual sharing value between different niches is zero.

Definition 6. For a population $S = \{s_1, s_2, ..., s_n\}$, the sharing scale of individual s_i in population is defined as:

$$m_i = \sum_{j=1}^{n} sh(d(s_i, s_j)), i = 1, 2, \cdots, n$$

Definition 7. The method to adjust the fitness-sharing value is defined as:

$$fitness'(s_i | E_c) = \frac{fitness(s_i | E_c)}{m_i}, n = 1, 2, \cdots, n.$$

Individual selection operator. Through adjusting fitness value, we use the optimal preserve policy to copy $\alpha|F|!$ oriental individuals to the next generation, where α is the scale of a niche.

The niche's adaptive value. For the niche's adaptive value
$NF(env_\alpha) = \Delta E(ECD(s | E_c)) / E(ECD(s | E_c))(s \in niche_\alpha)$, $E(ECD(s_i | E_c))$ is the expected value of the individuals' ECD in a niche. The smaller $E(ECD(s_i | E_c))$ are, the better niches are. $|\Delta E(ECD(s | E_c))|$ is the change rate of $E(ECD(s_i | E_c))$, which represents the individuals' evolution speed in a niche on one step length. The bigger $|\Delta E(ECD(s | E_c))|$ is, the faster the niche evolves.

We calculate that the niches with higher adaptive value should be counted in priority. Other niches which evolve slowly and whose value of $E(ECD(s_i | E_c))$ is big can be put aside for the moment. So unnecessary calculations load is reduced.

The following is the computing process of the optimal action sequence by the niche genetic algorithm.

Algorithm NGAOAS (Niche Genetic Algorithm for the Optimal Action Sequence).

(1) Initialize the counter $t = 0$
(2) Random generates the initial population $P(0)$, and the initial scale is $\alpha|F|!$ (α is the scale of the niches, $|F|!$ is the number of the niches)
(3) Initialize all the niches adaptive values $NF(niche) = +\infty$
(4) Select the optimal $|F|!/10$ niches based on $NF(env)$
(5) Process the crossover action between individuals $P'(t) \leftarrow Crossover[P(t)]$ by the probability β
(6) Process the individual mutation action by the probability γ : $P''(t) \leftarrow Mutation[P'(t)]$
(7) Process the individual reversal action by the probability σ : $P'''(t) \leftarrow Inversion[P''(t)]$
(8) Calculate the individual adaptive value $fitness(s | E_c)$
(9) Adjust the adaptive value by sharing scale to get an adjusted $fitness'(s)$

(10) Copy $\alpha|F|!$ individuals of the optimal adaptive value to the next generation, and get $P(t+1)$

(11) If t mod $10=0$, recalculate the adaptive value $NF(niche)$ of each evolving niche

(12) If $\exists niche, \exists s$, $s \in niche$, $fitness(s|E_c) = \max\{fitness(s'|E_c)\}$, $NF(niche)=0$ and $\forall niche$, $NF(niche) < \delta$, the algorithm ends. Otherwise, go to (4).

4 The Performance Analysis of NGAOAS

The keys of calculating the optimal action sequence with the genetic algorithm are selecting the proper parameters and holding good features. Before analyzing features of NGAOAS, we select parameters of NGAOAS by testing.

NGAOAS has four parameters. By experience we can assign the reference values to $\beta=0.3$, $\gamma=1$, $\sigma=0.03$. Notice that because the algorithm adapts the symbolic coding, the function of mutation operator is similar to the crossover operator in the general GA algorithm, while the reversal operator is real mutation operator. Therefore, we usually set the mutation operator's probability of NGAOAS according to GA experienced crossover probability value. Accordingly, we set the reversal operator's probability of NGAOAS according to GA algorithm's mutation experienced probability value.

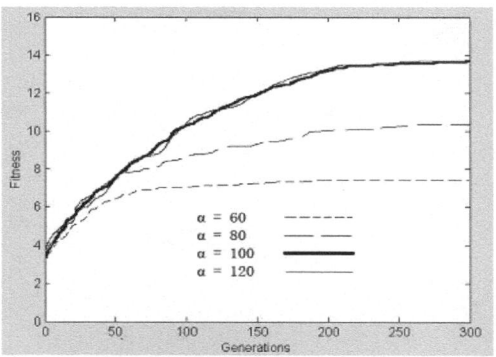

Fig. 1. Choose parameter α

We emphasize the parameter α , which decides the original population scale of NGAOAS. The selection of the original population scale affects the probability of finding the optimal result. An unsuitable original scale may cause premature problem. We select five network faults and 20 diagnosis actions, and each fault diagnosis needs average 10 actions. We do experiment on a stand-alone PC with PIV 2G CPU, 512M memory.

From the figure 1, we know that when the parameter α is bigger than 100, premature problem can be avoided. In order to reduce the memory space that the algorithm needs, we set $\alpha=100$ in the later test.

Furthermore, NGAOAS has some outstanding features.

Feature 1. The space complexity of NGAOAS is $O(|F|!)$, the memory space that the algorithm needs can be predicted.

Only $\alpha|F|!$ optimal individuals remain by the optimal reservation strategy. So the memory space that NGAOAS needs can be predicted and controlled under $\alpha|F|!$. We can avoid the case of insufficient memory before processing the algorithm. Besides, the space complexity of the algorithm is $O(|F|!)$, but not $O(length(s)!)$. Therefore, we reduce the space complexity effectively.

Feature 2. NGAOAS has better implicit parallelism.

Through the plain definition and assumptions for the problem, we study the background knowledge and adopt niche technology. Each niche is independent and seldom communicates with each other. Therefore, NGAOAS has better implicit parallelism than the normal genetic algorithm. We can make better use of the existing calculating grid, and the algorithm has better practicability.

We have tested the performance of NGAOAS on the single PC and grid environments according to feature 2. We select 5 network faults and 20 corresponding diagnosis actions. Each fault diagnosis needs 10 actions on average. Standalone environment is PIV 2G CPU, 512M memory. The grid environment is composed of Globus Toolkit 2.4 computation grid platform and 10 identical test PCs.

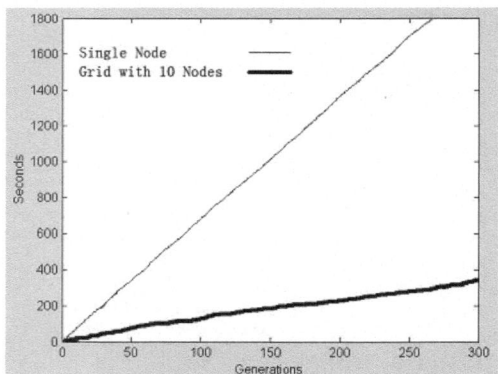

Fig. 2. Compare the performance under the circumstance between a single PC and the grid environment

From figure 2 we can find that performance of NGAOAS increases about 6 to 7 times in the grid environment. This proves that NGAOAS has better implicit parallelism.

Feature 3. The population is able to hold diversity, which can avoid premature problem.

NGAOAS defines the distance between individuals, adjusts the adaptive value with sharing scale and selects the optimal individuals according to adjusted adaptive value, which makes individuals in each niche achieve balanced development. NGAOAS takes the inner and global competitive stress all around, and thus avoids the convergence problem prematurely.

According to feature 3, we draw a comparison between the genetic algorithm without niche technology and NGAOAS. The test is processed in the similar environment as feature 2. We adapt $div(P(t)) = \sum_{s_i,s_j \in P(t)} d(s_i,s_j) \bigg/ \sum_{s_k,s_l \in P(0)} d(s_k,s_l)$ as the computation formula of the population's diversity.

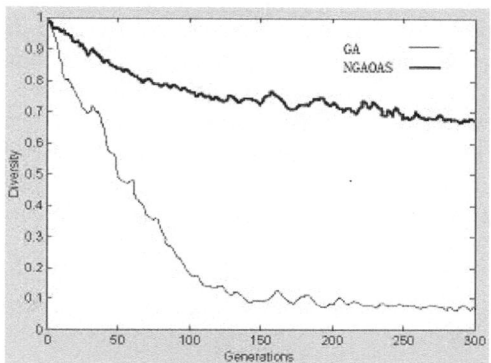

Fig. 3. The Diversity changing curve along with genetic generations

Figure3 is population's diversity changing curve along with the genetic generation.

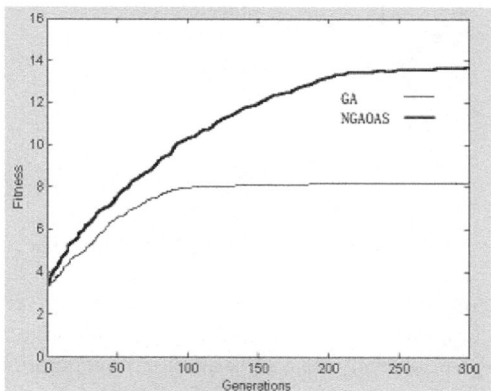

Fig. 4. The changing curve of adaptive value along with genetic generations

Figure 4 is the changing curve of the optimal individual's adaptive value in the same situation.

From figure 3 and figure 4, we can see that compared with general GA algorithm, NGAOAS can keep the adaptive values of individuals increasing and can retain better diversity along with the increase of genetic generations. While the population diversity of the general GA algorithm decreases quickly, the optimal individuals' adaptive values no longer increase after the 100th generation. Premature problem severely influences the optimal results or preferable results. Repeated experiments prove that NGAOAS has a moderate convergence speed, the probability of finding out the optimal results per 300 generations is 82%, and the rest 18% also generates preferable result, which is close to the optimal one. This effect is enough for the most situations.

Feature 4. NGAOAS introduces niche adaptive value to reduce calculation load, and thus speeds up convergence and avoid the premature problem at the same time.

NGAOAS introduces the concept of niche adaptive value according to the diagnosis decision problem. The niches' evolution speed is direct ratio to niches' adaptive value, so niches have a good potential of being optimized. The smaller current optimal ECD is, the better the niche is. The proper design of the niche adaptive value function ensures the generation of the optimal results, which decreases unnecessary calculation load at a certain extent and also speeds up the convergence without causing premature problem.

In the experiment, we choose a situation with less calculation load, in which we compare the previous convergence speed with the later introduction of the niche adaptive function. In the test environment, we select 6 network faults and 16 corresponding diagnosis actions. Each fault diagnosis needs 3 to 4 actions.

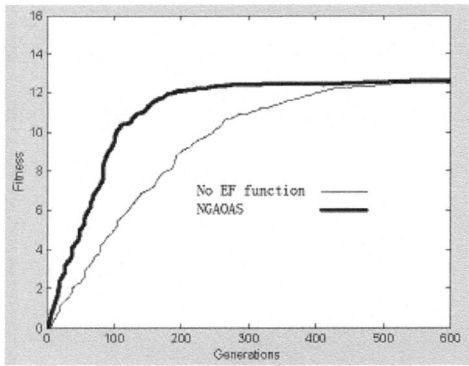

Fig. 5. Comparison of the convergence rates

In figure 5, the algorithm without niche adaptive value converges to a steady state after about 500 generations. After using the niche adaptive value function, the algorithm can converge to a steady state after about 200 generations. The convergence rate doubles. When there are more candidate faults, the convergence rate increases even more. From figure 5, we can also see that NGAOAS optimize the convergence rate without causing premature problem. It can find the optimal or preferable results at faster convergence rate.

At the same time, we do experiments in various of diagnosis environments and compare the diagnosis cost of the updated P/C algorithm with the one of NGAOAS.

As is shown in figure 6, when actions are dependent, compared with updated P/C algorithm, NGAOAS can acquire preferable diagnosis action sequence and the lower cost.

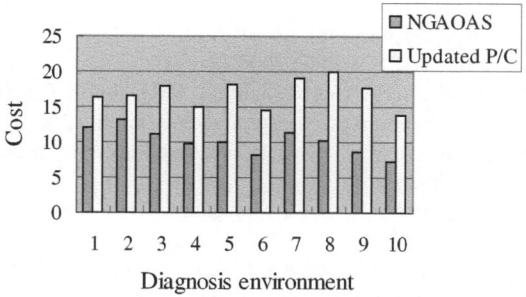

Fig. 6. Comparison of the optimal diagnosis costs

5 Summary and Conclusion

This paper presents an algorithm named NGAOAS to compute the minimum diagnosis cost in the condition of dependent actions. NGAOAS decomposes the candidate action sequences into $|F|$! niches, each of which corresponds to a single candidate faults permutation. An individual will compete with others in the same niche. At the same time NGAOAS will estimate whether a niche should continue evolving and when it should do evolution, which reduces calculation load as much as possible and get the result as soon as possible. The experiment proves NGAOAS has several outstanding features. Compared with updated P/C algorithm, NGAOAS can acquire preferable diagnosis action sequence and the less cost.

We describe the diagnosis problem in the condition of dependent actions. There are several other scenarios that we have not considered and that deserve further exploration. We described the diagnosis problem assuming there are independent faults or single fault, which simplifies the computations, but dependent faults and multiple faults may occur in real world. So [8] [9] introduce the model of exact 3-sets cover. These problems will be studied further. Another issue relates to the behavior of observations [10]. Observations can not solve the diagnosis problem, but they can exclude some faults and speed up the troubleshooting.

References

1. J.-F. Huard and A. A. Lazar. Fault isolation based on decision-theoretic troubleshooting. Technical Report 442-96-08, Center for Telecommunications Research, Columbia University, New York, NY 1996.
2. J.Baras, H.Li, and G..Mykoniatis. Integrated distributed fault management for communication networks. Technical Report 98-10, Center for Satellite and Hybrid Communication Networks, University of Maryland, College Park, MD,1998.
3. Breese J S, Heckerman D. Decision-Theoretic Troubleshooting: A Framework for Repair and Experiment. Proc. 12th Conf. on Uncertainty in Artificial Intelligence, Morgan Kaufmann Publishers, San Francisco, CA,1996:124-132

4. Skaanning C, Jensen F V, Kjærulff U. Printer Troubleshooting Using Bayesian Networks. Industrial and Engineering Applications of Artificial Intelligence and Expert Systems (IEA/AIE) 2000, New Orleans, USA, June, 2000
5. Vomlelová M. Decision Theoretic Troubleshooting. PhD Thesis, Faculty of Informatics and Statistics, University of Economics, Prague, Czech Republic, 2001.
6. Shi Zhong Zhi. Knowledge Discover. Beijing: Tsinghua university press, 2002. ISBN 7-302-05061-9:265~293(in Chinese)
7. Li Min Qiang, Kou Ji Song, Lin Dan, Li Shu Quan. Basic theory and application of Genetic Algorithm. Beijing: Science press, 2002.3, ISDN7-03-009960-5:233-244.
8. C.Skaanning, F.V.Jensen, U.Kjarulff, and A.L.Madsen. Acquisition and transformation of likelihoods to conditional probabilities for Bayesian networks. AAAI Spring Symposium, Stanford, USA,1999.
9. C.Skaanning,. A knowledge acquisition tool for Bayesian network troubleshooters. Proceedings of the Sixteenth Conference on Uncertainty in Artificial Intelligence, Stanford, USA, 2000.
10. M. Steinder and A. S. Sethi. Increasing robustness of fault localization through analysis of lost, spurious, and positive symptoms. IEEE. 2002

Diagnosis of Plan Execution and the Executing Agent

Nico Roos[1] and Cees Witteveen[2,3]

[1] Dept of Computer Science, Universiteit Maastricht,
P.O.Box 616, NL-6200 MD Maastricht
roos@cs.unimaas.nl
[2] Faculty EEMCS, Delft University of Technology,
P.O.Box 5031, NL-2600 GA Delft
witt@ewi.tudelft.nl
[3] Centre of Mathematics and Computer Science,
P.O. Box 94079, NL-1090 GB Amsterdam
C.Witteveen@cwi.tudelft.nl

Abstract. We adapt the Model-Based Diagnosis framework to perform (agent-based) plan diagnosis. In plan diagnosis, the system to be diagnosed is a plan, consisting of a partially ordered set of instances of actions, together with its executing agent. The execution of a plan can be monitored by making partial observations of the results of actions. Like in standard model-based diagnosis, observed deviations from the expected outcomes are explained qualifying some action instances that occur in the plan as behaving abnormally. Unlike in standard model-based diagnosis, however, in plan diagnosis we cannot assume that actions fail independently. We focus on two sources of dependencies between failures: dependencies that arise as a result of a malfunction of the executing agent, and dependencies that arise because of dependencies between action instances occurring in a plan. Therefore, we introduce causal rules that relate health states of the agent and health states of actions to abnormalities of other action instances. These rules enable us to introduce causal set and causal effect diagnoses that use the underlying causes of plan failing to explain deviations and to predict future anomalies in the execution of actions.

Keywords: planning, diagnosis, agent systems.

1 Introduction

In Model-Based Diagnosis (MBD) a model of a system consisting of components, their interrelations and their behavior is used to establish why the system is malfunctioning. Plans resemble such system specifications in the sense that plans also consist of components (action specifications), their interrelations and a specification of the (correct) behavior of each action. Based on this analogy, the aim of this paper is to adapt and extend a classical Model-Based Diagnosis (MBD) approach to the diagnosis of plans.

To this end, we will first formally model a plan consisting of a partially ordered set of actions as a system to be diagnosed, and subsequently we will describe how a diagnosis can be established using *partial observations* of a plan in progress. Distinguishing between normal and abnormal execution of actions in a plan, we will introduce sets of

U. Furbach (Ed.): KI 2005, LNAI 3698, pp. 161–175, 2005.

actions qualified as abnormal to explain the deviations between expected plan states and observed plan states. Hence, in this approach, a plan diagnosis is just a set of abnormal actions that is able to explain the deviations observed. Although plan diagnosis conceived in this way is a rather straightforward application of MBD to plans, we do need to introduce new criteria for selecting acceptable plan diagnoses: First of all, while in standard MBD usually subset-minimal diagnoses, or within them *minimum (cardinality)* diagnoses, are preferred, we also prefer *maximum informative* diagnoses. The latter type of diagnosis maximizes the exact similarity between predicted and observed plan states. Although maximum informative diagnoses are always subset minimal, they are not necessarily of minimum cardinality. More differences between MBD and plan diagnosis appear if we take a more detailed look into the reasons for choosing minimal diagnoses. The idea of establishing a minimal diagnosis in MBD is governed by the principle of *minimal change*: explain the abnormalities in the behavior observed by changing the qualification from normal to abnormal for as few system components as necessary. Using this principle is intuitively acceptable if the components qualified as abnormal are failing *independently*. However, as soon as *dependencies* exist between such components, the choice for minimal diagnoses cannot be justified. As we will argue, the existence of dependencies between failing actions in a plan is often the rule instead of an exception. Therefore, we will refine the concept of a plan diagnosis by introducing the concept of a *causal diagnosis*. To establish such a causal diagnosis, we consider both the executing agent and its plan as constituting the system to be diagnosed and we explicitly relate health states of the executing agent and subsets of (abnormally qualified) actions to the abnormality of other actions in the form of causal rules. These rules enable us to replace a set of dependent failing actions (e.g. a plan diagnosis) by a set of unrelated *causes* of the original diagnosis. This independent and usually smaller set of causes constitutes a causal diagnosis, consisting of a health state of an agent and an independent (possibly empty) set of failing actions. Such a causal diagnosis always generates a cover of a minimal diagnosis. More importantly, such causal diagnoses can also be used to predict failings of actions that have to be executed in the plan and thereby also can be used to assess the consequences of such failures for goal realizability.

This paper is organized as follows. Section 2 introduces the preliminaries of plan-based diagnosis, while Section 3 formalizes plan-based diagnosis. Section 4 extends the formalization to determining the agent's health state. Finally, we briefly discuss some computational aspects of (causal) plan diagnosis. In Section 6, we place our approach into perspective by discussing some related approaches to plan diagnosis. and Section 7 concludes the paper.

2   Preliminaries

Model Based Diagnosis.   In Model-Based Diagnosis (MBD) [4,5,12] a system S is modeled as consisting of a set $Comp$ of components and their relations, for each component $c \in Comp$ a set H_c of *health modes* is distinguished and for each health mode $h_c \in H_c$ of each component c a specific (input-output) behavior of c is specified. Given some input to S, its output is defined if the health mode of each component $c \in Comp$ is known. The diagnostic engine is triggered whenever, under the assumption that all components are functioning normally, there is a discrepancy between the output as predicted from the input observations, and the actually observed output. The result of MBD is a suitable assignment of health modes to the components, called a *diagnosis*, such that the actually observed output is *consistent* with this health mode qualification or can be *explained* by this qualification. Usually, in a diagnosis one requires the number of components qualified as abnormally to be minimized.

States and Partial States.   We consider plan-based diagnosis as a simple extension of the model-based diagnosis where the model is not a description of an underlying system but a *plan* of an agent. Before we discuss plans, we introduce a simplified state-based view on the world, assuming that for the planning problem at hand, the world can be simply described by a set $Var = \{v_1, v_2, \ldots, v_n\}$ of variables and their respective *value domains* D_i. A *state of the world* σ then is a value assignment $\sigma(v_i) = d_i \in D_i$ to the variables. We will denote a state simply by an element of $D_1 \times D_2 \times \ldots \times D_n$, i.e. an n-tuple of values. It will not always be possible to give a complete state description. Therefore, we introduce a *partial state* as an element $\pi \in D_{i_1} \times D_{i_2} \times \ldots \times D_{i_k}$, where $1 \leq k \leq n$ and $1 \leq i_1 < \ldots < i_k \leq n$. We use $Var(\pi)$ to denote the set of variables $\{v_{i_1}, v_{i_2}, \ldots, v_{i_k}\} \subseteq Var$ specified in such a state π. The value d_j of variable $v_j \in Var(\pi)$ in π will be denoted by $\pi(j)$. The value of a variable $v_j \in Var$ not occurring in a partial state π is said to be *unknown* (or unpredictable) in π, denoted by $\perp$. Including $\perp$ in every value domain D_i allows us to consider every partial state π as an element of $D_1 \times D_2 \times \ldots \times D_n$.

Partial states can be ordered with respect to their information content: Given values d and d', we say that $d \leq d'$ holds iff $d = \perp$ or $d = d'$. The containment relation $\sqsubseteq$ between partial states is the point-wise extension of $\leq$: π is said to be contained in π', denoted by $\pi \sqsubseteq \pi'$, iff $\forall j[\pi(j) \leq \pi(j')]$. Given a subset of variables $V \subseteq Var$, two partial states π, π' are said to be *V-equivalent*, denoted by $\pi =_V \pi'$, if for every $v_j \in V$, $\pi(j) = \pi'(j)$. We define the partial state π restricted to a given set V, denoted by $\pi \upharpoonright V$, as the state $\pi' \sqsubseteq \pi$ such that $Var(\pi') = V \cap Var(\pi)$.

An important notion for diagnosis is the notion of *compatibility* between partial states. Intuitively, two states π and π' are said to be compatible if there is no essential disagreement about the values assigned to variables in the two states. That is, for every j either $\pi(j) = \pi'(j)$ or at least one of the values $\pi(j)$ and $\pi'(j)$ is undefined. So we define π and π' to be compatible, denoted by $\pi \approx \pi'$, iff $\forall j[\pi(j) \leq \pi'(j)$ or $\pi'(j) \leq \pi(j)]$. As an easy consequence we have, using the notion of V-equivalent states, $\pi \approx \pi'$ iff $\pi =_{Var(\pi) \cap Var(\pi')} \pi'$. Finally, if π and π' are compatible states they can be *merged* into the $\sqsubseteq$-least state $\pi \sqcup \pi'$ containing them both: $\pi \sqcup \pi'(j) = max_{\leq}\{\pi(j), \pi'(j)\}$.

Goals.   An (elementary) goal g of an agent specifies a set of partial states an agent wants to bring about using a plan. Here, we specify each such a goal g as a constraint, that is a relation over some product $D_{i_1} \times \ldots \times D_{i_k}$ of domains.

We say that a goal g is satisfied by a partial state π, denoted by $\pi \models g$, if the relation g contains at least one tuple $(d_{i_1}, d_{i_2}, \ldots, d_{i_k})$ such that $(d_{i_1}, d_{i_2}, \ldots d_{i_k}) \sqsubseteq \pi$. We assume each agent to have a set G of such elementary goals $g \in G$. We use $\pi \models G$ to denote that all goals in G hold in π, i.e. for all $g \in G$, $\pi \models g$.

Actions and Action Schemes.   An *action scheme* or plan operator α is represented as a function that replaces the values of a subset $V_\alpha \subseteq Var$ by other values, dependent

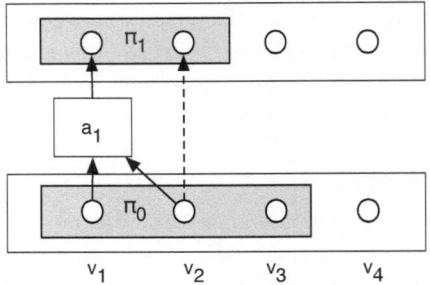

Fig. 1. Plan operators & states

upon the values of another set $V'_\alpha \supseteq V_\alpha$ of variables. Hence, every action scheme α can be modeled as a (partial) function $f_\alpha : D_{i_1} \times \ldots \times D_{i_k} \to D_{j_1} \times \ldots \times D_{j_l}$, where $1 \leq i_1 < \ldots < i_k \leq n$ and $\{j_1, \ldots, j_l\} \subseteq \{i_1, \ldots, i_k\}$. The variables whose value domains occur in $dom(f_\alpha)$ will be denoted by $dom_{Var}(\alpha) = \{v_{i_1}, \ldots, v_{i_k}\}$ and, likewise $ran_{Var}(\alpha) = \{v_{j_1}, \ldots, v_{j_l}\}$. Note that it is required that $ran_{Var}(\alpha) \subseteq dom_{Var}(\alpha)$. This functional specification f_α constitutes the *normal* behavior of the action scheme, denoted by f_α^{nor}.

Example 1. Figure 1 depicts two states σ_0 and σ_1 (the white boxes) each characterized by the values of four variables v_1, v_2, v_3 and v_4. The partial states π_0 and π_1 (the gray boxes) characterize a subset of values in a (complete) state. Action schemes are used to model state changes. The domain of the action scheme α is the subset $\{v_1, v_2\}$, which are denoted by the arrows pointing to α. The range of α is the subset $\{v_1\}$, which is denoted by the arrow pointing from α. Finally, the dashed arrow denotes that the value of variable v_2 is not changed by operator(s) causing the state change.   ∎

The correct execution of an action may fail either because of an inherent malfunctioning or because of a malfunctioning of an agent responsible for executing the action, or because of unknown external circumstances. In all these cases we would like to model the effects of executing such failed actions. Therefore, we introduce a set of *health modes* M_α for each action scheme α. This set M_α contains at least the normal mode nor, the mode ab indicating the most general abnormal behavior, and possibly several other specific fault modes. The most general abnormal behavior of action α is specified by the function f_α^{ab}, where $f_\alpha^{ab}(d_{i_1}, d_{i_2}, \ldots, d_{i_k}) = (\perp, \perp, \ldots, \perp)$ for every partial state $(d_{i_1}, d_{i_2}, \ldots, d_{i_k}) \in dom(f_\alpha)$.[1] To keep the discussion simple, in the sequel we distinguish only the health modes nor and ab.

Given a set $\mathcal{A}$ of action schemes, we will need to consider a set $A \subseteq inst(\mathcal{A})$ of *instances* of actions in $\mathcal{A}$. Such instances will be denoted by small roman letters a_i. If $type(a_i) = \alpha \in \mathcal{A}$, such an instance a_i is said to be of *type* α. If the context permits we will use "actions" and "instances of actions" interchangeably.

Plans. A plan is a tuple $P = \langle \mathcal{A}, A, < \rangle$ where $A \subseteq Inst(\mathcal{A})$ is a set of instances of actions occurring in $\mathcal{A}$ and $(A, <)$ is a partial order. The partial order relation $<$

---

[1] This definition implies that the behavior of abnormal actions is essentially unpredictable.

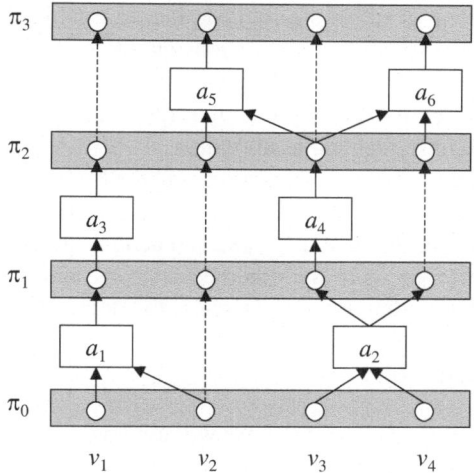

Fig. 2. Plans and action instances. Each state characterizes the values of four variables v_1, v_2, v_3 and v_4. States are changed by application of action instances.

specifies a precedence relation between these instances: $a < a'$ implies that the instance a must finish before the instance a' may start. We will denote the *transitive reduction* of $<$ by $\ll$, i.e., $\ll$ is the smallest subrelation of $<$ such that the transitive closure $\ll^+$ of $\ll$ equals $<$.

We assume that if in a plan P two action instances a and a' are independent, in principle they may be executed concurrently. This means that the dependency relation $<$ at least should capture all resource dependencies that would prohibit concurrent execution of actions. Therefore, we assume $<$ to satisfy the following *concurrency requirement*:

If $ran_{Var}(a) \cap dom_{Var}(a') \neq \varnothing$ then $a < a'$ or $a' < a$.[2]

That is, for concurrent instances, domains and ranges do not overlap.

Example 2. Figure 2 gives an illustration of a plan. Arrows relate the variables an action uses as inputs and the variables it produces as its outputs to the action itself. In this plan, the dependency relation is specified as $a_1 \ll a_3$, $a_2 \ll a_4$, $a_4 \ll a_5$, $a_4 \ll a_6$ and $a_1 \ll a_5$. Note that the last dependency has to be included because a_5 changes the value of v_2 needed by a_1. The action a_1 shows that not every variable occurring in the domain of an action need to be affected by the action. The actions a_5 and a_6 illustrate that concurrent actions may have overlapping domains. ∎

3   Standard Plan Diagnosis

Let us assume, for the moment, that each action instance can be viewed as an independent component of a plan. To each action instance a a health mode $m_a \in \{nor, ab\}$ can

---

[2] Note that since $ran_{Var}(a) \subseteq dom_{Var}(a)$, this requirement excludes overlapping ranges of concurrent actions, but domains of concurrent actions are allowed to overlap as long as the values of the variables in the overlapping domains are not affected by the actions.

be assigned and the result is called a *qualified* plan. In establishing which part of the plan fails, we are only interested in those actions qualifies as abnormal. Therefore, we define a qualified version P_Q of a plan $P = \langle \mathcal{A}, A, < \rangle$ as a tuple $P_Q = \langle \mathcal{A}, A, <, Q \rangle$, where $Q \subseteq A$ is the subset of instances of actions qualified as abnormal (and therefore, $A - Q$ the subset of actions qualified as normal).

Since a qualification Q corresponds to assigning the health mode ab to every action in Q and since $f_a^{ab}(d_{i_1}, d_{i_2}, \ldots, d_{i_k}) = (\bot, \bot, \ldots, \bot)$ for every action $a \in Q$ with $type(a) = \alpha$, the results of anomalously executed actions are unpredictable. Note that a "normal" plan P corresponds to the qualified plan $P_\varnothing$ and furthermore that in our context "undefined" is considered to be equivalent to "unpredictable".

3.1   Qualified Plan Execution

For simplicity, when a plan P is executed, we will assume that every action takes a unit of time to execute. We are allowed to observe the execution of a plan P at discrete times $t = 0, 1, 2, \ldots, k$ where k is the depth of the plan, i.e., the longest $<$-chain of actions occurring in P. Let $depth_P(a)$ be the depth of action a in plan $P = \langle \mathcal{A}, A, < \rangle$. Here, $depth_P(a) = 0$ if $\{a' \mid a' \ll a\} = \varnothing$ and $depth_P(a) = 1 + max\{depth_P(a') \mid a' \ll a\}$, else. If the context is clear, we often will omit the subscript P. We assume that the plan starts to be executed at time $t = 0$ and that concurrency is fully exploited, i.e., if $depth_P(a) = k$, then execution of a has been completed at time $t = k + 1$. Thus, all actions a with $depth_P(a) = 0$ are completed at time $t = 1$ and every action a with $depth_P(a) = k$ will be started at time k and will be completed at time $k + 1$. Note that thanks to the above specified concurrency requirement, concurrent execution of actions having the same depth leads to a well-defined result.

Let P_t denote the set of actions a with $depth_P(a) = t$, let $P_{>t} = \bigcup_{t'>t} P_{t'}$, $P_{<t} = \bigcup_{t'<t} P_{t'}$ and $P_{[t,t']} = \bigcup_{k=t}^{t'} P_k$. Execution of P on a given initial state σ_0 will induce a sequence of states $\sigma_0, \sigma_1, \ldots, \sigma_k$, where σ_{t+1} is generated from σ_t by applying the set of actions P_t to σ_t. Instead, however, of assuming total states and total state transitions, we define the (predicted) effect of the execution of plan P on a given (partial) state π at time $t \geq 0$, denoted by (π, t).

We say that $(\pi', t + 1)$ is (directly) generated by execution of P_Q from (π, t), abbreviated by $(\pi, t) \rightarrow_{Q;P} (\pi', t + 1)$, iff the following conditions hold:

1. $\pi' \upharpoonright ran_{Var}(a) = f_a^{nor}(\pi \upharpoonright dom_{Var}(a))$ for each $a \in P_t - Q$ such that $dom_{Var}(a) \subseteq Var(\pi)$, that is, the consequences of all actions a enabled in π can be predicted and occur in π'.[3]

2. $Var(\pi') \cap ran_{Var}(a) = \varnothing$ for each $a \in Q \cap P_t$, since the result of executing an abnormal action cannot be predicted (even if such an action is enabled in π);

3. $Var(\pi') \cap ran_{Var}(a) = \varnothing$ for each $a \in P_t$ with $dom_{Var}(a) \not\subseteq Var(\pi)$, that is, even if an action a is enabled in (the complete state) σ_t, if a is not enabled in $\pi \sqsubseteq \sigma_t$, the result is not predictable and therefore does not occur in π', since it is not possible to predict the consequences of actions that depend on values not defined in π.

---

[3] An action a is enabled in a state π if $dom_{Var}(a) \subseteq Var(\pi)$.

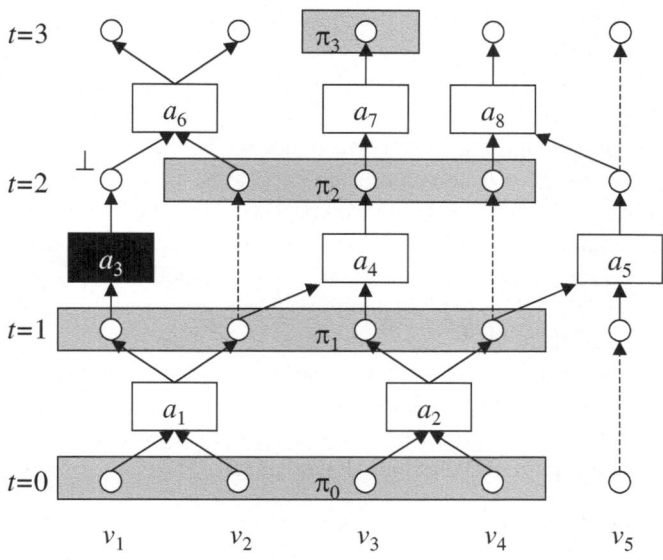

$t=3$

$t=2$

$t=1$

$t=0$

v_1 v_2 v_3 v_4 v_5

Fig. 3. Plan execution with abnormal actions

4. $\pi'(i) = \pi(i)$ for each $v_i \notin ran_{Var}(P_t)$, that is, the value of any variable not occurring in the range of an action in P_t should remain unchanged. Here, $ran_{Var}(P_t)$ is a shorthand for the union of the sets $ran_{Var}(a)$ with $a \in P_t$.

For arbitrary values of $t \le t'$ we say that (π', t') is (directly or indirectly) generated by execution of P_Q from (π, t), denoted by $(\pi, t) \to^*_{Q;P} (\pi', t')$, iff the following conditions hold:

1. if $t = t'$ then $\pi' = \pi$;
2. if $t' = t + 1$ then $(\pi, t) \to_{Q;P} (\pi', t')$;
3. if $t' > t + 1$ then there must exists some state $(\pi'', t' - 1)$ such that $(\pi, t) \to^*_{Q;P}$ $(\pi'', t' - 1)$ and $(\pi'', t' - 1) \to_{Q;P} (\pi', t')$.

Note that $(\pi, t) \to^*_{\varnothing;P} (\pi', t')$ denotes the normal execution of a normal plan $P_\varnothing$. Such a normal plan execution will also be denoted by $(\pi, t) \to^*_P (\pi', t')$.

Example 3. Figure 3 gives an illustration of an execution of a plan with abnormal actions. Suppose action a_3 is abnormal and generates a result that is unpredictable ($\bot$). Given the qualification $Q = \{a_3\}$ and the partially observed state π_0 at time point $t = 0$, we predict the partial states π_i as indicated in Figure 3, where $(\pi_0, t_0) \to^*_{Q;P} (\pi_i, t_i)$ for $i = 1, 2, 3$. Note that since the value of v_1 and of v_5 cannot be predicted at time $t = 2$, the result of action a_6 and of action a_8 cannot be predicted and π_3 contains only the value of v_3. ∎

3.2 Diagnosis

Suppose now that we have a (partial) observation $obs(t) = (\pi, t)$ of the state of the world at time t and an observation $obs(t') = (\pi', t')$ at time $t' > t \ge 0$ during the execution of the plan P. We would like to use these observations to infer the health states of the actions occurring in P. Assuming a normal execution of P, we can (partially) predict the state of the world at a time point t' given the observation $obs(t)$: if all actions behave normally, we predict a partial state $\pi'_\varnothing$ at time t' such that $obs(t) \to_P^* (\pi'_\varnothing, t')$.

There does not need to be a strict correspondence between the variables *predicted at time* t' and the variables *observed at time* t'. That is, $Var(\pi')$ and $Var(\pi'_\varnothing)$ need not to be identical sets. This means that to check whether the predicted state matches the observed state at time t', we have to verify whether the variables occurring in both $Var(\pi')$ and $Var(\pi'_\varnothing)$ have identical values, that is whether $\pi'(j) = \pi'_\varnothing(j)$ holds for all $v_j \in Var(\pi') \cap Var(\pi'_\varnothing)$. Therefore, these states match exactly if they are compatible i.e. $\pi' \approx \pi'_\varnothing$ holds.[4]

If this is not the case, the execution of some action instances must have gone wrong and we have to determine a qualification Q such that the predicted state π'_Q derived using Q is compatible with π'. Hence, we have the following straight-forward extension of the diagnosis concept in MBD to plan diagnosis (cf. [5]):

Definition 1. *Let* $P = \langle \mathcal{A}, A, < \rangle$ *be a plan with observations* $obs(t) = (\pi, t)$ *and* $obs(t') = (\pi', t')$, *where* $t < t' \leq depth(P)$ *and let* $obs(t) \to_{Q;P}^* (\pi'_Q, t')$ *be a derivation assuming a qualification* Q.

Then Q *is said to be a* plan diagnosis *of* $\langle P, obs(t), obs(t') \rangle$ *iff* $\pi' \approx \pi'_Q$.

Example 4. Consider again Figure 3 and suppose that we did not know that action a_3 was abnormal and that we observed $obs(0) = ((d_1, d_2, d_3, d_4), 0)$ and $obs(3) = ((d'_1, d'_3, d'_5), 3)$. Using the normal plan derivation relation starting with $obs(0)$ we will predict a state $\pi'_\varnothing$ at time $t = 3$ where $\pi'_\varnothing = (d''_1, d''_2, d''_3)$. If everything is ok, the values of the variables predicted as well as observed at time $t = 3$ should correspond, i.e. we should have $d'_j = d''_j$ for $j = 1, 3$. If, for example, only d'_1 would differ from d''_1, then we could qualify a_6 as abnormal, since then the predicted state at time $t = 3$ using $Q = \{a_6\}$ would be $\pi'_Q = (d''_3)$ and this partial state agrees with the predicted state on the value of v_3. ∎

Note that for all variables in $Var(\pi') \cap Var(\pi'_Q)$, the qualification Q provides an *explanation* for the observation π' made at time point t'. Hence, for these variables the qualification provides an *abductive diagnosis* [4] for the normal observations. For all observed variables in $Var(\pi') - Var(\pi'_Q)$, no value can be predicted given the qualification Q. Hence, by declaring them to be unpredictable, possible conflicts with respect to these variables if a normal execution of all actions is assumed, are resolved. This corresponds with the idea of a *consistency-based diagnosis* [12].

The following observation shows that we might easily trivialize plan diagnoses:

Observation 1. *If* $Q \subset A$ *is a plan diagnosis of* $\langle P, obs(t), obs(t') \rangle$, *then every superset* $Q' \supseteq Q$ *is also a plan diagnosis and in particular* A *is always a plan diagnosis.*

The reason is that (i) $Q' \supseteq Q$ implies $\pi'_{Q'} \sqsubseteq \pi'_Q$ where $\pi'_{Q'}$ and π'_Q are the predicted states using the qualifications Q and Q', respectively and (ii) $\pi'_Q \approx \pi'$ and $\pi'_{Q'} \sqsubseteq \pi'_Q$, using the definition of $\approx$, immediately imply that $\pi'_{Q'} \approx \pi'$, i.e., Q' is a diagnosis as

---

[4] See the definition in the preliminaries.

well. Since in particular $A \supseteq Q$ for every qualification Q, A is a diagnosis whenever there has been found any diagnosis.

Clearly then, the smaller a diagnosis is, the more values it will predict that are also actually observed in the resulting plan state. This, like in MBD, is a reason for us to prefer *subset-minimal* diagnoses and especially *minimum* diagnoses among the set of minimal diagnoses.

But there is a caveat: a minimum diagnosis only minimizes the number of abnormal actions to explain deviations; as important however for a diagnosis might be its *information content*, i.e. the exactness it provides in predicting the values of the variables occurring in the observed state π'. This means that besides *minimizing* the cardinality of abnormalities another criterion could be *maximizing* the exactness of the similarity by maximizing $|Var(\pi') \cap Var(\pi'_Q)|$ i.e. maximizing the number of variables having the same value in the predicted state and the observed state. Therefore, besides a minimum diagnosis we also define the notion of a *maximum informative diagnosis*:

Definition 2. *Given plan observations $\langle P, (\pi, t), (\pi', t') \rangle$, a qualification Q is said to be a* minimum plan diagnosis *if for every plan diagnosis Q' it holds that $|Q| \leq |Q'|$.*

Q is said to be a maximum informative plan-diagnosis *iff for all plan diagnoses Q^*, it holds that $|Var(\pi') \cap Var(\pi'_Q)| \geq |Var(\pi') \cap Var(\pi'_{Q^*})|$.*

Note that for every maximum informative diagnosis Q we have $Var(\pi') \cap Var(\pi'_Q) \subseteq Var(\pi') \cap Var(\pi'_\varnothing)$, where $obs(t) \rightarrow^*_{\varnothing;P} (\pi'_\varnothing, t')$ is the partial state derivation assuming a *normal plan* execution.

Also note that every maximum informative diagnosis is a minimal diagnosis. So both minimum plan diagnoses and maximum informative plan diagnoses are the result of different criteria for selecting minimal diagnoses, as the following example shows:

Example 5. To illustrate the difference between minimum plan diagnosis and maximum informative diagnosis, consider again the plan execution depicted in Figure 3. Given $obs(0)$ and $obs(3)$ and a deviation in the value of v_2 at time $t = 3$, there are three possible minimum diagnoses: $D_1 = \{a_1\}$, $D_2 = \{a_3\}$ and $D_3 = \{a_6\}$. D_2 and D_3 are also maximum-informative diagnoses.   ∎

4   Causes of Plan-Execution Failures

Unlike in classical MBD, minimum diagnosis and maximum-informative diagnosis need not provide the best explanation for the differences between observed effects of a plan execution and the predicted effects. The reason is that often in a plan instances of actions do not fail independently. For example, suppose that we have a plan for carrying luggage from a depot to a number of waiting planes. Such a plan might contain several instances of a drive action pertaining to the same carrier controlled by an agent. Suppose that an instance a_i of some drive action (type) α behaves abnormally because of malfunctioning of the carrier. Then it is reasonable to assume that other instances a_j of the same drive action that occur in the plan *after* a_i can be predicted to behave abnormally, too. Another possibility is that a number of instances of actions is related to the malfunctioning of an *agent* executing several actions in the plan. For example, in the luggage example, the carrier is controlled by a driving agent. If this agent itself is not functioning well, all driving actions as well as loading and unloading actions might be affected.

Such dependencies between action instances and between agent health states and action instances imply that sometimes qualifying an instance of an action as being abnormal implies that other instances of actions must be qualified a being abnormal, too. Minimum and information-maximum diagnosis do not take into account these dependencies between action failures. Therefore, we must take into consideration the underlying *causes* of a plan-execution failure.

4.1 Causal Rules

To be able to include a malfunctioning of an executing agent as a possible cause, we will consider a plan together with its executing agent as the system to be diagnosed. Here, an agent will be simply represented by a set H of specific health states. To identify causes of action failures, we use a set R of *causal rules* in combination with plan diagnosis. A causal rule is a rule that can appear in the following forms:

- $(\alpha_1, \alpha_2, \ldots, \alpha_k) \rightarrow \alpha_{k+1}$, where $k \geq 1$ and for $i = 1, 2 \ldots, k + 1$, $\alpha_i \in \mathcal{A}$ are action types. This type of rule relates the occurrence of a set of failed actions to the occurrence of a failed action implied by them. The intuitive meaning of these rules is that if during plan execution there are, for $i = 1, \ldots, k$, action instances a_i of type α_i that have been qualified as abnormal up to time t, then it is inferred that from time $t + 1$ on all instances of actions of type α_{k+1} will behave abnormally, too.
- $(h; \alpha_1, \alpha_2, \ldots, \alpha_k) \rightarrow \alpha_{k+1}$, where $k \geq 0$, $h \in H$ is a health state $(h \neq nor)$ of the plan executing agent and, for $i = 1, 2 \ldots, k + 1$, $\alpha_i \in \mathcal{A}$ are action types. This type of rule relates the occurrence of an agent abnormality h and a set of action abnormalities occurring at time t to the inference of a failed action at time $t+1$. The intuitive meaning of such a rule is that if during plan execution at some time $t' \leq t + 1$ the agent operates in some abnormal health states h and, for $i = 1, 2, \ldots, k$, there are action instances a_i of type α_i that have been qualified as abnormal up to time t, then it is inferred that from time $t+1$ on all instances of actions of type α_{k+1} that occur in the plan will behave abnormally, too.[5] If $k = 0$, this rule establishes a health state as a single cause for action failure.

The intuitive idea behind a causal diagnosis is to be able to explain a given plan diagnosis Q by a (usually smaller) set of qualifications (causes) Q' together with some health state h of the agent established at time t using the set of causal rules R. Using such a pair consisting of a health state and a qualification should enable us to generate, using the rules in R, a set containing Q.

To define the effect of applying R to a set of (unique) instances of actions occurring in a plan, we first construct the set $inst(R)$ of instance of actions with respect to given plan $P = \langle \mathcal{A}, A, < \rangle$ as follows:

---

[5] We allow abnormal health states to be detected at the same time that abnormal action consequences are generated.

- For every rule r of the form $(\alpha_1, \alpha_2, \ldots, \alpha_k) \rightarrow \alpha_{k+1} \in R$, $inst(R)$ contains an instance $(a_{i_1}, a_{i_2}, \ldots, a_{i_k}) \rightarrow a_{i_{k+1}}$ of r whenever there exists a $t \geq 0$ such that $\{a_{i_1}, a_{i_2}, \ldots, a_{i_k}\} \subseteq P_{\leq t}$ and $a_{i_{k+1}} \in P_{>t}$.
- For every rule r of the form $(h; \alpha_1, \alpha_2, \ldots, \alpha_k) \rightarrow \alpha_{k+1} \in R$, $inst(R)$ contains the instances $(h; a_{i_1}, a_{i_2}, \ldots, a_{i_k}) \rightarrow a_{i_{k+1}}$, whenever there exists a $t \geq 0$ such that $\{a_{i_1}, a_{i_2}, \ldots, a_{i_k}\} \subseteq P_{\leq t}$ and $a_{i_{k+1}} \in P_{>t}$.

For each $r \in inst(R)$, let $ante(r)$ denote the antecedent of r and $hd(r)$ denote the head of r. Furthermore, let $Ab \subseteq \{h\}$ be a set containing an abnormal agent health state h or be equal to the empty set (signifying a normal state of the agent) and let $Q \subseteq A$ be a qualification of instances of actions. We can now define a causal consequence of a qualification Q and a health state Ab using R as follows:

Definition 3. *An instance $a \in A$ is a causal consequence of a qualification $Q \subset A$ and the health state Ab using the causal rules R if*

1. *$a \in Q$ or*
2. *there exists a rule $r \in inst(R)$ such that (i) for each $a_i \in ante(r)$ either a_i is a causal consequence of Q or $a_i \in Ab$, and (ii) $a = hd(r)$.*

The set of causal consequences of Q using R and Ab is denoted by $C_{R,Ab}(Q)$.

We have a simple characterization of the set of causal consequences $C_{R,Ab}(Q)$ of a qualification Q and a health state Ab using a set of causal rules R:

Observation 2. $C_{R,Ab}(Q) = Cn_A(inst(R) \cup Q \cup Ab)$.

Here, $Cn_A(X)$ restricts the set $Cn(X)$ of classical consequences of a set of propositions X to the consequences occurring in A. To avoid cumbersome notation, we will omit the subscripts R and Ab from the operator C and use $C(Q)$ to denote the set of consequences of a qualification Q using a health state Ab and a set of causal rules R. We say that a qualification Q is *closed* under the set of rules R and an agent health state Ab if $Q = C(Q)$, i.e, Q is saturated under application of the rules R.

Proposition 1. *The operator C satisfies the following properties:*

1. *(inclusion): for every $Q \subseteq A$, $Q \subseteq C(Q)$*
2. *(idempotency): for every $Q \subseteq A$, $C(Q) = C(C(Q))$*
3. *(monotony): if $Q \subseteq Q' \subseteq A$ then $C(Q) \subseteq C(Q')$*

Proof. Note that $C(Q) = Cn(inst(R) \cup Q \cup Ab) \cap A$. Hence, monotony and inclusion follow immediately as a consequence of the monotony and inclusion of Cn. Monotony and inclusion imply $C(Q) \subseteq C(C(Q))$. To prove the reverse inclusion, let $Cn^*(Q) = Cn(instr(R) \cup Q \cup Ab)$. Then by inclusion and idempotency of Cn we have $C(C(Q)) = Cn^*(C(Q)) \cap A \subseteq Cn^*(Cn^*(Q)) \cap A = Cn^*(Q) \cap A = C(Q)$. $\square$

Thanks to Proposition 1 we conclude that every qualification can be easily extended to a closed set $C(Q)$ of qualifications. Due to the presence of causal rules, we require every diagnosis Q to be closed under the application of rules, that is in the sequel we restrict diagnoses to closed sets $Q = C(Q)$.

We define a causal diagnosis as a qualification Q such that its set of consequences $C(Q)$ constitutes a diagnosis:

Definition 4. *Let* $P = \langle \mathcal{A}, A, < \rangle$ *be a plan, R a set of causal rules and let* $obs(t)$ *and* $obs(t')$ *be two observations with* $t < t'$. *Then a qualification* $Q \subseteq A$ *is a causal diagnosis of* $(P, obs(t), obs(t'))$ *if* $C(Q) \cap P_{[t;t']}$ *is a diagnosis of* $(P, obs(t), obs(t'))$.

Like we defined a minimum diagnosis, we now define two kinds of minimum causal diagnoses: a minimum causal *set* diagnosis and a minimum causal *effect* diagnosis:

Definition 5. *Let* $P = \langle \mathcal{A}, A, < \rangle$ *be a plan and* $obs(t)$ *and* $obs(t')$ *with* $t < t'$ *be two observations.*

1. *A minimum causal set diagnosis is a causal diagnosis* Q *such that* $|Q| \leq |Q'|$ *for every causal diagnosis* Q' *of* P;
2. *A minimum causal effect diagnosis is a causal diagnosis* Q *such that* $|C(Q)| \leq |C(Q')|$ *for every causal diagnosis* Q'.

Maximum informative causal set and maximum informative causal effect diagnoses are defined completely analogous to the previous definitions using standard diagnosis.

The relationships between the different diagnostic concepts we have distinguished is partially summarized in the following proposition:

Proposition 2. *Let* $P = \langle \mathcal{A}, A, < \rangle$ *be a plan and* $obs(t)$ *and* $obs(t')$ *with* $t < t'$ *be two observations.*

1. $|Q| \leq |Q'|$ *for every minimum causal set diagnosis* Q *and minimum closed diagnosis* Q' *of* P;
2. $|Q| \leq |Q'|$ *for every minimum causal effect diagnosis* Q *and minimum closed diagnosis* Q' *of* P

Proof. Both properties follow immediately from the definitions and the inclusion property of C.   □

4.2   Causal Diagnoses and Prediction

Except for playing a role in establishing causal *explanations* of observations, (causal) diagnoses also can play a significant role in the *prediction* of future results (states) of the plan or even the attainability of the goals of the plan. First of all, we should realize that a diagnosis can be used to enhance observed state information as follows: Suppose that Q is a causal diagnosis of a plan P based on the observations $obs(t)$ and $obs(t')$ for some $t < t'$, let $obs(t) \rightarrow^*_{C(Q);P} (\pi'_Q, t')$ and let $obs(t') = (\pi', t')$. Since $C(Q)$ is a diagnosis, π' and π'_Q are compatible states. Hence, we can combine the information contained in both partial states by merging them into a new partial state $\pi'_{\sqcup} = \pi'_Q \sqcup \pi'$. This latter state can be seen as the partial state that can be obtained by direct observation at time t' as well as by making use of previous observations at time t and diagnostic information.

In the same way, we can use this information and the causal consequences $C(Q)$ to derive a prediction of the partial states derivable at a time $t'' > t'$:

Definition 6. *Let* Q *be a causal diagnosis of a plan* P *based on the observations* (π, t) *and* (π', t') *where* $t < t'$. *Furthermore, let* $obs(t) \rightarrow^*_{C(Q);P}(\pi'_Q, t')$ *and let* $obs(t') = (\pi', t')$. *Then, for some time* $t'' > t'$, (π'', t'') *is the partial state predicted using* Q *and the observations if* $(\pi'_Q \sqcup \pi', t') \rightarrow^*_{C(Q);P}(\pi'', t'')$.

In particular, if $t'' = depth(P)$, i.e., the plan has been executed completely, we can predict the values of some variables that will result from executing P and we can check which goals $g \in G$ will still be achieved by the execution of the plan, based on our current knowledge. That is, we can check for which goals $g \in G$ it holds that $\tau \models g$. So causal diagnosis might also help in evaluating which goals will be affected by failing actions.

4.3   Complexity Issues

It is well-known that the diagnosis problem is computationally intractable. The decision forms of both consistency-based and abductive based diagnosis are NP-hard ([2]). It is easy to see that standard plan diagnosis has the same order of complexity. Concerning (minimal) causal diagnoses, we can show that they are not more complex than establishing plan diagnoses if the latter problem is NP-hard. The reason is that in every case the verification of Q' being a causal diagnosis is as difficult as verifying a plan diagnosis under the assumption that the set $inst_P(R)$ is polynomially bounded in the size $\|P\|$ of the plan P.[6] Also note that subset minimality (under a set of rules $inst(R)$ of a set of causes can be checked in polynomial time.

5   Related Research

In this section we briefly discuss some other approaches to plan diagnosis. Like we use MBD as a starting point to plan diagnosis, Birnbaum et al. [1] apply MBD to *planning agents* relating health states of agents to *outcomes* of their planning activities, but not taking into account faults that can be attributed to actions occurring in a plan as a separate source of errors. However, instead of focusing upon the relationship between agent properties and outcomes of plan executions, we take a more detailed approach, distinguishing two separate sources of errors (actions and properties of the executing agents) and focusing upon the detection of anomalies during the plan execution. This enables us to predict the outcomes of a plan on beforehand instead of using them only as observations.

de Jonge et al. [6] propose another approach that directly applies model-based diagnosis to plan execution. Their paper focuses on agents each having an individual plan, and where conflicts between these plans may arise (e.g. if they require the same resource). Diagnosis is applied to determine those factors that are accountable for *future* conflicts. The authors, however, do not take into account dependencies between health modes of actions and do not consider agents that collaborate to execute a common plan.

Kalech and Kaminka [9,10] apply *social diagnosis* in order to find the cause of an anomalous plan execution. They consider hierarchical plans consisting of so-called *behaviors*. Such plans do not prescribe a (partial) execution order on a set of actions. Instead, based on its observations and beliefs, each agent chooses the appropriate behavior to be executed. Each behavior in turn may consist of primitive actions to be executed, or of a set of other behaviors to choose from. Social diagnosis then addresses the issue

---

[6] The reason is that computing consequences of Horn-theories can be achieved in a time linear in the size of $inst_P(R)$.

of determining what went wrong in the joint execution of such a plan by identifying the disagreeing agents and the causes for their selection of incompatible behaviors (e.g., belief disagreement, communication errors). This approach might complement our approach when conflicts not only arise as the consequence of faulty actions, but also as the consequence of different selections of sub-plans in a joint plan.

Lesser et al. [3,8] also apply diagnosis to (multi-agent) plans. Their research concentrates on the use of a *causal model* that can help an agent to refine its initial diagnosis of a failing *component* (called a *task*) of a plan. As a consequence of using such a causal model, the agent would be able to generate a new, situation-specific plan that is better suited to pursue its goal. While their approach in its ultimate intentions (establishing anomalies in order to find a suitable plan repair) comes close to our approach, their approach to diagnosis concentrates on specifying the exact causes of the failing of one single *component* (task) of a plan. Diagnosis is based on observations of a component without taking into account the consequences of failures of such a component w.r.t. the remaining plan. In our approach, instead, we are interested in applying MBD-inspired methods to *detect* plan failures. Such failures are based on observations during plan execution and may concern individual components of the plan, but also agent properties. Furthermore, we do not only concentrate on failing components themselves, but also on the consequences of these failures for the future execution of plan elements.

6   Conclusion

We have adapted model-based agent diagnosis to the diagnosis of plans and we have pointed out some differences with the classical approaches to diagnosis. We distinguished two types of diagnosis: minimum plan diagnosis and maximum informative diagnosis to identify (*i*) minimum sets of anomalously executed actions and (*ii*) maximum informative (w.r.t. to predicting the observations) sets of anomalously executed actions. Assuming that a plan is carried out by a single agent, anomalously executed action can be correlated if the anomaly is caused by some malfunctions in the agent. Therefore, (*iii*) causal diagnoses have been introduced and we have extended the diagnostic theory enabling the prediction of future failure of actions.

Current work can be extended in several ways. We mention two possible extensions: First of all, we could improve the diagnostic model of the executing agent. The causal diagnoses are based on the assumption that the agent enters an abnormal state at some time point and stays in that state until the agent is repaired. In our future work we wish to extend the model such that the agent might evolve through several abnormal states. The resulting model will be related diagnosis in Discrete Event Systems [7,11]. Moreover, we intend to investigate plan repair in the context of the agent's current (abnormal) state. Secondly, we would like to extend the diagnostic model with sequential observations and iterative diagnoses. Here, we would like to consider the possibilities of diagnosing a plan if more than two subsequent observations are made, the best way to detect errors in such cases and the construction of enhanced prediction methods.

Acknowledgements.   This research is supported by the Technology Foundation STW, applied science division of NWO and the technology programme of the Dutch Ministry of Economic Affairs.

References

1. L. Birnbaum, G. Collins, M. Freed, and B. Krulwich. Model-based diagnosis of planning failures. In *AAAI 90*, pages 318–323, 1990.
2. T. Bylander, D. Allemang, M. C. Tanner, and J. R. Josephson. The computational complexity of abduction. *Artif. Intell.*, 49(1-3):25–60, 1991.
3. N. Carver and V.R. Lesser. Domain monotonicity and the performance of local solutions strategies for cdps-based distributed sensor interpretation and distributed diagnosis. *Autonomous Agents and Multi-Agent Systems*, 6(1):35–76, 2003.
4. L. Console and P. Torasso. Hypothetical reasoning in causal models. *International Journal of Intelligence Systems*, 5:83–124, 1990.
5. L. Console and P. Torasso. A spectrum of logical definitions of model-based diagnosis. *Computational Intelligence*, 7:133–141, 1991.
6. F. de Jonge and N. Roos. Plan-execution health repair in a multi-agent system. In *PlanSIG 2004*, 2004.
7. R. Debouk, S. Lafortune, and D. Teneketzis. Coordinated decentralized protocols for failure diagnosis of discrete-event systems. *Journal of Discrete Event Dynamical Systems: Theory and Application*, 10:33–86, 2000.
8. B. Horling, B. Benyo, and V. Lesser. Using Self-Diagnosis to Adapt Organizational Structures. In *Proceedings of the 5th International Conference on Autonomous Agents*, pages 529–536. ACM Press, 2001.
9. M. Kalech and G. A. Kaminka. On the design ov social diagnosis algorithms for multi-agent teams. In *IJCAI-03*, pages 370–375, 2003.
10. M. Kalech and G. A. Kaminka. Diagnosing a team of agents: Scaling-up. In *AAMAS 2004*, 2004.
11. Y. Pencolé and M. Cordier. A formal framework for the decentralised diagnosis of large scale discrete event systems and its application to telecommunication networks. *Artif. Intell.*, 164(1-2):121–170, 2005.
12. R. Reiter. A theory of diagnosis from first principles. *Artificial Intelligence*, 32:57–95, 1987.

Automatic Abstraction of Time-Varying System Models for Model Based Diagnosis

Pietro Torasso and Gianluca Torta

Dipartimento di Informatica, Università di Torino, Italy
{torasso, torta}@di.unito.it

Abstract. This paper addresses the problem of automatic abstraction of component variables in the context of the MBD of Time-Varying Systems (i.e. systems where the behavioral modes of components can evolve over time); the main goal is to produce abstract models capable of deriving fewer and more general diagnoses when the current observability of the system is reduced and/or the system operates under specific operating conditions. The notion of indiscriminability among instantiations of a subset of components is introduced and constitutes the basis for a formal definition of abstractions which preserve all the distinctions that are relevant for diagnosis given the current observability and operating conditions of the system. The automatic synthesis of abstract models further restricts abstractions so that the temporal behavior of abstract components can be expressed in terms of a simple combination of the temporal behavior of their subcomponents. As a validation of our proposal, we present the results obtained with the automatic abstraction of a non-trivial model adapted from the spacecraft domain.

1 Introduction

System model abstraction has been successfully exploited in many approaches to model-based diagnosis (MBD). The pioneer work presented in [7] and recent improvements proposed e.g. in [9] and [2] mostly use human-provided abstractions in order to focus the diagnostic process and thus improve its efficiency. However, abstraction has another main benefit, namely to provide fewer and more concise abstract diagnoses when it is not possible to discriminate among detailed diagnoses (e.g. [4], [5]).

Recently, some authors have aimed at the same goal in a different way, namely automatic abstraction of static system models ([8], [11], [10]). The main idea behind these approaches is that, since the system is usually only partially observable, it can be convenient to abstract the domains of system variables ([10]) and possibly system components themselves ([8], [11]) without losing any information relevant for the diagnostic task. By using the abstracted model for diagnosis the number of returned diagnoses can be significantly reduced, and such diagnoses, by being "as abstract as possible", are more understandable for a human.

U. Furbach (Ed.): KI 2005, LNAI 3698, pp. 176–190, 2005.

The work presented in this paper aims at extending the results of [11] to the automatic abstraction of time-varying system models. Such models are able to capture the possible evolutions of the behavioral modes of system components over time; they thus have an intermediate expressive power between static models and fully dynamic models, where arbitrary status variables can be modelled [1]. The goal of automatically synthesizing an abstraction of a time-varying system is very ambitious since the abstraction has to specify not only the relations between the behavioral modes of the abstract component and the ones of its subcomponents, but also to automatically derive the temporal evolutions of the modes of the abstract component.

As the approaches mentioned above, our proposal relies on system observability as one of the two main factors for driving the automatic abstraction process. The other factor is represented by the fact that abstractions are tailored to specific *operating conditions* of the system (e.g. an engine system may be in operating conditions *shut down, warming up, cruising*); as already noted in the context of diagnosability analysis (e.g. [3]), indeed, different operating conditions imply differences in the system behavior so relevant that it is often impossible to draw general conclusions (such as *the system is diagnosable* or *components C and C' can be abstracted*) that hold for all of them.

Our proposal requires that abstractions lead to real simplifications of the system model while preserving all the information relevant for diagnosis: in order to exclude all the undesired abstractions we introduce a precise definition of *abstraction mapping*, and further drive the computation of abstractions through heuristics and enforcement of additional criteria.

The paper is structured as follows. In section 2 we formalize the notions of time-varying system model, diagnostic problem and diagnosis and in section 3 we give a precise definition of abstraction mapping. In section 4 we describe how the declarative notions introduced in 3 can be implemented and discuss correctness and complexity properties of the computational process. In section 5 we present the results obtained by applying our approach to the automatic abstraction of a non-trivial time-varying system model taken from the spacecraft domain. Finally, in section 6 we compare our work to related papers and conclude.

2   Characterising Time-Varying Systems

In order to formally define time-varying system models, we first give the definition of Temporal System Description which captures fully dynamic system models, and then derive the notion of Time-Varying System Description from it.

Definition 1. *A* Temporal System Description *(TSD) is a tuple (SV, DT, Δ) where:*

- *SV is the set of discrete system variables partitioned in INPUTS (system inputs), OPCONDS (operating conditions), STATUS (system status) and*

---

[1] An in-depth analysis of different types of temporal systems in reported in [1] where the main characteristics of time-varying systems are singled out.

INTVARS (internal variables). Distinguished subsets $COMPS \subseteq STATUS$ and $OBS \subseteq INTVARS$ represent system components and observables respectively. We will denote with $DOM(v)$ the finite domain of variable $v \in SV$; in particular, for each $C \in COMPS$, $DOM(C)$ consists of the list of possible behavioral modes *for C (an ok mode and one or more fault modes)*

- *DT (Domain Theory) is an acyclic set of Horn clauses defined over SV representing the* instantaneous *behavior of the system (under normal and abnormal conditions); we require that any instantiation of variables $STATUS \cup OPCONDS \cup INPUTS$ is consistent with DT*

- Δ *is the* System Transition Relation *mapping the values of the variables in $OPCONDS \cup STATUS$ at time* t *to the values of the variables in $STATUS$ at time* t+1 *(such a mapping is usually non deterministic). A generic element $\tau \in \Delta$ (named a* System Transition*) is then a pair $(OPC_t \cup S_t, S_{t+1})$ where OPC_t is an instantiation of $OPCONDS$ variables at time* t *and S_t, S_{t+1} are instantiations of $STATUS$ variables at times* t *and* t+1 *respectively.*

In the following, we will sometimes denote a System Transition $\tau = (OPC_t \cup S_t, S_{t+1})$ as $\tau = S_t \overset{OPC_t}{\to} S_{t+1}$. Given a system status S and an instantiation OPC of the $OPCONDS$ variables, the set of the possible successor states of S given OPC is defined as $\Pi(S, OPC) = \{S' \text{ s.t. } S \overset{OPC}{\to} S' \in \Delta\}$.

Operator Π can be easily extended to apply to a *set* $\mathcal{S} = \{S_1, \ldots, S_m\}$ of system states by taking the union of $\Pi(S_i, OPC)$, $i = 1, \ldots, m$.

Finally, let $\theta = (\tau_0, \ldots, \tau_{w-1})$ be a sequence of system transitions s.t. $\tau_i = S_i \overset{OPC_i}{\to} S_{i+1}$, $i = 0, \ldots, w-1$. We say that θ is a *(feasible) system trajectory*.

The following definition of *Time-Varying System Description* is derived by putting some restrictions on the notion of Temporal System Description.

Definition 2. *A* Time-Varying System Description *(TVSD) is a Temporal System Description TSD = (SV, DT, Δ) such that:*

- $STATUS = COMPS$ *(i.e. the only status variables are the ones representing the health conditions of system components)*
- Δ *can be obtained as $\Delta_1 \bowtie \ldots \bowtie \Delta_n$ where Δ_i (Component Transition Relation) represents the possible evolutions of the behavioral modes of component C_i. A generic element $\tau_i \in \Delta_i$ (Component Transition) is then a pair $(OPC \cup C_{i,t}(bm), C_{i,t+1}(bm'))$ where OPC is an instantiation of $OPCONDS$ variables at time* t *and $C_{i,t}(bm)$, $C_{i,t+1}(bm')$ are instantiations of variable C_i at times* t *and* t+1 *respectively.*

Since Δ can be partitioned in $\Delta_1, \ldots, \Delta_n$ we can easily extend the notion of trajectory to apply to any subsystem Γ involving some system components.

We are now ready to formalize the notions of *Temporal Diagnostic Problem* and *temporal diagnosis*; both of these notions apply to Temporal System Descriptions as well as to Time-Varying System Descriptions.

Definition 3. *A* Temporal Diagnostic Problem *is a tuple TDP = (TSD, S_0, w, σ) where:*

- TSD is a Temporal System Description
- S_0 is the set of possible initial states (i.e. system states at time 0)
- w is an integer representing the size of the time window $[0, \dots, w]$ over which the diagnostic problem is defined
- σ is a sequence $(\mathcal{X}_0, OPC_0, \mathcal{Y}_0, \dots, \mathcal{X}_w, OPC_w, \mathcal{Y}_w)$ where the $\mathcal{X}_i$s, the OPC_is and the $\mathcal{Y}_i$s are instantiations of the INPUTS, OPCONDS and OBS variables respectively at times $0, \dots, w$. Sequence σ then represents the available observed information about the system in time window $[0, \dots, w]$.

Given a temporal diagnostic problem TDP, we say that a system status S is *instantaneously consistent* at time t, $t \in \{0, \dots, w\}$ if:
$$DT \cup \mathcal{X}_t \cup OPC_t \cup \mathcal{Y}_t \cup S \nvdash \perp$$
Instantaneous consistency of S expresses the fact that the instantiation S of $STATUS$ variables is logically consistent with the current values of $INPUTS$, $OPCONDS$ and OBS variables, under constraints imposed by DT.

The following definition formalizes the notions of *belief state* and *temporal diagnosis*.

Definition 4. *Let TDP= (TSD, S_0, w, σ) be a Temporal Diagnostic Problem. We define the* belief state $\mathcal{B}_t$ *at time t (t = 0, \dots, w) recursively as follows:*
- $\mathcal{B}_0 = \{S_0$ s.t. $S_0 \in S_0$ and S_0 is instantaneously consistent at time 0\}
- $\forall t = 1 \dots w$, we have that $\mathcal{B}_t = \{S_t$ s.t. $S_t \in \Pi(\mathcal{B}_{t-1}, OPC_{t-1})$ and
  S_t *is instantaneously consistent at time t\}*
We say that any system status $S_w \in \mathcal{B}_w$ is a (temporal) diagnosis for TDP.

In order for S_t to belong to $\mathcal{B}_t$, then, it must be consistent with the current observed information about the system (i.e. $\mathcal{X}_t$, OPC_t and $\mathcal{Y}_t$) and with the prediction $\Pi(\mathcal{B}_{t-1}, OPC_{t-1})$ based on the past history.

Example 1. As a running example throughout the paper we will consider a very simple hydraulic system involving a pipe P1 connected in series with a valve V which in turn is in series with a pipe P2 (see the left part of figure 1).

The Time Varying System Description $TVSD_H$ of the hydraulic system is built by instantiating generic models of valve and pipe contained in a library of component models (figure 2 reports the domain theory DT and the transition relation Δ for both a valve and a pipe). For $TVSD_H$ we have that: $COMPS = \{P1, V, P2\}$, where P1 and P2 have three behavioral modes (ok, leaking and broken), while the valve V has six behavioral modes (ok, leaking, broken, stuck-open,

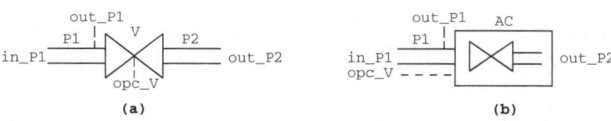

Fig. 1. A fragment of an hydraulic circuit at two levels of abstraction

$V(ok) \land in(f) \land opc(open) \Rightarrow out(f)$	$V(lk) \land in(f) \land opc(open) \Rightarrow out(rf)$
$V(ok) \land in(rf) \land opc(open) \Rightarrow out(rf)$	$V(lk) \land in(rf) \land opc(open) \Rightarrow out(rf)$
$V(ok) \land in(nf) \land opc(open) \Rightarrow out(nf)$	$V(lk) \land in(nf) \land opc(open) \Rightarrow out(nf)$
$V(ok) \land in(*) \land opc(closed) \Rightarrow out(nf)$	$V(lk) \land in(*) \land opc(closed) \Rightarrow out(nf)$
$V(so) \land in(f) \land opc(open) \Rightarrow out(f)$	$V(hb) \land in(f) \land opc(open) \Rightarrow out(rf)$
$V(so) \land in(rf) \land opc(open) \Rightarrow out(rf)$	$V(hb) \land in(rf) \land opc(open) \Rightarrow out(rf)$
$V(so) \land in(nf) \land opc(open) \Rightarrow out(nf)$	$V(hb) \land in(nf) \land opc(open) \Rightarrow out(nf)$
$V(so) \land in(f) \land opc(closed) \Rightarrow out(f)$	$V(hb) \land in(f) \land opc(closed) \Rightarrow out(rf)$
$V(so) \land in(rf) \land opc(closed) \Rightarrow out(rf)$	$V(hb) \land in(rf) \land opc(closed) \Rightarrow out(rf)$
$V(so) \land in(nf) \land opc(closed) \Rightarrow out(nf)$	$V(hb) \land in(nf) \land opc(closed) \Rightarrow out(nf)$
$V(sc) \land in(*) \land opc(open) \Rightarrow out(nf)$	$V(br) \land in(*) \land opc(open) \Rightarrow out(nf)$
$V(sc) \land in(*) \land opc(closed) \Rightarrow out(nf)$	$V(br) \land in(*) \land opc(closed) \Rightarrow out(nf)$

$P(ok) \land in(f) \Rightarrow out(f)$	$P(lk) \land in(f) \Rightarrow out(rf)$
$P(ok) \land in(rf) \Rightarrow out(rf)$	$P(lk) \land in(rf) \Rightarrow out(rf)$
$P(ok) \land in(nf) \Rightarrow out(nf)$	$P(lk) \land in(nf) \Rightarrow out(nf)$
$P(br) \land in(*) \Rightarrow out(nf)$	

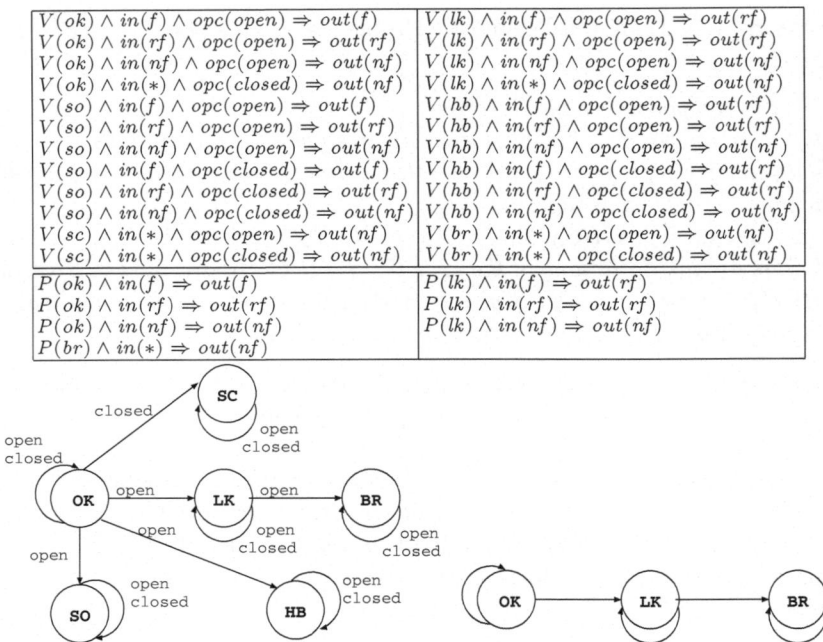

Fig. 2. Domain Theories and Transition Relations of a Valve and a Pipe

stuck-closed and half-blocked); $INPUTS = \{in_{P1}\}$ (i.e. the input of the pipe P1 is the only system input); $OPCONDS = \{opc_V\}$ (i.e. the op. cond. of the valve V specifying whether the valve should be open or closed); $INTVARS = \{out_{P1}, out_V, out_{P2}\}$; $OBS = \{out_{P1}, out_{P2}\}$ (i.e. the output of V is not observable).

Inputs and internal variables represent qualitative values of flows (flow, reduced-flow and no-flow). For lack of space the DT and the Δ of the system are omitted.

3   Abstractions Defined

As pointed out in the introduction, we drive automatic abstraction of time-varying system models mainly by exploiting system observability and by tailoring abstractions to specific operating conditions of the system.

The following definition, which introduces the notion of instantaneous indiscriminability among instantiations of subsets of $COMPS$, is explicitly based on both of the driving factors that we use for abstraction.

Definition 5. *Let Γ be a subsystem consisting of the subset $SCOMPS$ of $COMPS$ and OPC be an instantiation of $OPCONDS$. We say that two instantiations S_Γ, S'_Γ of $SCOMPS$ are OPC-indiscriminable[inst] iff for any instantiation $\mathcal{X}$ of $INPUTS$ and any instantiation $\mathcal{O}$ of $COMPS \backslash SCOMPS$ the following holds:*

$$tc_{DT}(\mathcal{X} \cup OPC \cup \mathcal{O} \cup S_\Gamma)|_{OBS} = tc_{DT}(\mathcal{X} \cup OPC \cup \mathcal{O} \cup S'_\Gamma)|_{OBS}$$

where $tc_{DT}(\mathcal{V})$ is the transitive closure of $\mathcal{V}$ given DT (i.e. the set of variables instantiations logically derivable from $\mathcal{V} \cup DT$) and operator $|$ means projection of a set of instantiations on a subset of the variables.

Note that the $OPC\text{--}indiscriminability^{inst}$ relation induces a partition into $OPC\text{--}indiscriminability^{inst}$ classes of the set of possible instantiations of $SCOMPS$.

Example 2. Let us consider a subsystem Γ of the hydraulic system of figure 1 consisting of V and P2. Let also assume that $OPC = \{opc_V(open)\}$, i.e. the valve has been commanded to be open. According to the notion of instantaneous indiscriminability introduced in definition 5, we have 3 indiscriminability classes for the instantiations of $\{V,P2\}$. In particular, class C1 = $\{$ $V(ok) \wedge P2(ok)$, $V(so) \wedge P2(ok)\}$, class C2 = $\{V(ok) \wedge P2(lk), V(so) \wedge P2(lk), V(hb) \wedge P2(lk),$ $V(hb) \wedge P2(ok), V(lk) \wedge P2(ok), V(lk) \wedge P2(lk)\}$ and class C3 including all the other assignments. It is easy to see that just on the basis of a single time point, the discriminability among different health states of the subsystem is very poor.

We now extend the notion of indiscriminability by taking into account the temporal dimension.

Definition 6. *Let Γ be a subsystem consisting of the subset $SCOMPS$ of $COMPS$ and OPC be an instantiation of $OPCONDS$. We say that two instantiations S_Γ, S'_Γ of $SCOMPS$ are OPC-indiscriminable$_k$ iff:*

- *S_Γ, S'_Γ are OPC-indiscriminable^{inst}*
- *given a trajectory $\theta_\Gamma = (S_{\Gamma,0}, \ldots, S_{\Gamma,k})$ of subsystem Γ s.t. $S_{\Gamma,t} = S_\Gamma$ for some $t \in \{0, \ldots, k\}$ there exists a trajectory $\theta'_\Gamma = (S'_{\Gamma,0}, \ldots, S'_{\Gamma,k})$ s.t. $S'_{\Gamma,t} = S'_\Gamma$ and $S_{\Gamma,i}$, $S'_{\Gamma,i}$ are OPC-indiscriminable^{inst} for $i = 0, \ldots, k$*
- *given a trajectory $\theta'_\Gamma = (S'_{\Gamma,0}, \ldots, S'_{\Gamma,k})$ of subsystem Γ s.t. $S'_{\Gamma,t} = S'_\Gamma$ for some $t \in \{0, \ldots, k\}$ there exists a trajectory $\theta_\Gamma = (S_{\Gamma,0}, \ldots, S_{\Gamma,k})$ s.t. $S_{\Gamma,t} = S_\Gamma$ and $S_{\Gamma,i}$, $S'_{\Gamma,i}$ are OPC-indiscriminable^{inst} for $i = 0, \ldots, k$*

When S_Γ, S'_Γ are *OPC-indiscriminable$_k$* for any k, we say that they are *OPC-indiscriminable$_\infty$*.

If we require that the second and third conditions of the definition above only hold for trajectories where S_Γ, S'_Γ are the first (last) status of the trajectory we obtain a weaker notion of indiscriminability that we denote as *OPC-indiscriminability$_k^{fut}$* (*OPC-indiscriminability$_k^{past}$*). It is possible to prove that the following property holds.

Property 1. Instantiations S_Γ, S'_Γ of $SCOMPS$ are *OPC-indiscriminable$_k$* iff they are both *OPC-indiscriminable$_k^{fut}$* and *OPC-indiscriminable$_k^{past}$*.

Example 3. The indiscriminability classes of the subsystem involving V and P2 change when we take into consideration the temporal dimension. In fact, if we consider $\{V(hb),P2(ok)\}$ and $\{V(lk),P2(lk)\}$, they are instantaneously indiscriminable (both belong to class C2) but are not *OPC-indiscriminable$_1^{fut}$*; in the same way the two assignments {V(lk),P2(br)} and {V(sc),P2(br)} are instantaneously indiscriminable, but are not $OPC\text{-}indiscriminable_1^{past}$, while {V(ok),P2(ok)} and {V(so),P2(ok)} are $OPC\text{-}indiscriminable_\infty$.

The final $OPC\text{-}indiscriminable_\infty$ classes for the considered subsystem {V,P2} when $OPC = \{opc_V(open)\}$ are C1' = {{V(ok),P2(ok)}, {V(so),P2(ok)}}, C2' = {{V(hb),P2(ok)}}, C3' = {{V(ok),P2(lk)}, {V(so),P2(lk)}, {V(hb),P2(lk)}, {V(lk),P2(ok)}, {V(lk),P2(lk)}}, C4' = {{V(ok),P2(br)}, {V(so),P2(br)}, {V(hb), P2(br)}, {V(lk),P2(br)}, {V(br),P2(ok)}, {V(br),P2(lk)}, {V(br), P2(br)}} and C5' = {{V(sc),P2(2ok)}, {V(sc),P2(lk)}, {V(sc),P2(br)}}.

It is easy to see that the temporal dimension reduces the degree of indiscriminability among mode assignment. Nevertheless, the $OPC\text{-}indiscriminable_\infty$ classes are far from being singletons even in this very simple subsystem.

We now introduce the notion of abstraction mapping; as it is common in structural abstractions, abstract components are recursively built bottom-up starting with the primitive components.

Definition 7. *Given a set $COMPS = \{C_1, \ldots, C_n\}$ of component variables and an instantiation OPC of OPCONDS, a components abstraction mapping $\mathcal{AM}$ of COMPS defines a set $COMPS_A = \{AC_1, \ldots, AC_m\}$ of discrete variables (abstract components) and associates with each $AC_i \in COMPS_A$ one or more $C_j \in COMPS$ (subcomponents of AC_i) s.t. each component in COMPS is the subcomponent of exactly one abstract component. Moreover, $\mathcal{AM}$ associates with each abstract component AC a definition def_{AC}, which is a characterization of the behavioral modes of AC in terms of the behavioral modes of its subcomponents. More precisely, an abstract component and its definition are built hierarchically as follows:*

- *if $C \in COMPS$, AC is a simple abstract component if its definition def_{AC} associates with each $abm \in DOM(AC)$ a formula $def_{abm} = C(bm_1) \vee \ldots \vee C(bm_k)$ s.t. $bm_1, \ldots, bm_k \in DOM(C)$; in the trivial case, AC has the same domain as C and $\forall bm \in DOM(C) : def_{bm} = C(bm)$*
- *Let AC', AC'' be two abstract components with disjoint sets of subcomponents $SCOMPS', SCOMPS'':AC$ is an abstract component with subcomponents $SCOMPS' \cup SCOMPS''$ if def_{AC} associates with each $abm \in DOM(AC)$ a definition def_{abm} which is a DNF s.t. each disjunct is in the form $def_{abm'} \wedge def_{abm''}$ where $abm' \in DOM(AC')$ and $abm'' \in DOM(AC'')$*

We require that, given the abstract component AC, its definition def_{AC} satisfies the following conditions:

1. *correctness: given $def_{abm} \in def_{AC}$, the set of instantiations of SCOMPS which satisfy def_{abm} is an OPC-indiscriminability$_\infty$ class*
2. *mutual exclusion: for any two distinct $def_{abm}, def_{abm'} \in def_{AC}$, and any instantiation **COMPS** of COMPS: **COMPS** $\cup \{def_{abm} \wedge def_{abm'}\} \vdash \perp$*
3. *completeness: for any instantiation **COMPS** of COMPS: **COMPS** $\vdash def_{abm_i}$ for some $i \in \{1, \ldots, |DOM(AC)|\}$*

The definition def_{AC} of AC thus specifies a relation between instantiations of the subcomponents of AC and instantiations (i.e. behavioral modes) of AC itself.

Example 4. Let us consider again the subsystem involving $\{V, P2\}$ and OPC $= opc_V(open)$. A *components abstraction mapping* $\mathcal{AM}$ may define a new abstract component AC whose subcomponents are V and P2 and whose behavioral modes are $\{abm1, \ldots, abm5\}$ (the behavioral modes correspond to the five $OPC\text{-}indiscriminable_\infty$ classes C1', ..., C5' reported above). In particular, the definition of the behavioral modes of AC are the following

- $def_{abm1} = V(ok) \wedge P2(ok) \vee V(so) \wedge P2(ok)$
- $def_{abm2} = V(hb) \wedge P2(ok)$
- $def_{abm3} = V(ok) \wedge P2(lk) \vee V(so) \wedge P2(lk) \vee V(hb) \wedge P2(lk) \vee V(lk) \wedge P2(ok) \vee$ $V(lk) \wedge P2(lk)$
- $def_{abm4} = V(ok) \wedge P2(br) \vee V(so) \wedge P2(br) \vee V(hb) \wedge P2(br) \vee V(lk) \wedge P2(br) \vee$ $V(br) \wedge P2(ok) \vee V(br) \wedge P2(lk) \vee V(br) \wedge P2(br)$
- $def_{abm5} = V(sc) \wedge P2(ok) \vee V(sc) \wedge P2(lk) \vee V(sc) \wedge P2(br)$

It is easy to verify that the definitions of abstract behavioral modes abm1, ..., abm5 satisfy the properties of correctness, mutual exclusion and completeness required by definition 7.

4   Computing Abstractions

The hierarchical way abstract components are defined in section 3 suggests that, after an initial step in which simple abstract components are built, the computational process can produce new abstract components incrementally, by merging only two components at each iteration. After some finite number of iterations, arbitrarily complex abstract components can be produced.

As already mentioned in section 1, however, our purpose is to produce abstractions that are not merely correct, but are also useful and meaningful. In the context of MBD, this essentially comes down to producing abstract components with a limited number of different behavioral modes. A proliferation of behavioral modes in the abstract components, indeed, has negative effects on both the efficiency of diagnosis and the understandability of abstract diagnoses. Given that the maximum number of behavioral modes of an abstract component AC built from AC', AC'' is clearly $|DOM(AC')| \cdot |DOM(AC'')|$, we chose to impose the limit $|DOM(AC)| \leq |DOM(AC')| + |DOM(AC'')|$; we call this condition the *limited domain condition*.

Figure 3 reports an high-level description of the abstraction process. The body of the `ForEach` loop at the beginning of the algorithm has the purpose of building a simple abstract component AC by possibly merging indiscriminable behavioral modes of the basic component C while taking into consideration the operating conditions of C. This step is performed by invoking `BuildSimpleAC()` and the system model is updated in order to replace C with AC.

The update of the Domain Theory DT_{temp} is performed by `ReviseDTSimple()` which essentially replaces the set of formulas DT_C where C occurs with formulas

Function Abstract($TVSD_D$, OPC)
    $\mathcal{AM} := \emptyset$
    $TVSD_{temp} := TVSD_D$
    ForEach $C \in COMPS_{temp}$
        $\langle AC, def_{AC} \rangle$ = BuildSimpleAC(C, OPC, DT_C, Δ_C)
        $COMPS_{temp} := COMPS_{temp} \setminus \{C\} \cup \{AC\}$
        DT_{temp} := ReviseDTSimple(C, DT_{temp}, DT_C, $\langle AC, def_{AC} \rangle$)
        Δ_{temp} := ReviseDeltaSimple(C, Δ_{temp}, Δ_C, $\langle AC, def_{AC} \rangle$)
        $\mathcal{AM} := \mathcal{AM} \cup \{\langle AC, def_{AC} \rangle\}$
    Loop
    $\mathcal{G}$:= BuildInfluenceGraph(DT_A)
    While (Oracle($TVSD_A$, $\mathcal{G}$) = $\langle C_i, C_j \rangle$)
        (DT_{loc}, $\mathcal{G}_{loc}$) := CutDT(C_i, C_j, $TVSD_{temp}$, $\mathcal{G}$)
        If (BuildAC(C_i, C_j, OPC, DT_{loc}, $\mathcal{G}_{loc}$, Δ_{C_i}, Δ_{C_j}) = $\langle AC, def_{AC} \rangle$)
            $COMPS_{temp} := COMPS_{temp} \setminus \{C_i, C_j\} \cup \{AC\}$
            DT_{temp} := ReviseDT(C_i, C_j, DT_{temp}, DT_{loc}, $\mathcal{G}_{loc}$, $\langle AC, def_{AC} \rangle$)
            Δ_{temp} := ReviseDelta(C_i, C_j, Δ_{temp}, Δ_{C_i}, Δ_{C_j}, $\langle AC, def_{AC} \rangle$)
            $\mathcal{G}$:= BuildInfluenceGraph(DT_{temp})
            $\mathcal{AM} := \mathcal{AM} \cup \{\langle AC, def_{AC} \rangle\}$
        EndIf
    Loop
    $TVSD_A := TVSD_{temp}$
    Return $\mathcal{AM}$, $TVSD_A$
EndFunction

Fig. 3. Sketch of the `Abstract()` function

that mention the new abstract component AC. Similarly, the System Transition Relation Δ_{temp} is updated by function `ReviseDeltaSimple()` which builds the Component Transition Relation Δ_{AC} of AC and replaces Δ_C with Δ_{AC}. Finally, the definition of AC is added to the abstraction mapping $\mathcal{AM}$.

Once the simple abstract components have been built, the algorithm enters the `While` loop for incrementally building complex abstract components. Function `Oracle()` selects the next two candidates C_i, C_j for abstraction according to strategies that will be discussed below; `Oracle()` uses a DAG $\mathcal{G}$ (automatically built by `BuildInfluenceGraph()`) which represents the influences among system variables: the variables appearing in the antecedent of a formula $\phi \in DT$ become parents of the variable occurring in the consequent of ϕ.

The call to function `CutDT()` has the purpose of isolating the portion DT_{loc} of DT_{temp} relevant for computing the influence of C_i, C_j on the values of observables. `CutDT()` also returns the Influence Graph $\mathcal{G}_{loc}$ associated with DT_{loc}.

Function `BuildAC()` then tries to compute the definition of a new abstract component AC starting from C_i and C_j. Unlike `BuildSimpleAC()`, `BuildAC()` may fail when AC turns out to have a number of behavioral modes larger than $|DOM(C_i)| + |DOM(C_j)|$. In case `BuildAC()` succeeds, the system model is updated in order to replace C_i and C_j with AC. As for the case of simple abstract components, the Domain Theory and the System Transition Relation are updated (by functions `ReviseDT()` and `ReviseDelta()` respectively).

Then, the definition of AC is added to the abstraction mapping $\mathcal{AM}$.

The whole process terminates when function `Oracle()` has no new suggestions for further abstractions.

The Oracle. Function `Oracle()` chooses, at each iteration, a pair of candidate components for abstraction. Since a search over the entire space of potential abstractions would be prohibitive, the function is based on a greedy heuristic that tries to achieve a good overall abstraction without doing any backtracking. In our experience it turned out that by following two simple principles significant results could be achieved.

First, `Oracle()` follows a *locality principle* for choosing C_i, C_j. This principle is often used when structural abstractions are built by human experts, in fact these abstractions tend to consist in hierarchies such that at each level of the hierarchy the new abstractions involve components that are structurally close to one another. The *locality principle* usually has the advantage of building abstract components that have a limited and meaningful set of behavioral modes. However, since there is no guarantee on the number of behavioral modes of the abstract component, the check of the *limited domain condition* must be deferred to a later time, in particular when function `BuildAC()` is called; there's thus no warranty that the components selected by `Oracle()` will end up being merged but just a good chance of it to happen. Two good examples of structural patterns that follow this principle are components connected in series and in parallel [2].

The second principle followed by `Oracle()` consists of preferring pairs of components that are connected to the same observables; the rationale behind this principle is that if at least one of the two components is observable separately from the other, it is more unlikely to find indiscriminable instantiations of the two components.

In the current implementation, `Oracle()` terminates when it can't find any pair of components which are connected in series or parallel and influence the same set of observables.

Building Abstract Components. Figure 4 shows the pseudo-code of function `BuildSimpleAC()` which has the purpose of building a simple abstract component. First of all, $DOM(C)$ is partitioned into $OPC\text{-}indiscriminability^{inst}$ classes by function `FindIndiscriminableInst()`. The function is essentially an algorithm for computing static indiscriminability classes as the one described in [11]; the only difference is that the values of $OPCONDS$ variables are constrained to match parameter OPC, so that in general it is possible to merge more behavioral modes in the same indiscriminability class.

The call to `FindIndiscriminableFut()` has the purpose of partitioning $DOM(C)$ into $OPC\text{-}indiscriminability^{fut}$ classes. Function `FindIndiscriminableFut()` starts with the $OPC\text{-}indiscriminability^{inst}$ classes of partition $\mathcal{P}_{inst}$ and gradually refines them into $OPC\text{-}indiscriminability_1^{fut}$ classes, $OPC\text{-}indiscriminability_2^{fut}$ classes, etc. until a fixed point is reached.

---

[2] Note that the structural notions of vicinity, seriality and parallelism can be naturally transposed in terms of relationships in the influence graph $\mathcal{G}$.

Function BuildSimpleAC(C, OPC, DT_C, Δ_C)
    $\mathcal{P}^{inst}$ = FindIndiscriminableInst(C, OPC, DT_C)
    $\mathcal{P}^{fut}$ = FindIndiscriminableFut(C, OPC, Δ_C, $\mathcal{P}_{inst}$)
    $\mathcal{P}^{past}$ = FindIndiscriminablePast(C, OPC, Δ_C, $\mathcal{P}_{inst}$)
    $\mathcal{P} = \mathcal{P}^{fut} \circ \mathcal{P}^{past}$
    AC = NewComponentName()
    $def_{AC} = \emptyset$
    ForEach $\pi \in \mathcal{P}$
        abm = NewBehavioralModeName()
        $def_{abm} = \bigvee_{bm \in \pi} C(bm)$
        $def_{AC} = def_{AC} \cup \{(abm, def_{abm})\}$
    Loop
    Return $\langle AC, def_{AC} \rangle$
EndFunction

Fig. 4. Sketch of the `BuildSimpleAC()` function

It is important to note that partition $\mathcal{P}^{inst}$ is used by `FindIndiscriminable` `Fut()` at each iteration, since it is required by definition 6 that all the states on trajectories starting from OPC-$indiscriminability^{fut}$ states are OPC-$indiscriminable^{inst}$. Function `FindIndiscriminablePast()` is similar to `FindIndiscriminableFut()`, except that it partitions $DOM(C)$ into OPC-$indiscriminability^{past}$ classes by considering backwards transitions on Δ_C.

The partition $\mathcal{P}$ into OPC-$indiscriminability_\infty$ classes is computed by merging $\mathcal{P}^{fut}$ and $\mathcal{P}^{past}$ according to property 1.

The classes of $\mathcal{P}$ form the basis for building abstract behavioral modes definitions: associating exactly one abstract behavioral mode to each of them guarantees that the mutual exclusion, completeness and correctness conditions given in definition 7 are automatically satisfied. For this reason, the definitions of the abstract behavioral modes are generated by considering an indiscriminability class π at a time and building a behavioral mode definition for π.

We do not report the algorithm for function `BuildAC()` because of lack of space. The function exhibits only a few differences w.r.t. `BuildSimpleAC()`: first of all, partitions $\mathcal{P}^{inst}$, $\mathcal{P}^{fut}$, $\mathcal{P}^{past}$ and $\mathcal{P}$ are computed on $DOM(C_i) \times DOM(C_j)$ instead of $DOM(C)$; second, since the number of classes in the partition $\mathcal{P}$ corresponds to the number of abstract behavioral modes to be eventually generated, the function fails (and returns `null`) if $|\mathcal{P}|$ is larger than $|DOM(C_i)| + |DOM(C_j)|$; finally, the generated abstract behavioral modes are DNFs of terms in the form $C_i(bm_i) \wedge C_j(bm_j)$.

Example 5. Let us consider now the abstractions performed by the algorithm on the $TVSD_H$ system when the operating condition of the valve V is open. The first step performed by `Abstract()` is the creation of simple abstract modes for each component P1, V and P2. It is worth noting that the model of P1 and P2 does not change since the behavior of the pipe is not influenced by any operating condition. On the contrary, the simple abstract model of V is different from the

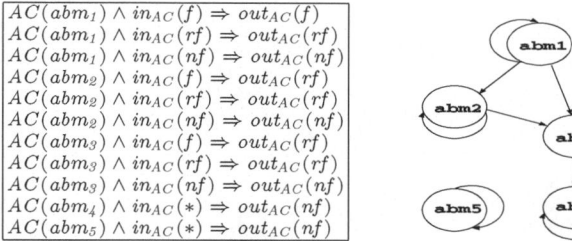

$$
\begin{array}{l}
AC(abm_1) \wedge in_{AC}(f) \Rightarrow out_{AC}(f) \\
AC(abm_1) \wedge in_{AC}(rf) \Rightarrow out_{AC}(rf) \\
AC(abm_1) \wedge in_{AC}(nf) \Rightarrow out_{AC}(nf) \\
AC(abm_2) \wedge in_{AC}(f) \Rightarrow out_{AC}(rf) \\
AC(abm_2) \wedge in_{AC}(rf) \Rightarrow out_{AC}(rf) \\
AC(abm_2) \wedge in_{AC}(nf) \Rightarrow out_{AC}(nf) \\
AC(abm_3) \wedge in_{AC}(f) \Rightarrow out_{AC}(rf) \\
AC(abm_3) \wedge in_{AC}(rf) \Rightarrow out_{AC}(rf) \\
AC(abm_3) \wedge in_{AC}(nf) \Rightarrow out_{AC}(nf) \\
AC(abm_4) \wedge in_{AC}(*) \Rightarrow out_{AC}(nf) \\
AC(abm_5) \wedge in_{AC}(*) \Rightarrow out_{AC}(nf)
\end{array}
$$

Fig. 5. Model of the Abstract Component built from V and P2

detailed model of V, in particular the transition relation changes because of the influence of the operating condition; however the simple abstract model of V has six modes (despite the restriction on the operating condition, all the original behavioral modes of the valve are discriminable over time).

Now Oracle() has to select two candidate components for abstraction. It selects V and P2 because they are connected in series and no observable exists that is reachable from just one of them (this is not the case for P1 and V since out_{P1} is observable).

When BuildAC() is invoked on components V and P2, the abstraction is successful since the abstract component involves just 5 abstract behavioral modes (the limited domain condition is satisfied). The inferred definition of the abstract component is exactly the same as the one reported in example 4. The DT and Δ of the abstract component are reported in figure 5. No attempt is made to create an abstract component starting from P1 and AC because of the heuristics used by Oracle() (i.e. out_{P1} is observable)

Correctness and Complexity. We now state some results concerning algorithm Abstract() [3]. The first result just states that the algorithm builds abstractions according to the definition given in section 3 while the second one makes explicit the correspondence between detailed and abstract diagnoses.

Property 2. The $\mathcal{AM}$ returned by Abstract() is a components abstraction mapping according to definition 7.

Property 3. Let $\mathcal{AM}$ and $TVSD_A$ be the components abstraction mapping and system description obtained from $TVSD_D$ and OPC by applying Abstract(). Given a Temporal Diagnostic Problem $TDP_D = (TVSD_D, \mathcal{S}_0, w, \sigma)$ and the corresponding abstract Temporal Diagnostic Problem $TDP_A = (TVSD_A, \mathcal{S}_{A,0}, w, \sigma)$, D_D is a diagnosis for TDP_D iff its abstraction D_A according to $\mathcal{AM}$[4] is a diagnosis for TDP_A.

---

[3] The proofs are not reported because of lack of space.
[4] Each instantiation **COMPS** of $COMPS$ corresponds to exactly one instantiation **COMPS$_A$** of $COMPS_A$; we say that **COMPS$_A$** is the *abstraction* of **COMPS** according to $\mathcal{AM}$.

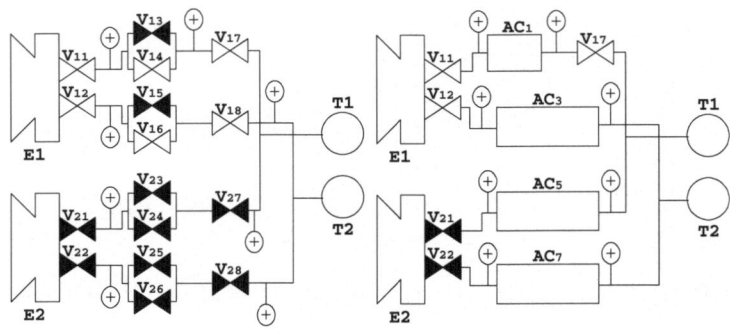

Fig. 6. The Propulsion Sub-System (left) and its abstraction (right)

As concerns the computational complexity of `Abstract()`, it turns out that it is polynomial in the number $|\mathcal{SV}|$ of the System Variables in $TVSD_D$; however, some of the operations in the `While` loop can be exponential in $|\mathcal{SN}_{loc}|$, where $\mathcal{SN}_{loc}$ denotes the set of all source nodes in $\mathcal{G}_{loc}$. This exponential complexity is taken under control by the *locality principle* used by `Oracle()`; indeed, when C_i and C_j are close, $\mathcal{G}_{loc}$ tends to be small (this is particularly true when C_i, C_j are connected in series or parallel).

5   Applying Abstraction to a Propulsion Sub-system

In order to show the applicability of our approach to realistic domains, we report the main results of the abstraction on the model of a part of a propulsion system of a spacecraft (derived from [6]). In figure 6 (left part) the detailed model is reported where V_{ij} represent valves whose Domain Theory and Transition Relation are the same as the ones reported in figure 2 while T_i represent tanks and E_i engines. Apart from the outputs of the two engines, the observable parameters are marked with $\oplus$ symbols.

The only system components that have operating conditions are the valves. We analyze the case where V_{11}, V_{12}, V_{14}, V_{16}, V_{17} and V_{18} are open, while all the other valves (depicted in black) are closed. For sake of simplicity, pipes are not modeled as components (i.e. we assume they cannot fail).

Figure 6 (right part) schematically shows the abstractions performed by our algorithm. In particular, abstract component AC_1 is the abstraction of the sub-system involving V_{13} and V_{14} connected in parallel; abstract component AC_3 is the abstraction of a subsystem involving V_{18} and the abstract component AC_2 that has been derived by abstracting V_{15} and V_{16}; AC_5 is the abstraction of a subsystem involving V_{23}, V_{24} and V_{27} (in an intermediate step of the algorithm V_{23} and V_{24} have been abstracted in AC_4); similar remarks hold for AC_7.

The model of the propulsion system is interesting for analyzing the behavior of `Oracle()` in selecting the components to be abstracted. As said above the current version of `Oracle()` exploits heuristics for selecting components that are connected in parallel or in series. By following this criterion Oracle() identifies that V_{13} and V_{14} are connected in parallel, so they are suitable candidates for abstraction. The abstraction is successful and returns AC_1 which has just 9 behavioral modes (out of 36 possible behavioral modes that are obtained by the Cartesian product of the modes of the two valves), its Δ involves 27 state transitions (out of the 289 obtained as the Cartesian product of the Δs of V_{13} and V_{14} or the 70 transitions if we take into account the restrictions imposed by the operating conditions). After the update of the system model, the Oracle() is able to detect that V_{17} is connected in series with AC_1 and therefore they could be candidates for abstraction. However, since the output of V_{17} (and input of AC_1) is observable, Oracle() does not suggest the abstraction of V_{17} and AC_1.

Oracle() continues its search and the next two selected components are V_{15} and V_{16} which are successfully abstracted in AC_2. Then, Oracle() selects V_{18} and AC_2, since they are connected in series and no observable parameter exists in between the two; this step leads to the construction of AC_3. Abstract components AC_5 and AC_7 are built in a similar way; the main difference is that the involved valves have different operating conditions. In all these cases abstraction provides a significant reduction in the number of behavioral modes (15 for AC_3, 3 for AC_5 and AC_7, just a fraction of the possible 216 behavioral modes), in the number of transitions of Δ (for AC_3 36 versus the 4913 possible transitions or 700 transition if we take into account operating conditions), in the number of propositional formulas in DT (in case of AC_3 54 instead of 108). Even more dramatic reductions are obtained with AC_5 and AC_7 whose Δs involve just 3 transitions and whose DTs involve 9 formulas.

6   Conclusions

So far, only a few methods have been proposed for automatic model abstraction in the context of MBD and all of them deal with static system models. The work described in [10] aims at automatic abstraction of the domains of system variables given a desired level of abstraction τ_{targ} for the domains of some variables and the system observability τ_{obs}. In [11] the authors consider the automatic abstraction of both the domains of system variables and of the system components, given the system observability expressed as the set of observables OBS_{AV} and their granularity Π. Both of these approaches guarantee that the abstract models they build preserve all the relevant diagnostic information, i.e. there is no loss of discrimination power when using the abstract model so that the ground model can potentially be substituted by the abstract one. This is different from the goal of [7] and its improvements, where the abstract model is viewed as a focusing mechanism, and is used in conjunction with the detailed model in order to improve efficiency.

As [10] and [11], the approach described in this paper guarantees that the information relevant for diagnosis is preserved, as shown by the possibility of converting back and forth between detailed and abstract diagnoses (property 3).

However, our approach applies to a significantly broader class of systems, namely time-varying systems; for this reason, we tailor our abstractions not only to the system observability, but also to specific operating conditions of the system.

The abstraction is clearly connected with the diagnosability assessment task ([3]), since both of them strongly depend on the indiscriminability of elements of the model. Note however that the two tasks are actually dual problems: diagnosability assessment is based on finding elements that are indiscriminable in *at least* one situation, while abstraction is based on finding elements that are indiscriminable in *all* situations; moreover, while diagnosability is a system wide property that either holds or does not hold, a system can be abstracted in several different ways, so abstraction can be viewed more as a creative process whose goal is to simplify the system model in a useful and meaningful way.

The application of our approach to the propulsion sub-system has shown its ability at automatically synthesizing abstract components with complex behavior. Note that these results have been obtained under very demanding requirements: the algorithm had to identify combinations of behavioral modes indiscriminable regardless of the input vectors over unlimited time windows.

The problem of the abstraction of fully dynamic system models is an even more challenging problem, since it requires not only the abstraction of component variables but also of other system status variables that interact with component variables in determining the temporal evolution of the system. We believe that this paper has laid solid ground for attacking this problem.

References

1. Brusoni, V., Console, L., Terenziani, P., Theseider Dupré, D.: A Spectrum of Definitions for Temporal Model-Based Diagnosis. Artificial Intelligence **102** (1998) 39–79
2. Chittaro, L., Ranon, R.: Hierarchical Model-Based Diagnosis Based on Structural Abstraction. Artificial Intelligence **155**(1–2) (2004) 147–182
3. Cimatti, A., Pecheur, C., Cavada, R.: Formal Verification of Diagnosability via Symbolic Model Checking. Proc. IJCAI (2003) 363–369
4. Console, L., Theseider Dupré, D.: Abductive Reasoning with Abstraction Axioms. LNCS **810** (1994) 98–112
5. Friedrich, G.: Theory Diagnoses: A Concise Characterization of Faulty Systems. Proc. IJCAI (1993) 1466–1471
6. Kurien, J., Nayak, P. Pandurang: Back to the Future for Consistency-Based Trajectory Tracking. Proc. AAAI00 (2000) 370–377
7. Mozetič, I.: Hierarchical Model-Based Diagnosis. Int. Journal of Man-Machine Studies **35**(3) (1991) 329–362
8. Out, D.J., van Rikxoort, R., Bakker, R.: On the Construction of Hierarchic Models. Annals of Mathematics and AI **11** (1994) 283–296
9. Provan, G.: Hierarchical Model-Based Diagnosis. Proc. DX (2001) 167–174
10. Sachenbacher, M., Struss, P.: Task-Dependent Qualitative Domain Abstraction. Artificial Intelligence **162** (1–2) (2004) 121–143
11. Torta, G., Torasso, P.: Automatic Abstraction in Component-Based Diagnosis Driven by System Observability. Proc. IJCAI (2003) 394–400

Neuro-Fuzzy Kolmogorov's Network for Time Series Prediction and Pattern Classification

Yevgeniy Bodyanskiy[1], Vitaliy Kolodyazhniy[1], and Peter Otto[2]

[1] Control Systems Research Laboratory,
Kharkiv National University of Radioelectronics,
14, Lenin Av., Kharkiv, 61166, Ukraine
bodya@kture.kharkov.ua, kolodyazhniy@ukr.net
[2] Department of Informatics and Automation,
Technical University of Ilmenau,
PF 100565, 98684 Ilmenau, Germany
peter.otto@tu-ilmenau.de

Abstract. In the paper, a novel *Neuro-Fuzzy Kolmogorov's Network* (NFKN) is considered. The NFKN is based on and is the development of the previously proposed neural and fuzzy systems using the famous Kolmogorov's superposition theorem (KST). The network consists of two layers of neo-fuzzy neurons (NFNs) and is linear in both the hidden and output layer parameters, so it can be trained with very fast and simple procedures: the gradient-descent based learning rule for the hidden layer, and the recursive least squares algorithm for the output layer. The validity of theoretical results and the advantages of the NFKN are confirmed by experiments.

1 Introduction

According to the Kolmogorov's superposition theorem (KST) [1], *any* continuous function of d variables can be *exactly* represented by superposition of continuous functions of one variable and addition:

$$f(x_1,...,x_d) = \sum_{l=1}^{2d+1} g_l \left[\sum_{i=1}^{d} \psi_{l,i}(x_i) \right], \tag{1}$$

where $g_l(\bullet)$ and $\psi_{l,i}(\bullet)$ are some continuous univariate functions, and $\psi_{l,i}(\bullet)$ are independent of f. Aside from the exact representation, the KST can be used as the basis for the construction of parsimonious universal approximators, and has thus attracted the attention of many researchers in the field of soft computing.

Hecht-Nielsen was the first to propose a neural network implementation of KST [2], but did not consider how such a network can be constructed. Computational aspects of approximate version of KST were studied by Sprecher [3], [4] and Kůrková [5]. Igelnik and Parikh [6] proposed the use of spline functions for the construction of Kolmogorov's approximation. Yam *et al* [7] proposed the multi-resolution approach to fuzzy control, based on the KST, and proved that the KST representation can be

U. Furbach (Ed.): KI 2005, LNAI 3698, pp. 191–202, 2005.

realized by a two-stage rule base, but did not demonstrate how such a rule base could be created from data. Lopez-Gomez and Hirota developed the Fuzzy Functional Link Network (FFLN) [8] based on the fuzzy extension of the Kolmogorov's theorem. The FFLN is trained via fuzzy delta rule, whose convergence can be quite slow. A novel KST-based universal approximator called Fuzzy Kolmogorov's Network (FKN) with simple structure and training procedure with high rate of convergence was proposed in [9 – 11]. The training of the FKN is based on the alternating linear least squares technique for both the output and hidden layers. However, this training algorithm may require a large number of computations in the problems of high dimensionality.

In this paper we propose an efficient and computationally simple learning algorithm, whose complexity depends linearly on the dimensionality of the input space. The network considered in this paper (*Neuro-Fuzzy Kolmogorov's Network* – NFKN) is based on the FKN architecture and uses the gradient descent-based procedure for the tuning of the hidden layer weights, and recursive least squares method for the output layer.

2 Network Architecture

The NFKN (Fig. 1) is comprised of two layers of neo-fuzzy neurons (NFNs, Fig. 2) [12] and is described by the following equations:

$$\hat{f}(x_1,\ldots,x_d) = \sum_{l=1}^{n} f_l^{[2]}(o^{[1,l]}), \quad o^{[1,l]} = \sum_{i=1}^{d} f_i^{[1,l]}(x_i), \quad l = 1,\ldots,n, \tag{2}$$

where n is the number of hidden layer neurons, $f_l^{[2]}(o^{[1,l]})$ is the l-th nonlinear synapse in the output layer, $o^{[1,l]}$ is the output of the l-th NFN in the hidden layer, $f_i^{[1,l]}(x_i)$ is the i-th nonlinear synapse of the l-th NFN in the hidden layer.

The equations for the hidden and output layer synapses are

$$f_i^{[1,l]}(x_i) = \sum_{h=1}^{m_1} \mu_{i,h}^{[1]}(x_i) w_{i,h}^{[1,l]}, \quad f_l^{[2]}(o^{[1,l]}) = \sum_{j=1}^{m_2} \mu_{l,j}^{[2]}(o^{[1,l]}) w_{l,j}^{[2]}, \tag{3}$$

$$l = 1,\ldots,n, \quad i = 1,\ldots,d,$$

where m_1 and m_2 is the number of membership functions (MFs) per input in the hidden and output layers respectively, $\mu_{i,h}^{[1]}(x_i)$ and $\mu_{l,j}^{[2]}(o^{[1,l]})$ are the MFs, $w_{i,h}^{[1,l]}$ and $w_{l,j}^{[2]}$ are tunable weights.

Nonlinear synapse is a single input-single output fuzzy inference system with crisp consequents, and is thus a universal approximator [13] of univariate functions (see Fig. 3). It can provide a piecewise-linear approximation of any functions $g_l(\bullet)$ and $\psi_{l,i}(\bullet)$ in (1). So the NFKN, in turn, can approximate any function $f(x_1,\ldots,x_d)$.

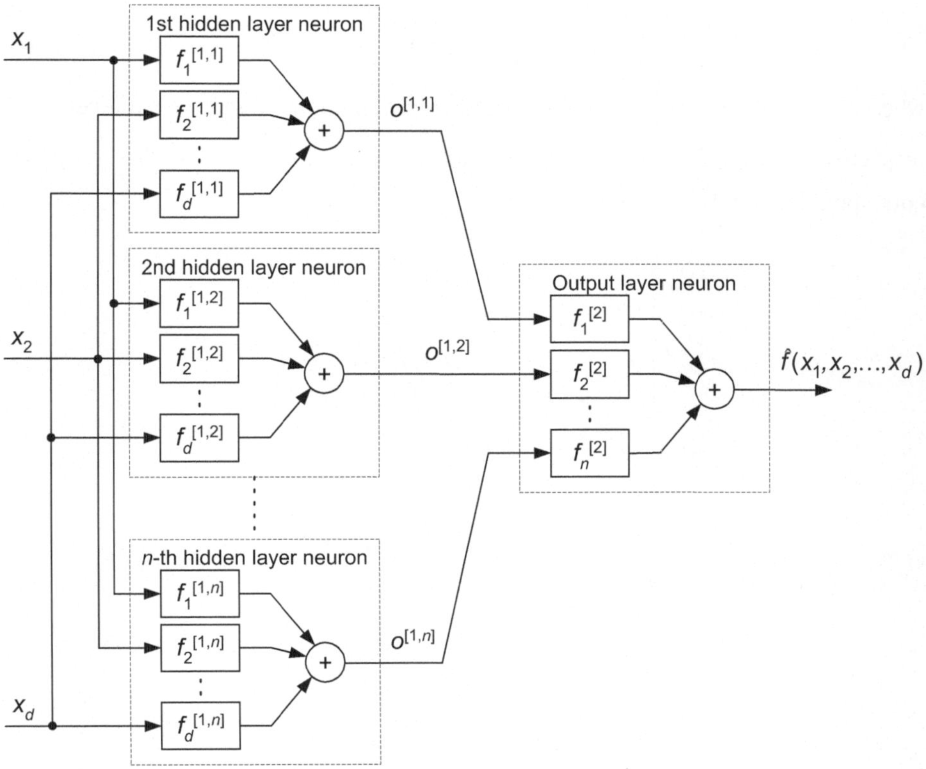

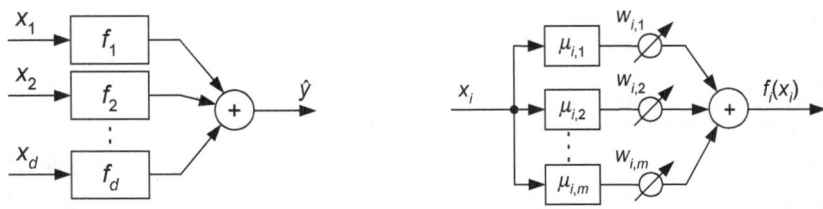

Fig. 1. NFKN architecture with d inputs and n hidden layer neurons

Fig. 2. Neo-fuzzy neuron (*left*) and its nonlinear synapse (*right*)

The output of the NFKN is computed as the result of two-stage fuzzy inference:

$$\hat{y} = \sum_{l=1}^{n} \sum_{j=1}^{m_2} \mu_{l,j}^{[2]} \left[\sum_{i=1}^{d} \sum_{h=1}^{m_1} \mu_{i,h}^{[1]}(x_i) w_{i,h}^{[1,l]} \right] w_{l,j}^{[2]} . \tag{4}$$

The description (4) corresponds to the following two-level fuzzy rule base:

$$\text{IF } x_i \text{ IS } X_{i,h} \text{ THEN } o^{[1,1]} = w_{i,h}^{[1,1]} d \text{ AND}\dots\text{AND } o^{[1,n]} = w_{i,h}^{[1,n]} d ,$$
$$i = 1,\dots,d , \quad h = 1,\dots,m_1 , \tag{5}$$

$$\text{IF } o^{[1,l]} \text{ IS } O_{l,j} \text{ THEN } \hat{y} = w_{l,j}^{[2]} n \,, \quad l = 1,\ldots,n \,, \quad j = 1,\ldots,m_2 \,, \tag{6}$$

where $X_{i,h}$ and $O_{l,j}$ are the antecedent fuzzy sets in the first and second level rules, respectively. Each first level rule contains n consequent terms $w_{i,h}^{[1,1]} d,\ldots,w_{i,h}^{[1,n]} d$, corresponding to n hidden layer neurons.

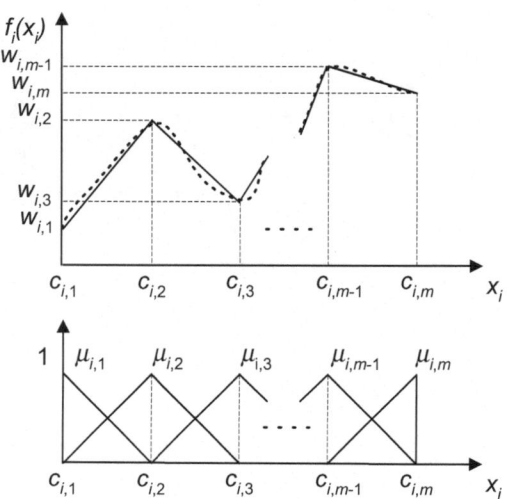

Fig. 3. Approximation of a univariate nonlinear function by a nonlinear synapse

Total number of rules is

$$N_R^{FKN} = d \cdot m_1 + n \cdot m_2 \,, \tag{7}$$

i.e., it depends *linearly* on the number of inputs d.

The rule base is complete, as the fuzzy sets $X_{i,h}$ in (5) completely cover the input hyperbox with m_1 MFs per input variable. Due to the linear dependence (7), this approach is feasible for input spaces with high dimensionality d without the need for clustering techniques for the construction of the rule base. Straightforward grid-partitioning approach with m_1 MFs per input requires $(m_1)^d$ fuzzy rules, which results in combinatorial explosion and is practically not feasible for $d > 4$.

To simplify computations and to improve the transparency of fuzzy rules, MFs of each input are shared between all the neurons in the hidden layer (Fig. 4). So the set of rules (5), (6) in an NFKN can be interpreted as a grid-partitioned fuzzy rule-base

$$\text{IF } x_1 \text{ IS } X_{1,1} \text{ AND} \ldots \text{AND } x_d \text{ IS } X_{d,1} \text{ THEN } \hat{y} = \hat{f}(c_{1,1}^{[1]},\ldots,c_{d,1}^{[1]}),$$

$$\vdots \tag{8}$$

$$\text{IF } x_1 \text{ IS } X_{1,m_1} \text{ AND} \ldots \text{AND } x_d \text{ IS } X_{d,m_1} \text{ THEN } \hat{y} = \hat{f}(c_{1,m_1}^{[1]},\ldots,c_{d,m_1}^{[1]}),$$

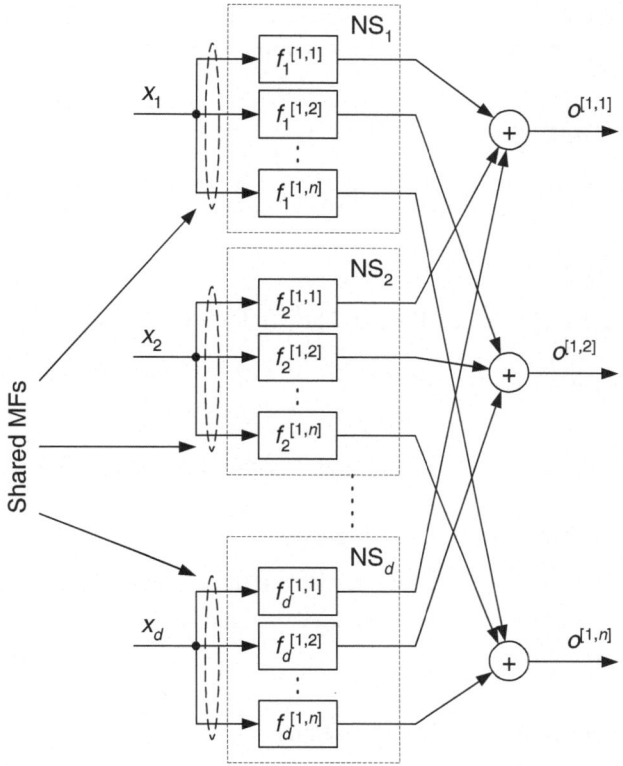

Fig. 4. Representation of the hidden layer of the NFKN with shared MFs

with total of $(m_1)^d$ rules, whose consequent values are equal to the outputs of the NFKN computed on all the possible d-tuples of prototypes of the input fuzzy sets $c_{1,1}^{[1]}, \ldots, c_{d,m_1}^{[1]}$ [9].

3 Learning Algorithm

The weights of the NFKN are determined by means of a batch-training algorithm as described below. A training set containing N samples is used. The minimized error function is

$$E(t) = \sum_{k=1}^{N} [y(k) - \hat{y}(t,k)]^2 = [Y - \hat{Y}(t)]^T [Y - \hat{Y}(t)], \tag{9}$$

where $Y = [y(1), \ldots, y(N)]^T$ is the vector of target values, and
$\hat{Y}(t) = [\hat{y}(t,1), \ldots, \hat{y}(t, N)]^T$ is the vector of network outputs at epoch t.

Since the nonlinear synapses (3) are linear in parameters, we can employ recursive least squares (RLS) procedure for the estimation of the output layer weights. Re-write (4) as

$$
\hat{y} = W^{[2]^T} \varphi^{[2]}(o^{[1]}), \quad W^{[2]} = \left[w_{1,1}^{[2]}, w_{1,2}^{[2]}, \ldots, w_{n,m_2}^{[2]} \right]^T,
$$
$$
\varphi^{[2]}(o^{[1]}) = \left[\mu_{1,1}^{[2]}(o^{[1,1]}), \mu_{1,2}^{[2]}(o^{[1,1]}), \ldots, \mu_{n,m_2}^{[2]}(o^{[1,n]}) \right]^T.
$$
(10)

Then the RLS procedure will be

$$
\begin{cases}
W^{[2]}(t,k) = W^{[2]}(t,k-1) + P(t,k)\varphi^{[2]}(t,k)(y(k) - \hat{y}(t,k)), \\
P(t,k) = P(t,k-1) - \dfrac{P(t,k-1)\varphi^{[2]}(t,k)\varphi^{[2]^T}(t,k)P(t,k-1)}{1 + \varphi^{[2]^T}(t,k)P(t,k-1)\varphi^{[2]}(t,k)},
\end{cases}
$$
(11)

where $k = 1, \ldots, N$, $P(t,0) = 10000 \cdot I$, and I is the identity matrix of corresponding dimension.

Introducing after that the regressor matrix of the hidden layer $\Phi^{[1]} = \left[\varphi^{[1]}(x(1)), \ldots, \varphi^{[1]}(x(N)) \right]^T$, we can obtain the expression for the gradient of the error function with respect to the hidden layer weights at the epoch t:

$$
\nabla_{W^{[1]}} E(t) = -\Phi^{[1]^T} \left[Y - \hat{Y}(t) \right],
$$
(12)

and then use the well-known gradient-based technique to update these weights:

$$
W^{[1]}(t+1) = W^{[1]}(t) - \gamma(t) \frac{\nabla_{W^{[1]}} E(t)}{\left\| \nabla_{W^{[1]}} E(t) \right\|},
$$
(13)

where $\gamma(t)$ is the adjustable learning rate, and

$$
W^{[1]} = \left[w_{1,1}^{[1,1]}, w_{1,2}^{[1,1]}, \ldots, w_{d,m_1}^{[1,1]}, \ldots, w_{d,m_1}^{[1,n]} \right]^T,
$$
$$
\varphi^{[1]}(x) = \left[\varphi_{1,1}^{[1,1]}(x_1), \varphi_{1,2}^{[1,1]}(x_1), \ldots, \varphi_{d,m_1}^{[1,1]}(x_d), \ldots, \varphi_{d,m_1}^{[1,n]}(x_d) \right]^T,
$$
$$
\varphi_{i,h}^{[1,l]}(x_i) = a_l^{[2]}(o^{[1,l]}) \mu_{i,h}^{[1,l]}(x_i),
$$
(14)

and $a_l^{[2]}(o^{[1,l]})$ is determined as in [9 – 11]

$$
a_l^{[2]}(o^{[1,l]}) = \frac{w_{l,p+1}^{[2]} - w_{l,p}^{[2]}}{c_{l,p+1}^{[2]} - c_{l,p}^{[2]}},
$$
(15)

where $w_{l,p}^{[2]}$ and $c_{l,p}^{[2]}$ are the weight and center of the p-th MF in the l-th synapse of the output layer, respectively. The MFs in an NFN are chosen such that only two adjacent MFs p and $p+1$ fire at a time [12].

Thus, the NFKN is trained via a two-stage optimization procedure without any nonlinear operations, similar to the ANFIS learning rule for the Sugeno-type fuzzy inference systems [14]. In the forward pass, the output layer weights are adjusted. In the backward pass, adjusted are the hidden layer weights. An epoch of training is considered 'successful' if the error measure (e.g. the root mean squared error (RMSE)) on the training set is reduced in comparison with the previous epoch. Only successful epochs are counted. If the error measure is not reduced, the training cycle (forward and backward passes) is repeated until the error measure is reduced or the maximum number of cycles per epoch is reached. Once it is reached, the algorithm is considered to converge, and the parameters from the last successful epoch are saved as the result of training. Otherwise the algorithm is stopped when the maximum number of epochs is reached.

The number of tuned parameters in the hidden layer is $S_1 = d \cdot m_1 \cdot n$, in the output layer $S_2 = n \cdot m_2$, and total $S = S_1 + S_2 = n \cdot (d \cdot m_1 + m_2)$.

Hidden layer weights are initialized deterministically using the formula [9 – 11]

$$w_{h,i}^{[1,l]} = \exp\left\{ -\frac{i\left[m_1(l-1)+h-1\right]}{d(m_1 n-1)} \right\}, h=1,\ldots,m_1, i=1,\ldots,d, l=1,\ldots,n, \qquad (16)$$

broadly similar to the parameter initialization technique proposed in [6] for the Kolmogorov's spline network based on rationally independent random numbers.

Further improvement of the learning algorithm can be achieved through the adaptive choice of the learning rate $\gamma(t)$ as was proposed for the ANFIS.

4 Experimental Results

To verify the theoretical results and compare the performance of the proposed network to the known approaches, we have carried out three experiments. The first two were prediction and emulation of the Mackey-Glass time series [15], and the third one was classification of the Wisconsin breast cancer (WBC) data [16]. In all the experiments, the learning rate was $\gamma(t) = 0.05$, maximum number of cycles per epoch was 50. In the first and second experiments, the error measure was root mean squared error, in the third one – the number of misclassified patterns.

The Mackey-Glass time series was generated by the following equation:

$$\frac{dy(t)}{dt} = \frac{0.2\, y(t-\tau)}{1+ y^{10}(t-\tau)} - 0.1\, y(t), \qquad (17)$$

where τ is time delay. In our experiments, we used $\tau = 17$. The values of the time series (17) at each integer point were obtained by means of the fourth-order Runge-Kutta method. The time step used in the method was 0.1, initial condition $y(0) = 1.2$, and $y(t)$ was derived for t = 0,...,5000.

4.1 Mackey-Glass Time Series Prediction

In the first experiment, the values $y(t-18), y(t-12), y(t-6)$, and $y(t)$ were used to predict $y(t+85)$. From the generated data, 1000 values for $t = 118,...,1117$ were used as the training set, and the next 500 for $t = 1118,...,1617$ as the checking set. The NFKN used for prediction had 4 inputs, 9 neurons in the hidden layer with 3 MFs per input, and 1 neuron in the output layer with 9 MFs per synapse (189 adjustable parameters altogether). The training algorithm converged after 43 epochs.

The NFKN demonstrated similar performance as multilayer perceptron (MLP) with 2 hidden layers each containing 10 neurons. The MLP was trained for 50 epochs with the Levenberg-Marquardt algorithm. Because the MLP weights are initialized randomly, the training and prediction were repeated 10 times.

Root mean squared error on the training and checking sets (trnRMSE and chkRMSE) was used to estimate the accuracy of predictions. For the MLP, median values of 10 runs for the training and checking errors were calculated. The results are shown in Fig. 5 and Table 1.

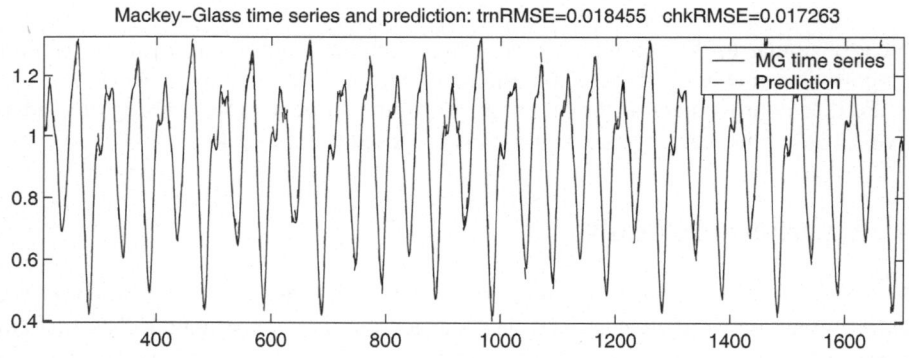

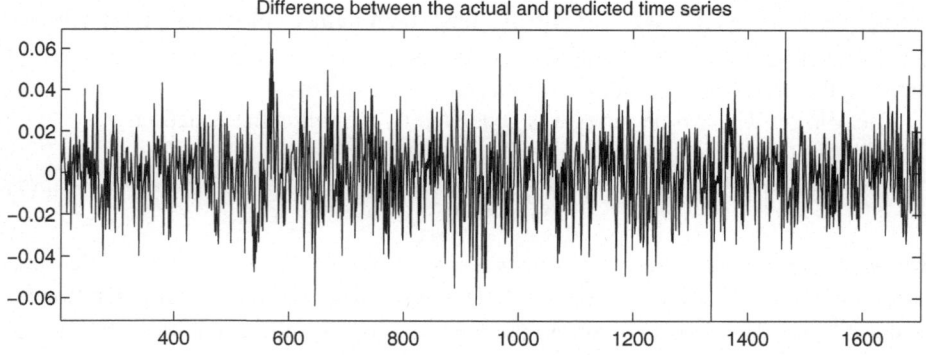

Fig. 5. Mackey-Glass time series prediction

Table 1. Results of Mackey-Glass time series prediction

Network	Parameters	Epochs	trnRMSE	chkRMSE
MLP 10-10-1	171	50	0.022	0.0197
NFKN 9-1	189	43	0.018455	0.017263

With a performance similar to that of the MLP in combination with the Levenberg-Marquardt training method, the NFKN with its hybrid learning algorithm requires much less computations because no matrix inversions for the tuning of the synaptic weights are required.

4.2 Mackey-Glass Time Series Emulation

The second experiment consisted in the emulation of the Mackey-Glass time series by the NFKN. The values of the time series were fed into the NFKN only during the training stage. 3000 values for $t = 118,...,3117$ were used as the training set. The NFKN had 17 inputs corresponding to the delays from 1 to 17, 5 neurons in the hidden layer with 5 MFs per input, and 1 neuron in the output layer with 7 MFs per synapse (total 460 adjustable parameters). The NFKN was trained to predict the value of the time series one step ahead.

The training procedure converged after 28 epochs with the final value of $RMSE_{TRN}=0.0027$, and the last 17 values of the time series from the training set were fed to the inputs of the NFKN. Then the output of the network was connected to its inputs through the delay lines, and subsequent 1000 values of the NFKN output were computed. As can be seen from Fig. 6, the NFKN captured the dynamics of the real time series very well. The difference between the real and emulated time series becomes visible only after about 500 time steps. The emulated chaotic oscillations remain stable, and neither fade out nor diverge. In such a way, the NFKN can be used for long-term chaotic time series predictions.

4.3 Classification of the Wisconsin Breast Cancer Data

In this experiment, the NFKN was tested on the known WBC dataset [16], which consists of 699 cases (*malignant* or *benign*) with 9 features. We used only 683 cases, because 16 had missing values. The dataset was randomly shuffled, and then divided into 10 approximately equally-sized parts to perform 10-fold cross-validation. The NFKN that demonstrated best results had 9 inputs, only 1 neuron in the hidden layer with 4 MFs per input, and 1 synapse in the output layer with 7 MFs (total 43 adjustable weights). It provided high classification accuracy (see Table 2).

Typical error rates the on the WBC data with the application of other classifiers range from 2% to 5% [17, 18]. But it is important to note that most neuro-fuzzy systems rely on time-consuming learning procedures, which fine-tune the membership functions [17, 19, 20]. In the NFKN, no tuning of MFs is performed, but the classification results are very good.

Table 2. Results of the WBC data classification

Run	Epochs	Training set errors	Checking set errors
1	6	1.7915%	2.8986%
2	5	1.9544%	2.8986%
3	6	2.1173%	0%
4	12	0.97561%	4.4118%
5	8	1.9512%	0%
6	7	1.7886%	2.9412%
7	6	1.7886%	2.9412%
8	6	1.7886%	2.9412%
9	9	1.4634%	2.9412%
10	8	1.626%	2.9412%
Min	5	0.97561%	0%
Max	12	2.1173%	4.4118%
Average	7.3	1.7245%	2.4915%

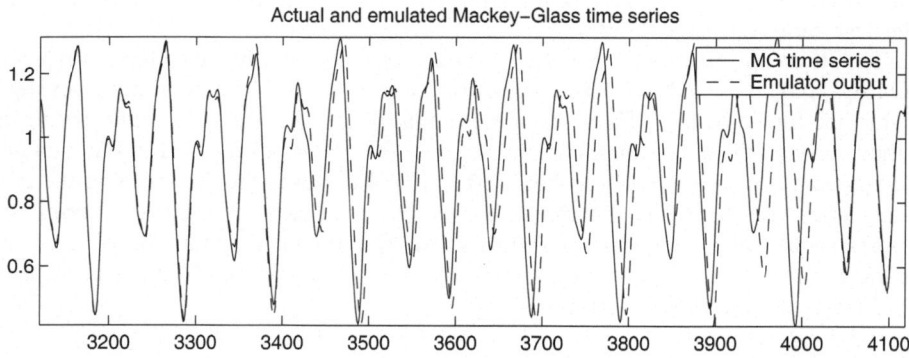

Actual and emulated Mackey–Glass time series

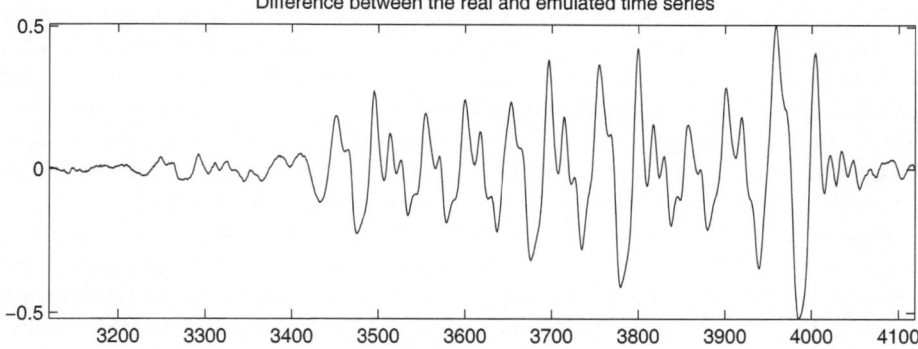

Difference between the real and emulated time series

Fig. 6. Mackey-Glass time series emulation

5 Conclusion

A new simple and efficient neuro-fuzzy network (NFKN) and its training algorithm were considered. The NFKN contains the neo-fuzzy neurons in both the hidden and output layers and is not affected by the curse of dimensionality because of its two-level structure. The use of the neo-fuzzy neurons enabled us to develop fast training procedures for all the parameters in the NFKN.

Two-level structure of the rule base helps the NFKN avoid the combinatorial explosion in the number of rules even with a large number of inputs (17 in the experiment with chaotic time series emulation).

Further work will be directed to the improvement of the convergence properties of the training algorithm by means of the adaptation of the learning rate, and to the development of a multiple output modification of the NFKN for the classification problems with more than two classes.

We expect that the proposed neuro-fuzzy network can find real-world applications, because the experimental results presented in the paper are quite promising.

References

1. Kolmogorov, A.N.: On the representation of continuous functions of many variables by superposition of continuous functions of one variable and addition. Dokl. Akad. Nauk SSSR **114** (1957) 953-956
2. Hecht-Nielsen, R: Kolmogorov's mapping neural network existence theorem. Proc. IEEE Int. Conf. on Neural Networks, San Diego, CA, Vol. 3 (1987) 11-14
3. Sprecher, D.A.: A numerical implementation of Kolmogorov's superpositions. Neural Networks **9** (1996) 765-772
4. Sprecher, D.A.: A numerical implementation of Kolmogorov's superpositions II. Neural Networks **10** (1997) 447–457
5. Kůrková, V.: Kolmogorov's theorem is relevant. Neural Computation 3 (1991) 617-622
6. Igelnik, B., and Parikh, N.: Kolmogorov's spline network. IEEE Transactions on Neural Networks **14** (2003) 725-733
7. Yam, Y., Nguyen, H. T., and Kreinovich, V.: Multi-resolution techniques in the rules-based intelligent control systems: a universal approximation result. Proc. 14th IEEE Int. Symp. on Intelligent Control/Intelligent Systems and Semiotics ISIC/ISAS'99, Cambridge, Massachusetts, September 15-17 (1999) 213-218
8. Lopez-Gomez, A., Yoshida, S., and Hirota, K.: Fuzzy functional link network and its application to the representation of the extended Kolmogorov theorem. International Journal of Fuzzy Systems **4** (2002) 690-695
9. Kolodyazhniy, V., and Bodyanskiy, Ye.: Fuzzy Neural Network with Kolmogorov's Structure. Proc. 11th East-West Fuzzy Colloquium, Zittau, Germany (2004) 139-146
10. Kolodyazhniy, V., and Bodyanskiy, Ye.: Fuzzy Kolmogorov's Network. Proc. 8th Int. Conf. on Knowledge-Based Intelligent Information and Engineering Systems (KES 2004), Wellington, New Zealand, September 20-25, Part II (2004) 764-771
11. Kolodyazhniy, V., Bodyanskiy, Ye., and Otto, P.: Universal Approximator Employing Neo-Fuzzy Neurons. Proc. 8th Fuzzy Days, Dortmund, Germany, Sep. 29 – Oct. 1 (2004) CD-ROM

12. Yamakawa, T., Uchino, E., Miki, T., and Kusanagi, H.: A neo fuzzy neuron and its applications to system identification and prediction of the system behavior. Proc. 2nd Int. Conf. on Fuzzy Logic and Neural Networks "IIZUKA-92", Iizuka, Japan (1992) 477-483
13. Kosko, B.: Fuzzy systems as universal approximators. Proc. 1st IEEE Int. Conf. on Fuzzy Systems, San Diego, CA (1992) 1153-1162
14. Jang, J.-S. R.: Neuro-Fuzzy Modeling: Architectures, Analyses and Applications. PhD Thesis. Department of Electrical Engineering and Computer Science, University of California, Berkeley (1992)
15. Mackey, M. C., and Glass, L.: Oscillation and chaos in physiological control systems. Science **197** (1977) 287-289
16. UCI Machine Learning Repository, http://www.ics.uci.edu/~mlearn/MLRepository.html
17. Nauck, D., Nauck, U., and Kruse, R.: Generating classification rules with the neuro-fuzzy system NEFCLASS. Proc. NAFIPS'96, Berkeley (1996)
18. Duch, W.: Datasets used for classification: comparison of results. Computational Intelligence Laboratory, Nicolaus Copernicus University, Torun, Poland.
http://www.phys.uni.torun.pl/kmk/projects/datasets.html
19. Sun, C.-T., and Jang, J.-S. R.: Adaptive network based fuzzy classification, Proc. of the Japan-U.S.A. Symposium on Flexible Automation (1992)
20. Sun, C.-T., and Jang, J.-S. R.: A neuro-fuzzy classifier and its applications. Proc. IEEE Int. Conf. on Fuzzy Systems, San Francisco (1993)

Design of Optimal Power Distribution Networks Using Multiobjective Genetic Algorithm

Alireza Hadi and Farzan Rashidi

Azad University of Tehran-South Branch, Tehran, Iran
f.rashidi@ece.ut.ac.ir

Abstract. This paper presents solution of optimal power distribution networks problem of large-sized power systems via a genetic algorithm of real type. The objective is to find out the best power distribution network reliability while simultaneously minimizing the system expansion costs. To save an important CPU time, the constraints are to be decomposing into active constraints and passives ones. The active constraints are the parameters which enter directly in the cost function and the passives constraints are affecting the cost function indirectly as soft limits. The proposed methodology is tested for real distribution systems with dimensions that are significantly larger than the ones frequently found in the literature. Simulation results show that by this method, an optimum solution can be given quickly. Analyses indicate that proposed method is effective for large-scale power systems. Further, the developed model is easily applicable to n objectives without increasing the conceptual complexity of the corresponding algorithm and can be useful for very large-scale power system.

1 Introduction

Investments in power distribution systems constitute a significant part of the utilities' expenses. They can account for up to 60 percent of capital budget and 20 percent of operating costs [1]. For this reason, efficient planning tools are needed to allow planners to reduce costs. Computer optimization algorithms improve cost reduction compared to system design by hand. Therefore, in recent years, a lot of mathematical models and algorithms have been developed. The optimal power distribution system design has been frequently solved by using classical optimization methods during the last two decades for single stage or multistage design problems. The basic problem has been usually considered as the minimization of an objective function representing the global power system expansion costs in order to solve the optimal sizing and/or locating problems for the feeder and/or substations of the distribution system [2]. Some mathematical programming techniques, such as Mix-Integral Program(MIP)[3,4], Branch and Bound [5], Dynamic Program(DP)[6], Quadratic MIP, AI/Expert system approach[7,8], Branch Exchange[9], etc., were extensively used to solve this problem. Some of the above models considered a simple linear objective function to represent the actual non-linear expansion costs, what leads to unsatisfactory representations of the real costs. Mixed-integer models achieved a

U. Furbach (Ed.): KI 2005, LNAI 3698, pp. 203–215, 2005.
© Springer-Verlag Berlin Heidelberg 2005

better description of such costs by including fixed costs associated to integer 0-1 variables, and linear variable costs associated to continuous variables [10]. The most frequent optimization techniques to solve these models have been the branch and bound ones. These kinds of methods need large amounts of computer time to achieve the optimal solution, especially when the number of integer 0-1 variables grows for large power distribution systems [11]. Some models such as DP contained the non-linear variable expansion costs and used dynamic programming that requires excessive CPU time for large distribution systems [12]. AI/Expert based methods propose a distribution planning heuristic method, that is faster than classical ones but the obtained solutions are sometimes local optimal solutions [13]. This model takes into account linearized costs of the feeders, that is, a linear approximation of the non-linear variable costs.

Genetic Algorithms (GAs) have solved industrial optimization problems in recent years, achieving suitable results. Genetic algorithms have been applied to the optimal multistage distribution network and to the street lighting network design [14, 15]. In these works, a binary codification was used, but this codification makes difficult to include some relevant design aspects in the mathematical models. Moreover, most papers about optimal distribution design do not include practical examples of a true multi objective optimization of real distribution networks of significant dimensions (except in [16]), achieving with detail the set of optimal multi objective non dominated solutions [17, 18] (true simultaneous optimization of the costs and the reliability).

In this paper, because the total objective function is very non-linear, a genetic algorithm with real coding of control parameters is chosen in order to well taking into account this nonlinearity and to converge rapidly to the global optimum. To accelerate the processes of the optimization, the controllable variables are decomposed to active constraints and passive constraints. The active constraints, which influence directly the cost function, are included in the GA process. The passive constraints, which affect indirectly this function, are maintained in their soft limits using a conventional constraint, only, one time after the convergence of GA. Novel operators of the genetic algorithm is developed, that allow for getting out of local optimal solutions and obtaining the global optimal solution or solutions very close to such optimal one.

This algorithm considers the true non-linear variable cost. It also obtains the distribution nodes voltages and an index that gives a measure of the distribution system reliability. The proposed algorithm allows for more flexibility, taking into account several design aspects as, for example, different conductor sizes, different sizes of substations, and suitable building of additional feeders for improving the network reliability. This methodology is tested for real distribution systems with dimensions that are significantly larger than the ones frequently found in the literature. Simulation results show that by this method, an optimum solution can be given quickly. Further analyses indicate that this method is effective for large-scale power systems. CPU times can be reduced by decomposing the optimization constraints of the power system to active constraints manipulated directly by the genetic algorithm, and passive constraints maintained in their soft limits using a conventional constraint power distribution network.

2   Power Distribution Problems

The problem that we want to solve is a simplified form that it disposes a series of sources (substations) and a series of drains or demand nodes (demand centers). To each demand node it is associated a determined demand of boosts and each source has a maximum limit of the power supply. Moreover, it is known the possible routes for the construction of electric line to transport the powers from the sources to the demand nodes are known. Each line possesses a cost that depends principally on its length (basic fixed costs) and the power value that transports (fundamentally variable costs). The model contains an objective function that represents the costs corresponding to the lines and substations of the electric distribution system. Multiobjective optimization can be defined as the problem of finding [10]:

A vector of decision variables, which satisfies constraints and optimizes a vector function whose elements represent the objective functions. These functions form a mathematical description of performance criteria, which are usually in conflict with each other. Hence, the term "optimizes" means finding such a solution, which would give the values of all the objective functions acceptable to the designer.

The standard power distribution problem can be written in the following form,

$$
\begin{aligned}
&\text{Minimize } f(x) && \text{(the objective function)}\\
&\text{subject to : } h(x) = 0 && \text{(equality constraints)} && (1)\\
&\text{and } g(x) \le 0 && \text{(inequality constraints)}
\end{aligned}
$$

Where $x = [x_1, x_2, ..., x_n]$ is the vector of the control variables, that is those which can be varied (set of routes, set of nodes, power flow, fixed cost of a feeder to be built and etc.). The essence of the optimal power distribution problem resides in reducing the objective function and simultaneously satisfying the objective function (equality constraints) without violating the inequality constraints.

2.1   Objective Function

The most commonly used objective in the power distribution problem formulation is the optimization of the total economic cost and the reliability of the distribution network. In this paper, the multi objective design model is a nonlinear mixed-integer one for the optimal sizing and location of feeders and substations, which can be used for single stage or for multistage planning under a pseudo dynamic methodology. The vector of objective functions to be minimized is $f(x) = [f_1(x), f_2(x)]$, where $f_1(x)$ is the objective function of the global economic costs and $f_2(x)$ is a function related with the reliability of the distribution network. The objective function $f_1(x)$ can be considered as below [10]:

$$
f_1(x) = \sum_{(i,j)\in N_F}\left\{ (CF_{ij})(Y_{ij}) + (CV_{ij})(X_{ij})^2 \right\} + \sum_{k\in N_S}\left\{ (CF_k)(Y_k) + (CV_k)(X_k)^2 \right\} \tag{2}
$$

Where:

N_F : is the set of routes (between nodes) associated with lines in the network.

N_S : is the set of nodes associated with substations in the network.

(i, j): is the Route between nodes i and j.

X_k: is the Power flow, in kVA, supplied from node associated with a substation size.

X_{ij}: is the power flow, in kVA, carried through route associated with a feeder size.

CV_{ij}: is the Variable cost coefficient of a feeder in the network, on route (i, j).

CF_{ij}: is the Fixed cost of a feeder to be built, on route (i, j).

CV_k: is the Variable cost coefficient of a substation in the network, in the node k.

CF_k: is the Fixed cost of a substation to be built, in the node k.

Y_k: is equal to 1, if substation associated with node $k \in N_S$ is built. Otherwise, it is equal to 0.

Y_{ij}: is equal to 1, if feeder associated with route $(i, hj) \subset N_F$ is built. Otherwise, it is equal to 0.

The objective function $f_2(x)$ is:

$$f_2(x) = \sum_{(i,j) \in N_F} (u_{ij})(X_f)_{(i,j)} \tag{3}$$

Where:

N_F: is the set of "proposed" routes (proposed feeders) connecting the network nodes with the proposed substation.

$(X_f)_{(i,j)}$: is the power flow, in kVA, carried through the route $f \in N_F$, that is calculated for a possible failure of an existing feeder (usually in operation) on the route $(i, j) \in N_F$ In this case, reserve feeders are used to supply the power demands.

(u_{ij}): is the constants obtained from other suitable reliability constants in the distribution network, including several reliability related parameters such as failure rates and repair rates for distribution feeders, as well as the length of the corresponding feeders on routes or $(i, j) \in N_F$. The simultaneous minimization of the two objective functions is subject to technical constraints, which are [10]:

1. The first Kirchhoff's law (in the existed nodes of the distribution system).
2. The relative restrictions to the power transport capacitance limits of the lines.
3. The relative restrictions to the power supply capacitance limits of the substations.

The presented model corresponds to a mathematical formulation of non linear mixed-integer programming, which includes the true non linear costs of the lines and substations of the distribution system, to search the optimal solution from an economic viewpoint.

3   Evolutionary Algorithms in Power Distribution Networks

Evolutionary algorithms (EAs) are computer-based problem solving systems, which are computational models of evolutionary processes as key elements in their design and implementation. Genetic Algorithms (GAs) are the most popular and widely used of all evolutionary algorithms. They have been used to solve difficult problems with objective functions that do not possess well properties such as continuity, differentiability, satisfaction of the Lipschitz Condition, etc. These algorithms maintain and manipulate a population of solutions and implement the principle of survival of the fittest in their search to produce better and better approximations to a solution [19]. This provides an implicit as well as explicit parallelism that allows for the exploitation of several promising areas of the solution space at the same time. The implicit parallelism is due to the schema theory developed by Holland, while the explicit parallelism arises from the manipulation of a population of points. The power of the GA comes from the mechanism of evolution, which allows searching through a huge number of possibilities for solutions. The simplicity of the representations and operations in the GA is another feature to make them so popular. There are three major advantages of using genetic algorithms for optimization problems [20].

1. GAs do not involve many mathematical assumptions about the problems to be solved. Due to their evolutionary nature, genetic algorithms will search for solutions without regard for the specific inner structure of the problem. GAs can handle any kind of objective functions and any kind of constraints, linear or nonlinear, defined on discrete, continuous, or mixed search spaces.

2. The ergodicity of evolution operators makes GAs effective at performing global search. The traditional approaches perform local search by a convergent stepwise procedure, which compares the values of nearby points and moves to the relative optimal points. Global optima can be found only if the problem possesses certain convexity properties that essentially guarantee that any local optimum is a global optimum.

3. GAs provide a great flexibility to hybridize with domain-dependent heuristics to make an efficient implementation for a specific problem.

The implementation of GA involves some preparatory stages. Having decoded the chromosome representation (genotype) into the decision variable domain (phenotype), it is possible to assess the performance, or fitness, of individual members of a population [21]. This is done through an objective function that characterises an individual's performance in the problem domain. During the reproduction phase, each individual is assigned a fitness value derived from its raw performance measure given by objective function.

Once the individuals have been assigned a fitness value, they can be chosen from population, with a probability according to their relative fitness, and recombined to produce the next generation. Genetic operators manipulate the genes. The recombination operator is used to exchange genetic information between pairs of individuals. The crossover operation is applied with a probability p_x when the pairs are chosen for breeding. Mutation causes the individual genetic representation to be changed according to some probabilistic rule.

Mutation is generally considered as a background operator that ensures that the probability of searching a particular subspace of the problem space is never zero. This has the effect of tending to inhibit the possibility of converging to a local optimum [22-25]. The traditional binary representation used for the genetic algorithms creates some difficulties for the optimization problems of large size with high numeric precision. According to the problem, the resolution of the algorithm can be expensive in time. The crossover and the mutation can be not adapted. For such problems, the genetic algorithms based on binary representations have poor performance. The first assumption that is typically made is that the variables representing parameters can be represented by bit strings. This means that the variables are discretized in an a priori fashion, and that the range of the discretization corresponds to some power of two. For example, with 10 bits per parameter, we obtain a range with 1024 discrete values. If the parameters are actually continuous then this discretization is not a particular problem. This assumes, of course, that the discretization provides enough resolution to make it possible to adjust the output with the desired level of precision. It also assumes that the discretization is in some sense representative of the underlying function.

Therefore, one can say that a more natural representation of the problem offers more efficient solutions. Then one of the greater improvements consists in the use of real numbers directly. Evolutionary Programming algorithms in Economic Power Dispatch provide an edge over common GA mainly because they do not require any special coding of individuals. In this case, since the desired outcome is the operating point of each of the dispatched generators (a real number), each of the individuals can be directly presented as a set of real numbers, each one being the produced power of the generator it concerns [26].

Our Evolutionary programming is based on the completed Genocop III system [27], developed by Michalewicz, Z. and Nazhiyath, G. Genecop III for constrained numerical optimization (nonlinear constraints) is based on repair algorithms. Genocop III incorporates the original Genocop system [28] (which handles linear constraints only), but also extends it by maintaining two separate populations, where a development in one population influences evaluations of individuals in the other population. The first population Ps consists of so-called search points, which satisfy linear constraints of the problem, the feasibility (in the sense of linear constraints) of these points is maintained by specialized operators (as in Genocop). The second population, P_r, consists of fully feasible reference points. These reference points, being feasible, are evaluated directly by the objective function, whereas search points are "repaired" for evaluation.

3.1 The Proposed GA for Design of Optimal Power Distribution Network

The Genecop (for GEnetic algorithm for Numerical Optimization of Constrained Problem) system [28] assumes linear constraints only and a feasible starting point (a feasible population).

3.1.1 Initialization

Let $Pg^k = \left[Pg_1^k, Pg_2^k, ..., Pg_{ng}^k \right]$ be the trial vector that presents the kth individual, $k = 1, 2, ..., P$ of the population to be evolved, where ng is the population size. The elements of the Pg^k are the desired values. The initial parent trial vectors, Pg^k, $k = 1, ..., P$, are generated randomly from a reasonable range in each dimension by setting the elements of Pg^k as:

$$pg_i^k = U\left(pg_{i,\min}, pg_{i,\max}\right) \qquad \text{for } i = 1, 2, ..., ng \qquad (4)$$

Where $U(pg_{i,\min}, pg_{i,\max})$ denotes the outcome of a uniformly distributed random variable ranging over the given lower bounded value and upper bounded values of the active power outputs of generators. A closed set of operators maintains feasibility of solutions. For example, when a particular component x_i of a solution vector X is mutated, the system determines its current domain dom(x_i) (which is a function of liner constraints and remaining value of the solution vector X) and the new value xi is taken from this domain (either with flat probability distribution for uniform mutation, or other probability distributions for non-uniform and boundary mutations). In any case the offspring solution vector is always feasible. Similarly, arithmetic crossover, $aX+(1-a)Y$, of two feasible solution vectors X and Y yields always a feasible solution (for $0 \leq a \leq 1$) in convex search spaces.

3.1.2 Offspring Creation

By adding a Gaussian random variation with zero mean, a standard derivation with zero mean and a standard derivation proportional to the fitness value of the parent trial solution, each parent Pg^k, $k=1,...,P$, creates an offspring vector, Pg^{k+1}, that is:

$$pg^{k+1} = pg^k + N(0, \sigma_k^2) \tag{5}$$

Where $N(0, \sigma_k^2)$ designates a vector of Gaussian random variables with mean zero and standard deviation σ_k, which is given according to:

$$\sigma_k = r \frac{f}{f_{\min}} (pg_{i,\min} - pg_{i,\max}) + M \tag{6}$$

Where f is the objective function value to be minimized in (2) associated with the control vector $f_{\min}$ represents the minimum objective function value among the trial solution; r is a scaling factor; and M indicates an offset.

3.1.3 Stopping Rule

As the stopping rule of maximum generation or the minimum criterion of f value in Equations (2, 3) is satisfied, the evolution process stops, and the solution with the highest fitness value is regarded as the best control vector.

3.2 Replacement of Individuals by Their Repaired Versions

The question of replacing repaired individuals is related to so-called Lamarckian evolution, which assumes that an individual improves during its lifetime and that the resulting improvements are coded back into the chromosome. In continuous domains, Michalewicz, Z. and Nazhiyath, G. indicated that an evolutionary computation technique with a repair algorithm provides the best results when 20% of repaired individuals replace their infeasible originals.

4 Computational Results

The new evolutionary algorithm has been applied intensively to the multi objective optimal design of several real size distribution systems. A compatible PC has been used (CPU Pentium 1.6GHz and 256Mb of RAM) with the operating system WinXp

Table 1. Characteristics of the distribution system

Number of existing demand nodes	*43*
Number of proposed demand nodes	*121*
Number of total demand nodes	*164*
Number of existing lines	*43*
Number of proposed lines	*133*
Number of total lines	*176*
Number of existing substations	*2*
Number of proposed substations	*0*
Number of total substations	*2*
Number of total 0-1 variables	*532*

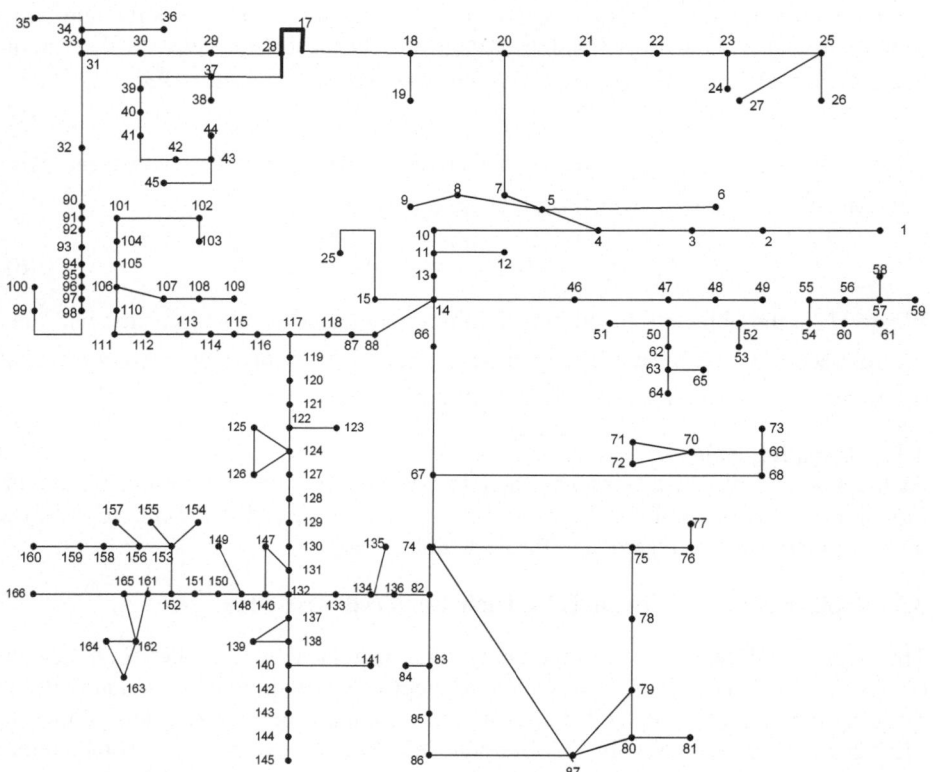

Fig.1. Existing and proposed feeders of the distribution network

Table 2. Results of the multi objective optimization program

Generations	Cost (Rials)	Reliability
75	42511.4367	197.17357

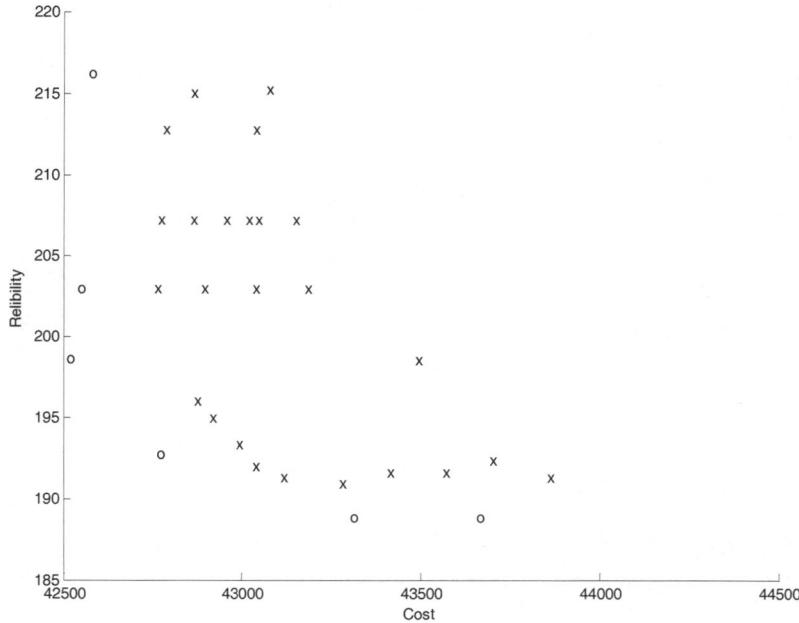

Fig. 2. Evolution of the non-dominated solution curve

to run MATLAB. Table I. shows the characteristics of the example of optimal design, indicating the dimensions of the used distribution system and the mathematical aspects. Notice the number of 0-1 variables of the distribution network indicating that the complexity of the optimization and the dimensions of this network is significantly larger than most of the ones usually described in technical papers [20].

In this study we use a 20 kV electric net, which is a portion of Iran's power distribution network and the network is represented in the Figure (1). The complete network, the existing feeders (thickness lines) and the proposed feeders (other lines). The existing network presents two substations at nodes 1 and 17 with the capacitance of 40MVA. The sizes of proposed substation are 40MVA and 10MVA. The existing lines of the network possess the following conductor sizes: 3x75Al, 3x118Al, 3x126Al and 3x42Al. Also in the design of the network, it is proposed 4 different conductor sizes for the construction of new lines: 3x75Al, 3x118Al, 3x126Al and 3x42Al.

We have used a multi objective toolbox that is written to work in MATLAB environment, which is called MOEA. In Table 2, we have shown the results of the multi objective optimization program that have leaded us to the final non dominated solutions curve. This table provides the objective function values ("cost" in millions of Rials, and "Reliability", function of expected energy not supplied, in kWh) of the ideal solutions, the number of generations (Gen.) and the objective function values of the best topologically meshed network solution for the distribution system.

The multi objective optimization finishes when the stop criteria is reached, that is when the upper limit of the defined generation number reached. The used crossover rate is 0.3 and the mutation rate is 0.02. The population size is 180 individuals and the

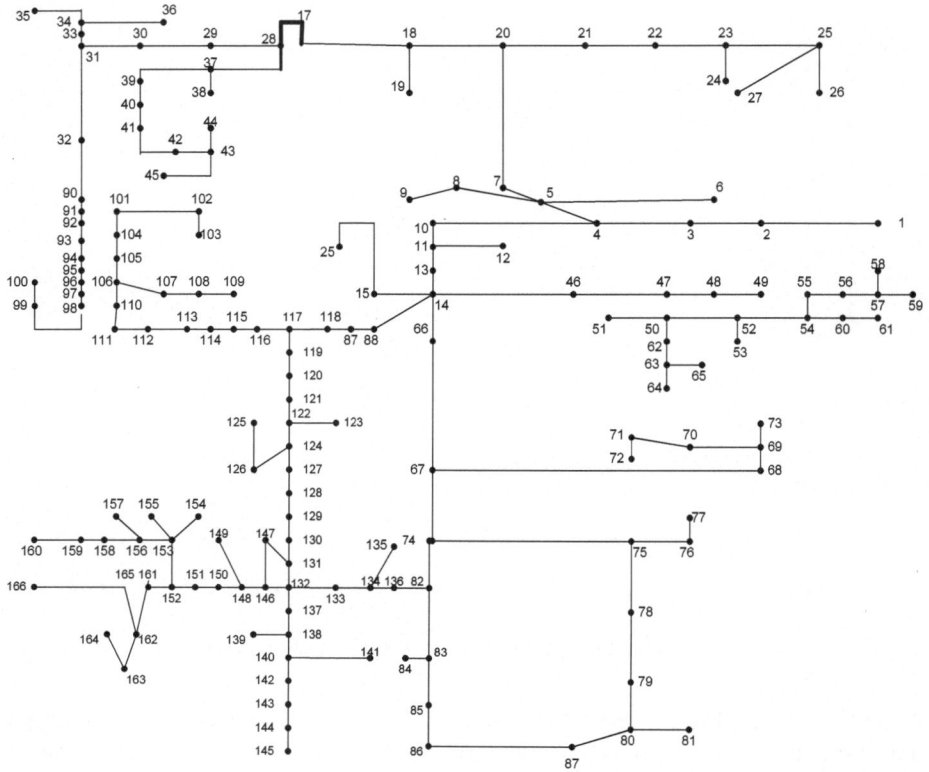

Fig. 3. Solution of multi objective optimal design model

process uses roulette wheel selection method; it reuses the previous population in the new generation. Figure (2), shows the non-dominated solution curve during the optimization design. The trade off between two solutions is very easy and subjective. The curve shows that when the cost improves the reliability decreases and vice versa.

After analyzing the curve of the final non-dominated solutions, the planner can select the definitive non-dominated solution, taking into account simultaneously the most satisfactory values of the two objective functions. We believe that the set of non-dominated solutions is the best set that can be offered to the planner in order to select the best satisfactory solution from such set. Finally, as mentioned before, Figure (1), shows the existing and future proposed distribution network. Figure (3), shows the final selected multi objective non-dominated solution in this paper, corresponding to the non dominated one that represents the topologically meshed distribution system in radial operating state with the best reliability achieved by the complete multi objective optimal design. The comparison of the multi objective algorithm of this paper (evolutionary algorithm) with other existing ones (for distribution systems of significant dimensions) has not been possible since such existing algorithms are not able to consider the characteristics of our mathematical model used in this paper.

5   Conclusions

This paper was reported the use of multiobjective evolutionary algorithm to solve optimal power distribution network problem with coordination of reliability and economy objectives. The multi objective optimization model contemplates the following aspects: Non-linear objective function of economic costs associated to the electric distribution network and a new objective function representing the reliability of the electric network to effects of the optimal design. These objective functions are subject to the mathematical restrictions that were mentioned. For the optimal design localizations and optimal sizes of electric lines as a future substations, proposed for the planner. Reliability objective of the electric power distribution system evaluated by means of a fitting function to indicate a meshing made topologies (or radial) electric network, keeping in mind, in general, faults (errors) of any order in the electric lines (studious for the first-order faults in the obtained computational results) to effects of the optimal design of distribution network. As a study case, a portion of Iran's power distribution network was selected. The simulation results show that for large-scale system a genetic algorithm with real code can give a best result with reduced time. To save an important CPU time, the constraints are to be decomposing into active constraints and passives ones. The active constraints are the parameters which enter directly in the cost function and the passives constraints are affecting the cost function indirectly. The developed model is easily applicable to n objectives without increasing the conceptual complexity of the corresponding algorithm. The proposed method can be useful for very large-scale power system.

Reference

1.   H. Lee Willis, H. Tram, M.V. Engel, and L. Finley. "Optimization applications to power distribution", IEEE Computer Applications in Power, 8(4):12-17, October 1995.
2.   Gonen, T., and B.L. Foote, "Distribution-System Planning Using Mixed-Integer Programming", Proceeding IEE, vol.128, Pt. C, no.2, pp. 70-79, March 1981.
3.   R.N. Adams and M.A. Laughton, "Optimal planning of power networks using mixed-integer programming Part I static and time-phaseed network synthesis", Proc. IEE, v01.121(2), pp. 139-147, February 1974.
4.   G.L.Thompson, D.L.Wall, "A Branch and Bound model for Choosing Optimal Substation Locations", IEEE Trans. PAS, pp 2683-2688, May 1981.
5.   R.N. Adams and M.A. Laughton, "A dynamic programming network flow procedure for distribution system planning", Presented at the IEEE Power Industry Computer Applications Conference (PICA), June 1973.
6.   M.Ponnaviakko, K.S.Parkasa Rao, S.S.Venkata, "Distribution System Planning through a Quadratic Mixed Integer Programming Approach", IEEE Trans. PWRD, pp 1157-1 163, October 1987.
7.   Brauner, G. and Zobel, M. "Knowledge Based Planning of Distribution Networks", IEEE Trans Power Delivery, vol.5, no.3, pp. 1514-1519, 1990.
8.   Chen, J. and Hsu, Y. "An Expert System for Load Allocation in Distribution Expansion Planning", IEEE Trans Power Delivery, vol.4, no.3, pp 1910-1917, 1989.

9. S.K.Goswami, "Distribution System Planning Using Branch Exchange Technique", IEEE Trans. On Power System, v01.12, no.2, pp.718-723, May 1997. D.E. Goldberg, Genetic Algorithm in Search, Optimization and Machine Learning, Addison Wesley, 1989.

10. I. J. Ramírez-Rosado and J. L. Bernal-Agustín, "Genetic algorithms applied to the design of large power distribution systems," IEEE Trans. On Power Systems, vol. 13, no. 2, pp. 696–703, May 1998.

11. I.J. Ramfrez-Rosado and T. Giinen, "Pseudodynamic Planning for Expansion of Power Distribution Systems", IEEE Transactions on Power Systems, Vol. 6, No. 1, February 1991, pp. 245-254.

12. Partanen, J., "A Modified Dynaniic Programming Algorithm for Sizing Locating and Timing of Feeder Reinforcements", IEEE Tram. on Power Delivery, Vol. 5, No. 1, January 1990, pp. 277-283.

13. Ramirez-Rosado, I.J., and JosB L. Bemal-Agustfn, "Optimization of the Power Distribution Netwoik Design by Applications of Genetic Algorithms", International Journal of Power and Energy Systems, Vol. 15, No. 3, 1995, pp. 104-110.

14. I.J. Ramirez-Rosado and R.N. Adams, "Multiobjective Planning of the Optimal Voltage Profile m Electric Power Distribution Systems", International Journal for Computation and Mathematics in Electrical and Electronic Engineering, Vol. 10, No. 2, 1991, pp. 115-128.

15. J. z Zhu, "Optimal reconfiguration of electrical distribution network using the refined genetic algorithm", Electric Power System Research 62 , 37-42, 2002

16. J. Partanen, "A modified dynamic programming algorithm for sizing locating and timing of feeder reinforcements", IEEE Trans. on Power Delivery, vol. 5, No. 1, pp. 277–283, Jan. 1990.

17. T. Gönen and I. J. Ramírez-Rosado, "Optimal multi-stage planning of power distribution systems", IEEE Trans. on Power Delivery, vol. PWRD-2, no. 2, pp. 512–519, Apr. 1987.

18. S.Sundhararajan, A.Pahwa, "Optimal Selection of Capacitors for Radial Distribution System Using A Genetic Algorithm", IEEE Trans. On Power System, vo1.9, no.3, pp. 1499-1507, August 1994.

19. Goldberg D., Genetic Algorithms in Search, "Optimization and Machine Learning", Addison Wesley 1989,ISBN: 0201157675

20. Coello Coello, C.A.,VanVeldhuizen, D.A., Lamont, G.B., "Evolutionary Algorithms for Solving Multi-Objective Problems", Genetic Algorithms and Evolutionary Computation. Kluwer Academic Publishers, NewYork, NY 2002

21. Obayashi, S., "Pareto genetic algorithm for aerodynamic design using the Navier-Stokes equations", In Quagliarella, D., P´eriaux, J., Poloni, C., Winter, G., eds.: Genetic Algorithms and Evolution Strategies in Engineering and Computer Science. John Wiley & Sons, Ltd., Trieste, Italy (1997) 245–266

22. Zitzler, E., Thiele, L.: Multiobjective evolutionary algorithms: A comparative case study and the strength pareto approach. IEEE Transactions on Evolutionary Computation 3, 257–271, 1999

23. Bentley, P.J.,Wakefield, J.P. "Finding acceptable solutions in the pareto-optimal range using multiobjective genetic algorithms. Proceedings of the Second Online World Conference on Soft Computing in Engineering Design and Manufacturing (WSC2) 5, pp. 231–240, 1998

24. Deb, K., Mohan, M., Mishra, S. "A fast multi-objective evolutionary algorithm for finding well-spread pareto-optimal solutions", Technical Report 2003002, Kanpur Genetic Algorithms Laboratory (KanGAL), Indian Institute of Technology Kanpur, Kanpur, PIN 208016, India,2003

25. Parmee, I.C.,Watson,A.H., "Preliminary airframe design using co-evolutionary multiobjective genetic algorithms", Proceedings of the Genetic and Evolutionary Computation Conference 1999, pp.1657–1665, 1999

26. K. De Jong: Lecture on Coevolution., "Theory of Evolutionary Computation", H.-G. Beyer et al. (chairs), Max Planck Inst. for Comput. Sci. Conf. Cntr., Schloß Dagstuhl, Saarland, Germany, 2002

27. Z. Michalewicz, G. Nazhiyzth, "Genocop III: A Co-evolutionary Algorithm for Numerical Optimization Problems with Nonlinear Constraints", Proceedings of the Second IEEE ICEC, Perth, Australia, 1995.

28. Z. Michalewicz, C. Jaikow, "Handling Constraints in Genetic Algorithms", Proceedings of Fourth ICGA, Morgan Kauffmann, pp. 151-157, 1999

New Stability Results for Delayed Neural Networks

Qiang Zhang

Liaoning Key Lab of Intelligent Information Processing,
Dalian University, Dalian, 116622, China
zhangq30@hotmail.com

Abstract. By constructing suitable Lyapunov functionals and combing with matrix inequality technique, some new sufficient conditions are presented for the global asymptotic stability of delayed neural networks. These conditions contain and improve some of the previous results in the earlier references.

1 Introduction

In recent years, the stability of a unique equilibrium point of delayed neural networks has extensively been discussed by many researchers[1]-[16]. Several criteria ensuring the global asymptotic stability of the equilibrium point are given by using the comparison method, Lyapunov functional method, M-matrix, diagonal dominance technique and linear matrix inequality approach. In [1]-[12], some sufficient conditions are given for the global asymptotic stability of delayed neural networks by constructing Lyapunov functionals. In [13]-[16], several results for asymptotic stability of neural networks with constant or time-varying delays are obtained based on the linear matrix inequality approach. In this paper, some new sufficient conditions to the global asymptotic stability for delayed neural networks are derived. These conditions are independent of delays and impose constraints on both the feedback matrix and delayed feedback matrix. The results are also compared with the earlier results derived in the literature.

2 Stability Analysis for Delayed Neural Networks

The dynamic behavior of a continuous time delayed neural networks can be described by the following state equations:

$$x_i^{'}(t) = -c_i x_i(t) + \sum_{j=1}^{n} a_{ij} f_j(x_j(t)) + \sum_{j=1}^{n} b_{ij} f_j(x_j(t - \tau_j)) + J_i, \quad i = 1, 2, \cdots, n \tag{1}$$

or equivalently

$$x'(t) = -Cx(t) + Af(x(t)) + Bf(x(t - \tau)) + J \tag{2}$$

U. Furbach (Ed.): KI 2005, LNAI 3698, pp. 216–221, 2005.

where $x(t) = [x_1(t), \cdots, x_n(t)]^T \in R^n, f(x(t)) = [f_1(x_1(t)), \cdots, f_n(x_n(t))]^T \in R^n, f(x(t - \tau)) = [f_1(x_1(t - \tau_1)), \cdots, f_n(x_n(t - \tau_n))]^T \in R^n. \ C = diag(c_i > 0)$ (a positive diagonal matrix), $A = \{a_{ij}\}$ is referred to as the feedback matrix, $B = \{b_{ij}\}$ represents the delayed feedback matrix, while $J = [J_1, \cdots, J_n]^T$ is a constant input vector and the delays τ_j are nonnegative constants. In this paper, we will assume that the activation functions $f_i, i = 1, 2, \cdots, n$ satisfy the following condition (H)

$$0 \le \frac{f_i(\xi_1) - f_i(\xi_2)}{\xi_1 - \xi_2} \le L_i \ , \forall \xi_1, \xi_2 \in R, \xi_1 \ne \xi_2.$$

This type of activation functions is clearly more general than both the usual sigmoid activation functions and the piecewise linear function (PWL): $f_i(x) = \frac{1}{2}(|x + 1| - |x - 1|)$ which is used in [4].

The initial conditions associated with system (1) are of the form

$$x_i(s) = \phi_i(s), s \in [-\tau, 0], \ i = 1, 2, \cdots, n, \ \tau = \max_{1 \le j \le n} \{\tau_j\}$$

in which $\phi_i(s)$ is continuous for $s \in [-\tau, 0]$.

Assume $x^* = (x_1^*, x_2^*, \cdots, x_n^*)^T$ is an equilibrium of the Eq.(1), one can derive from (1) that the transformation $y_i = x_i - x_i^*$ transforms system (1) or (2) into the following system:

$$y_i'(t) = -c_i y_i(t) + \sum_{j=1}^{n} a_{ij} g_j(y_j(t)) + \sum_{j=1}^{n} b_{ij} g_j(y_j(t - \tau_j)), \ \forall i \qquad (3)$$

where $g_j(y_j(t)) = f_j(y_j(t) + x_j^*) - f_j(x_j^*)$, or,

$$y'(t) = -Cy(t) + Ag(y(t)) + Bg(y(t - \tau)) \qquad (4)$$

respectively. Note that since each function $f_j(\cdot)$ satisfies the hypothesis (H), hence, each $g_j(\cdot)$ satisfies

$$g_j^2(\xi_j) \le L_j^2 \xi_j^2$$

$$\xi_j g_j(\xi_j) \ge \frac{g_j^2(\xi_j)}{L_j}, \ \forall \xi_j \in R.$$

$$g_j(0) = 0 \qquad (5)$$

To prove the stability of x^* of the Eq.(1), it is sufficient to prove the stability of the trivial solution of Eq.(3) or Eq.(4).

In this paper, we will use the notation $A > 0$ (or $A < 0$) to denote the matrix A is a symmetric and positive definite (or negative definite) matrix. The notation A^T and A^{-1} means the transpose of and the inverse of a square matrix A.

In order to obtain our results, we need establishing the following lemma:

Lemma 1. *For any vectors $a, b \in R^n$, the inequality*

$$2a^T b \le a^T X a + b^T X^{-1} b \qquad (6)$$

holds, in which X is any matrix with $X > 0$.

Proof. Since $X > 0$, we have

$$a^T X a - 2a^T b + b^T X^{-1} b = (X^{\frac{1}{2}} a - X^{-\frac{1}{2}} b)^T (X^{\frac{1}{2}} a - X^{-\frac{1}{2}} b) \geq 0$$

From this, we can easily obtain the inequality (6).

Now we will present several results for the global stability of Eq.(1).

Theorem 1. *The equilibrium point of Eq.(1) is globally asymptotically stable if there exist a positive definite matrix Q and a positive diagonal matrix D such that*

$$- 2L^{-1}DC + DA + A^T D + Q^{-1} + DBQB^T D < 0 \qquad (7)$$

holds.

Proof. We will employ the following positive definite Lyapunov functional:

$$V(y_t) = 2 \sum_{i=1}^{n} d_i \int_0^{y_i(t)} g_i(s) ds + \int_{t-\tau}^{t} g^T(y(s)) Q^{-1} g(y(s)) ds$$

in which $Q > 0$ and $D = diag(d_i)$ is a positive diagonal matrix. The time derivative of $V(y_t)$ along the trajectories of the system (4) is obtained as

$$\begin{aligned}
V'(y_t) &= 2g^T(y(t)) D(-Cy(t) + Ag(y(t)) + Bg(y(t-\tau))) \\
&\quad + g^T(y(t)) Q^{-1} g(y(t)) - g^T(y(t-\tau)) Q^{-1} g(y(t-\tau)) \\
&= -2g^T(y(t)) DCy(t) + 2g^T(y(t)) DAg(y(t)) + 2g^T(y(t)) DBg(y(t-\tau)) \\
&\quad + g^T(y(t)) Q^{-1} g(y(t)) - g^T(y(t-\tau)) Q^{-1} g(y(t-\tau)) \qquad (8)
\end{aligned}$$

Let $B^T Dg(y(t)) = a, g(y(t-\tau)) = b$. By Lemma 1, we have

$$2g^T(y(t)) DBg(y(t-\tau)) \leq g^T(y(t)) DBQB^T Dg(y(t)) + g^T(y(t-\tau)) Q^{-1} g(y(t-\tau)) \qquad (9)$$

Using the inequality (5), we can write the following:

$$- 2g^T(y(t)) DCy(t) \leq -2g^T(y(t)) L^{-1} DCg(y(t)) \qquad (10)$$

Substituting (9), (10) into (8), we get

$$\begin{aligned}
V'(y_t) &\leq g^T(y(t))(-2L^{-1}DC + DA + A^T D + Q^{-1} + DBQB^T D) g(y(t)) \\
&< 0
\end{aligned}$$

Therefore, the proof is completed.

Corollary 1. *The equilibrium point of Eq.(1) is globally asymptotically stable if there exist a positive definite matrix Q, a positive diagonal matrix D and a symmetric matrix K such that the following conditions hold.*

$$\begin{aligned}
&1) \quad DA + A^T D + K < 0 \\
&2) \quad - 2L^{-1}DC - K + Q^{-1} + DBQB^T D \leq 0 \qquad (11)
\end{aligned}$$

Note that the above condition is a sufficient condition for the inequality (7) holds.

Remark 1. In (11), if we let $L = C = Q = I$ and $K > 0$, then we can easily obtain the main result in [6]. As discussed in [6], therefore, the conditions given in [7]-[10] can be considered as the special cases of the result in Theorem 1 above.

3   An Example

In this section, we will give the following example showing that the conditions presented here hold for different classes of feedback matrices than those given in [11]. To compare our results with those given in [11] , we first restate the results of [11].

Theorem 2. *[11]: If 1) A has nonnegative off-diagonal elements, 2) B has nonnegative elements; and 3) $-(A+B)$ is row sum dominant, then the system defined by (1) has a globally asymptotically stable equilibrium point for every constant input when the activation function is described by a PWL function and $c_i = 1$.*

Example 1. Consider the following delayed neural networks

$$x_1'(t) = -x_1(t) - 0.1f(x_1(t)) + 0.2f(x_2(t)) + 0.1f(x_1(t - \tau_1)) + 0.1f(x_2(t - \tau_2)) - 3$$
$$x_2'(t) = -x_2(t) + 0.2f(x_1(t)) + 0.1f(x_2(t)) + 0.1f(x_1(t - \tau_1)) + 0.1f(x_2(t - \tau_2)) + 2$$
$$(12)$$

where the activation function is described by a PWL function $f_i(x) = 0.5(|x + 1| - |x - 1|)$. The constant matrices are

$$A = \begin{bmatrix} -0.1 & 0.2 \\ 0.2 & 0.1 \end{bmatrix}, B = \begin{bmatrix} 0.1 & 0.1 \\ 0.1 & 0.1 \end{bmatrix}, C = I$$

where A has nonnegative off-diagonal elements and B has nonnegative elements. Let $L = D = Q = I$ in Theorem 1 above, we can check that

$$-2L^{-1}DC + DA + A^T D + Q^{-1} + DBQB^T D$$
$$= A + A^T + BB^T - I$$
$$= \begin{bmatrix} -1.18 & 0.42 \\ 0.42 & -0.78 \end{bmatrix} < 0$$

Hence, Eq.(12) is globally stable.
The matrix $-(A + B)$ is obtained as

$$-(A + B) = \begin{bmatrix} 0 & -0.3 \\ -0.3 & -0.2 \end{bmatrix}$$

It can easily be seen that $-(A + B)$ is not row sum dominant. Therefore, the condition given in [11] does not hold. The simulation is shown in Fig.1.

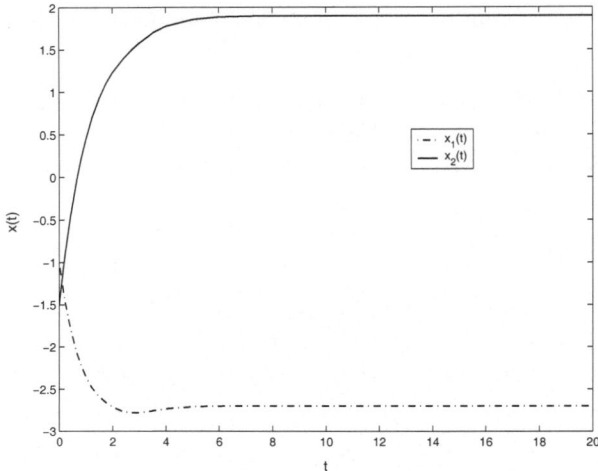

Fig. 1. The solution trajectory of Eq.(12)

References

1. Zhang, Q., Ma, R., Xu, J.: Stability of Cellular Neural Networks with Delay. Electron. Lett. **37** (2001) 575–576
2. Zhang, Q., Ma, R., Chao, W., Jin, X.: On the Global Stability of Delayed Neural Networks. IEEE Trans.Automatic Control **48** (2003) 794–797
3. Zhang, Q., Wei, X.P. Xu, J.: Global Asymptotic Stability of Hopfield Neural Networks with Transmission Delays. Phys.Lett.A **318** (2003) 399–405
4. Chua, L.O., Yang, L.: Cellular Neural Networks: Theory and Applications. IEEE Trans.Circuits Syst.I **35** (1988) 1257–1290
5. Arik, S.: Stability Analysis of Delayed Neural Networks. IEEE Trans.Circuits Syst.I **47** (2000) 1089–1092
6. Arik, S.: An Improved Global Stability Result for Delayed Cellular Neural Networks. IEEE Trans.Circuits Syst.I **49** (2002) 1211–1214
7. Arik, S.: On the Global Asymptotic Stability of Delayed Cellular Neural Networks. IEEE Trans.Circuits Syst.I **47** (2000) 571–574
8. Arik, S.: An Analysis of Global Asymptotic Stability of Delayed Cellular Neural Networks. IEEE Trans. Neural Networks **13** (2002) 1239–1242
9. Liao, T.L., Wang, F.C.: Global Stability for Cellular Neural Networks with Time Delay. IEEE Trans. Neural Networks **11** (2000) 1481–1484
10. Cao, J.: Global Stability Conditions for Delayed CNNs. IEEE Trans.Circuits Syst.I **48** (2001) 1330–1333
11. Roska, T., Wu, C.W., Balsi, M., Chua, L.O.: Stability and Dynamics of Delay-Type General and Cellular Neural Networks. IEEE Trans.Circuits Syst. **39** (1992) 487–490
12. Zhang, J., Jin, X.: Global Stability Analysis in Delayed Hopfiled Neural Network Models. Neural Networks **13** (2000) 745–753
13. Liao, X., Chen, G., Sanchez, E.N.: LMI-Based Approach for Asymptotically Stability Analysis of Delayed Neural Networks. IEEE Trans.Circuits Syst.I **49** (2002) 1033–1039

14. Arik, S.: Global Asymptotic Stability of A Larger Class of Neural Networks with Constant Time Delay. Phys.Lett.A **311** (2003) 504–511
15. Liao, X., Wong, K.W., Yang, S.: Stability Analysis for Delayed Cellular Neural Networks Based on Linear Matrix Inequality Approach. International Journal of Bifurcation and Chaos **14** (2004) 3377–3384
16. Singh, V.: A Generalized LMI-Based Approach to the Global Asymptotic Stability of Delayed Cellular Neural Networks. IEEE Trans. Neural Networks **15** (2004) 223–225

Metaheuristics for Late Work Minimization in Two-Machine Flow Shop with Common Due Date

Jacek Blazewicz[1], Erwin Pesch[2], Malgorzata Sterna[1], and Frank Werner[3]

[1] Institute of Computing Science, Poznan University of Technology,
Piotrowo 3A, 60-965 Poznan, Poland
{Jacek.Blazewicz, Malgorzata.Sterna}@cs.put.poznan.pl
[2] Institute of Information Systems, FB 5 - Faculty of Economics, University of Siegen,
Hoelderlinstrasse 3, 57068 Siegen, Germany
Erwin.Pesch@uni-siegen.de
[3] Faculty of Mathematics, Otto-von-Guericke-University,
PSF 4120, 39016, Magdeburg, Germany
Frank.Werner@mathematik.uni-magdeburg.de

Abstract. In this paper, metaheuristic approaches for the weighted late work minimization in the two-machine flow shop problem with a common due date ($F2 \mid d_j=d \mid Y_w$) are presented. The late work performance measure estimates the quality of a solution with regard to the duration of the late parts of jobs not taking into account the quantity of the delay for the fully late activities. Since, the problem mentioned is known to be NP-hard, three trajectory based methods, namely simulated annealing, tabu search and variable neighborhood search were designed and compared to an exact approach and a list scheduling algorithm.

1 Introduction

Due date based performance measures are important objective functions, which find many practical applications (cf. e.g. [3], [10], [22]), since they can be used, for example, to describe the solution feasibility or the quality of service in a system. The late work performance measure can be applied in all cases in which a solution is estimated with regard to the amount of the late units, not to the quantity of their delay. For example, in control systems [2], the accuracy of control processes depends on the amount of information on the system provided. Data exposed by sensing devices after a predefined time window are lost and can be modeled as the late work. In the modern flexible manufacturing environment, the late work may be used e.g. for representing delayed parts of customer orders realized after a given due date or production tasks which could not be scheduled within a given time horizon [25]. In the agriculture, the late work models the amount of wasted crops, which has not been harvested in time [23], as well as the amount of not spread fertilizers or pesticides, which influences the quantity of these crops.

The late work criteria are relatively new performance measures, which were first considered in the parallel [2], [4] and single [23], [24] machine environments. Then shop scheduling problems with the late work performance measures were investigated, starting with the two-machine cases with a common due date [5], [6],

U. Furbach (Ed.): KI 2005, LNAI 3698, pp. 222–234, 2005.

[7], [8], [25], which appeared to be binary NP-hard. (It is worth to be mentioned that the late work concept can be considered as a special case of the imprecise computation model [20].)

In this paper, we continue the research [7], [8] on the problem $F2 \mid d_j=d \mid Y_w$, where a set of jobs $\{J_1, J_2, \ldots, J_n\}$ have to be processed on two dedicated machines M_1, M_2. Each job has to be performed first by the machine M_1 and then by M_2 for p_{1j} and p_{2j} time units, respectively. Each machine can process only one job at any time and, analogously, each job can be executed by only one machine at any time. We look for a non-preemptive schedule minimizing the late work in the system, i.e. minimizing the amount of work executed after a given common due date d. Denoting with C_{ij} the completion time of the job J_j on the machine M_i and with w_j the weight of this activity, the late work Y_j is determined as (cf. Fig. 1):

$$Y_j = \sum_{i=1,2} \min\{\max\{0, C_{ij} - d\}, p_{ij}\} \tag{1}$$

The criterion / objective value, estimating the quality of a complete schedule for the whole set of n jobs, is given by:

$$Y_w = \sum_{j=1}^{n} w_j Y_j \tag{2}$$

The problem $F2 \mid d_j=d \mid Y_w$ stated above is binary NP-hard [13], since a polynomial transformation was constructed from the set partition problem and a dynamic programming method (DP) with a pseudo-polynomial time complexity ($O(n^2d^4)$) was proposed [7]. This exact approach is important mainly from a theoretical point of view, for it determines the complexity status of the case under consideration. On the other hand, DP finds optimal solutions in reasonable time only for small problem instances. Thus, to solve the problem efficiently, heuristic approaches have to be proposed.

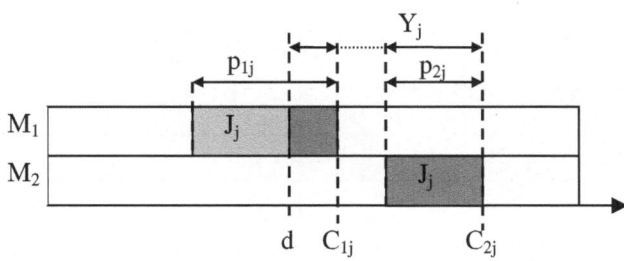

Fig. 1. The late work parameter Y_j in the two-machine flow shop environment

First, the list scheduling method was implemented for the problem under consideration and compared to the dynamic programming and enumerative algorithms [8]. The list approach [16] is a scheduling technique commonly used especially for practical applications, because of its ease of implementation and the low time complexity. This constructive procedure builds a feasible solution by scheduling jobs on available machines, one by one, in a certain order determined by a given priority dispatching rule. The job selection rule influences the execution order of jobs, so applying different strategies makes it possible to construct different feasible solutions of the problem under consideration. In consequence, the list scheduling algorithm can be also used for constructing a set of feasible solutions of different quality in a reasonable amount of time. The computational experiments showed an extremely high efficiency of the list scheduling algorithm from the run time and the solution quality points of view. This heuristic constructed solutions with a criterion value of only 2.5% worse than the optimum on average. Taking into account these encouraging results, it was interesting, whether further improvement of the solution quality could be obtained by extending the list scheduling approach with an additional search engine, i.e. by proposing metaheuristic methods. Summing up, the theoretical studies on the problem $F2 \mid d_j=d \mid Y_w$ have been completed with extended computational experiments. They concerned two exact approaches (the dynamic programming and enumerative ones) and the list scheduling method [8], as well as more advanced search techniques presented in this paper.

2   Metaheuristic Approaches

Within this research, we proposed three trajectory-based metaheuristics [9], [11] for the problem $F2 \mid d_j=d \mid Y_w$: simulated annealing (SA), tabu search (TS) and variable neighborhood search (VNS). Similarly to the dynamic programming approach [7] and the list scheduling algorithm [8], the metaheuristics are based on the specific structure of an optimal solution of the problem under consideration. It has been proven [7], [25] that in such a schedule, the set of early jobs, executed before a common due date, have to be sequenced in Johnson's order [19], which is optimal from the schedule length point of view. Thus, any method solving the problem $F2 \mid d_j=d \mid Y_w$ has to select the first late job in the system and to divide the remaining activities into two sets of early and late activities (cf. Fig. 2).

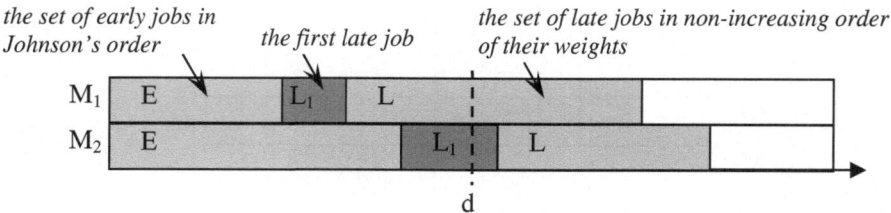

Fig. 2. The general structure of an optimal solution for the problem $F2 \mid d_j=d \mid Y_w$ (with the first late job partially late on the machine M_2)

Early jobs are scheduled by Johnson's algorithm (by ordering jobs with $p_{1j} \leq p_{2j}$ in non-decreasing order of p_{1j} and jobs with $p_{1j} > p_{2j}$ in non-increasing order of p_{2j}), while the late activities are executed in non-increasing order of their weights w_j. The crucial element of any method solving the problem is taking the decision whether a particular job is processed early or late in a schedule.

Initial Solution. For all metaheuristic approaches proposed, an initial solution is determined either by Johnson's algorithm (applied as a simple heuristic for $F2 \mid d_j=d \mid Y_w$) or by the list scheduling algorithm. As we have mentioned, the latter method constructs a solution by scheduling jobs, selected according to a given priority dispatching rule [16], one by one on the machines. Based on the results obtained within the previous research, the maximum weight rule has been applied, for it allowed constructing schedules of the highest quality [7]. Since two methods of determining an initial solution are available, it is possible for the metaheuristics to start the solution space exploration from two different points, which may lead to final schedules of a different quality.

Neighborhood Structures. As it was mentioned, a solution of the problem is described as a sequence (E, L_1, L), where E denotes the set of early jobs (scheduled in Johnson's order), L_1 is the first late job (executed partially late either on M_1 or on M_2) and L denotes the set of late jobs (performed in non-increasing order of their weights). In order to explore the solution space two neighborhood structures are proposed based on job shifts (N^1) and jobs interchanges (N^2).

In the first neighborhood, N^1, a new solution is generated by selecting a late job in a current schedule and shifting it to the set of early activities. Since all early jobs have to be scheduled in Johnson's order and to ensure that the selected activity will be executed totally early, it might be necessary to move some early jobs preceding it in Johnson's sequence after the common due date d. On the other hand, after these modifications, some late jobs succeeding the chosen one in Johnson's order might need shifting before d to complete a schedule. This move idea for the neighborhood N^1 is sketched below:

```
Step 1: move a job J_j selected from L∪{L_1} to the set E and
        calculate Johnson's schedule for E;
Step 2: if the job J_j is late in the new subschedule for E,
        then move to the set L some early jobs J_i /selected in non-
        decreasing order of (p_{1i}+p_{2i})w_i values or at random/, until J_j
        becomes early or there are no more early jobs in E to be
        moved to L;
Step 3: add to the subschedule for E some jobs J_k from the set
        L∪{L_1} succeeding J_j in Johnson's order /selected in non-
        increasing order of (p_{1k}+p_{2k})w_k values or at random/, until
        the last job in E exceeds the common due date d becoming L_1;
Step 4: remove the job included to E as the last one and re-
        include it into L, then try all jobs from L as the first
        late job and select as L_1 the job for which the best
        criterion value has been obtained;
Step 5: schedule all late jobs J_k from the set L in non-
        increasing order of their w_k values.
```

In the neighborhood N^2, a new schedule is obtained by choosing a pair of jobs - one from the set of late ones and one from the set of early ones - and interchanging them between these sets. A solution modification is performed in a similar way as for the neighborhood N^1 according to the following move scheme:

```
Step 1:move a job J_j selected from the set E to the set L and
       disable it (i.e. exclude it from the further analysis, such
       that J_j cannot become early in Step 2);
Step 2:apply move N^1 to a job J_i selected from the set L∪{L_1}
```

The schedule modification in the neighborhood N^1 as well as in N^2 requires the application of Johnson's algorithm, which takes O(nlogn) time (where n denotes the number of jobs). In consequence, both types of moves presented are performed in O(nlogn) time.

Termination Condition. The SA, TS and VNS methods are trajectory-based metaheuristics, i.e. they start the exploration of the solution space from an initial schedule, and, at each iteration, they construct a new schedule by picking up one solution from a given neighborhood. In the research reported, the search is terminated after exceeding a certain run time limit or after exceeding a given number of iterations without improvement in the schedule quality.

Simulated Annealing. The SA method proposed is based on the classical framework of this metaheuristic (cf. e.g. [9], [11], [18]). The algorithm starts the search from an initial solution with an initial temperature, whose value is set as the multiplied total processing time. At each iteration, the method constructs a new solution from a current one according to the neighborhood definition N^1 or N^2, in which jobs for shifting or interchanging are selected at random or according to their weighted processing times. If the new schedule improves the criterion value, then it is accepted, otherwise it replaces the previous solution with a probability depending on the criterion value deterioration and the temperature value. The temperature is decreased at the end of an iteration according to the geometrical reduction scheme (the preliminary tests showed that the arithmetic reduction scheme is much less efficient for the problem under consideration). The search finishes after reaching the termination condition or the (nearly) zero temperature value.

Tabu Search. The TS method implemented follows the classical scheme of this approach (cf. e.g. [9], [11], [12], [14]). In this algorithm, a new solution is selected from the whole neighborhood generated for a current schedule by applying move N^1 or N^2, i.e. by shifting particular late jobs early or by interchanging particular pairs of late and early jobs. It is possible to generate the whole set of neighbor solutions or the restricted one by selecting jobs for the analysis at random or according to their weighted processing times. In the restricted neighborhood N^1 not all late jobs are made early, while in the restricted N^2, only some pairs of early and late activities are considered. The best schedule from the neighborhood, which is not forbidden, i.e. which does not have the tabu status, becomes the starting point for the next iteration. The move leading to this point is stored in the tabu list, which contains the index/indices of the job/jobs which the move concerned. The tabu list is managed according to the FIFO (First In First Out) rule and its length is determined with regard to the number of jobs. The method stops after reaching the termination condition or when there is no solution in the neighborhood without the tabu status.

Variable Neighborhood Search. The VNS method is a strategy using a local search algorithm and dynamically changing neighborhood structures (cf. e.g. [9], [12], [15], [21]). The algorithm picks at random a solution from the neighborhood generated for a current solution. The schedule selected becomes the starting point for a local search method. The role of a local search procedure is played by the simulated annealing (VNS-SA) or the tabu search algorithm (VNS-TS). If a resulting solution improves the criterion value, then it is accepted, otherwise the local search is restarted from another neighbor solution generated according to a different neighborhood definition. The VNS method applies repeatedly the neighborhood N^1 and N^2 (restricted to two different sizes) to generate the starting points for the local search procedure. The variable neighborhood algorithm finishes after reaching the termination conditions. Since VNS calls SA or TS as subprocedures, the number of iterations without improvement (after which VNS should stop) has to be rather small.

3 Computational Experiments

Computational experiments performed within the research were devoted to comparing the efficiency of particular metaheuristic methods, as well as to evaluating the quality of their solutions with regard to the optimal schedules. Moreover, the test results analysis made it possible to determine some specific features of the approaches proposed.

The main computational experiments were obviously preceded by a careful tuning process [1], [17], during which we tested the metaheuristics efficiency for different values of control parameters in order to determine their best settings.

3.1 Comparison of Metaheuristic Approaches

The first stage of the computational experiments was designed for comparing particular metaheuristics (SA, TS, VNS-SA, VNS-TS) and for comparing them to the list scheduling algorithm (LA). Since only heuristic methods were applied, big problem instances could be included into the test set containing 20 instances with:

- the number of jobs equal to 20 and 200 (two instances per each number of jobs), as well as to 50, 80, 110 and 140 (four instances per each number of jobs),
- the task processing times generated from the range [2, 40],
- the job weights taken from the range [1, 5],
- the due date equal to 40% of the half of the total job processing time.

For each test instance, particular metaheuristic methods were run four times: starting from Johnson's schedule (used as a heuristic solution for the problem under consideration) or from a list schedule and with a limited number of iterations without improvement or the run time limit as a termination condition. The values of the remaining control parameters were determined during the tuning process. (Obviously the time or iteration limit was not applied to the list scheduling algorithm, which is not a metaheuristic approach. Thus, double bars for LA exist in Fig. 3 and Fig. 4 for the sake of clarity.)

The test results (cf. Fig. 3 and Fig. 4) showed that even a very good list schedule can be still improved without a huge time effort, just by performing a systematic search in the solution space. The metaheuristics improved the list scheduling solution in 70% of all tests (14 instances among 20) by 2.82‰ on average. The smallest improvement was equal to 0.2‰, while the maximum one was equal to 6.8‰. Because the list scheduling algorithm generates schedules of a very high quality, nearly optimal ones [8], the further improvement achieved by the metaheuristic methods should be considered as notable, especially, with regard to their rather simple structures and the short run times.

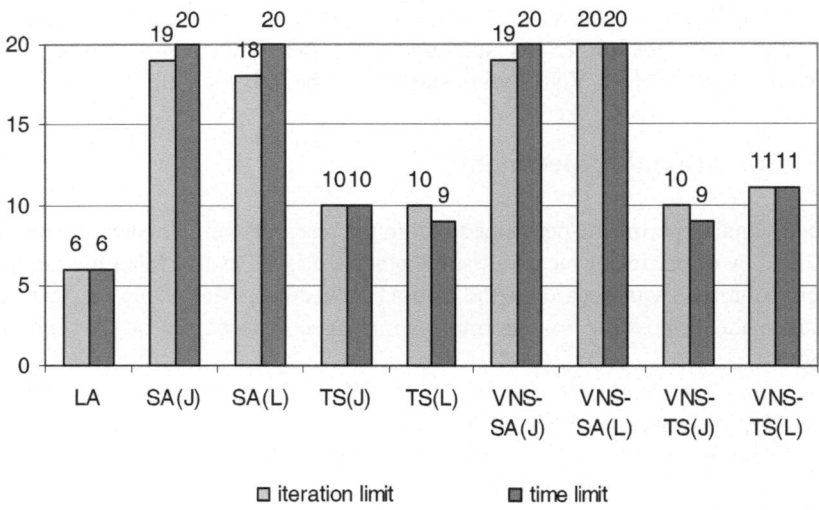

Fig. 3. The number of test instances (among 20 instances) for which a particular method constructed the solution of the best criterion value, starting from Johnson's schedule (J) or a list schedule (L), for the number of iterations without improvement and 5-second time limit as a termination condition

The variable neighbourhood search method with simulated annealing appeared to be the best choice for the problem under consideration. It constructed the best schedules for all test instances independently of the termination condition applied. The superiority of VNS-SA resulted from the high efficiency of the simulated annealing algorithm. It was possible to achieve the same solution quality level by extending the SA's run time to 5 seconds or by restarting it from different initial solutions within the VNS framework. The tabu search approach appeared to be much less efficient. In this case introducing a time limit as a termination condition, as well as multiple restarting TS within VNS did not result in a significant solution improvement. Taking into account the fact that the initial schedule was close to the optimum, searching the solution space by a single modification of a current schedule (as in SA) appeared to be a better approach than generating the whole neighborhood (as in TS), containing schedules mostly worse than the current one.

The efficiency of the metaheuristic depends not only on its search engine, but also on the quality of the initial solution. TS and TS-VNS generated much better schedules starting from the list solution than from Johnson's one with the worse criterion value (cf. Fig. 4). According to the very high efficiency of SA and VNS-SA, in their cases this influence is almost invisible when the performance measure value is considered; but it can be detected looking at the number of the best schedules constructed during the experiments (cf. Fig.3). Summing up, starting the search from a better (list) schedule the metaheuristics were usually able to obtain the higher final solution quality than starting from Johnson's schedule with a worse criterion value.

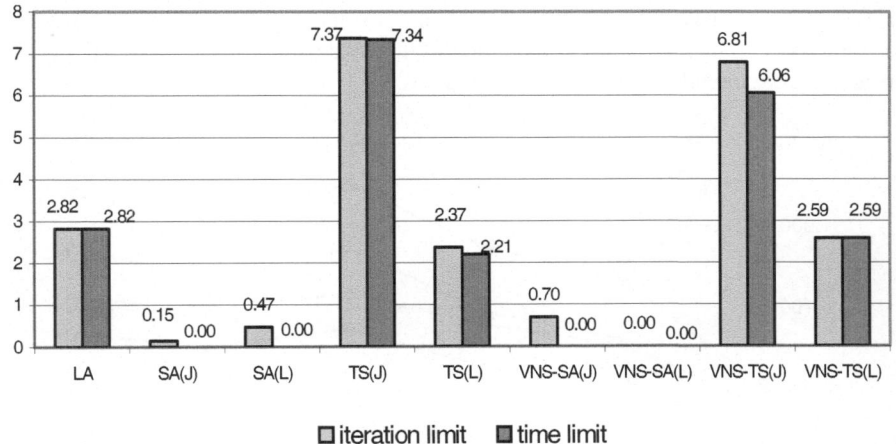

□ iteration limit   ■ time limit

Fig. 4. The average distance to the best solution in [‰] for particular methods starting from Johnson's schedule (J) or a list schedule (L), for the number of iterations without improvement and a 5-second time limit as a termination condition (restricted to instances for which a particular method did not find the best solution)

The simulated annealing algorithm appeared to be extremely fast (cf. Fig. 5). For this reason, a 5-second time limit enforced the method to search the solution space for the time of 3 orders of magnitude longer than the time resulting from the termination condition based on the number of iterations without improvement. The long time limit allowed SA to leave the local optimum in the number of iterations, which would be not possible for the latter termination condition.

These results showed that forcing a fast metaheuristic to the long-lasting search might result in an additional objective value improvement. A similar effect can be obtained by restarting a method from different initial solutions, as it was performed within the variable neighborhood algorithm.

The tabu search approach was much more time consuming because of the necessity of generating the neighborhood of a current schedule (instead of a single neighbor solution as for SA). For the same reason, VNS-TS appeared to be the slowest approach. It is interesting, that for one test instance VNS-TS generated a schedule worse than TS. Within the variable neighborhood search, the tabu search starts from a solution taken at random from the neighborhood of a current schedule, which might be worse than this latter one. For this reason, a single run of TS, starting from a good schedule, could result in a higher solution quality than a few runs of TS within VNS initialized with worse schedules.

3.2 Comparison of Metaheuristics to an Exact Approach

At the second stage of the computational experiments, metaheuristic approaches were compared to an exact enumerative method (EM). Algorithm EM constructs all possible sets of early jobs and completes them with all possible first late jobs in $O(n2^n)$ time. The earlier research results showed that the enumerative algorithm with its huge time complexity is more time efficient than the pseudopolynomial dynamic programming approach for the problem under consideration [8]. However, taking into account the exponential time complexity of the exact approach, 16 small problem instances were included into the test set with:

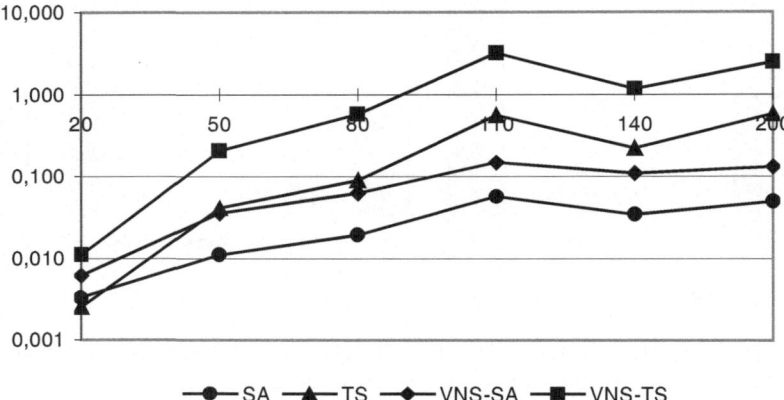

Fig. 5. The average run time in seconds for particular number of jobs n and the number of iterations without improvement as a termination condition with the logarithmic time axis

- the number of jobs equal from 5 to 20 (with a unitary increment),
- task processing times generated from the range [2, 20],
- job weights from the range [1, 5],
- the deadline equal to 40% of the half of the total processing time.

For each problem instance the optimal solution was compared to the schedules constructed by the particular metaheuristics (SA, TS, VNS-SA, VNS-TS) and the list scheduling (LA) as well as Johnson's algorithm (JA) (cf. Fig. 6 and Fig. 7).

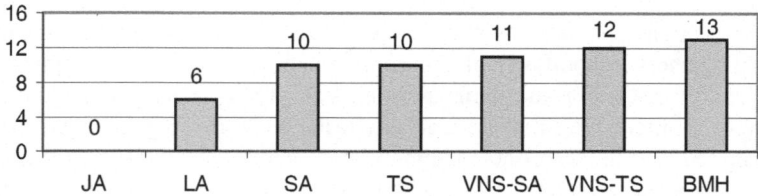

Fig. 6. The number of optima found for 16 instances for particular methods (BMH denotes the best metaheuristic result)

As one could expect, Johnson's algorithm (JA) appeared to be the weakest method. It found no optimum for the test set generated and constructed schedules of the quality of nearly 11% worse than the optimum (because of this huge difference JA is not depicted in Fig. 7). On the other hand, taking into account the negligible short run time and the simplicity of this procedure, the results obtained for JA can be considered as quite satisfying (especially, bearing in mind the fact that this algorithm has been designed for a different problem F2 || C_{max} [19]).

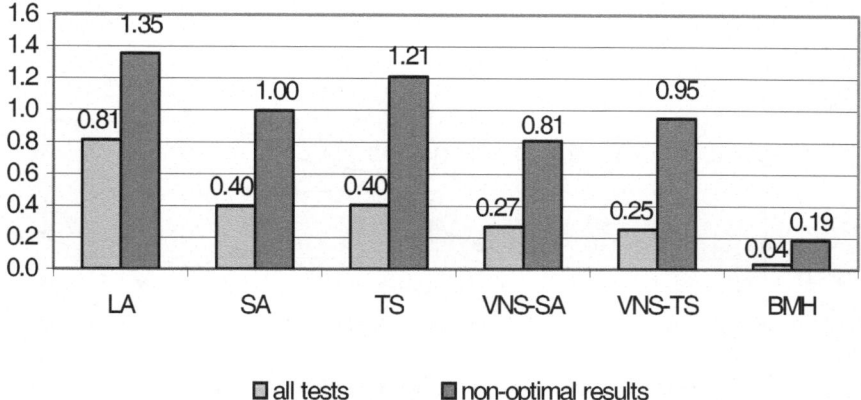

Fig. 7. The distance to the optimal objective value for particular methods and the best metaheuristic in [%] for all tests and for these tests for which a particular algorithm did not find the optimum

The list scheduling algorithm was able to construct an optimal solution for only 6 instances among 16. Moreover, it found an optimum mostly for small instances with 6 up to 9 jobs. The simulated annealing and tabu search approaches reached the optimal criterion value for 10 instances, while applying these procedures within the variable neighbourhood search framework increased this number to 11 and 12, respectively. In general, the best metaheuristic solution (BMH) had the same criterion value as the exact one for 13 instances. BMH differed from the optimum by only 0.19%, while for the list scheduling one this value equals to 1.35%. Similarly as in the previous stage of the experiments, the highest performance in terms of the schedule quality was observed for VNS-SA. The multiple restarting of SA within VNS made it possible to additionally improve the schedules constructed by simulate annealing, whose quality was already quite good. The tabu search method appeared to be less efficient than SA, even if applied within the variable neighborhood search algorithm.

These results showed that the specificity of the problem under consideration made the search difficult. Taking into account the fact that all early jobs have to be scheduled in Johnson's order, the crucial decision is the selection of activities performed before the common due date d. In an initial (list) solution, the early jobs are selected based on their weights in order to increase the schedule quality already at the beginning of the search. When particular jobs differ only slightly in their processing times and weights, a further improvement of the criterion value is difficult and it can be obtained only by finding a very specific sequence of jobs.

The run time comparison confirms the superiority of SA over TS applied as a stand-alone method as well as within the variable neighborhood approach (cf. Fig. 8). Moreover, the computational effort required by the enumerative algorithm confirms the necessity of applying heuristic procedures for hard problems. For the exact approach the exponential explosion could be observed already for 18 jobs, while the metheuristic run time increased very slowly.

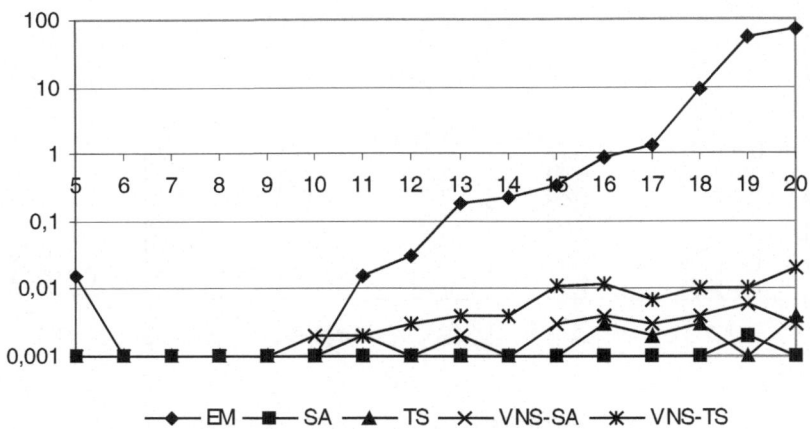

Fig. 8. The run time in milliseconds for particular methods

4 Conclusions

The work presented completed the study on the two-machine flow-shop scheduling problem with a common due date and the weighted late work criterion (F2 | d_j=d | Y_w), which consists of a theoretical and a computational part.

The NP-hardness proof and the dynamic programming formulation [7] have been completed with the computational experiments performed for the DP algorithm, the simple list scheduling heuristic [8] and the more advanced metaheuristic approaches. The simulated annealing, the tabu search and the variable neighborhood search methods proposed have been based on the specific features of the problem under consideration.

The computational experiments showed the superiority of the simulated annealing technique in terms of the time efficiency and the solution quality. An additional improvement could be achieved be embedding this algorithm within a variable neighborhood search framework.

The test results confirmed the advantages of the metaheuristic approaches. They improved the solution quality with respect to the list scheduling algorithm besides the fact that this procedure generates nearly optimal schedules. Applying a systematic search through the solution space made it possible to decrease the objective value, even when starting from very good initial solutions. Moreover, the computational time requirements for this kind of optimization techniques are extremely small compared with exact (exponential) algorithms.

Acknowledgements. We would like to thank to Michal Glowinski for his effort in implementing and testing the approaches investigated within the presented research. The fourth author has been supported by INTAS (project 03-51-5501).

References

1. Barr, R.S., Golden, B.L., Kelly, J.P., Resende, M.G.C., Stewart JR., W.R.: Designing and Reporting on Computational Experiments with Heuristic Methods. Journal of Heuristics 1 (1995) 9-32

2. Blazewicz, J.: Scheduling Preemptible Tasks on Parallel Processors with Information Loss. Recherche Technique et Science Informatiques 3/6 (1984) 415-420

3. Blazewicz, J., Ecker, K., Pesch, E., Schmidt, G., Weglarz, J.: Scheduling Computer and Manufacturing Processes. 2nd edn. Springer-Verlag, Berlin Heidelberg New York (2001)

4. Blazewicz, J., Finke, G.: Minimizing Mean Weighted Execution Time Loss on Identical and Uniform Processors. Information Processing Letters 24 (1987) 259-263

5. Blazewicz, J., Pesch, E., Sterna, M., Werner, F.: Revenue Management in a Job-shop: a Dynamic Programming Approach. Preprint Nr. 40/03. Otto-von-Guericke-University, Magdeburg (2003)

6. Blazewicz, J., Pesch, E., Sterna, M., Werner, F.: Open Shop Scheduling Problems with Late Work Criteria. Discrete Applied Mathematics 134 (2004) 1-24

7. Blazewicz, J., Pesch, E., Sterna, M., Werner, F.: The Two-Machine Flow-Shop Problem with Weighted Late Work Criterion and Common Due Date. European Journal of Operational Research 165/2 (2005) 408-415

8. Blazewicz, J., Pesch, E., Sterna, M., Werner, F.: Flow Shop Scheduling with Late Work Criterion – Choosing the Best Solution Strategy. Lecture Notes in Computer Science 3285 (2004) 68-75

9. Blum, Ch., Roli, A.: Metaheuristics in Combinatorial Optimization: Overview and Conceptual Comparison. ACM Computing Surveys 35/3 (2003) 268-308

10. Brucker, P.: Scheduling Algorithms. 2nd edn. Springer-Verlag, Berlin Heidelberg New York (1998)

11. Crama, Y., Kolen, A., Pesch, E.: Local Search in Combinatorial Optimization. Lecture Notes in Computer Science 931 (1995) 157-174

12. Dorndorf, U., Pesch, E.: Variable Depth Search and Embedded Schedule Neighbourhoods for Job Shop Scheduling, Proceedings of the 4th International Workshop on Project Management and Scheduling (1994) 232-235

13. Garey, M.R., Johnson, D.S.: Computers and Intractability. W.H. Freeman and Co., San Francisco (1979)

14. Glover, F., Laguna, M.: Tabu Search. Kluwer Academic Publishers, Boston (1997)

15. Hansen, P., Mladenović, N.: Variable Neighbour Search. Principles and Applications. European Journal of Operational Research 130 (2001) 449-467

16. Haupt, R.: A Survey of Priority Rule–Based Scheduling. OR Spektrum 11 (1989) 3-16

17. Hooker, J.N.: Testing Heuristics: We Have It All Wrong. Journal of Heuristics 1 (1995) 33-42

18. Kirkpatrick, S., Gelatt, C.D., Vecchi, M.P.: Optimization by Simulated Annealing. Science 220/4598 (1983) 671-680

19. Johnson, S.M.: Optimal Two- and Three-Stage Production Schedules. Naval Research Logistics Quarterly 1 (1954) 61-68

20. Leung, J.Y.T.: Minimizing Total Weighted Error for Imprecise Computation Tasks and Related Problems. In: Leung, J.Y.T. (Ed.): Handbook of Scheduling: Algorithms, Models, and Performance Analysis. CRC Press, Boca Raton (2004) Chapter 34, 1-16

21. Pesch, E., Glover, F.: TSP Ejection Chains. Discrete Applied Mathematics 76 (1997) 165-181

22. Pinedo, M., Chao, X.: Operation Scheduling with Applications in Manufacturing and Services. Irwin/McGraw-Hill, Boston (1999)
23. Potts, C.N., Van Wassenhove, L.N.: Single Machine Scheduling to Minimize Total Late Work. Operations Research 40/3 (1991) 586-595
24. Potts, C.N., Van Wassenhove, L.N.: Approximation Algorithms for Scheduling a Single Machine to Minimize Total Late Work. Operations Research Letters 11 (1991) 261-266
25. Sterna, M.: Problems and Algorithms in Non-Classical Shop Scheduling. Scientific Publishers of the Polish Academy of Sciences, Poznan (2000)

An Optimal Algorithm for Disassembly Scheduling with Assembly Product Structure

Hwa-Joong Kim[1], Dong-Ho Lee[2], and Paul Xirouchakis[1]

[1] Institute of Production and Robotics (STI-IPR-LICP),
Swiss Federal Institute of Technology (EPFL), Lausanne, CH-1015, Switzerland
{hwa-joong.kim, paul.xirouchakis}@epfl.ch
[2] Department of Industrial Engineering, Hanyang University,
Sungdong-gu, Seoul 133-791, Korea
leman@hanyang.ac.kr

Abstract. This paper considers the problem of determining the quantity and timing of disassembling used or end-of-life products in order to satisfy the demand of their parts or components over a finite planning horizon. We focus on the case of single product type without parts commonality, i.e., assembly product structure. The objective is to minimize the sum of setup, disassembly operation, and inventory holding costs. Several properties of optimal solutions are derived, and then a branch and bound algorithm is developed based on the Lagrangian relaxation technique. Results of computational experiments on randomly generated test problems show that the algorithm finds the optimal solutions in a reasonable amount of computation time.

1 Introduction

Disassembly has become a major issue for countries and companies due to the obligation to the environment and the society and its profitability incurred from remanufacturing used or end-of-life products. In the mean time, disassembly scheduling has become an important problem in the area of operations planning in disassembly systems. In general, disassembly scheduling is defined as the problem of determining the quantity and timing of disassembling (used or end-of-life) products in order to satisfy the demand of their parts or components over a planning horizon. In general, disassembly scheduling is known to have more complex computational complexity than ordinary production planning in assembly systems because of the difference in the number of demand sources [2].

Previous research has addressed several cases of the disassembly scheduling problem and these can be classified according to two aspects: (single or multiple) product types and (with or without) parts commonality [5]. Here, the parts commonality implies that products or subassemblies share their parts/components. For the case of single product type without parts commonality, Gupta and Taleb [2] suggest a reversed form of the material requirement planning (MRP) algorithm without explicit objective function. Lee *et al.* [8] consider the case with resource capacity constraints and suggest an integer programming model. Recently, Lee and Xirouchakis [7] suggest a heuristic algorithm for the objective of minimizing various costs related with disassembly systems. As a variant of the problem with single product type, the case of

U. Furbach (Ed.): KI 2005, LNAI 3698, pp. 235–248, 2005.

single product type with parts commonality is considered. Taleb *et al.* [11] suggest another MRP-like heuristic algorithm for the objective of minimizing the number of products to be disassembled, and Neuendorf *et al.* [9] suggest Petri-net based algorithm for the problem of Taleb *et al.* [11]. For the case with multiple product types and parts commonality, Taleb and Gupta [10] suggest a two-phase heuristic for two independent objectives of minimizing the number of products to be disassembled and minimizing the product disassembly costs, and Kim *et al.* [4] suggest a heuristic algorithm based on a linear programming relaxation approach for the objective of minimizing the sum of setup, disassembly operation, and inventory holding costs, and perform a case study. Recently, Lee *et al.* [6] suggest integer programming models for all cases and they require excessive computation time.

This paper focuses on the case of single product type without parts commonality, i.e., assembly product structure, and suggests an optimal solution algorithm for the objective of minimizing the sum of setup, disassembly operation, and inventory holding costs. Although Lee *et al.* [6] tried to obtain the optimal solutions using an integer programming approach, it is not efficient in that excessive amounts of computation time are required. Instead, a branch and bound algorithm, after deriving several optimal properties, is suggested using the Lagrangian relaxation technique. Here, the Lagrangian relaxation technique is used to obtain a good initial solution and good lower and upper bounds.

2 Problem Description

In the disassembly structure of single product type without parts commonality, the root item represents the (used or end-of-life) product to be ordered and disassembled, and the leaf items are the parts or components to be demanded and not to be disassembled further. A parent item denotes an item that has more than one child and a child item denotes an item that has only one parent. Figure 1 shows an example of the disassembly product structure. In the figure, item 1 denotes the root item and items 6 to 12 denote leaf items. The number in parenthesis represents the yield of the item obtained when its parent is disassembled.

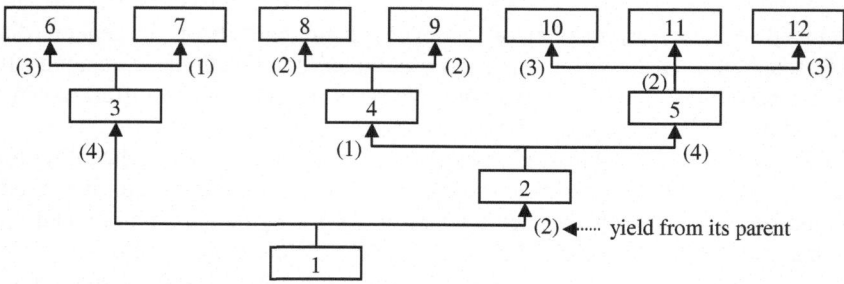

Fig. 1. Disassembly product structure: an example

Now, the disassembly scheduling problem considered in this paper is defined as follows: *for a given disassembly structure, the problem is to determine the quantity and timing of ordering the root item and disassembling each parent item in order to satisfy the demand of leaf items over a planning horizon.* The objective is to minimize the sum of setup, disassembly operation, and inventory holding costs. The setup cost implies the cost required for preparing the corresponding disassembly operation. It is assumed that the setup cost occurs in a period if any disassembly operation is performed in that period. In general, the setup cost is significant since most disassembly operations are performed manually [3]. The disassembly operation cost is the cost proportional to the labor or machine processing time required for performing the corresponding disassembly operation, and the inventory holding cost occurs if items are stored to satisfy future demand, and they are computed based on the end-of-period inventory. Other assumptions made in this problem are summarized as follows: (a) (used or end-of-life) products can be obtained whenever ordered; (b) demand of leaf items is given and deterministic; (c) backlogging is not allowed and hence demands are satisfied on time; (d) parts/components are perfect in quality, i.e., no defectives are considered.

The problem is formulated as an integer programming model. In the model, without loss of generality, all items are numbered with integers 1, 2, ... i_l, ... N, where 1 and i_l denote the indices for the root item and the first leaf item, respectively. The notations used are summarized below.

Parameters

s_i setup cost of parent item i

p_i disassembly operation cost of parent item i

h_i inventory holding cost of item i

d_{it} demand of leaf item i in period t

a_{ij} yield of item j obtained by disassembling one unit of item i ($i < j$)

I_{i0} initial inventory of item i

$\varphi(i)$ parent of item i

$H(i)$ set of child items of parent item i

M arbitrary large number

Decision variables

X_{it} disassembly quantity of item i in period t

Y_{it} = 1 if there is a setup for item i in period t, and 0 otherwise

I_{it} inventory level of item i at the end of period t

Now, the integer program is given as follows.

[P1] Minimize $\sum_{i=1}^{i_l-1}\sum_{t=1}^{T} s_i \cdot Y_{it} + \sum_{i=1}^{i_l-1}\sum_{t=1}^{T} p_i \cdot X_{it} + \sum_{i=2}^{N}\sum_{t=1}^{T} h_i \cdot I_{it}$

subject to

$$I_{it} = I_{i,t-1} + a_{\varphi(i),i} \cdot X_{\varphi(i),t} - X_{it} \qquad \text{for } i = 2, 3, \dots i_l-1 \text{ and } t = 1, 2, \dots T \quad (1)$$

$$I_{it} = I_{i,t-1} + a_{\varphi(i),i} \cdot X_{\varphi(i),t} - d_{it} \qquad \text{for } i = i_l, i_l+1, \dots N \text{ and } t = 1, 2, \dots T \quad (2)$$

$$X_{it} \leq M \cdot Y_{it} \qquad \text{for } i = 1, 2, \dots i_l-1 \text{ and } t = 1, 2, \dots T \quad (3)$$

$$Y_{it} \in \{0,1\} \qquad \text{for } i = 1, 2, \dots i_l-1 \text{ and } t = 1, 2, \dots T \quad (4)$$

$$X_{it} \geq 0 \text{ and integer} \qquad\qquad \text{for } i = 1, 2, \dots i_l-1 \text{ and } t = 1, 2, \dots T \qquad (5)$$

$$I_{it} \geq 0 \text{ and integer} \qquad\qquad \text{for } i = 2, 3, \dots N \text{ and } t = 1, 2, \dots T \qquad (6)$$

The objective function denotes the sum of setup, disassembly operation, and inventory holding costs. Constraints (1) and (2) define the inventory flow conservation of non-root items at the end of each period. Note that no inventory flow conservation constraint is needed for the root item since its surplus-inventory results in unnecessary cost increase. Constraint (3) guarantees that a setup cost in a period is incurred if any disassembly operation is performed in that period.

3 Problem Analysis

In this section, we analyse the problem considered in this paper. First, the original model [P1] is transformed into another one so that the Lagrangean relaxation technique can be applied more effectively. Second, the properties of optimal solutions are derived which are used to reduce the search space in the branch and bound (B&B) algorithm suggested in this paper.

3.1 Problem Reformulation

The integer program [P1] can be reformulated by substituting and eliminating inventory variables. The equation used is given below.

$$I_{it} = I_{i0} + \sum_{j=1}^{t} \left(a_{\varphi(i),i} \cdot X_{\varphi(i),j} - Q_{ij} \right) \qquad \text{for } i = 2, 3, \dots N \text{ and } t = 1, 2, \dots T,$$

where $Q_{it} = X_{it}$ for $i = 2, 3, \dots i_l-1$ and $t = 1, 2, \dots T$, and $Q_{it} = d_{it}$ for $i = i_l, i_l+1, \dots N$ and $t = 1, 2, \dots T$. Then, the objective function of [P1] becomes

$$\sum_{i=1}^{i_l-1} \sum_{t=1}^{T} s_i \cdot Y_{it} + \sum_{i=1}^{i_l-1} \sum_{t=1}^{T} c_{it} \cdot X_{it} + R.$$

Here, $R = \sum_{i=i_l}^{N} \sum_{t=1}^{T} h_i \cdot \left\{ I_{i0} - \sum_{j=1}^{t} d_{ij} \right\}$, a constant that can be eliminated from further consideration, and c_{it} is defined as follows:

$$c_{it} = p_i + (T-t+1) \cdot \sum_{k \in H(i)} h_k \cdot a_{ik} \qquad \text{for } i = 1 \text{ and } t = 1, 2, \dots T,$$

$$c_{it} = p_i + (T-t+1) \cdot \sum_{k \in H(i)} h_k \cdot a_{ik} - (T-t+1) \cdot h_i \quad \text{for } i = 2, 3, \dots i_l-1 \text{ and } t = 1, 2, \dots T.$$

Also, a new demand constraint is added to the reformulation since it is needed in the Lagrangean relaxation technique used when obtaining lower bounds. To explain the new demand constraint, let $L(i)$ and $P(i, j)$ denote the set of leaf items among successors of parent item i and the path from parent item i to leaf item j, respectively. For example, $L(2) = \{8, 9, 10, 11, 12\}$ and $P(2, 12) = 2 \rightarrow 5 \rightarrow 12$ in Figure 1. Then, the new demand constraint can be represented as

$$\sum_{j=1}^{t} X_{ij} \geq \max_{e \in L(i)} \left\{ \sum_{j=1}^{t} D_{ij}^{e} - A_{i0}^{e} + I_{i0} \right\} \qquad \text{for } i = 1, 2, \dots i_l-1 \text{ and } t = 1, 2, \dots T \quad (7)$$

where D_{it}^e and A_{i0}^e denote the transformed demand and the transformed initial inventory of item i in period t associated with the demand of leaf item e. More specifically,

$$D_{it}^e = \frac{D_{kt}^e}{a_{ik}}, \; D_{et}^e = d_{et} \qquad \text{for } k \in H(i) \cap P(i, e), \text{ and}$$

$$A_{i0}^e = \frac{A_{k0}^e}{a_{ik}} + I_{i0}, \; A_{e0}^e = I_{e0} \qquad \text{for } k \in H(i) \cap P(i, e).$$

In the new demand constraint, the maximum term is inserted since the demand of each item in $L(i)$ can be satisfied by disassembling item i to the maximum amount so that the demands of all leaf items in $L(i)$ are satisfied.

Then, the alternative formulation without inventory variables is given as follows.

[P2] Minimize $\displaystyle\sum_{i=1}^{i_l-1}\sum_{t=1}^{T} s_i \cdot Y_{it} + \sum_{i=1}^{i_l-1}\sum_{t=1}^{T} c_{it} \cdot X_{it}$

subject to

$$\sum_{j=1}^{t} a_{ik} \cdot X_{ij} \geq \sum_{j=1}^{t} Q_{kj} - I_{k0} \qquad \text{for } i = 1, 2, \dots i_l{-}1,\, k \in H(i), \text{ and } t = 1, 2, \dots T \; (6')$$

and (3) – (5) and (7).

In the reformulation [P2], the new constraint (7) does not affect the optimal solution since all demands can be satisfied with constraint (6'). However, it is added to the reformulation since the bounding method in the B&B algorithm is based on the Lagrangean relaxation of constraint (6') and hence there is no way to satisfy demands of leaf items in the relaxed problem without constraint (7).

3.2 Properties of Optimal Solutions

This section presents the properties that characterize the optimal solutions. As stated earlier, the properties are helpful to reduce the search space. Note that the properties can be applied to model [P2] since [P1] and [P2] are mathematically equivalent. Before presenting the properties, we make the following assumption:

$$h_i \leq \sum_{k \in H(i)} h_k \cdot a_{ik} \qquad \text{for } i = 2, 3, \dots i_l{-}1 \qquad (8)$$

This assumption is not very restrictive since the inventory holding cost is directly related to the item value. That is, the item value may increase as the disassembly operations progress.

Proposition 1 is an extension of the well-known zero-inventory property for lot sizing problem.

Proposition 1. *For the disassembly scheduling problem [P1], there exists an optimal solution that satisfies*

$$\min_{k \in H(i)} \left\{ \left\lfloor \frac{I_{k,t-1}}{a_{ik}} \right\rfloor \right\} \cdot X_{it} = 0 \qquad \text{for } i = 1, 2, \dots i_l{-}1 \text{ and } t = 1, 2, \dots T \qquad (9)$$

where $\lfloor \bullet \rfloor$ *gives the largest integer that is less than or equal to* $\bullet$.

Proof. Suppose that there is an optimal solution such that $X_{iv} > 0$ and $I_{k,v-1} \geq a_{ik}$ for all $k \in H(i)$, i.e.,

$$\min_{k \in H(i)} \left\{ \left\lfloor \frac{I_{k,v-1}}{a_{ik}} \right\rfloor \right\} \cdot X_{iv} > 0 .$$

We show that this leads to a contradiction. Let u be the last setup period prior to period v ($> u$), such that $X_{iu} > 0$ and $X_{it} = 0$ for $t = u+1, u+2, \ldots v-1$. Then, the current solution can be improved in two ways.

(i) Consider an alternative solution in which $X'_{iu} = X_{iu} - \Delta$ and the others are the same, which satisfies (9) for item i in period v, i.e.,

$$\min_{k \in H(i)} \left\{ \left\lfloor \frac{I'_{k,v-1}}{a_{ik}} \right\rfloor \right\} = 0 , \text{ where } \Delta = \min_{k \in H(i)} \left\{ \left\lfloor \frac{I_{k,v-1}}{a_{ik}} \right\rfloor \right\}$$

Then, the inventory level is changed as follows:

$$I'_{it} = I_{it} + \Delta \qquad \text{for } t = u, u+1, \ldots v-1,$$

$$I'_{kt} = I_{kt} - a_{ik} \cdot \Delta \qquad \text{for } k \in H(i) \text{ and } t = u, u+1, \ldots v-1, \text{ and}$$

$$I'_{it} = I_{it} \qquad \text{for the others.}$$

Therefore, the alternative solution increases the inventory holding cost of item i and decreases the inventory holding costs of its leaf items defined as

$$\sum_{t=u}^{v-1} h_i \cdot \Delta - \sum_{k \in H(i)} \sum_{t=u}^{v-1} h_k \cdot a_{ik} \cdot \Delta = \sum_{t=u}^{v-1} (h_i - \sum_{k \in H(i)} h_k \cdot a_{ik}) \cdot \Delta \le 0,$$

while setup and disassembly operation costs are not affected. Note that for the root item, the first term in the above equation can be eliminated. Therefore, the alternative solution reduces the total cost because of the assumption (8).

(ii) Consider another alternative solution such that $X'_{iu} = X_{iu} + X_{iv}$, $X_{iv} = 0$, and the others are the same, which satisfies (9) for item i in period v. Then, the inventory level is changed as follows:

$$I'_{it} = I_{it} - X_{iv} \qquad \text{for } t = u, u+1, \ldots v-1,$$

$$I'_{kt} = I_{kt} + a_{ik} \cdot X_{iv} \qquad \text{for } k \in H(i) \text{ and } t = u, u+1, \ldots v-1, \text{ and}$$

$$I'_{it} = I_{it} \qquad \text{for the others.}$$

Therefore, the alternative solution results in the decrease of the inventory holding cost of item i, the increase of the inventory holding costs of its leaf items, and the decrease of setup costs of item i, while the disassembly operation cost is not affected by the change. The total cost change is defined as

$$\sum_{k \in H(i)} \sum_{t=u}^{v-1} h_k \cdot a_{ik} \cdot X_{iv} - \sum_{t=u}^{v-1} h_i \cdot X_{iv} - s_i ,$$

Therefore, we can see that the alternative solution improves the current one if

$$s_i \ge \sum_{t=u}^{v-1} (\sum_{k \in H(i)} h_k \cdot a_{ik} \cdot X_{iv} - h_i) \cdot X_{iv} .$$

From (i) and (ii), we can see that the current solution can be improved and hence is not optimal. This completes the proof. $\qquad\square$

Proposition 1 implies that we can search a limited set of feasible solutions when finding the optimal solutions. Here, the subset is characterized by the set of solutions such that item i is disassembled in period t, i.e., $X_{it} > 0$, if and only if its child items satisfy

$$\min_{k \in H(i)} \left\{ \left\lfloor I_{k,t-1} / a_{ik} \right\rfloor \right\} = 0.$$

Proposition 2 given below is an extension of the nested property of Crowston and Wagner [1] for the lot sizing problems, which describes the condition that if an item is assembled in a period, its parent item should be also assembled in that period.

Proposition 2. *For the disassembly scheduling problem [P1], there exists an optimal solution that satisfies the following condition: $X_{it} > 0$ implies that at least one of its child items satisfies $Q_{kt} > 0$ for $i = 1, 2, \dots i_t-1$ and $k \in H(i)$.*

Proof. Suppose that there is an optimal solution that does not satisfy the condition. Then, we show that this leads to a contradiction. Suppose that in the optimal solution, $X_{iu} > 0$ and $Q_{ku} = 0$ for all $k \in H(i)$, i.e., parent item i is disassembled in period u while any child item of item i is not disassembled or demanded in that period. Let v (> u) be the next disassembling or demanding period of child item $k \in H(i)$.

Consider an alternative solution defined as: $X'_{iv} = X_{iv} + X_{iu}$, $X'_{iu} = 0$, and the others are the same. Then, the inventory level is changed as follows:

$$I'_{it} = I_{it} + X_{iu} \qquad \text{for } t = u, u+1, \dots v-1,$$

$$I'_{kt} = I_{kt} - a_{ik} \cdot X_{iu} \qquad \text{for } k \in H(i) \text{ and } t = u, u+1, \dots v-1, \text{ and}$$

$$I'_{it} = I_{it} \qquad \text{for the others.}$$

Therefore, the alternative solution results in the decrease of the inventory holding costs of the child items and the increase of the inventory holding cost of the parent item. This can be formally described as

$$\sum_{j=u}^{v-1} h_i \cdot X_{iu} - \sum_{k \in H(i)} \sum_{j=u}^{v-1} h_k \cdot a_{ik} \cdot X_{iu} = \sum_{j=u}^{v-1} (h_i - \sum_{k \in H(i)} h_k \cdot a_{ik}) \cdot X_{iu} \leq 0$$

while it results in no change of the setup and the disassembly operation costs. Here, the first term can be eliminated for the root item. From this, we can see that the alternative solution can be improved and hence is not optimal. This completes the proof. □

An immediate outcome of Proposition 2 is that the optimal solution satisfies the property that $X_{it} = 0$ if $X_{kt} = 0$ for all $k \in H(i)$.

4 Branch and Bound Algorithm

This section presents the B&B algorithm that can give the optimal solutions. First, the branching scheme is explained with the definition of the subproblem at each node of the B&B tree. Second, the Lagrangean relaxation based methods to obtain the lower and upper bounds are explained. As in the ordinary B&B algorithm, each node of the B&B tree can be deleted from further consideration (fathomed) if the lower bound at the node is greater than or equal to the incumbent solution, i.e., the smallest upper bound of all nodes generated so far.

4.1 Branching

We first define the B&B tree that can generate all possible solutions. To explain the B&B tree, an example with three parent items and two planning periods is shown in Figure 2. Each level corresponds to the setup variable Y_{it}, and the level number increases from the last parent item i_l-1 to the root item 1 and for each item, from the first period to the last. (See Figure 2.) Two branches emanate from each node, which are associated with 1 (left side) and 0 (right side) of the corresponding setup variable, respectively. By doing this, we can generate all possible solutions. In this branching scheme, the variables related with the first period, i.e., Y_{i1} for $i = 1, 2, \ldots i_l-1$, are set to 1 to satisfy the demand requirements, and from Proposition 2, Y_{it} is set to 0 if $Y_{kt} = 0$ for all $k \in H(i)$. These can reduce the number of branches.

As stated earlier, each node in the B&B tree corresponds to a partial solution associated with nodes: from that node to the root. Suppose that the current node is associated with item i' and period t', the corresponding partial solution is Y_{it}^F for $t =1, 2, \ldots T$ if $i = i'+1, i'+2, \ldots i_l-1$ and $t = 1, 2, \ldots t'$ if $i = i'$, and the last setup period be period l in the current partial solution, such that $Y_{it}^F =1$ and $Y_{it}^F = 0$ for $t = l+1, l+2, \ldots t'$. Then, the solution value of the partial solution after branching becomes

$$V = \sum_{t=1}^{l-1}(s_{i'} \cdot Y_{i't}^F + c_{i't} \cdot X_{i't}^F) + \sum_{i=i'+1}^{i_l-1}\sum_{t=1}^{T}(s_i \cdot Y_{it}^F + c_{it} \cdot X_{it}^F),$$

where the disassembly quantity X_{it}^F corresponding to the fixed setup schedule Y_{it}^F can be obtained as (using Proposition 1)

$$X_{it}^F = \left\lceil \max_{k \in H(i)}(\sum_{j=1}^{v-1}Q_{kj} - I_{k0})/a_{ik} - \max_{k \in H(i)}\left\lceil(\sum_{j=1}^{t-1}Q_{kj} - I_{k0})/a_{ik}\right\rceil\right\rceil.$$

In the current subproblem, items i', $i'+1$, $i'+2$, ... i_l-1 can be regarded as leaf items since their disassembly schedules are fixed using the above method. The new leaf items are defined as follows:

$$E^- = \{i', i'+1, \ldots e^-\} \qquad \text{for } t = 1, 2, \ldots l-1$$
$$E^+ = \{i'+1, i'+2, \ldots e^+\} \qquad \text{for } t = l, l+1, \ldots T$$

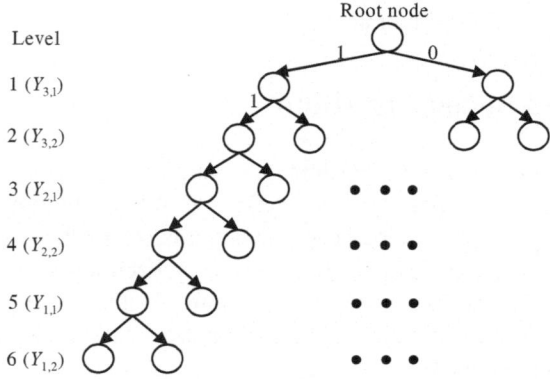

Fig. 2. The branch and bound tree: an example

where $e^- = \max_{\text{all } i}\{i < \min_{k \in H(i')} k\}$, and $e^+ = \max_{\text{all } i}\{i < \min_{k \in H(i'+1)} k\}$ if $i' < i_l-1$ and N, otherwise. For example, in Figure 1, if $i' = 4$, $t' = 4$, and $l = 4$ in the current node, $E^- = \{4, 5, 6, 7\}$ for $t = 1, 2, 3$, and $E^+ = \{5, 6, 7, 8, 9\}$ for $t = 4, 5, \dots T$, where $e^- = 7$ and $e^+ = 9$. Then, using the new leaf items defined above, $L(i)$, the set of leaf items among successors of parent item i, is modified as $L(i) \subset E^-$ for $t = 1, 2, \dots l-1$ and $L(i) \subset E^+$ for $t = l, l+1, \dots T$.

Based on the above terms and the problem [P2], the current subproblem can be formulated as follows.

[CP] Minimize $V + \sum\limits_{i=1}^{i'-1} \sum\limits_{t=1}^{T} (s_i \cdot Y_{it} + c_{it} \cdot X_{it}) + \sum\limits_{t=l}^{T} (s_{i'} \cdot Y_{i't} + c_{i't} \cdot X_{i't})$

subject to

$$\sum_{j=1}^{t} a_{ik} \cdot X_{ij} \geq \sum_{j=1}^{t} Q_{kj} - I_{k0} \qquad \text{for } i = 1, 2, \dots i'-1, k \in H(i), \text{ and } t = 1, 2, \dots T \quad (6'')$$

$$\sum_{j=1}^{t} X_{ij} \geq \max_{e \in L(i)} \left\{ \sum_{j=1}^{t} D_{ij}^e - A_{i0}^e + I_{i0} \right\} \qquad \text{for } i = 1, 2, \dots i'-1 \text{ and } t = 1, 2, \dots T \quad (7')$$

$$X_{it} \leq M \cdot Y_{it} \qquad \text{for } i = 1, 2, \dots i'-1 \text{ and } t = 1, 2, \dots T \quad (3')$$

$$Y_{it} \in \{0,1\} \qquad \text{for } i = 1, 2, \dots i'-1 \text{ and } t = 1, 2, \dots T \quad (4')$$

$$X_{it} \geq 0 \text{ and integer} \qquad \text{for } i = 1, 2, \dots i'-1 \text{ and } t = 1, 2, \dots T \quad (5')$$

$$\sum_{j=l}^{t} a_{i'k} \cdot X_{i'j} \geq \sum_{j=1}^{t} Q_{kj} - \sum_{j=1}^{l-1} a_{i'k} \cdot X_{i'j}^F - I_{k0} \quad \text{for } k \in H(i') \text{ and } t = l, l+1, \dots T \quad (6''a)$$

$$\sum_{j=l}^{t} X_{i'j} \geq \max_{e \in L(i')} \left\{ \sum_{j=1}^{t} D_{i'j}^e - A_{i'0}^e + I_{i'0} \right\} - \sum_{j=1}^{l-1} X_{i'j}^F \qquad \text{for } t = l, l+1, \dots T \quad (7'a)$$

$$X_{i't} \leq M \cdot Y_{i't} \qquad \text{for } t = l, t'+1, \dots T \quad (3'a)$$

$$Y_{i't} \in \{0,1\} \qquad \text{for } t = l, t'+1, \dots T \quad (4'a)$$

$$X_{i't} \geq 0 \text{ and integer} \qquad \text{for } t = l, t'+1, \dots T \quad (5'a)$$

$$Y_{i'l} = 1 \text{ and } Y_{i't} = X_{i't} = 0 \quad \text{for } t = l+1, l+2, \dots t' \quad (10)$$

For node selection (or branching), the depth-first rule is used. In the rule, if the current node is not fathomed, the next node to be considered is the left node among its child nodes, i.e., the child node on the left is considered before the right. When a node is fathomed, we go back on the path from this node toward the root node until we find the first node with a child node that has not been considered yet, i.e., backtracking.

4.2 Bounding

As stated earlier, the lower and upper bounds are obtained using the Lagrangean relaxation technique. In this section, we explain the method to obtain the relaxed problem at each node of the B&B tree, which gives the lower bounds directly from its optimal solution as well as the upper bounds after modifying its optimal solution into the one that satisfies the constraints of the original problem. Note that the relaxed problem can be decomposed into the single item lot-sizing problems and hence can be solved with a polynomial-time algorithm.

Obtaining the lower bounds

At each node of the B&B tree, the relaxed problem is obtained from relaxing (6″) and (6″a) from the subproblem [CP] and can be represented as follows.

[LR] $LB(\lambda) = \min\left\{ \sum_{i=1}^{i'-1}\sum_{t=1}^{T}(s_i \cdot Y_{it} + v_{it} \cdot X_{it}) + \sum_{t=l}^{T}(s_{i'} \cdot Y_{i't} + v_{i't} \cdot X_{i't}) + R' \right\}$

$\quad$ subject to $\quad \lambda_{ikt} \geq 0 \qquad$ for $i = 1, 2, \dots i'-1,\ k \in H(i),$ and $t = 1, 2, \dots T$

$\qquad\qquad\qquad \lambda_{i'kt} \geq 0 \qquad$ for $k \in H(i')$ and $t = l, l+1, \dots T$

$\qquad\qquad$ and (3′), (4′), (5′), (7′), (3′a), (4′a), (5′a), (7′a), and (10).

In the objective function,

$$v_{1t} = c_{1t} - \sum_{k \in H(1)}\sum_{j=t}^{T} a_{1k} \cdot \lambda_{1kj} \qquad\qquad \text{for } t = 1, 2, \dots T$$

$$v_{it} = c_{it} - \sum_{k \in H(i)}\sum_{j=t}^{T} a_{ik} \cdot \lambda_{ikj} + \sum_{j=t}^{T}\lambda_{\varphi(i),ij} \qquad \text{for } i = 2, 3, \dots i'-1 \text{ and } t = 1, 2, \dots T$$

$$v_{i't} = c_{i't} - \sum_{k \in H(i')}\sum_{j=t}^{T} a_{i'k} \cdot \lambda_{i'kj} + \sum_{j=t}^{T}\lambda_{\varphi(i'),i'j} \qquad \text{for } t = l, l+1, \dots T, \text{ and}$$

$$R' = V + \sum_{i=1}^{i'-1}\sum_{k \in H(i)\cap L(i)}\sum_{t=1}^{T}\lambda_{ikt}\cdot\sum_{j=1}^{t}Q_{kj} - \sum_{i=1}^{i'-1}\sum_{k \in H(i)}\sum_{t=1}^{T}\lambda_{ikt}\cdot I_{k0} + \sum_{t=1}^{l-1}\lambda_{\varphi(i'),i't}\cdot\sum_{j=1}^{t}X_{i'j}^{F}$$

$$+ \sum_{k \in H(i')}\sum_{t=l}^{T}\lambda_{i'kt}\cdot\left\{\sum_{j=1}^{t}Q_{kj} - \sum_{j=1}^{l-1}a_{i'k}\cdot X_{i'j}^{F} - I_{k0}\right\}.$$

Here, λ_{itk} is the Lagrangean multipliers corresponding to (6″) and (6″a), and R' is a constant and hence can be eliminated from further consideration.

The relaxed problem [LR] can be decomposed into the following mutually independent problems [SP$_i$], $i = 1, 2, \dots i'$:

[SP$_i$] Minimize $\sum_{t=1}^{T} s_i \cdot Y_{it} + \sum_{t=1}^{T} v_{it} \cdot X_{it}$

$\quad$ subject to $\quad \lambda_{ikt} \geq 0 \qquad$ for $k \in H(i)$ and $t = 1, 2, \dots T$

$\qquad\qquad$ and (3′), (4′), (5′), and (7′)

[SP$_{i'}$] Minimize $\sum_{t=l}^{T}(s_{i'} \cdot Y_{i't} + v_{i't} \cdot X_{i't})$

$\quad$ subject to $\quad \lambda_{i'kt} \geq 0 \qquad$ for $k \in H(i')$ and $t = l, l+1, \dots T$

$\qquad\qquad$ and (3′a), (4′a), (5′a), (7′a), and (10)

Each of the problems [SP$_i$] can be considered as the single item lot-sizing problem that is the same as that of Wagelmans *et al.* [12] except for the maximum terms in the constraints. Therefore, a lower bound can be obtained by solving problems [SP$_i$] using the algorithm suggested by Wagelmans *et al.* [12], and the best one can be obtained with the Lagrangean dual problem given below:

$$LB(\lambda^*) = \max_{\lambda} LB(\lambda)$$

The Lagrangean multipliers are updated using the subgradient optimization method. In this method, the followings:

$$\lambda_{ikt}^{w+1} = \max\left[0, \quad \lambda_{ikt}^{w} - \alpha_w \cdot \left\{\sum_{j=1}^{t}(a_{ik} \cdot X_{ij}^* - Q_{kj}^*) + I_{k0}\right\}\right]$$

for $i = 1, 2, \ldots i'-1$, $k \in H(i)$, and $t = 1, 2, \ldots T$

$$\lambda_{i'kt}^{w+1} = \max\left[0, \quad \lambda_{i'kt}^{w} - \alpha_w \cdot \left\{\sum_{j=l}^{t} a_{i'k} \cdot X_{i'j}^* - \sum_{j=1}^{t} Q_{kj}^* + \sum_{j=1}^{l-1} a_{i'k} \cdot X_{i'j}^F + I_{k0}\right\}\right]$$

for $k \in H(i')$ and $t = l, l+1, \ldots T$

where λ_{itk}^{w} and X_{it}^* denote the multiplier values and the optimal solution of the relaxed problem [LR] at iteration w, respectively. Also, α_w denotes the step sizes at iteration w, updated by

$$\alpha_w = \delta_w \frac{\overline{UB} - LB(\lambda^w)}{\sum_{i=1}^{i'-1}\sum_{k \in H(i)}\sum_{t=1}^{T}\left\{\sum_{j=1}^{t}(a_{ik} \cdot X_{ij}^* - Q_{kj}^*) + I_{k0}\right\}^2 + \sum_{k \in H(i')}\sum_{t=l}^{T}\left\{\sum_{j=l}^{t}a_{i'k} \cdot X_{i'j}^* - \sum_{j=1}^{t}Q_{kj}^* + \sum_{j=1}^{l-1}a_{i'k} \cdot X_{i'j}^F + I_{k0}\right\}^2}$$

where $\overline{UB}$ is the best upper bound and δ_w $(0 < \delta_w \leq 2)$ is a constant, which is initially set to 2 and is halved if $LB(\lambda^w)$ has not been improved in a given number of iterations. Finally, the algorithm is terminated after a given number of iterations, i.e., when the iteration count (w) reaches a given limit (W).

Obtaining the upper bounds

To obtain the upper bound at each node of the B&B tree, the decomposed problem [SP$_i$] is modified so that a feasible solution for the original problem [CP] is obtained. To do this, the following constraints are used instead of (7′) and (7′a) in [SP$_i$].

$$\sum_{j=1}^{t} X_{ij} \geq \max_{k \in H(i)} \frac{\sum_{j=1}^{t} Q_{kj} - I_{k0}}{a_{ik}} \qquad \text{for } i = 1, 2, \ldots i'-1 \text{ and } t = 1, 2, \ldots T$$

$$\sum_{j=l}^{t} X_{i'j} \geq \max_{k \in H(i')} \frac{\sum_{j=1}^{t} Q_{kj} - \sum_{j=1}^{l-1} a_{i'k} \cdot X_{i'j}^F - I_{k0}}{a_{i'k}} \qquad \text{for } t = l, l+1, \ldots T$$

The constraints given above are the modifications of (6″) and (6″a), in which the maximum terms are added. The modified problem can be also considered as the single item lot-sizing problem, and hence solved easily. Then, by solving the subproblem recursively from parent item i' to the root item 1, we can obtain a feasible solution for the original problem [CP]. Note that in the B&B algorithm, the initial solution is obtained by solving the modified problem after setting $i' = i_f-1$, $t' = 0$, and $l = 0$. Now, after obtaining the solution of the modified problem, the upper bound is calculated as

$$UB = V + \sum_{i=1}^{i'-1}\sum_{t=1}^{T}(s_i \cdot Y_{it} + c_{it} \cdot X_{it}) + \sum_{t=l}^{T}(s_{i'} \cdot Y_{i't} + c_{i't} \cdot X_{i't}).$$

5 Computational Experiments

This section presents the test results on the B&B algorithm suggested in this paper. For this purpose, we generated 60 test problems randomly using the disassembly structure given in Figure 1, i.e., 10 problems for each combination of two levels of the amount of setup cost (low and high) and three levels of the number of periods (10, 20, and 30). In the problem set, inventory holding costs were generated from $DU(5, 10)$ for leaf items and $U(0.2, 0.4) \cdot \sum_{k \in H(i)} h_k \cdot a_{ik}$ for parent item i. Here, $DU(a, b)$ and $U(a, b)$ are discrete and continuous uniform distributions with the range $[a, b]$, respectively. Also, disassembly operation costs were generated from $DU(50, 100)$, and setup costs were generated from $DU(500, 1000)$ and $DU(5000, 10000)$ for the cases of low and high setup cost, respectively. Finally, the demand in each period was generated from $DU(50, 200)$ and the initial inventory was set to 0 without loss of generality.

Table 1. Test results for the B&B algorithm

(a) Case of low setup costs

Problem	Number Period					
	10		20		30	
	$n_N{}^{\dagger}$	CPU	n_N	CPU	n_N	CPU
1	65	0.11	419	0.59	121	0.75
2	55	0.11	83	0.34	103	0.73
3	1	0.00	1	0.00	1	0.00
4	151	0.13	175	0.40	473	1.28
5	59	0.11	1	0.00	1	0.01
6	1	0.04	139	0.39	889	1.91
7	71	0.11	215	0.43	439	1.23
8	11	0.10	5	0.28	1	0.01
9	47	0.10	127	0.37	621	1.54
10	1	0.00	181	0.41	2053	3.41

$\dagger$ number of nodes generated in the B&B algorithm.

(b) Case of high setup costs

Problem	Number Period					
	10		20		30	
	$n_N{}^{\dagger}$	CPU	n_N	CPU	n_N	CPU
1	3787	1.02	–	–	–	–
2	569	0.24	8831	6.41	–	–
3	1157	0.38	1531913	1077.77	–	–
4	1	0.00	465	0.65	54287	82.63
5	1761	0.51	1089983	791.05	–	–
6	5911	1.40	678445	492.43	–	–
7	2289	0.67	2662965	1936.69	–	–
8	941	0.33	–	–	–	–
9	4767	1.20	–	–	–	–
10	735	0.25	25699	17.25	178353	255.62

Two performance measures were used in the test: number of nodes generated in the B&B algorithm and CPU seconds. Here, we set the time limit as 3600 seconds for the B&B algorithm due to the computational burden. Also, the parameters for the Lagrangean relaxation technique were set after a preliminary test. They are as follows. the iteration limit W was set to 1000 and 3 for the initial problem and subproblems, respectively; and δ_w was set to 2 initially and halved if the lower bound has not been improved in 10 and 1 iterations for the initial problem and subproblems, respectively. The B&B algorithm was coded in C and run on a Pentium processor (800 MHz).

The test results are summarized in Table 1. It can be seen from the table that the B&B algorithm solved quickly up to the problems with 30 periods for the case of low setup costs and 10 periods for the case of high setup costs. For these cases, the numbers of nodes generated were very small, which implies that the Lagrangean lower bounds suggested in this paper works well. As anticipated, the CPU seconds increase as the problem complexity (in terms of the number of periods) increases. Nevertheless, for the case of low setup costs, the number of periods has slightly impact on the performance of the B&B algorithm. Therefore, we can see that the amount of setup costs has highly impact on the performance and the lower and upper bounds obtained from solving the subproblems may not be close to the optimal solution value.

6 Concluding Remarks

In this paper, we considered the problem of determining the quantity and timing of disassembling used or end-of-life products in order to satisfy the demand of their parts or components over a finite planning horizon. Among various cases of the problem, we focus on the case of single product type with assembly structure for the objective of minimizing the sum of setup, disassembly operation, and inventory holding costs. Several optimal properties are derived which are used to develop the Lagrangean relaxation based B&B algorithm suggested in this paper. Test results showed that the B&B algorithm gives the optimal solutions for most of the test problems.

Acknowledgements

The financial support from the Swiss National Science Foundation under contract number 2000-066640 is gratefully acknowledged.

References

1. Crowston, W. B., Wagner, M. H.: Dynamic Lot Size Models for Multi-Stage Assembly Systems, Management Science, Vol. 20 (1973) 14–21.
2. Gupta, S. M., Taleb, K. N.: Scheduling Disassembly, International Journal of Production Research, Vol. 32 (1994) 1857–1886.
3. Kang, J.-G., Lee, D.-H., Xirouchakis, P., Lambert, A. J. D.: Optimal Disassembly Sequencing with Sequence Dependent Operation Times based on the Directed Graph of Assembly States, Journal of the Korean Institute of Industrial Engineers, Vol. 28 (2002) 264–273.

4. Kim, H.-J., Lee, D.-H., Xirouchakis, P., Züst, R.: Disassembly Scheduling with Multiple Product Types, CIRP Annals – Manufacturing Technology, Vol. 52 (2003) 403–406.
5. Lee, D.-H., Kang, J.-G., Xirouchakis, P.: Disassembly Planning and Scheduling: Review and Further Research, Proceedings of the Institution of Mechanical Engineers: Journal of Engineering Manufacture – Part B, Vol. 215 (2001) 695–710.
6. Lee, D.-H., Kim, H.-J., Choi, G., Xirouchakis, P.: Disassembly Scheduling: Integer Programming Models, Proceedings of the Institution of Mechanical Engineers: Journal of Engineering Manufacture – Part B, 218 (2004) 1357–1372.
7. Lee, D.-H., Xirouchakis, P.: A Two-Stage Heuristic for Disassembly Scheduling with Assembly Product Structure, Journal of the Operational Research Society, Vol. 55 (2004) 287–297.
8. Lee D.-H., Xirouchakis, P., Züst, R.: Disassembly Scheduling with Capacity Constraints, CIRP Annals – Manufacturing Technology, Vol. 51 (2002) 387–390.
9. Neuendorf, K.-P., Lee, D.-H., Kiritsis, D., Xirouchakis, P.: Disassembly Scheduling with Parts commonality using Petri-nets with Timestamps, Fundamenta Informaticae, Vol. 47 (2001) 295–306.
10. Taleb, K. N., Gupta, S. M.: Disassembly of Multiple Product Structures, Computers and Industrial Engineering, Vol. 32 (1997) 949–961.
11. Taleb, K. N., Gupta, S. M., Brennan, L.: Disassembly of Complex Product Structures with Parts and Materials Commonality, Production Planning & Control, Vol. 8 (1997) 255–269.
12. Wagelmans, A., Hoesel, S. V., Kolen, A.: Economic Lot Sizing: an $O(n \log n)$ Algorithm that Runs in Linear Time in the Wagner-Whitin Case, Operations Research, Vol. 40 (1992) 145–156.

Hybrid Planning Using Flexible Strategies

Bernd Schattenberg, Andreas Weigl, and Susanne Biundo

Dept. of Artificial Intelligence,
University of Ulm, D-89069 Ulm, Germany
firstname.lastname@informatik.uni-ulm.de

Abstract. In this paper we present a highly modular planning system architecture. It is based on a proper formal account of hybrid planning, which allows for the formal definition of (flexible) planning strategies. Groups of modules for flaw detection and plan refinement provide the basic functionalities of a planning system. The concept of explicit strategy modules serves to formulate and implement strategies that orchestrate the basic modules. This way a variety of fixed plan generation procedures as well as novel flexible planning strategies can easily be implemented and evaluated. We present a number of such strategies and show some first comparative performance results.

1 Introduction

Hybrid planning – the combination of hierarchical task network (HTN) planning with partial order causal link (POCL) techniques – turned out to be most appropriate for complex real-world planning applications [1,2,3]. Here, the solution of planning problems often requires the integration of planning from first principles with the utilization of predefined plans to perform certain complex tasks.

Up to now the question of how suitable strategies for hybrid planning, which actually exploit the merge of the two planning paradigms, should look like has rarely been addressed. Nor have existing strategies been systematically compared or evaluated in this context.

In this paper we present a highly modular architecture that can serve as an experimental platform for the implementation of hybrid planning systems and respective strategies as well as for their evaluation and comparison. Based on a formal account of hybrid planning, which provides formal definitions of flaws and plan modification steps, and the relation between both, elementary modules for flaw detection and plan modification are specified. These modules implement not only the basic functionalities of the system described in previous work [2], but also for any hybrid planning system which operates on the plan space. In order to design the behavior (strategy) of a particular planning system, the concept of a so-called *strategy module* is provided. As strategy modules are able to organize the interplay between flaw detection and plan modification in various ways, total freedom is given for the design of planning strategies. This does enable the empirical evaluation of search control for hybrid planning and in addition allows for the definition of novel *flexible planning strategies*, which are independent

U. Furbach (Ed.): KI 2005, LNAI 3698, pp. 249–263, 2005.

from the actual planning methodology. These *flexible* planning strategies act opportunistically instead of following a fixed plan generation schema.

The rest of the paper is organized as follows. In Section 2 we present the formal framework on which our hybrid planning approach relies. Basic planning system components and the design of strategies are presented in Section 3. In Section 4 we introduce various such strategies and show some comparative experimental results in Section 5.

2 Basic Definitions

Our hybrid planning formalism relies on a logical language $L = \{P, C, V, O, T\}$. P and C are non-empty, finite sets of predicate and constant symbols. V is a finite set of variables. O denotes a non-empty finite set of *operator* symbols, T a finite set of *task* symbols. All sets are assumed to be disjoint and each element of P, O, and T is assigned a natural number denoting its arity.

An *operator schema* $o(\bar{\tau}) = (\mathrm{prec}(o(\bar{\tau})), \mathrm{add}(o(\bar{\tau})), \mathrm{del}(o(\bar{\tau})))$ specifies the *preconditions* and the *positive* and *negative effects* of that operator ($o \in O$, $\bar{\tau} = \tau_1, \ldots, \tau_n$ with $\tau_i \in \{C \cup V\}$ for $1 \leq i \leq n$ and n being the arity of o). Preconditions and effects are sets of literals and atoms over $P \cup C \cup V$, respectively. A ground instance of an operator schema is called an *operation*. A *state* is a finite set of ground atoms and an operation $o(\bar{c})$ is *applicable* in a state s iff for the positive literals in the precondition of o: $\mathrm{prec}^{\oplus}(o(\bar{c})) \subseteq s$ and for the negative literals: $\mathrm{prec}^{\ominus}(o(\bar{c})) \cap s = \emptyset$. The result of applying $o(\bar{c})$ in state s is a state $s' = (s \cup \mathrm{add}(o(\bar{c}))) \setminus \mathrm{del}(o(\bar{c}))$. Operators are also called *primitive tasks*.

Abstract actions are represented by *complex tasks*. In hybrid planning, complex tasks, like primitive ones, show preconditions and effects. They are defined by *task schemata* $t(\bar{\tau}) = (\mathrm{prec}(t(\bar{\tau})), \mathrm{add}(t(\bar{\tau})), \mathrm{del}(t(\bar{\tau})))$ for $t \in T$. A *decomposition method* $m = (t(\bar{\tau}), d)$ relates a complex task $t(\bar{\tau})$ to a *task network* or *partial plan* d. This task network can be seen as a pre-defined *implementation* of $t(\bar{\tau})$.

A task network over L is a tuple $(TE, \prec, VC, CL)$ with TE being a set of *task expressions* $te = l : t(\tau_1, \ldots, \tau_m)$, where l represents a unique label, $t \in T \cup O$, and $\tau_i \in V \cup C$ for $1 \leq i \leq m$. $\prec$ is a set of *ordering constraints* imposing a partial order on TE. VC denotes a set of *variable constraints* (equations and inequations), which *codesignate* and *non-codesignate* variables in TE with variables or constants. CL is a set of *causal links* $te_i \xrightarrow{\phi} te_j$ with $te_i, te_j \in TE$ and ϕ being a literal with $\sigma(\phi) \in \sigma(\mathrm{prec}(te_j))$ and $\sigma(\phi) \in \sigma(\mathrm{add}(te_i))$, if ϕ is positive, and $\sigma(|\phi|) \in \sigma(\mathrm{del}(te_i))$, if ϕ is negative. σ is a VC-compatible variable substitution, i.e. a substitution consistent with the variable constraints. A causal link indicates that a task $(l : t_i(\bar{\tau}))$ *establishes* a precondition of a task $(l' : t_j(\bar{\tau}'))$.

Definition 1 (Planning Problems). *A planning problem $(d, \mathsf{T}, \mathsf{M})$ consists of an initial task network d, a set of task and operator schemata T, and a set M of methods to implement the complex tasks in T.*

Given a planning problem $(d, \mathsf{T}, \mathsf{M})$, a hybrid planning system transforms d into a task network that is a *solution* to the problem.

Definition 2 (Solutions). *A partial plan $P = (TE, \prec, VC, CL)$ over L is a solution to a planning problem $(d, \mathsf{T}, \mathsf{M})$ iff the following conditions hold.*

1. *for all $(l : t(\bar{\tau})) \in TE \ : \ t \in O$ (the plan is primitive)*
2. *for all $te \in TE \ : \ te \not\prec^* te$, where $\prec^*$ denotes the transitive closure of $\prec$ (the partial order on task expressions is acyclic)*
3. *for all $v \in V \ : \ VC \not\models v \neq v$ (VC is consistent)*
4. *for all $v \in V$ occuring in TE, there exists $c \in C$ such that $VC \models v = c$ (the plan is fully instantiated)*
5. *for all $(te_i \xrightarrow{\phi} te_j) \in CL \ : \ te_j \not\prec^* te_i$ (CL is compatible with $\prec$)*
6. *for all $te \in TE$, there exists a set $\{te_i \xrightarrow{\phi_i} te | 1 \leq i \leq k\} \subseteq CL$ such that $prec(te) \subseteq \bigcup_{1 \leq i \leq k} \phi_i$ (all preconditions of each task are established)*
7. *for all $(te_i \xrightarrow{\phi} te_j) \in CL$, there exists no te_k with $te_k \not\prec^* te_i$ and $te_j \not\prec^* te_k$, and $\sigma(\phi) \in \sigma(del(te_k))$, if ϕ is positive, or $\sigma(|\phi|) \in \sigma(add(te_k))$ if ϕ is negative for a VC-compatible variable substitution σ (all tasks are applicable)*

In order to obtain a solution, an initial task network is stepwise refined by adding new tasks and constraints – causal links, ordering and variable constraints – and by replacing abstract tasks through appropriate partial plans as specified by the given decomposition methods. The following plan modifications are used to perform these plan refinements.

Definition 3 (Plan Modifications). *A plan modification m is a pair $(mod, E^{\oplus} \cup E_{\ominus})$ with mod denoting the modification class. $E^{\oplus}$ and $E_{\ominus}$ are elementary additions and deletions of plan components, respectively. The set of all plan modifications is denoted by $\mathcal{M}$ and grouped into subsets $\mathcal{M}_{mod}$ for given classes mod.*

For a partial plan $P = (TE, \prec, VC, CL)$ over $L = \{P, C, V, O, T\}$ and a planning problem $(d, \mathsf{T}, \mathsf{M})$, the following classes of plan modifications do exist:

1. $(\texttt{InsertTask}, \{\oplus te, \oplus(te \xrightarrow{\phi} te'), \oplus(v_1 = \tau_1), \ldots, \oplus(v_k = \tau_k)\})$ with $te = l : t(\bar{\tau}) \notin TE, t \in \mathsf{T}$, being a new task expression to be added, $te' \in TE$, and $\sigma'(\phi) \in \sigma'(add(te))$ for positive literals ϕ, $\sigma'(|\phi|) \in \sigma'(del(te))$ for negative literals, and $\sigma'(\phi) \in \sigma'(prec(te'))$ with σ' being a VC'-compatible substitution for $VC' = VC \cup \{v_i = \tau_i | 1 \leq i \leq k\}$.
2. $(\texttt{AddOrdConstr}, \{\oplus(te_i \prec te_j)\})$ for $te_i, te_j \in TE$
3. $(\texttt{AddVarConstr}, \{\oplus(v = \tau)\})$ for codesignating and $(\texttt{AddVarConstr}, \{\oplus(v \neq \tau)\})$ for non-codesignating variables $v \in V$ with terms $\tau \in V \cup C$.
4. $(\texttt{AddCLink}, \{\oplus(te_i \xrightarrow{\phi} te_j), \oplus(v_1 = \tau_1), \ldots, \oplus(v_k = \tau_k)\})$, with $te_i, te_j \in TE$. The codesignations represent variable substitutions, such that there exists a VC'-compatible substitution σ' for $VC' = VC \cup \{(v_1 = \tau_1), \ldots, (v_k = \tau_k)\}$ for which $\sigma'(\phi) \in \sigma'(add(te_i))$ for positive literals ϕ, $\sigma'(|\phi|) \in \sigma'(del(te_i))$ for negative literals, and $\sigma'(\phi) \in \sigma'(prec(te_j))$.

5. Given an abstract task expression $te = l : t(\bar{\tau})$ in TE, $l \in T$ and a method $m = (t, (TE_m, \prec_m, VE_m, CL_m))$ in M:

$$(\texttt{ExpandTask}, \{\ominus te\} \cup \{\oplus te_m | te_m \in TE_m\} \cup$$
$$\{\oplus(te_{m1} \prec te_{m2}) | (te_{m1} \prec_m te_{m2})\} \cup$$
$$\{\oplus(te_{m1} \xrightarrow{\phi} te_{m2}) | (te_{m1} \xrightarrow{\phi} te_{m2}) \in CL_m\}$$
$$\{\oplus(v = \tau) | (v = \tau) \in VC_m\} \cup \{\oplus(v \neq \tau) | (v \neq \tau) \in VC_m\} \cup$$
$$\{\ominus(te' \prec te), \oplus(te' \prec te_m) | (te' \prec te), te_m \in TE_m\} \cup$$
$$\{\ominus(te \prec te'), \oplus(te_m \prec te') | (te \prec te'), te_m \in TE_m\} \cup$$
$$\{\ominus(te \xrightarrow{\phi} te'), \oplus(te_m \xrightarrow{\phi} te') | (te \xrightarrow{\phi} te') \in CL,$$
$$\qquad te_m \in TE_m, |\phi| \in \text{add}(te_m) \cup \text{del}(te_m)\} \cup$$
$$\{\ominus(te' \xrightarrow{\phi} te), \oplus(te' \xrightarrow{\phi} te_m) | (te' \xrightarrow{\phi} te) \in CL,$$
$$\qquad te_m \in TE_m, \phi \in \text{prec}(te_m)\}$$

During an expansion the abstract task is replaced by the decomposition network with all its sub-tasks being ordered between the predecessors and successors of the abstract task and with all the causalities re-distributed among the appropriate sub-tasks[1]. In this paper we omit the use of axiomatic state refinements as presented in previous work [2], thereby restricting task networks occuring in methods to those in which all preconditions and effects of the abstract task explicitly occur.

Plan modifications are the canonical generators for a refinement-based planner: starting from an initial task network, the current plan is checked against the solution criteria, and while not all of them are met, plan modifications are applied. If no applicable modification exists, backtracking is performed. In order to make the search more systematic and efficient, we want the algorithm to focus on those plan modification steps which are appropriate to overcome the deficiencies in the current plan. To this end, we make these deficiencies explicit by representing solution criteria violations as *flaws* which literally "point" to critical components of the plan. Flaws will be used to trigger the selection of appropriate plan modifications.

Definition 4 (Flaws). *A flaw* f *is a pair* $(flaw, E)$ *with flaw indicating the flaw class and* E *being a set of plan components the flaw refers to. The set of flaws is denoted by* $\mathcal{F}$ *with subsets* $\mathcal{F}_{flaw}$ *for given labels flaw.*

The set $\mathcal{F}$ of flaw classes in a partial plan $(TE, \prec, VC, CL)$ over L addresses the solution criteria and consists of the following sub-sets:

1. $(\texttt{AbstrTask}, \{te\})$ with $te = l : t(\bar{\tau}) \in TE, t \in T$ being an abstract task expression
2. $(\texttt{OrdIncons}, \{te_1, \dots, te_k\})$ with $te_i \in TE, te_i \prec^* te_i, 1 \leq i \leq k$, i.e. if $\prec^*$ defines a cyclic partial order

---

[1] If the causal links cannot be re-distributed unambiguously, e.g., if there are two tasks in the expansion network which carry the appropriate precondition, one expanding modification is generated for each such permutation.

3. $(\texttt{VarIncons}, \{v\})$ with $v \in V$ being a variable for which $VC \models v \neq v$ holds
4. $(\texttt{OpenVarBind}, \{v\})$ with $v \in V$ being a variable occurring in TE and there exists no constant $c \in C$ with $VC \models v = c$
5. $(\texttt{OrdIncons}, \{te_1, te_2\})$ with $te_1, te_2 \in TE$ being causally linked task expressions, say $te_1 \xrightarrow{\phi} te_2 \in CL$, for which $te_2 \prec^* te_1$ does hold
6. $(\texttt{OpenPrec}, \{te, \phi\})$ with $\phi \in \operatorname{prec}(te), te \in TE$ denotes a not fully supported task, i.e. for the subset of te-supporting causal links $\{te_i \xrightarrow{\phi_i} te | 1 \leq i \leq k\} \subseteq CL$ we find $\bigcup_{1 \leq i \leq k} \phi_i \subset \operatorname{prec}(te)$
7. $(\texttt{Threat}, \{te_i \xrightarrow{\phi} te_j, te_k\})$ with $te_k \not\prec^* te_i$ or $te_j \not\prec^* te_k$ and there exists a VC-compatible substitution σ such that $\sigma(\phi) \in \sigma(\operatorname{del}(te_k))$ for positive literals ϕ and $\sigma(|\phi|) \in \sigma(\operatorname{add}(te_k))$ for negative literals

3   Strategic Hybrid Planning

Based on the formal notions of plan modifications and flaws, arbitrary planning strategies can be defined. A planning strategy specifies *how* and *when* the flaws in a partial plan are eliminated through appropriate plan modification steps. As a prerequisite, we therefore need to define the conditions under which a plan modification can *in principle* eliminate a given flaw.

Definition 5 (Suitable Modifications). *A class of plan modifications* $\mathcal{M}_m \subseteq \mathcal{M}$ *is suitable for a class of flaws* $\mathcal{F}_f \subseteq \mathcal{F}$ *iff there exists a partial plan* P, *which contains a flaw* $\texttt{f} \in \mathcal{F}_f$, *and a modification* $\texttt{m} \in \mathcal{M}_m$ *such that the refined plan* $P' = app(\texttt{m}, P)$ *does not contain* $\texttt{f}$.

From the definition of modifications it follows that an eliminated flaw cannot be re-introduced, i.e. once $\texttt{f}$ has been eliminated from P it does not appear in any successor plan obtained by the application of a sequence of plan modifications.

Relying on the above definitions the following "advice" or trigger function $\alpha : 2^{\mathcal{F}} \rightarrow 2^{\mathcal{M}}$ relates flaws and suitable modifications.(The suitability proofs are omitted for lack of space.)

Definition 6 (Modification Triggering Function). *Flaws in a partial plan can be removed by triggering the application of suitable plan modification steps according to the following function:*

$$\alpha(\mathcal{F}_x) = \begin{cases} \mathcal{M}_{\texttt{ExpandTask}} & \textit{if } \mathcal{F}_x = \mathcal{F}_{\texttt{AbstrTask}} \\ \mathcal{M}_{\texttt{AddCLink}} \cup \mathcal{M}_{\texttt{InsertTask}} \cup \mathcal{M}_{\texttt{ExpandTask}} & \textit{if } \mathcal{F}_x = \mathcal{F}_{\texttt{OpenPrec}} \\ \mathcal{M}_{\texttt{ExpandTask}} \cup \mathcal{M}_{\texttt{AddOrdConstr}} \cup \mathcal{M}_{\texttt{AddVarConstr}} & \textit{if } \mathcal{F}_x = \mathcal{F}_{\texttt{Threat}} \\ \mathcal{M}_{\texttt{AddVarConstr}} & \textit{if } \mathcal{F}_x = \mathcal{F}_{\texttt{OpenVarBind}} \\ \emptyset & \textit{if } \mathcal{F}_x = \mathcal{F}_{\texttt{OrdIncons}} \\ \emptyset & \textit{if } \mathcal{F}_x = \mathcal{F}_{\texttt{VarIncons}} \end{cases}$$

This definition clearly displays the flexibility of hybrid planning. Supplying a precondition literal for a task te can not only be done by adding a causal link or inserting a "producing" task appropriately; it can also be done by decomposing a complex task – not ordered "behind" te – that produces the desired effect through tasks occurring in its expansion network. In a similar fashion, threats of causal links can not only be addressed through the usual placing of threatening tasks outside the scope of the causal links or by decoupling variable constraints. If abstract tasks are involved in the threat situation, hybrid planning can also make use of task decomposition for producing "overlapping" task networks that may offer less strict promotion or demotion opportunities, since the causalities in the expansion network are typically linked from and to several of the introduced subtasks. As a side effect, the variable constraints of such a network may also rule out the threat. Ordering cycles and variable inconsistencies obviously cannot be resolved by adding constraints. Therefore, they do not trigger any modification.

As an important feature of our approach, the *explicitly stated* trigger function allows to completely separate the computation of flaws from the computation of modifications, and in turn both computations can be separated from search related issues. The system architecture relies on this separation and exploits it in two ways: module invocation and interplay are specified through α while the explicit reasoning about search can be performed on the basis of flaws and modifications without taking their actual computation into account. The issued flaws can *only* be addressed by the assigned modification generators; if none can solve the flaw, the system has to backtrack. Hence, we can map flaw and modification classes directly onto groups of modules which are responsible for their computation, thereby encapsulating both the detection of solution criteria violations and the computation of possible modifications.

Definition 7 (Detection Modules). *A detection module x is a function which given a partial plan P returns all flaws of type x that are in P:*

$$f_x^{det} : \mathcal{P} \to 2^{\mathcal{F}_x}$$

It ranks these flaws according to its local priorities. $\mathcal{F}_{\texttt{OpenPrec}}$ flaws are, e.g., prioritized according to the number of literals in the tasks' preconditions.

Definition 8 (Modification Modules). *A modification module y is a function which computes all modification steps of type y addressing the flaws in a partial plan:*

$$f_y^{mod} : \mathcal{P} \times 2^{\mathcal{F}_x} \to 2^{\mathcal{M}_y} \text{ for } \mathcal{M}_y \subseteq \alpha(\mathcal{F}_x)$$

These modules also prioritize their answers with local preferences. Priorities for modifications in $\mathcal{M}_{\texttt{InsertTask}}$ correlate with the number of available task schemata and implied variable constraints, for example. Furthermore, the modification generators store information about which flaw a modification refers to, so that the strategy is able to look up corresponding flaw/modification pairs if required.

Based on the detected flaws and proposed modifications, a separate search strategy chooses a plan modification step to remove these flaws.

Definition 9 (Strategy Modules). *A strategy module z is a function which selects plan modifications for application to a current plan. It is defined by the projection*

$$f_z^{strat} : \mathcal{P} \times 2^{\mathcal{F}} \times 2^{\mathcal{M}} \to \mathcal{M} \cup \epsilon$$

Strategies discard a current plan P if any flaw remains un-addressed by the associated modification modules, i.e., if for any f_x^{det} and $f_{y_1}^{mod}, \ldots, f_{y_n}^{mod}$ with $\mathcal{M}_{y_1} \cup \ldots \cup \mathcal{M}_{y_n} = \alpha(\mathcal{F}_x)$:

$$\bigcup_{1 \leq i \leq n} f_{y_i}^{mod}(P, f_x^{det}(P)) = \emptyset$$

The following generic algorithm implements the stepwise refinement of partial plans and is used as the core component of any planning system to be implemented within our architecture:

plan(P, T, M):
$F \leftarrow \emptyset$
for all f_x^{det} **do**
  $F \leftarrow F \cup f_x^{det}(P)$
if $F = \emptyset$ **then**
  return P
$M \leftarrow \emptyset$
for all $F_x = F \cap \mathcal{F}_x$ with $F_x \neq \emptyset$ **do**
  answered $\leftarrow$ **false**
  for all f_y^{mod} with $\mathcal{M}_y \subseteq \alpha(\mathcal{F}_x)$ **do**
    $M' \leftarrow f_y^{mod}(P, F_x)$
    if $M' \neq \emptyset$ **then**
      $M \leftarrow M \cup M'$
      answered $\leftarrow$ **true**
  if answered = **false then**
    return fail
return plan(apply(P, $f_z^{strat}(P, F, M)$), T, M)

Please note, that the algorithm does not depend on the deployed modules, since the options to address existing flaws by appropriate plan modifications is defined via the α function. The call of the strategy function z is of course the backtracking point of the system.

4   Search Strategies

The following adaptations of strategies taken from the literature illustrate the potential of our framework. Most planning algorithms follow a fixed preference relation on flaws and/or plan modifications. For example, a hybrid variant of the HTN-based UMCP strategy [4] prefers task decomposition steps until at the primitive plan level:

$$f_{UMCP}^{strat}(P, F, M) = \mathtt{m} \in \begin{cases} select(M \cap \mathcal{M}_{\mathtt{ExpandTask}}) \text{ if } F \cap \mathcal{F}_{\mathtt{AbstrTask}} \neq \emptyset \\ M \text{ otherwise} \end{cases}$$

The task selection function for decomposition *select* returns expansion refinements according to heuristics like *fewest alternatives first* which returns a modification from the smallest answering set of an abstract task flaw. Additional heuristic selection functions can be called, e.g. selecting expansions with a minimal number of added tasks, variables, or constraints.

Another example for a successful hybrid planning strategy is that of the OCLh system [5]: *Expand then Make Sound.* Our EMS variant alternates task expansions with other modifications such that a task expansion is only performed if no other flaws are issued (only expand the next abstract task if the network is sound):

$$f_{EMS}^{strat}(P, F, M) = \mathtt{m} \in \begin{cases} select(M \cap \mathcal{M}_{\mathtt{ExpandTask}}) \text{ if } F \cap (\mathcal{F} \setminus \mathcal{F}_{\mathtt{AbstrTask}}) = \emptyset \\ M \text{ otherwise} \end{cases}$$

Regarding the modification selection for decomposition, it relies on the local preferences of the modification module.

Although widely considered to be much more related to UMCP, the Shop planning strategy turns out to be very close to OCLh when mirrored in our hybrid framework. Shop follows the paradigm to generate plans by expanding abstract tasks in the order in which they are to be executed, thereby always updating the current state from the initial situation until the point of task execution [6]. This can be translated into solving all but the abstract task flaws, as far as these open precondition, etc. flaws concern those primitive tasks which are ordered before the first complex task. Let T_{left} be the set of such "first" complex tasks and T_{before} the tasks ordered before them.

$$f_{SHOP}^{strat}(P, F, M) = \mathtt{m} \in \begin{cases} select(M \cap \mathcal{M}_{\mathtt{ExpandTask}}) \\ \quad \text{if } F \cap (\mathcal{F} \setminus \mathcal{F}_{\mathtt{AbstrTask}}) = \emptyset \\ \quad \text{for } t_f \text{ in } \mathtt{f}, \mathtt{f} \in F, t_f \in T_{before} \text{ and} \\ \quad t_m \text{ in } \mathtt{m}, \mathtt{m} \in M, t_m \in T_{left} \\ M \text{ otherwise} \end{cases}$$

The main difference between more generalized techniques in the spirit of least commitment planning and the above *fixed* strategies is the exploitation of flaw/modification information: the "classical" strategies are *flaw-dependent* as they primarily rely on a flaw type preference schema, and *modification-dependent* as their second criterion is usually the modification type.

Least Committing First is a generalized variant of *Least Cost Flaw Repair* [7], a flexible strategy which selects modifications that address flaws for which the smallest number of modifications has been found (this corresponds to a LCFR with uniform modification costs). Thanks to our framework's formal basis and uniform representation, we can easily apply this heuristic to hybrid planning as well, thereby combining it with the principles from the UMCP *fewest alternatives first* task expansion selection.

Definition 10 (Least Committing First). *Let* $\mathtt{f} \in \mathcal{F}$ *be a flaw and* $\mathtt{m}_1, \ldots, \mathtt{m}_n \in \mathcal{M}$ *an arbitrary set of modifications. We define the committing level* f *of such a flaw as follows:*

$$f(\mathtt{f}, \{\mathtt{m}_1, \ldots, \mathtt{m}_n\}) = \begin{cases} 0 & \text{for } n = 0 \\ 1 + f(\mathtt{f}, \{\mathtt{m}_1, \ldots, \mathtt{m}_{n-1}\}) & \text{for } \mathtt{m}_n \text{ answering } \mathtt{f} \\ f(\mathtt{f}, \{\mathtt{m}_1, \ldots, \mathtt{m}_{n-1}\}) & \text{otherwise} \end{cases}$$

The Least Committing First *strategy selects from the set of modifications, answering flaws with a minimal f value.*

$$f_{LCF}^{strat}(P, F, M) = \mathtt{m} \in \{\mathtt{m}_\mathtt{f} | \mathtt{f} \in \min(f(\mathtt{f}, M))$$

It can easily be seen, that this is a *flexible* strategy, since it does not depend on the actual types of issued flaws and modifications: it just compares answer set sizes in order to keep the branching low. As an interesting historic note, many classical strategies have been formulated in terms of this more general flaw repair costs, their algorithmic procedure however jams these principles into a fixed preference schema like the ones presented above, and uses them only as a tie-break rule for equal preferences.

Going one step further in this direction, the problem of modification selection can also be tackled by comparing the triggering flaws and identifying commonly affected parts of the plan.

Let $\mathtt{f} \in \mathcal{F}$ be a flaw structure $(l, \{e_1, \ldots, e_n\})$. We define the *HotSpot value* g of such a flaw, given alternatives $\mathtt{f}_1, \ldots, \mathtt{f}_m$, as follows:

$$g(\mathtt{f}, \{\mathtt{f}_1, \ldots, \mathtt{f}_m\}) = \begin{cases} 0 & \text{if } m = 0 \\ g(\mathtt{f}, \{\mathtt{f}_1, \ldots, \mathtt{f}_{m-1}\}) & \text{if } \mathtt{f} = \mathtt{f}_m \\ g(\mathtt{f}, \{\mathtt{f}_1, \ldots, \mathtt{f}_{m-1}\}) \\ \quad + |\{e_1, \ldots, e_n\} \cap \{e_{m_1}, \ldots, e_{m_n}\}| & \text{otherwise} \end{cases}$$

Let us look deeper into the structure of our plan components which typically consist of "smaller" elements, say $e_1, \ldots, e_v$: A task can be decomposed into the task itself, its argument variables, and its literal sets; causal threats into the producer, the consumer, the threatening task, and the protected condition, etc. With this decomposition, we can extend g to a heuristic h of finer granularity.

$$h(\mathtt{f}, \{\mathtt{f}_1, \ldots, \mathtt{f}_m\}) = \begin{cases} 0 & \text{if } m = 0 \\ h(\mathtt{f}, \{\mathtt{f}_1, \ldots, \mathtt{f}_{m-1}\}) & \text{if } \mathtt{f} = \mathtt{f}_m \\ h(\mathtt{f}, \{\mathtt{f}_1, \ldots, \mathtt{f}_{m-1}\}) \\ \quad + |\{e_1, e_{1_1} \ldots, e_{n_v}\} \cap \{e_{m_1}, e_{m_{1_1}} \ldots, e_{m_{n_v}}\}| & \text{otherwise} \end{cases}$$

The HotSpot values g and h determine the number of overlapping elements within flaws, e.g. an abstract task with two arguments and with an open precondition increases the heuristic value g of the two respective flaws by 1 and the h value by 3 (assuming that the two argument variables appear in the precondition). The rationale behind this calculations is that a plan might have "weak points" at which numerous deficiencies can be found, and that refinements for fixing these problems typically interfere with each other. We therefore advise the system to solve isolated sub-problems first – a very strong notion of least commitment.

Definition 11 (HotSpot Avoidance Strategy). *The HotSpot Avoidance strategy chooses those modifications which try to solve the flaws with the lowest HotSpot h values.*

$$f^{strat}_{HotSpot}(P, F, M) = select_g(\{\mathtt{m_f} | \mathtt{f} \in \min(h(\mathtt{f}, F))\})$$

The selection function uses the heuristic g to prioritize the modifications with minimal h value. This hot spot avoidance seems counter-intuitive, since one would expect that concentrating on "common issues" would be comparably effective like it has turned out to be the case in constraint satisfaction. The DPLL search algorithm, e.g., gains its performance through dealing with the most constrained variables first. Early experiments have shown, however, that a hot spot *seeking* heuristic performs significantly worse than the presented one. We believe, that this is due to hidden dependencies between the modifiaction steps on different levels of plan development: the influence of modification choices on later choices are not trivially visible – a HotSpot-like heuristic on the competing modifications for a given HotSpot candidate shows, e.g., that typically these modifications show no commonalities, while the consequences in the next planning cycle, however, are dramatically which in turn tends to lead to a high number of backtrackings at an early stage. In contrast, our hot spot avoidance strategy tries to solve those problems in the plan first, which are not subject to competing solution proposals and are therefore very likely to be solved by any of the few proposals. As a side effect, these modifications typically reduce the number of modification proposals for the hot spots indirectly by constraining involved step orderings or variable assignments, etc.

5   Experimental Results

For a first evaluation of strategies, we have chosen a former planning competition benchmark for HTN systems, the UMTranslog domain as it has been shipped with UMCP. This domain deals with transportation and logistic problems on a structured territory: packages have to be delivered to customers by various means. The abstraction hierarchy on actions structures the transportation variants: between cities in the same and in different regions, using transportation hubs, handling valuable and hazardous packages, etc. Our system builds upon a sorted logic, so we translated the type predicates into a corresponding sort hierarchy; additional type annotations for indicating physical properties, etc., are represented via multiple super-sorts (cf. Fig. 1). Between network constraints are mapped on respective causal links, while most of the initially and finally constraints were translated into preconditions and effects of the corresponding abstract task.

The planning problems we considered consisted of one to five abstract transportation tasks for delivering a package to a chosen destination provided with enough transportation means in the initial state to do so in parallel. Solution plan length varied from 3 to 6 times the number of initial tasks.

The competing search strategy classes were the fixed preference based f_{EMS}^{strat} (which shows basically the same behaviour as our f_{SHOP}^{strat} implementation) and f_{UMCP}^{strat} against the flexible f_{LCF}^{strat} and $f_{HotSpot}^{strat}$. We used an adaption of the breadth-first plan selection heuristic as defined in [8]. The actual plan selection implementation, however, turned out to have a minor influence on the overall performance. Please note, that a direct comparison between our results and those in the original literature is neither intended nor adequate, because their original focus lay on different domains and plan generation paradigms. They have rather to be viewed as representatives of specific search schemata.

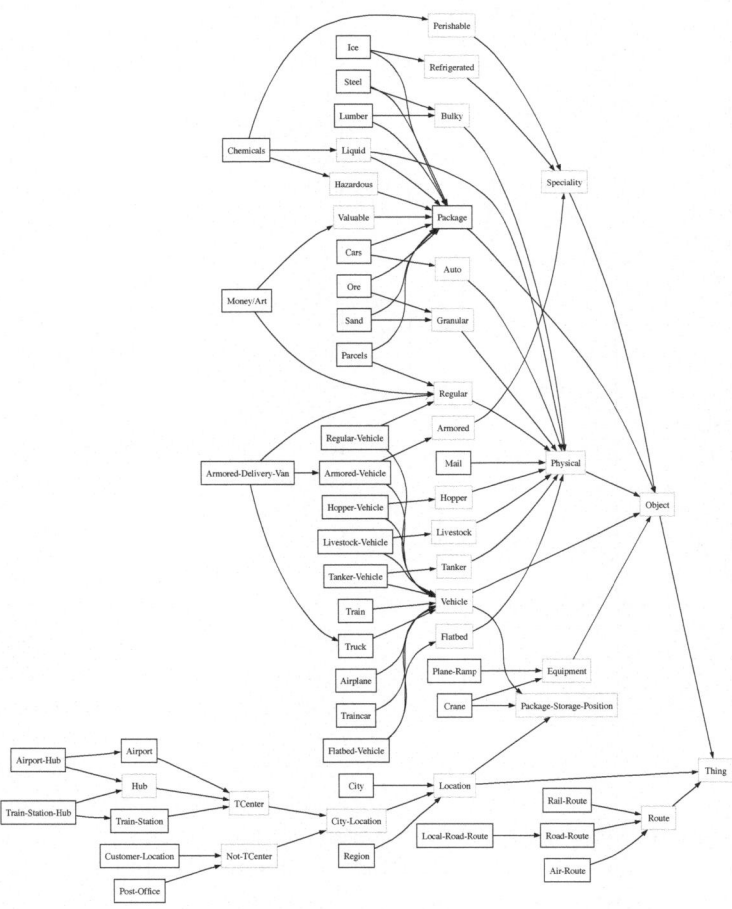

Fig. 1. The sort hierarchy used in the UMTranslog domain. Grey boxes indicate abstract sorts for which no instances are allowed.

Table 1 shows the average results for 10 runs per problem in terms of the number of examined plans, the number of discarded plans (backtracking), and the ratios of flaws and modifications selected to the ones issued (i.e. options which have been pruned by the strategy). The latter indicates how much the strategies' modification selection pruned the search space. The last two columns show the longest path generated during search and the number of plan modification steps that constitute the first solution.

Table 1. Experimental results

f^{strat}	size	bt	% det	% mod	max	sol
Initial Tasks: 1						
UMCP	5	0	26.3	1.5	6	6
EMS	6	0	42.9	4.1	6	6
LCF	5	0	26.3	1.4	6	6
HotSpot	5	0	66.7	6.3	6	6
Initial Tasks: 2						
UMCP	17	1	17.9	0.7	12	12
EMS	22	4	35.9	7.0	16	16
LCF	18	4	17.6	13.1	16	16
HotSpot	17	1	82.9	8.2	12	12
Initial Tasks: 3						
UMCP	32	3	16.1	0.4	19	19
EMS	45.2	5.4	25.8	4.3	23	23
LCF	32	6	15	1.8	23	23
HotSpot	32	3	76.9	9.9	19	19
Initial Tasks: 4						
UMCP	49.3	5.6	18.8	0.3	28	27
EMS	75	4.4	22.7	4.2	28	28
LCF	49.2	5.6	20.3	2.8	27	27
HotSpot	50	6	50.9	8.2	28	27
Initial Tasks: 5						
UMCP	157	43.4	18.9	0.3	40	40
EMS	142.2	15.6	17.0	3.8	43	43
LCF	75	13	19.5	3.3	40	40
HotSpot	82	14.8	50.2	8.5	40	40

The results indicate that a flexible strategy guides the system reasonably well through the search space. The fixed strategies seem not to run into purely branching related problems, their cut-down of detections and modifications appears to work much better than for HotSpot. In the runs on 3 initial tasks, e.g., f_{EMS}^{strat} generates plans which have on average only 1.75 relevant flaws and 4.72 modifications, while f_{LCF}^{strat} plans even show 2.2 flaws and 5.36 modifications. Deeper analysis of their logs however reveals, that f_{LCF}^{strat} and $f_{HotSpot}^{strat}$ prune much better in the early cycles (differences in the search space development are depicted in Figure 2). The low number of HotSpot's detection pruning results from choosing symmetric variable assignments which leave a lot of them un-selected for application after a solution has been found.

The results support our claim: letting the modules perform their planning-specific computation focuses search on a set of qualitatively good refinements, which carry enough information for a de-coupled strategy to properly judge the alternatives. It has also to be stressed that the participating strategies and modules did not undergo any kind of heuristic fine tuning of parameters and alike. What we report are results merely obtained by simple selection principles which are still open to all kind of performance enhancements. Furthermore, the results have been obtained for problems which are solvable by HTN-only systems to meet literature's standards on common ground. We note that despite performing competitive in the structurally better understood "HTN-part", the flexible strategies – and in particular $f_{HotSpot}^{strat}$ – improve their performance significantly on partially specified initial task networks with goal conditions.

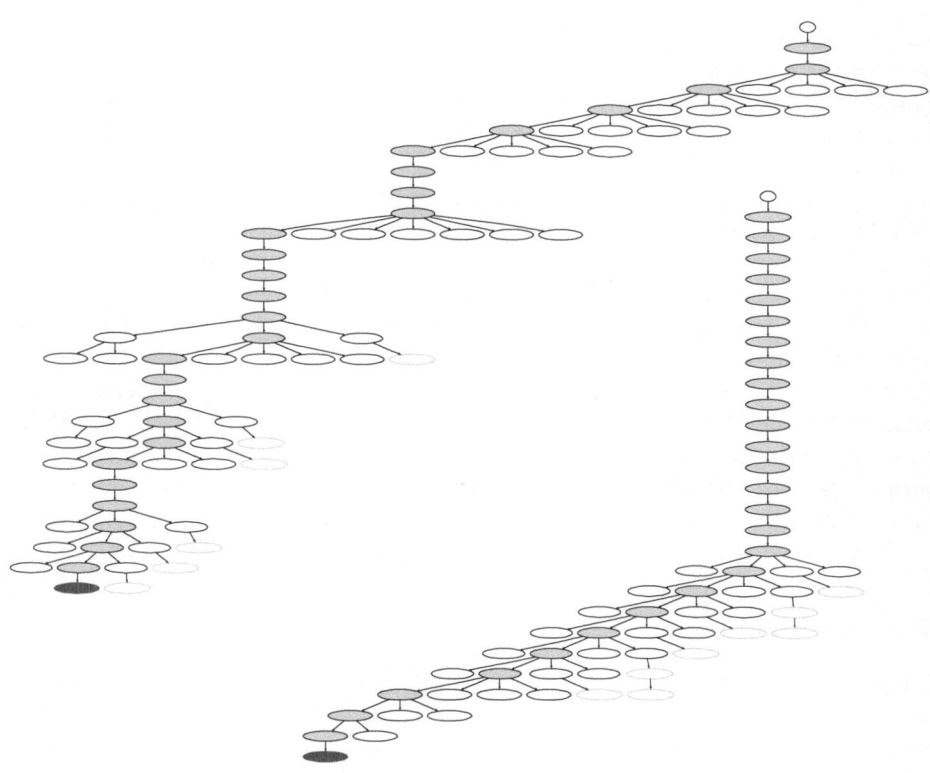

Fig. 2. An example for the structure of the examined search spaces with the f_{LCF}^{strat} (lower right corner) and f_{EMS}^{strat} (upper left corner) strategy

6 Related Work

Our approach relates to a number of proposed planning architectures. In the O-Plan system [9], e.g., a blackboard is modified by integrated detection and modification modules writing their highest ranked flaws on the agenda of a search controller which then selects one entry by triggering the respective knowledge source to perform its highest prioritized modification. The strategy's options are formulated in terms of delegating plan manipulations to the knowledge sources. Thus, it lacks the sort of flexibility that is the main characteristic of our approach.

Representatives for application frameworks or class-library toolboxes for building planning applications are Aspen [10] and work in the context of the Planform project [11]. They provide planning-specific data infrastructure, supportive inference mechanisms, and algorithmic templates, in order to facilitate rapid planning application development "out-of-the-box". I.e., they deliver pieces of a planning software independently from the architecture, while our view is centered on the architecture which allows for flexible planning system designs.

The approach by Yang et al. [12] addresses challenges in search-based applications from the software engineering point of view by proposing planning specific design patterns. The most important challenge is a reusable search engine, which our approach clearly provides. The problem of "obscurity of module boundary" does not apply to our system, because modules are strictly defined along flaw detection and plan modification.

From a technical point of view, the idea of mapping hybrid planning onto search through the space of plan refinements has been discussed in [13]. In this view, the algorithm uses reduction schemata where available and cares about primitive actions otherwise. Abstract conditions are closed by specific "phantom establishers" that are identified at a later stage. Conflict detection and resolution can only be done at the primitive level. Castillo et al. also use a flaw-based approach to hybrid planning, focusing on the acting entities in plan execution [1]. Specific abstraction-aware causal links manage causality issues during task expansions, thereby resulting in a co-ordinated combination of POCL and HTN methods. As opposed to both approaches, our methods substantially rely on a proper formal foundation of hybrid planning and make use of the explicit flaw and modification structures. It is these features that provide the capability to perform any kind of flexible strategy in our framework.

7   Conclusion

We have presented a novel architecture for planning systems. It relies on a formal account of hybrid planning, which allows to decouple flaw detection, modification computation, and search control. Planning capabilities, like HTN and POCL, can easily be combined by orchestrating respective elementary modules via an appropriate strategy module. The implemented system can be employed as a platform to implement and evaluate various planning methods and strategies. It can be easily extended to additional functionality, e.g. integrated scheduling [14] and probabilistic reasoning [15,16], without implying changes to the deployed modules – in particular flexible strategy modules – and without jeopardizing system consistency through interfering activity.

First experiments indicated that flexible planning strategies offer good results for hybrid planning problems. Although this analysis has to be refined by further experimentation with other domains and problems, it is a first step towards a systematic examination of planning algorithm characteristics. Furthermore, it opens the perspective for the development of novel sets of planning strategies which might be more widely applicable and it encourages the creation of better informed, paradigm-independent heuristics in the fashion of $f_{HotSpot}^{strat}$.

References

1. Castillo, L., Fdez-Olivares, J., González, A.: On the adequacy of hierarchical planning characteristics for real-world problem solving. In: Proc. of 6th European Conference on Planning (ECP-01). (2001)
2. Biundo, S., Schattenberg, B.: From abstract crisis to concrete relief – A preliminary report on combining state abstraction and HTN planning. In Cesta, A., Borrajo, D., eds.: Proc. of 6th European Conference on Planning (ECP-01). (2001)

3. Estlin, T.A., Chien, S.A., Wang, X.: An argument for a hybrid HTN/operator-based approach to planning. In Steel, S., Alami, R., eds.: Proc. of 4th European Conference on Planning (ECP-97). Volume 1348 of LNAI., Springer (1997) 182–194
4. Tsuneto, R., Nau, D., Hendler, J.: Plan-refinement strategies and search-space size. In Steel, S., Alami, R., eds.: Proc. of 4th European Conference on Planning (ECP-97). LNCS, Springer (1997) 414–426
5. McCluskey, T.: Object transition sequences: A new form of abstraction for HTN planners. In Chien, S., Kambhampathi, R., Knoblock, C., eds.: Proc. of 5th International Conference on Artificial Intelligence Planning Systems (AIPS-00), AAAI (2000) 216–225
6. Nau, D., Cao, Y., Lotem, A., Munoz-Avila, H.: SHOP: Simple hierarchical ordered planner. In Dean, T., ed.: Proc. of 16th International Joint Conference on Artificial Intelligence (IJCAI-99), Morgan Kaufmann (1999) 968–975
7. Joslin, D., Pollack, M.: Least-cost flaw repair: A plan refinement strategy for partial-order planning. In Hayes-Roth, B., Korf, R., eds.: Proc. of 12th National Conference on Artificial Intelligence (AAAI-94), AAAI (1994) 1004–1009
8. Gerevini, A., Schubert, L.: Accelerating partial-order planners: Some techniques for effective search control and pruning. Journal of Artificial Intelligence Research (JAIR) 5 (1996) 95–137
9. Tate, A., Drabble, B., Kirby, R.: O-Plan2: An architecture for command, planning and control. In Zweben, M., Fox, M., eds.: Intelligent Scheduling. Morgan Kaufmann (1994) 213–240
10. Fukunaga, A., Rabideau, G., Chien, S., Yan, D.: Towards an application framework for automated planning and scheduling. In: Proc. of 1997 Int. Symp. on AI, Robotics & Automation for Space. (1997)
11. PLANFORM: An open environment for building planners. Project web site at http://scom.hud.ac.uk/planform/ (2001)
12. Yang, Q., Fong, P., Kim, E.: Design patterns for planning systems. In Simmons, R., Veloso, M., Smith, S., eds.: Proc. of 4th International Conference on Artificial Intelligence Planning Systems (AIPS-98) Workshop on Knowledge Engineering and Acquisition for Planning: Bridging Theory and Practice, AAAI (1998) 104–112
13. Kambhampati, S., Mali, A., Srivastava, B.: Hybrid planning for partially hierarchical domains. In Rich, C., Mostow, J., eds.: Proc. of 15th National Conference on Artificial Intelligence (AAAI-98), AAAI (1998) 882–888
14. Schattenberg, B., Biundo, S.: On the identification and use of hierarchical resources in planning and scheduling. In Ghallab, M., Hertzberg, J., Traverso, P., eds.: Proc. of 6th International Conference on Artificial Intelligence Planning Systems (AIPS-02), AAAI (2002) 263–272
15. Biundo, S., Holzer, R., Schattenberg, B.: Dealing with continuous resources in AI planning. In: Proceedings of the 4th International Workshop on Planning and Scheduling for Space (IWPSS'04), ESA-ESOC, Darmstadt, Germany, European Space Agency Publications Division (2004) 213–218
16. Biundo, S., Holzer, R., Schattenberg, B.: Project planning under temporal uncertainty. In Castillo, L., Borrajo, D., Salido, M.A., Oddi, A., eds.: Planning, Scheduling, and Constraint Satisfaction: From Theory to Practice. Volume 117 of Frontiers in Artificial Intelligence and Applications. IOS Press (2005) 189–198

Controlled Reachability Analysis in AI Planning: Theory and Practice

Yacine Zemali

ONERA / DCSD - Centre de Toulouse,
2 avenue Edouard Belin,
BP 4025, 31055 Toulouse cedex 4 - France
yacine.zemali@gmx.net

Abstract. Heuristic search has been widely applied to classical planning and has proven its efficiency. Even GraphPlan can be interpreted as a heuristic planner. Good heuristics can generally be computed by solving a relaxed problem, but it may be difficult to take into account enough constraints with a fast computation method: The relaxed problem should not make too strong assumptions about the independence of subgoals. Starting from the idea that state-of-the-art heuristics suffer from the difficulty to take some interactions into account, we propose a new approach to control the amount and nature of the constraints taken into account during a reachability analysis process. We formalize search space splitting as a general framework allowing to neglect or take into account a controlled amount of dependences between sub-sets of the reachable space. We show how this reachability analysis can be used to compute a range of heuristics. Experiments are presented and discussed.

Keywords: Classical Planning, GraphPlan, Heuristics, Search Algorithms, Search Space Splitting, STRIPS.

1 Introduction

A classical planning task is defined by an initial state I, a final state G and a set of actions A. The goal is to find the action sequence leading from I to G. The formalism used to describe states and actions is PDDL [1], a refinement from STRIPS [2]. This language is a subset of the first order logic and it allows to manipulate non-instantiated predicates and objects. We call an instantiated predicate a fact. States of the world are sets of facts. The set of all possible facts for a planning task is noted F. An action $a \in A$ is defined by the usual triple $\langle pre(a), add(a), del(a) \rangle$.

One of the most efficient planning methods is heuristic planning [3]. Heuristics can be computed by considering a relaxed problem that is much quicker to solve (e.g. by neglecting interactions between actions or between facts).

GraphPlan [4] performs an approximate reachability analysis by neglecting some constraints: it solves a relaxed problem by assuming that all actions can be applied simultaneously (except for *mutex* actions). In terms of heuristic search,

U. Furbach (Ed.): KI 2005, LNAI 3698, pp. 264–278, 2005.

the planning graph assesses a lower bound of the length of the shortest parallel plan [5]. The search algorithm employed is a version of IDA^* [6]. HSP [7] computes its heuristic by assuming that subgoals are independent: It does not take any interaction into account. HSPr [5] takes some negative interactions into account while computing its heuristic (a_1 and a_2 have a negative interaction if $del(a_i) \cap (add(a_j) \cup pre(a_j)) \neq \varnothing$, $(i, j) \in \{(1, 2), (2, 1)\}$). It uses a WA^* algorithm performing a backward search from the goals to find a valid plan. FF [8] uses a relaxed GraphPlan (without mutex propagation nor backtrack) to compute an approximation of the length of the solution, i.e. the length of a solution to the relaxed problem: This heuristic takes into account some positive interactions, but no negative effects. The search is performed using an Enforced Hill Climbing algorithm. YAHSP [9] uses FF heuristic, but searches for a solution by using an elaborated version of the Enforced Hill Climbing Algorithm preserving the completeness. Fast Downward [10] computes its heuristic by translating STRIPS problems into causal graphs. This method allows to take into account some interactions that are neglected by others heuristics. In particular more dead-ends are detected. Recent work by Haslum et al. [11] presents refinements of two existing admissible heuristics (pattern database and h^m). The search algorithm employed is A^*.

Although heuristic methods are the most efficient approaches, it seems that they have reached a limit [3]. The main problem of existing approaches is their lack of flexibility: They use a fixed relaxation scheme, and the problems are always relaxed the same way. The same relaxation scheme is used on problems having different structures. This phenomenon explains why some particular planners are good on some particular domains (i.e. problem families). Our idea is to propose a more flexible relaxation scheme in order to adapt the amount and nature of neglected interactions to the structure of the problem. We propose and formalize search space splitting as a general framework allowing to neglect or take into account a controlled part of the problem constraints. Some planning systems build their search space by explicitly visiting instantiated reachable states. Some others, like GraphPlan, develop a leveled planning graph in which each level contains an aggregation of several states. Search space splitting is a compromise between those two extremes: Some states are fully instantiated, while some others are grouped together.

In section 2, we formalize the search space splitting. Section 3 presents the use of our theoretical core as a flexible reachability analysis scheme. In section 4, we show how we have used the search space splitting to build a heuristic planner for sequential planning. Section 5 presents some experiments and results. Section 6 offers concluding remarks.

2   Search Space Splitting: A Theoretical Core

2.1   Intuitions and Motivations

In GraphPlan, the number of levels of the planning graph may be a very poor estimate of the exact length of the solution: In the building phase, GraphPlan groups together incompatible facts into sets (levels) that contain a number of incoherences. In terms of reachability analysis, the successive action layers encode an estimate of an optimal parallel plan.

On the contrary, a naive tree exploration (FSS: Forward State Search) will separate each individual state during the tree construction. The depth of the tree (when the goals are reached) is the exact length of an optimal solution. In terms of reachability analysis, the successive action nodes between the root and the goals encode an optimal plan.

GraphPlan's approximation is too coarse. A tree exploration would compute a perfect heuristic, but in an exponential time. By reintroducing a tree structure into a disjunctive planning graph, we aim at combining the advantages of GraphPlan (i.e. a compact structure) and a tree exploration (i.e. accuracy of the reachability analysis). In a splitted planning graph, facts are grouped into different nodes at each level k instead of one undistinguished set of facts L_k (like in GraphPlan). For that purpose, it is necessary to split the levels into different nodes constitutive of the tree structure.

Splitting the planning graph allows to better approximate the solution plans. For instance, two incoherent sets of facts can be splitted into two distinct nodes to take more constraints into account. In consequence, some actions will not be introduced as early as in a GraphPlan-like building phase. Therefore, the depth of the planning graph (when it reaches the goals) is increased, getting closer to the optimal plan length.

2.2   Definitions and Graph Construction

An elaborated version of the following definitions, properties and theorems can be found in [12] (proofs are also detailed). The splitted planning graph is a leveled tree structure. The nodes can be either sets of facts (fact nodes) or sets of actions (action nodes). Each level contains a fact layer (set of fact nodes) and an action layer (set of action nodes), except the first level which only contains the root I. Mutex relations are not explicitly considered when building the graph.

Definition 1 (Fact node). *We call $Nf_i^j \in 2^F$ the jth fact node of the level i. We call $Nf_i = \{Nf_i^1, \ldots, Nf_i^{nf_i}\}$ the set of fact nodes at level i; nf_i is the number of fact nodes at level i.*

Each fact node Nf_i^j from level i is connected to one or several action nodes at level $i + 1$. All actions having their preconditions satisfied in Nf_i^j are introduced in the planning graph (in one or several action nodes). We distinguish branching actions and non-branching actions.

Definition 2 (Non-branching actions set). *The set $\mathcal{A}^{GPr} \subseteq A$ contains the actions which are applied in a relaxed way: By ignoring delete lists.*

Definition 3 (Branching actions set). *The set $\mathcal{A}^{FSS} \subseteq A$ contains the actions which are applied with their delete lists.*

Remark: $\mathcal{A}^{FSS} \cap \mathcal{A}^{GPr} = \emptyset$ and $\mathcal{A}^{FSS} \cup \mathcal{A}^{GPr} = A$.

Like in a relaxed GraphPlan, non-branching actions can be grouped together in the same action node. When a branching action is introduced in the graph, it leads to the creation of a new action node. Branching and non-branching actions are determined prior to the building phase.

Definition 4 (Splitting strategy). *A splitting strategy* $\mathcal{S}_{split}$ *is a function from* A *to* $\mathbb{B}$:

$$\mathcal{S}_{split} : A \longrightarrow \mathbb{B} \text{ with } \begin{cases} \forall \, a \in A : \\ \mathcal{S}_{split}(a) = 1 \text{ if } a \in \mathcal{A}^{FSS} \\ \mathcal{S}_{split}(a) = 0 \text{ if } a \in \mathcal{A}^{GPr} \end{cases} \tag{1}$$

The splitting strategy is the "branching rule" of the tree structure introduced in the planning graph. An action contained in $\mathcal{A}^{FSS}$ will introduce a new branch in the planning graph when applied. The splitting level t is the percentage of splitting actions ($|\mathcal{A}^{FSS}| = t.|A|$).

Definition 5 (Action node). *We call* $Na_i^j \in 2^A$ *the* j^{th} *action node of the level* i. *There exists two types of action nodes:*

- *nodes* Na_i^j *such that* $Na_i^j = \{a\}$ *with* $\mathcal{S}_{split}(a) = 1$. *Such nodes are called splitting action nodes;*
- *nodes* $Na_{i'}^{j'}$ *such that* $|Na_{i'}^{j'}| \geq 1$ *and* $\forall a \in Na_{i'}^{j'}$, $\mathcal{S}_{split}(a) = 0$.

We call $Na_i = \{Na_i^1, \ldots, Na_i^{na_i}\}$ *the set of action nodes at level* i; na_i *is the number of action nodes at level* i.

Each action node Na_i^j from level i is connected to a fact node Nf_i^j at level i. Nf_i^j results from the application of actions contained in Na_i^j (the application function depends on the type of action node: With or without relaxation).

Definition 6 (Splitted planning graph). *Given a planning task* $\mathcal{P} = \langle A, I, G \rangle$, *a splitting strategy* $\mathcal{S}_{split}$ *and* $k \in \mathbb{N}$. *We call the* k-*order splitted planning graph* $\mathcal{G}_k^s$ *the couple* $\langle \mathcal{N}_k, \mathcal{A}_k \rangle$. $\mathcal{N}_k$ *is the set of nodes;* $\mathcal{A}_k$ *is the set of arcs:*

$$\mathcal{N}_k = \bigcup_{i=0,\ldots,k} (Na_i \cup Nf_i) \tag{2}$$

$$\mathcal{A}_k = \bigcup_{i=0,\ldots,k} (Ap_i \cup Aa_i) \tag{3}$$

$Ap_i \subseteq Na_i \times Nf_{i-1}$ *is the set of precondition arcs between action nodes from level* i *and fact nodes from level* $i - 1$.
$Aa_i \subseteq Na_i \times Nf_i$ *is the set of application arcs between action nodes from level* i *and fact nodes from level* i.

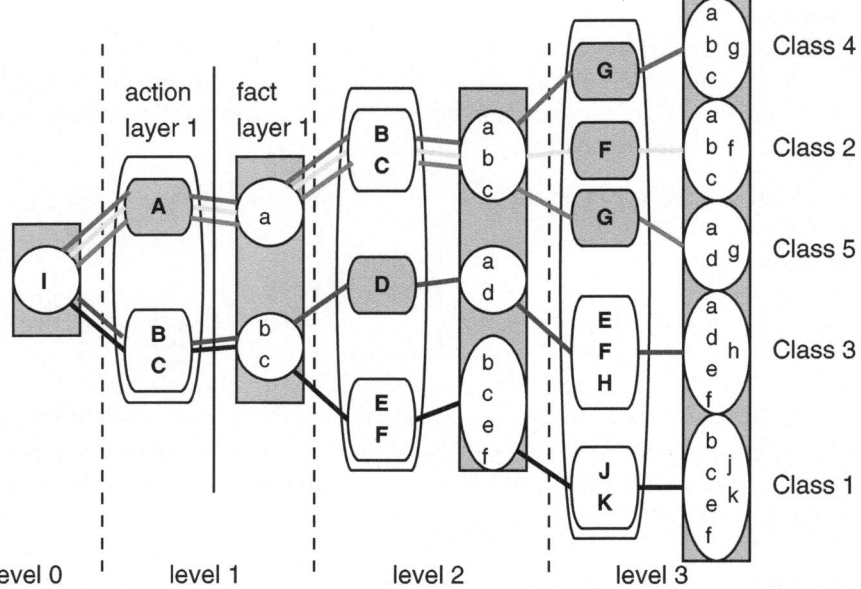

Fig. 1. A splitted planning graph

Figure 1 shows a splitted planning graph (splitting action nodes are in grey). The time complexity to build such a structure using a naive algorithm would be exponential. For this reason, we have implemented a memorization mechanism to prune useless branches. This technique allows avoiding to build several times the same sub-graph: During the building phase, each fact node is stored. If it happens that a new fact node is equal to an already memorized fact node, the corresponding action nodes are not built. A fact node is described by a vector $v \in \{0, 1\}^{|F|}$ ($v[i]$ gives the truth value of the ith fact). The vectors are stored in a decision tree (DT). Read and write operations in the DT are in constant time ($|F|$). Thanks to this mechanism, we can also ensure that the construction process terminates (i.e. the graph stabilizes: There exists a level k such that all succeeding levels are equivalent).

2.3   Time Complexity

Complexities are based on the number of node creations.

Worst case: In this case, the DT is useless: Each node has to be computed. Worst case complexity is in $O((t.|A|)^k)$.

Best case: The algorithm builds only one node per level. This case happens when all potential nodes to be computed in a given level where already built and memorized in a previous level. In this case the complexity is linear (the algorithm will build k nodes for a k-order splitted planning graph).

Average: We consider that only some actions are applicable on each fact node. Practically speaking, fact nodes which are close to the root do not contain enough facts for a large amount of actions to be applicable. On the contrary, fact nodes which are close to the leaves contain more facts and more actions will then be applicable. In consequence, the branching factor tends to rise with the depth. Similarly, the number of already visited fact nodes rises with the depth. Thanks to our memorization mechanism, an already visited fact node will not lead to an action node. This feature tends to decrease the branching factor with depth.

The average complexity is too difficult to be computed analytically. We rather remark that the graphs built by our algorithm are incomplete trees. We can reasonably compare the average time complexity of our tree construction to well known complexities. For instance, the A^* algorithm also builds an incomplete tree (its search space). Its complexity is exponential in the worst case, but often polynomial in practice.

3   Reachability Analysis and Heuristics

3.1   The Concept of Class

We have introduced the concept of class to facilitate the formal proof of the reachability analysis encoded by the graph.

Definition 7 (Action class). *We call action class $Ca_j \subset 2^A$ the ordered set of action nodes which are connected on a path between the root and the jth leaf in $\mathcal{G}_k^s$. In other words, an actions class is a set containing k action nodes, each node belonging to a distinct level of the splitted planning graph:*

$$Ca_j = \bigcup_{i=1\ldots k} \{Na_i^{j_i}\} \tag{4}$$

$$with \quad \begin{cases} \forall i \in [1,k], j_i \in [1, nf_i] \\ j_1 = 1 \\ j_k = j \\ \\ (Na_{i-1}^{j_{i-1}}, Na_i^{j_i}) \in Ca_j^2 \Leftrightarrow \begin{cases} \exists Nf_{i-1}^{j_{i-1}} \in Nf_{i-1} \ s.t.: \\ (Na_{i-1}^{j_{i-1}}, Nf_{i-1}^{j_{i-1}}) \in Aa_{i-1} \\ (Na_i^{j_i}, Nf_{i-1}^{j_{i-1}}) \in Ap_i \end{cases} \end{cases}$$

Notation: *for simplicity reasons, we substitute subscripts $j_1, \ldots, j_k$ by a unique subscript j. With this notation, the jth facts class is written as follows:*

$$Ca_j = \{Na_1^j, \ldots, Na_k^j\} \ with \ \forall i \in [1,k], Na_i^j \overset{def}{=} Na_i^{j_i}$$

Definition 8 (Fact class). *We call fact class $Cf_j \subset 2^F$ the ordered set of fact nodes which are connected on a path between the root and the jth leaf in $\mathcal{G}_k^s$.*

$$Cf_j = \bigcup_{i=0\ldots k} \{Nf_i^{j_i}\} \tag{5}$$

Notation: $Cf_j = \{Nf_1^j, \ldots, Nf_k^j\} \ with \ \forall i \in [0,k], Nf_i^j \overset{def}{=} Nf_i^{j_i}$

With those definitions, each node Nf_i^j of a fact class Cf_j assesses a potentially reachable set of facts at level i. The action classes assess potential plans able to produce the corresponding fact nodes. As a result, if the goals G belong to a fact class Cf_j, then the corresponding action class Ca_j approximates a solution plan. Figure 1 shows the representation of several classes on a graph.

In this context, the splitting strategy consists in determining which actions will create new classes. A new class is introduced in the graph each time a splitting action is applied, i.e. each branch creation corresponds to a class.

3.2   Reachability

In terms of reachability analysis, the splitted planning graph allows to estimate the minimal number of levels to obtain the goals. In terms of plan assessment, we have achieved a theoretical work to prove that the solution plan, if it exists, is included into one of the action classes.

Definition 9 (Inclusion). *Given a sequential plan $P = \langle a_1, \ldots, a_l \rangle$, and an action class $Ca_j = \{Na_i^j\}_{i=1..k}$. P is included in Ca_j (we note: $P \sqsubseteq Ca_j$) if there exists an integer $y \leq k$, and a function $f : [1, l] \to [1, y]$ such that:*

$$\begin{cases} f(1) = 1 \\ f(l) = y \\ \forall x \in [1, l[, \ (f(x+1) = f(x)) \vee (f(x+1) = f(x) + 1) \\ \forall x \in [1, l], \ a_x \in Na_{f(x)}^j \end{cases}$$

More intuitively, $P \sqsubseteq Ca_j$ means that actions contained in P are also contained in the sets $\{Na_i^j\}_{i=1..k}$ with respect to their ordering in P, i.e. two actions a_x and a_y from the plan P, such that $x < y$, are respectively contained in two (or one) sets $Na_{x'}^j$ and $Na_{y'}^j$ such that $x' \leq y'$.

Theorem 1 (Sequential plans). *Let $P = \langle a_1, \ldots, a_l \rangle$ be a sequential plan of length l with:*

$$\begin{cases} l \in [1, k] \\ \forall i \in [1, l], \ a_i \in A \\ pre(a_1) \subseteq I \end{cases}$$

Then, there exists an integer $j \leq na_k$ such that the action class Ca_j contains P, i.e. $\exists\, j \leq na_k \big/ P \sqsubseteq Ca_j$ (Proof is in [12]).

The previous theorem ensures that any solution plan of length $l \leq k$ is included in the k-order planning graph.

Theorem 2 (Stabilization and solution existence). *If $\mathcal{G}_k^s$ is stabilized without any of its fact nodes containing the goals G, then $\mathcal{P}$ does not have any solution (Proof is in [12]).*

Definition 10 (Splitted reachability). *Let $B \subseteq F$ be a set of facts. We call splitted reachability of the set B, the minimal order of a splitted planning graph containing B. We note this reachability measure $R^s(\mathcal{S}_{split}, B)$.*

$$R^s(\mathcal{S}_{split}, B) = \min_{B \subseteq Nf_i^j,\ j \in [1, nf_i]} i \in \mathbb{N} \tag{6}$$

If there does not exist any fact node containing B, then $R^s(\mathcal{S}_{split}, B) = \infty$.

Property 1 (Reachability and splitting level). *Let $(t, t') \in [0, 100]^2$ be two splitting levels such that $t < t'$. Let $\mathcal{S}_{split}$ and $\mathcal{S}'_{split}$ be two splitting strategies built respectively using t and t' (and the same splitting criterion, i.e. same ranking of the actions). Let $B \subseteq F$ be a set of facts. We have:*

$$R^s(\mathcal{S}_{split}, B) \leq R^s(\mathcal{S}'_{split}, B) \tag{7}$$

Proof is in [12].

In other words, a rise of the splitting level leads to a more accurate reachability analysis (for the same splitting criterion).

3.3   Heuristics

If the goals G are contained in a fact node Nf_i^j, we know that the action class Ca_j contains an estimate of the solution plan (other classes may contain other estimates). We call such a plan a semi-relaxed plan (partial relaxation). We know that at least one action of each action node in Ca_j has to belong to the solution plan if this one exists. We can simply use the number of action nodes in Ca_j to estimate the length of the optimal sequential solution (this is the depth of the planning graph when goals are reached). This measure is an admissible heuristic, we call it h^*. We have $h^*(G) = R^s(\mathcal{S}_{split}, G)$. With small splitting levels, this heuristic is too coarse: We consider that only one action per node is in the assessment of the solution, but this is generally not the case. With greater levels, computed heuristics are more accurate and still admissible (we still assume that only one action per node is in the plan, but this is more often the case when lots of actions are splitting actions). Unfortunately, it may be more difficult to compute them in a reasonable time.

Therefore, we also compute a non-admissible heuristic – h – which is more accurate than h^* (for the same splitting level). We give a hint of the computation mechanism for h (the detailed mechanism is in [12]): We use an iterative calculus to estimate the number of actions useful to achieve the sub-goals at each level. This calculus starts from the last level with the set of sub-goals G. Then we estimate the average number of actions involved in the last level (in class Ca_j). We use this information (and others) to estimate a new set of sub-goals for the previous level, and so on.

4   Resulting Heuristic Planner

4.1   Architecture

We have implemented a heuristic planner based on the h^* and h heuristics. This planner visits the state space forward, starting from the initial state I. The search algorithms used are: A^*, IDA^*, and an elaboration of Hill Climbing. Our Hill Climbing algorithm can detect local minima and dead-ends (detection mechanism is based on decision trees). In such cases, the algorithm randomly restarts its search. Thus, we have six configurations (each of them is a couple heuristic/search algorithm). Two of those configurations are optimal planners (use of h^* with A^* and IDA^*).

4.2   Elaboration of h^*

At each call of the function h^*, the algorithm assesses the reachability from lots of sets of facts. During the successive calls, the algorithm may assess several times the same sets. To avoid this, we store intermediate estimates obtained during h^* computations as follows: When the goals are reached in a class Cf_j, with $G \subseteq Nf_k^j$, we have $h^*(Nf_k^j) = 0$. For each set Nf_i^j of class Cf_j, we can deduce $h^*(Nf_i^j) = k - i$. Once again, we use a DT to store those intermediate results. Each set is described by a vector which is stored in a DT where the leaves contain the h^* intermediate estimates. An advantage of this mechanism is that it does not recompute h^* from scratch: Suppose that during a heuristic computation, the graph-building algorithm reaches a node Nf_a^b. If a value $h^*(Nf_a^b)$ is already stored, the algorithm does not need to visit this node's successors because it is aware of the fact that $h^*(Nf_a^b)$ levels would be necessary to reach the goals. Furthermore, the retrieved value is used as a bound for the graph construction.

This mechanism is very useful when the successive splitted planning graphs share a lot of nodes. In particular, this case happens when the computed heuristics are accurate: The successive states visited by the search algorithm (the successive roots of the planning graphs) are located in the same (virtual) fact class. The search algorithm then focuses in one direction. With too small splitting levels, there are too much changes of direction for the intermediate storage to be used efficiently. With greater levels, stored results are reused. We have empirically verified that a rise of the splitting level leads to a better use of the storage mechanism.

Thanks to this technique, the optimal configurations of our planner behave like a dynamic programming algorithm when used with a full splitting: The first graph construction computes all required heuristic values and the next graph constructions consist in value retrieval in the DT.

4.3   Splitting Criterion

In practice, the splitting strategy is deduced from the splitting level t and from a ranking of the actions. Each action is assigned a mark. The $t.|A|$ best ranked actions are marked as splitting actions (they belong to $\mathcal{A}^{FSS}$). The policy used to rank the actions is called the splitting criterion.

The ranking can be achieved manually or automatically. The ranking used in this paper is based on the minimization of the interference between actions. At each level, facts are grouped according to their level of mutual coherence: Facts created by actions that are not mutex with each other can be grouped together. This splitting criterion tends to enforce actions with a big interference level (i.e. the number of actions with which a given action is permanently mutex) to be in $\mathcal{A}^{FSS}$. Practically speaking, each action is automatically ranked according to its interference level.

A number of algorithms allow to compute invariants features such as permanent mutex relations in a planning problem [13,14]. We have designed a simple algorithm to assess the interference levels. It considers facts as resources [15]: If two actions delete the same fact, they are mutex because they try to consume the same resource. Our algorithm takes advantage of this by coloring the facts. This marking allows to know if two facts need competitive resources to be produced. Starting form the initial state, a propagation process (not detailed here) assesses permanent mutex relations between facts. From the mutex between facts, we deduce the mutex between actions to compute the interference levels.

5   Experiments and Results

We have used our planner to compare splitting criteria and splitting levels in terms of the quality of the resulting reachability analysis and in terms of the benefit obtained from using the corresponding heuristics at search stage. We have also studied the impact of the search space splitting on solutions quality. Our framework allows to study a continuum between the resolution with a relaxed GraphPlan heuristic (no splitting) and with a dynamic programming approach. With the intermediate splitting levels, more or less constraints are taken into account in the heuristic computation. We have investigated three classical domains: Gripper, Blocksworld, and Mystery which are representative of the available STRIPS benchmarks (in terms of difficulty). Experiments were carried out on a 1 Ghz machine with a time limit of 300 s for each instance.

5.1   Impact of the Splitting Criterion

We have tested five splitting criteria:

1. Criterion based on the minimization of the interference level. The resulting splitting strategies tend to avoid to group mutex facts together;
2. Criterion based on the maximization of the interference level. Mutex facts are preferentially grouped together;
3. Criterion based on non-invertible actions. The non-invertible actions are often responsible for dead-ends when they are relaxed [16]. With such strategies, dead-ends are better detected;
4. Random ranking of the actions: Criterion used for comparison purpose;
5. Manual ranking of the actions: Criterion used to test specific behaviors.

The best criterion, in terms of combined computational time and accuracy of the reachability analysis, is the one based on the minimization of the interference level (mutex avoidance). The worst criterion is – as expected – the one based on the maximization of the interference level.

5.2 Splitting Level and Performances

The theory suggests that a rise of the splitting level leads to a better heuristic accuracy. We have empirically verified the influence of the splitting level on the accuracy: The search algorithm visits the search space more efficiently with greater splitting levels. But the gain, in terms of computational time, is not systematic: Even if the search space is visited in a more efficient way, elevated splitting levels also lead to a more complex heuristic computation. The memorization mechanisms tend to limit this rise of complexity. The observed phenomena are quite heterogeneous: On some instances, we have noticed better performances with high splitting levels. This case happens when the cost of the successive graph constructions is compensated by the accuracy of the heuristic: Less states are visited. On the contrary, we have observed cases where the accuracy does not compensate the extra-cost of the splitting process.

To understand the influence of the splitting level on the performances, we have performed a range of experiments with five configurations of our planner. We have run our planner on 85 problem instances (20 Gripper, 35 Blocksworld, 30 Mystery). For each instance, we have tested 21 splitting levels (from 0% to 100% with an increment of 5%). In total, we have run 8925 planning tasks (5 configurations*85 instances*21 levels).

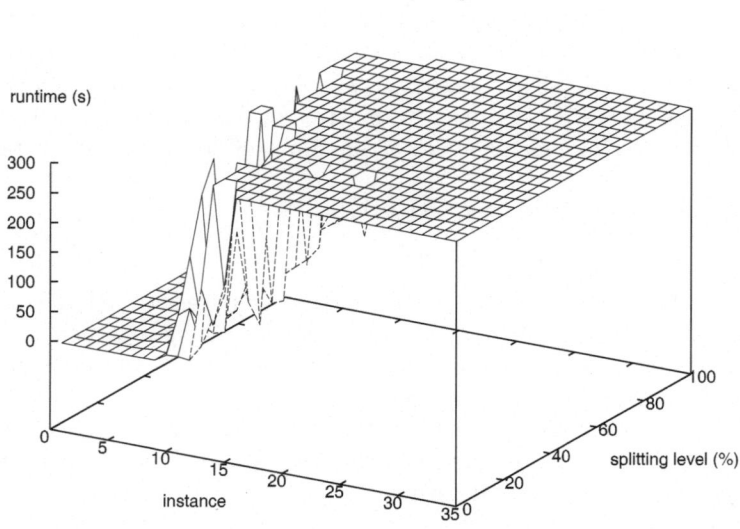

Fig. 2. CPU runtimes for Blocksworld domain (Hill Climbing)

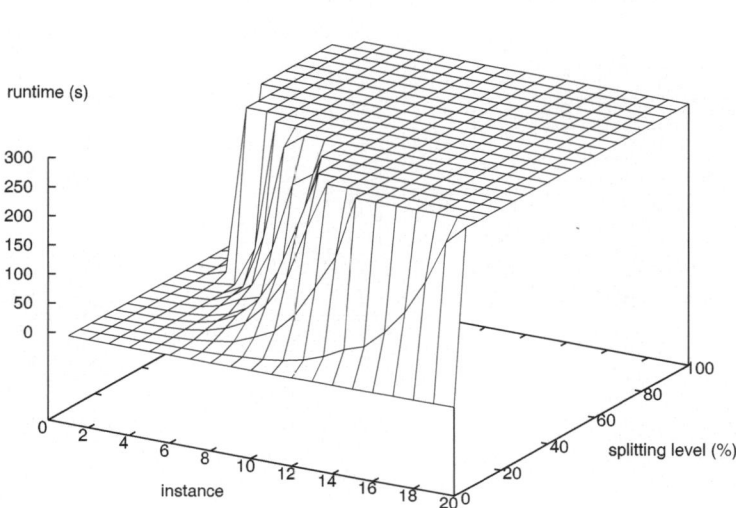

Fig. 3. CPU runtimes for Gripper domain (Hill Climbing)

For example, Fig. 2 shows CPU runtimes for the instances 1 to 35 of Blocksworld domain (and a varying splitting level). We see that our planner can solve 17 of the 35 instances by using the Hill Climbing algorithm and h heuristic (non-solved instances show a 300 s runtime). We see that several intermediate splitting levels give better results than full or null splitting. Figure 3 shows CPU runtimes for the instances 1 to 20 of Gripper domain. In this case, a rise of the splitting level almost always leads to greater runtimes.

Tables 1 and 2 give a summary of our experiments in terms of performances. For each couple configuration/domain we have computed four percentages.

Intermediate splitting: Percentage of solved instances with a minimal runtime in comparison with full splitting and no splitting approaches.

No splitting: Percentage of solved instances with a minimal runtime in comparison with full splitting and intermediate splitting. This configuration is equivalent to the use of a relaxed GraphPlan heuristic.

Full splitting: Percentage of solved instances with a minimal runtime in comparison with no splitting and intermediate splitting. Equivalent to dynamic programming with h^*.

Total: Percentage of solved instances (the sum of the three splitting can rise above the total because some instances can be counted several times).

We have found a lot of instances where an intermediate splitting is more efficient than an extreme splitting (full or null). In particular, we have found 3 couples (configuration, domain) where the intermediate splitting performs better than the no splitting approach. Additionally, we have found 11 couples where the intermediate splitting performs better than the full splitting. In one case (Blocksworld, A^*/h^*), the intermediate splitting approach performs better than

Table 1. Solved instances with optimal approaches (in a minimal time)

	Grip.	Bloc.	Myst.
$[A^*/h^*]$ int. split.	5%	29%	13%
$[A^*/h^*]$ no split.	10%	20%	43%
$[A^*/h^*]$ full split.	10%	9%	0%
$[A^*/h^*]$ total solv.	20%	38%	50%
$[IDA^*/h^*]$ int. split.	0%	17%	10%
$[IDA^*/h^*]$ no split.	0%	9%	47%
$[IDA^*/h^*]$ full split.	20%	37%	0%
$[IDA^*/h^*]$ total solv.	20%	43%	54%

Table 2. Solved instances with sub-optimal approaches (in a minimal time)

	Grip.	Bloc.	Myst.
$[A^*/h]$ int. split.	5%	9%	3%
$[A^*/h]$ no split.	15%	46%	43%
$[A^*/h]$ full split.	0%	0%	0%
$[A^*/h]$ total solv.	15%	46%	50%
$[IDA^*/h]$ int. split.	5%	11%	13%
$[IDA^*/h]$ no split.	0%	34%	33%
$[IDA^*/h]$ full split.	15%	6%	0%
$[IDA^*/h]$ total solv.	20%	43%	44%
$[H.\ Climb./h]$ int. split.	5%	17%	13%
$[H.\ Climb./h]$ no split.	100%	37%	37%
$[H.\ Climb./h]$ full split.	0%	0%	0%
$[H.\ Climb./h]$ total solv.	100%	49%	50%

both full and no splitting approaches. We have also observed that the intermediate splitting is not very efficient on the Gripper domain, compared to the Blocksworld or Mystery domains which are more difficult. It seems that the intermediate splitting is more adapted to complicated domains (because those domains contain more interactions). In the same way, the intermediate splitting generally overcomes the full splitting (even when equivalent to dynamic programming) on tough domains, but not on the Gripper domain.

5.3 Splitting Level and Solution Quality

By using sub-optimal configurations, we have noticed that the splitting level behaves like a parameter allowing to tune the quality of the solution. Table 3 shows such an example (instance prob07 of the Gripper domain): Thanks to the splitting, the length of the solution decreases from 63 to 57 steps. Table 4 shows the percentages of solved instances (for each domain and for each sub-optimal configuration) where we have observed an improvement of the solution's

Table 3. Instance prob07 of Gripper domain (Hill Climbing / h): quality improvement

t	0	5	10	15	20	25	30	35	40
solution length	63	61	61	61	61	61	59	59	57

Table 4. Quality improvement

Domain	A^*/h	IDA^*/h	$H.\ Climb./h$
Gripper	0%	50%	100%
Blocks	19%	20%	88%
Mystery	0%	0%	31%

quality with the rise of the splitting level. Results are more eloquent with the Hill Climbing. In particular, the Blocksworld and Gripper domains are very responsive to the splitting level. With A^* and IDA^*, the impact of the splitting level is limited (we can neglect the 50% for the Gripper/IDA^*/h because it only concerns 4 instances). One explanation is that the solutions provided by those configurations are often close to the optima (but those configurations also require more computational time than the Hill Climbing).

6   Concluding Remarks

We have shown that the splitting level and the splitting criterion have a strong impact on the planning process: Accuracy of the heuristics and quality of the solutions can be improved. Concerning the overall performances (heuristic computation + search stage), results are less clear. It is not surprising that the introduction of a branching factor in the graph leads to a more complicated building phase. Yet, we have shown that the gain (heuristics accuracy) can significantly reduce the search time in certain cases. This result is very promising: Splitting can relax the problem enough to break the complexity while providing an informative heuristic.

We have presented an automatic splitting criterion which appears to be efficient in the general case. However, on the Blocksworld domain, we have observed the existence of better criteria. On this domain, our criterion is equivalent to a random ranking of the actions (each action has the same interference level). We have found hand coded criteria that give better results. So this proves the existence of better strategies.

A good splitting strategy is crucial: It can enable to take into account some vital information during the reachability analysis while preserving a tractable computational time. Thus, we are working on a new automatic splitting strategy which is more problem-dependent than the mutex avoidance. The splitting criterion is deduced during a learning phase on small size instances. During this process, several splitting actions are tried. The best actions are then inserted in the final splitting strategy. The splitting level is also deduced from this preliminary step. Other splitting strategies could be imagined to take advantage of the flexibility of our planning framework.

Current research also includes: Enabling our planner to deal with optimization criteria richer than the plan length (use of PDDL 2.1 [17]); Using a regression search in order to build the planning graph only once.

References

1. McDermott, D., Committee, AIPS-98 Planning Competition.: PDDL -The Planning Domain Definition Language Version 1.2. (1998)
2. Fikes, R.E., Nilsson, N.J.: STRIPS: A new approach to the application for theorem proving to problem solving. In: Advance Papers IJCAI-71, Edinburgh, UK (1971) 608–620

3. Hoffmann, J.: Local search topology in planning benchmarks: An empirical analysis. In: Proc. IJCAI-01, Seattle, WA, USA (2001) 453–458
4. Blum, A.L., Furst, M.L.: Fast planning through planning graph analysis. Artificial Intelligence **90** (1997) 281–300
5. Bonet, B., Geffner, H.: Planning as heuristic search: New results. In: Proc. ECP-99, Durham, UK (1999) 360–372
6. Korf, R.E.: Depth-first iterative deepening: an optimal admissible tree search. Artificial Intelligence **27** (1985) 97–109
7. Bonet, B., Geffner, H.: HSP: Heuristic search planner. In: Proc. AIPS-98: Planning Competition, Pittsburgh, PA, USA (1998)
8. Hoffmann, J., Nebel, B.: The FF planning system: Fast plan generation through heuristic search. JAIR **14** (2001) 253–302
9. Vidal, V.: A lookahead strategy for heuristic search planning. In: Proc. ICAPS-04, Whistler, BC, Canada (2004) 150–159
10. Helmert, M.: A planning heuristic based on causal graph analysis. In: Proc. ICAPS-04, Whistler, BC, Canada (2004) 161–170
11. Haslum, P., Bonet, B., Geffner, H.: New admissible heuristics for domain-independent planning. In: Proc. AAAI-05, Pittsburgh, PA, USA (2005) 1163–1168
12. Zemali, Y.: Search space splitting in planning: a theoretical study and its use for heuristics computation. PhD thesis, Supaéro, Toulouse, France (2004)
13. Fox, M., Long, D.: The automatic inference of state invariants in TIM. JAIR **9** (1998) 367–421
14. Gerevini, A., Schubert, L.K.: Discovering state constraints in DISCOPLAN: Some new results. In: Proc. AAAI-00, Austin, TX, USA (2000) 761–767
15. Meiller, Y., Fabiani, P.: Tokenplan: a planner for both satisfaction and optimization problems. AI Magazine **22** (2001) 85–87
16. Hoffmann, J.: Local search topology in planning benchmarks: A theoretical analysis. In: Proc. AIPS-02, Toulouse, France (2002) 379–387
17. Fox, M., Long, D.: PDDL2.1: An extension to PDDL for expressing temporal planning domains. JAIR **20** (2003) 61–124

Distributed Multi-robot Localization Based on Mutual Path Detection

Vazha Amiranashvili and Gerhard Lakemeyer

Computer Science V, RWTH Aachen, Ahornstr. 55,
D-52056 Aachen, Germany
vazha@i5.informatik.rwth-aachen.de, gerhard@cs.rwth-aachen.de
http://www-kbsg.informatik.rwth-aachen.de

Abstract. This paper presents a new algorithm for the problem of multi-robot localization in a known environment. The approach is based on the mutual refinement by robots of their beliefs about the global poses, whenever they detect each other's paths. In contrast to existing approaches in the field the detection of robots (e.g. by cameras) by each other is not required any more. The only requirement is the ability of robots to communicate with each other. In our approach the robots try to detect the paths of other robots by comparing their own perception sensor information to that "seen" by other robots. This way the relative poses of the robots can be determined, which in turn enables the exchange of the beliefs about the global poses. Another advantage of our algorithm is that it is not computationally more complex than the conventional single robot algorithms. The results obtained by simulations show a substantial reduction of the localization time in comparison to single robot approaches.

1 Introduction

The self-localization of robots in known environments has been recognized as one of the key problems in the autonomous mobile robotics [4,2]. Indeed, over the last decade a number of very successful probabilistic algorithms has been developed for the solution of this problem in the single robot case [10,8,20,24,14,3,26]. All these algorithms have the same probabilistic basis. They all estimate posterior distributions over the robot's poses given the sensor data acquired so far by the robot and certain independence assumptions. The posterior distributions (called beliefs further on) are then in many cases estimated by an efficient Monte Carlo technique known as "particle filters" [5,1,17,21].

It is natural to try to extend the mentioned approaches to the multi-robot case. Indeed, there is a number of works dealing with this problem [12,16,22,9]. For example in [22] the multi-robot localization is solved by introducing one central Kalman filter for all robots. It is interesting that no information about the relative poses of the robots is used - only the sensor measurements of external landmarks. On the other hand the method of [9] uses the information about the relative poses of the robots to mutually refine the beliefs of the robots about the

U. Furbach (Ed.): KI 2005, LNAI 3698, pp. 279–290, 2005.
© Springer-Verlag Berlin Heidelberg 2005

global poses. The robots determine the relative poses by detecting each other with cameras. The method naturally extends the Monte Carlo algorithms for single robots to multi-robot systems.

In this context also the so called SLAM (simultaneous localization and mapping) algorithms for multi-robot systems should be mentioned [25,15,11]. In [25] each robot uses a combination of the incremental maximum likelihood and gradient search methods for the concurrent localization and mapping. The algorithm can then be extended to the multi-robot case, but with restrictive assumptions that the robots start at nearby locations such that their range scans considerably overlap and the initial relative poses of the robots are known. In [11] these restrictions were removed by solving the so called revisiting problem for multiple robots, which implies determining whether the current position of the robot lies within an area (map) already explored by the same or other robots. By locating a robot within a visited area it is possible to merge the maps of different robots.

In the approach introduced in this paper we also solve the revisiting problem in order to determine the relative poses of the robots. However in contrast to [11] we deal with a global environment (map) known to all robots. In our approach the robots while moving compare their current laser range scans to the ones possibly "seen" by other robots and if the scans match with a high probability, it is possible to determine the relative locations of the robots. The latter is possible because the robots in addition keep track of their local poses just like in SLAM approaches. After the relative poses are determined the robots can exchange their beliefs about their global poses. This way globally better localized robots can assist others to refine their beliefs about the global pose. Here we use a technique similar to that of [9] to fuse sample sets representing beliefs of the robots.

On a more intuitive level, what our approach does is let the robots share with all other robots whatever they have "seen" so far, which then can be used to accelerate the convergence of the localization method of the individual robot, if it happens to come accross an area which someone else has visited in the past. While it may seem paradoxical at first, a robot could in principle achieve the same effect by simulating other robots in its own map. This is so because we assume that the map is aleady known and hence obtaining sensor data from other robots (real or simulated) does not provide new information about the environment. Why then is our approach interesting for multi-robot scenarios when each robot could do all the work itself locally? For one, it seems natural to make use of other robots' work whereever possible. Also, using real data from real robots seems preferable as maps are hardly ever exact representations of the environment. But perhaps the main reason against the simulation idea is computational. Due to our efficient representation of scans the overhead of communicating, storing, and accessing the scans is fairly small. Simulating other robots, on the other hand, is costly and often not feasible given the limited computing resources of today's robots. By using real robots, we are essentially distributing the work to obtain scan data along certain paths. The result can then be used to speed up the convergence of the local particle filter. At the beginning the chances of a robot visiting an area which somebody else has been to may be low, but over time the likelyhood increases and the opportunistic (and cheap) gathering of what others have seen pays off, as we will see in the discussion of our experimental results.

The rest of the paper is organized as follows. In the following section we will discuss Monte Carlo techniques for localization. In Section 3 we describe our approach to the solution of the revisiting problem in the context of a known environment and present our MCL algorithm for multi-robot systems. Section 4 presents some experimental results illustrating our approach. Finally, Section 5 concludes the work.

2   Monte Carlo Localization

Monte Carlo localization (MCL) is a special case of probabilistic state estimation, applied to robotic localization. We assume that robots have odometry and perception sensors (wheel encoders and laser range finders in our case), which give us the information about the motion of the robot and the environment respectively. Our goal is to estimate the conditional probability distribution over all possible poses of the robot given all the sensor data that the robot gathers up to the time of the estimation. Let $l = (x, y, \theta)$ denote a robot location, where x, y and θ are the Cartesian coordinates in the global coordinate system and the orientation (with respect to the x-axis of this system) of a robot respectively. We seek to estimate $bel(l_{0:t}) \triangleq p(l_{0:t}|d_{0:t}, m)$ the belief that at time t the robot has made the path $l_{0:t}$ and its current position is l_t. Here $d_{0:t} = \{a_{0:t-1}, s_{1:t}\}$ is the set of all odometry measurements $a_{0:t-1}$ and perceptions of the environment $s_{1:t}$, while for $i < j$ $x_{i:j} \triangleq \{x_i, x_{i+1}, ..., x_j\}$, m is the map of the environment. We model $bel(l_{0:t})$ by the Bayesian network represented in Figure 1.

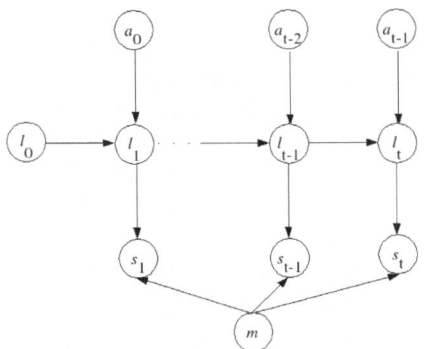

Fig. 1. Bayesian network used as a model for MCL

According to the model of Figure 1 we can rewrite $bel(l_{0:t})$ as follows

$$bel(l_{0:t}) = p(s_t|l_t, m) \cdot p(l_t|l_{t-1}, a_{t-1}) \cdot bel(l_{0:t-1}), \tag{1}$$

where $p(s_t|l_t, m)$ and $p(l_t|l_{t-1}, a_{t-1})$ are the so called perception and motion models respectively [10]. The model of Figure 1 makes an important assumption, which states that the future states do not depend on the past ones given the current state. This assumption is known as the "Markov assumption" or "static world assumption" and is also made in this work.

The MCL is an efficient way to compute (1). It is based on the SIR (sampling/importance resampling) algorithm [23], which is a version of "particle filters" [5]. It presents the belief function $bel(l_{0:t})$ by a set of weighted samples, or particles. Each particle is a tuple of the form $\langle w, l_{0:t} \rangle$, where $w \geqslant 0$ and if there are N particles in a set, $\sum_{n=1}^{N} w_n = 1$.

The MCL algorithm is presented below.

Table 1. MCL-Algorithm

1. $S_{0:t-1} \triangleq \{\langle w_{0:t-1}^{(i)}, l_{0:t-1}^{(i)} \rangle | 1 \leq i \leq N\}$; // Input sample set.

2. a_{t-1};   // Current odometry data.

3. s_t;   // Current perception data.

4. $S_{0:t} = \varnothing$;   // Output sample set.

5. **for** $i = 1$ **to** N **do**

6.   sample $l_{0:t-1} \sim S_{0:t-1}$ according to $w_{0:t-1}^{(i)}$, $1 \leq i \leq N$;

7.   sample $l_t \sim p(l_t|l_{t-1}, a_{t-1})$;

8. $w_{0:t}^{(i)} = p(s_t|l_t, m)$;

9. $S_{0:t} = S_{0:t} \cup \{\langle w_{0:t}^{(i)}, \{l_{0:t-1}, l_t\} \rangle\}$;

10.**return** $S_{0:t}$;

Note that the algorithm assumes that the map m is known. On the other hand, in the SLAM problem the map is unknown and should itself be determined. Recently, a Monte Carlo technique called Rao-Blackwellised particle filtering has been applied for the solution of the SLAM problem [6,19]. Without going into details, we only mention how the MCL- algorithm should be modified in order to solve the SLAM problem by the Rao-Blackwellised particle filtering. It suffices to replace the assignment in line 8 of the algorithm by $w_{0:t}^{(i)} = p(s_t|l_t, \widehat{m}(l_{0:t-1}, s_{1:t-1}))$, where $\widehat{m}$ is a function, which constructs maps by merging the perception data $s_{1:t-1}$ according to the poses $l_{0:t-1}$, that is, in our case $\widehat{m}$ places the laser range scans s_i at the locations l_i, $1 \leqslant i \leqslant t - 1$.

In the following section we will make use of a simplified version of the mentioned SLAM algorithm to solve the problem of tracking a robot's local pose when the initial global pose of the robot is unknown.

3   Multi-robot Localization

As we have already mentioned in the introduction, our approach to the multi-robot localization is based on determining the relative poses of robots. This can be done by enabling the robots to detect (and identify) each other by cameras or laser sensors. A good example of such an approach is [9]. This method has one substantial restriction however - a robot can reliably detect another one only when the distance between them is relatively short (about 3 meters for cameras). This means that if the robots localize themselves in a relatively large area, they either have to stay near to each other or the exchange of beliefs about the global poses can occur relatively rarely. In both cases the effect of mutually refining the beliefs will be minimized. We propose to compare the current perception data (laser scans) of one robot with the perception data acquired by another robot while moving. If the data match with high probability then the first robot is probably located on the path made by the second one. In addition, each robot keeps track of its own local poses (which is the pose relative to the coordinate system centered at the robot's initial pose) and annotates the corresponding perception data with this pose. After getting this annotated local pose from the matched perception data, the first robot can easily determine its relative pose to the second robot and fuse its belief about the global pose with that of the second robot. This scenario requires only data communication between robots and no direct detection of robots by each other. As it was already mentioned in the introduction, if a robot matches its perception sensor data with that of other robots it does not in principle get any new information about the map. Rather it reuses the computations of the global pose made by other robots. For example, a robot could itself simulate several other robots running in the environment and get the same results, however in this case it would not be able to reuse the computations of those simulated robots because these computations are made by the robot itself.

3.1   Overview

The problem of matching perception data from different robots is a kind of revisiting problem. A robot should decide, whether it is at a location previously visited by another robot. To realize the mentioned scenario we introduce two functions $F : S \longrightarrow \mathfrak{P}(\mathfrak{S} \times S \times \mathbb{N})$ and $G : S \times L \longrightarrow [0, 1]$, S, L and $\mathfrak{S}$ are the sets of all possible laser range scans poses and sample sets over poses respectively, $\mathbb{N}$ is the set of natural numbers and $\mathfrak{P}$ denotes the power set. Function G is defined as $G(s, \Delta l) = \int_{l'-\Delta l}^{l'+\Delta l} p(s|l, m)dl / \int p(s|l, m)dl$, where Δl is a small interval in the space of robot poses and $l' = \underset{l \in L}{argmax}\{p(s|l, m)\}$. The function returns the probability that a given scan s represents some limited area $[l' - \Delta l, l' + \Delta l]$ on the map (note that G is NOT a probability distribution). Intuitively G is a measure of "ambiguity" of a laser range scan s. Higher values of $G(s, \Delta l)$ indicate that s is less "ambiguous", that is s can probably be observed only within a small connected area in the map, while lower values mean that s is probably observed at several unconnected areas.

F is a kind of global repository where robots can store information on their own paths and retrieve that on the paths of other robots. Initially, $F(s) = \varnothing$ for all $s \in S$. As the robots start moving they estimate their local poses by means of a simplified version of the Rao-Blackwellised particle filter algorithm mentioned in Section II. The simplification here is that $\widehat{m}(l_{0:t-1}, s_{1:t-1}) = \widehat{m}(l_{t-1}, s_{t-1})$, i.e. we use only the most recent laser range scan as a map. If the j-th robot estimates a sample set S_t representing its local pose and reads a laser range scan s_t at time t then it sets $F(s_t) = F(s_t) \cup \{(S_t, s_t, j)\}$. In addition for all $(S', s', k) \in F(s_t)$ s.t. $k \neq j$ robot j first checks whether it should fuse its belief about the global pose with that of robot k (this could be done e.g. by comparing the entropies of both distributions) and whether the probability of a false positive path match is low, i.e. $1 - G(s_t, \Delta l) < a$, where Δl and a are some small (but not infinitely small) interval and threshold value respectively. The latter condition checks whether the local beliefs S_t and S' represent a same small area. This condition can be easily extended to longer pieces of paths. Let $S_{t_0:t}$, $t_0 < t$ be sample sets representing the recent local pose estimations of robot j. Then we should check whether $\prod_{i=t_0}^{t} (1 - G(s_i, \Delta l)) < a$ and for any t_1, t_2 s.t. $t_0 \leq t_1 < t_2 \leq t$ there exist t'_1, t'_2 s.t. $(S_{t'_1}, s_{t'_1}, k) \in F(s_{t_1})$ and $(S_{t'_2}, s_{t'_2}, k) \in F(s_{t_2})$ and either $t_2 - t_1 = t'_2 - t'_1$ for all t_1, t_2 or $t_2 - t_1 = t'_1 - t'_2$ for all t_1, t_2. That is we check whether consecutive pose estimations in a path of robot j correspond to consecutive pose estimations (forwards of backwards) in a path of robot k. We call the latter condition "path matching of robot j with respect to robot k" for further reference.

If both mentioned conditions are met robot j first estimates its local pose S'_t with respect to the coordinate system centered at robot k (this could be done e.g. by matching the scans s_t and s' with scan matching techniques like [18], because the poses of robots j and k are quite close to each other, or again using the MCL algorithm assuming $\widehat{m}((0,0,0), s')$ as a map). After that it is easy to determine the relative poses of robots j and k and fuse their beliefs.

3.2   Implementation

We use the perception model $p(s|l, m)$ to implement F and G. We note also that we should modify laser range scans before we can use them. The reason for this is that our implementation assumes learning F and G with a reasonable number of data points from the set of all possible laser range scans in a given map and because "raw " scans show no "structure", we would need the number of points exponential in the number of dimensions. Our approach is to transform a scan into a lower dimensional space and at the same time reveal the structure carried in the scan. We apply the multi-resolution analysis to scans. More precisely, we first compute the center of gravity of a scan and then get a modified scan by connecting the points of the old scan to the center of gravity. We then apply the Haar wavelet transform [13] to the modified scan. The power of the multi-resolution analysis is that it makes it possible to represent the most important information about the structure of a scan in a considerably smaller dimensional space (our experience shows that the first 10-12 elements of the Haar wavelet transform of 100 dimensional laser range scans in a given map carry more than 99% of the information). The mentioned transform can be carried out in $O(n)$ time, where n is the number of readings in a scan and is very "robust" in the sense that a small difference between poses will not cause a big change in the transformed scans read at these poses as it would be the case with "raw" scans. Figure 2 illustrates this.

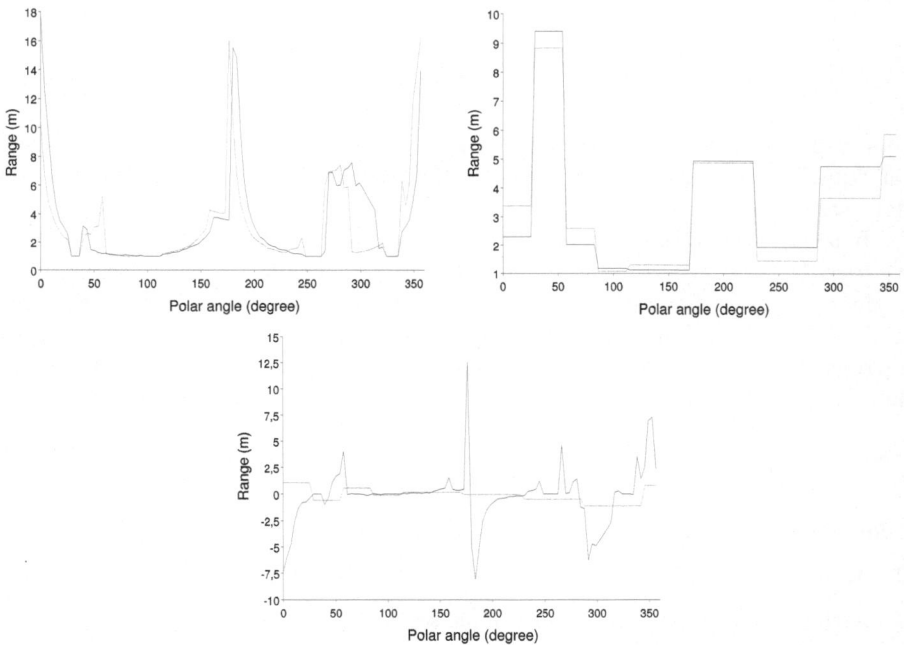

Fig. 2. Two laser range scans (represented by black and gray curves) recorded in a map. "Raw" scans are represented in the upper left image and the transformed scans in the upper right one. The gray scan was recorded at a relative pose -0.54 m, -0.03 m, -6 deg. with respect to the black one. The black and gray curves in the lower middle image represent the differences between the "raw" and transformed scans respectively.

We learn the functions F and G in a way similar to [26]. First, we randomly generate a sufficient number of poses in the map (one million in our current implementation) and then by means of ray-tracing and sampling from the perception model $p(s|l, m)$ compute "noisy" scans corresponding to the poses. Each scan is then transformed as mentioned above and a kd-tree is built over the transformed scan vectors. Initially, we store at each leaf of the tree the set of poses corresponding to the scan vectors making up the leaf. As a next step we construct a density tree (a kd-tree representing a piecewise-constant approximation of a probability distribution) for each leaf of the scan kd-tree out of the

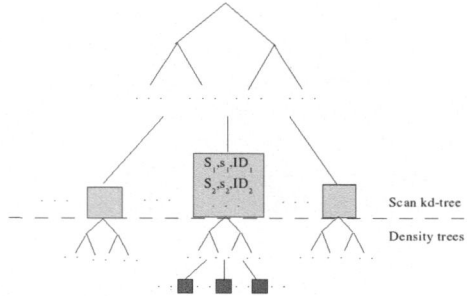

Fig. 3. The kd-tree representation of function F

sets of poses stored at that leaf. The resulting data structure can also be used for "dual" sampling in the Mixture-MCL algorithm [26]. Figure 3 illustrates the data structure.

It is easy to see that the mentioned computations can be done in $O(m \cdot n + n \cdot log(n))$ time, where m is the number of dimensions of "raw" scans and n is the number of generated scans. The function F can be easily computed and updated just by finding an appropriate leaf in the scan kd-tree and storing the current estimate of the local pose, (raw) scan and robot ID (refer to Figure 3). The function G can also be easily computed. In each density tree at the leaves of

Table 2. Multi-Robot MCL

1.**for each** robot i **do**

2.   estimate the current global pose sample set S_i by MCL;

3.   estimate the current local pose sample set S_i'

   by the simplified Rao-Blackwellised particle filter algorithm;

4. $F(s_i) = F(s_i) \cup \{(S_i', s_i, i)\}$,

   where s_i is the current laser range scan;

5.   **for each** $(S_j', s_j, j) \in F(s_i), j \neq i$ **do**

6.   **if** path matching of robot i with respect to robot j

      and j is localized better than i **do**

7.   localize robot i in a small map defined by $\widehat{m}((0, 0, 0), s_j)$

      and compute sample set S_{ij} representing

      the relative pose of robot i with respect to robot j;

8.   refine S_i by setting $S_i = S_i \cap S_i''$,

      where S_i'' is a sample set resulting from translating the

      global pose samples of robot j by relative poses in S_{ij};

the scan kd-tree we find a leaf with the maximum probability. This probability will be the value of G for a scan corresponding to the given leaf in the scan kd-tree. We mention also that the fusion of the beliefs about the global pose from different robots can also be done using density trees. This technique is the same as in [9].

Our algorithm for the multi-robot localization is summarized below.It is easy to see, that the run-time complexity of the algorithm (after the mentioned data structure has been constructed) is not significantly higher than that of single robot approaches, because the path matching does not depend on the number of samples and is done very efficiently by looking up in a binary tree. In the algorithm we implicitly assume that the robots communicate their sensor data and beliefs among themselves. Using WLAN 802.11b e.g. sample sets containing 10000 samples can be transfered in fractions of seconds. Finally the algorithm is robust in the sense that in case of communication failure the robots just do the single robot localization before they can communicate again, after which they simply continue running the multi-robot algorithm.

4   Experimental Results

We carried out a number of simulations to see how well the algorithm works in practice. We used the robot simulation software of the Aachen RoboCup team

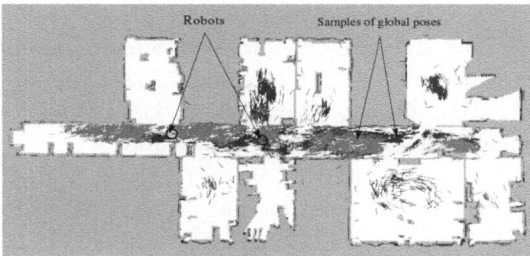

Fig. 4. Two robots in an occupancy grid map of LuFG V at RWTH Aachen. Current global pose sample sets of dark and bright robots are shown as dark and bright segments respectively.

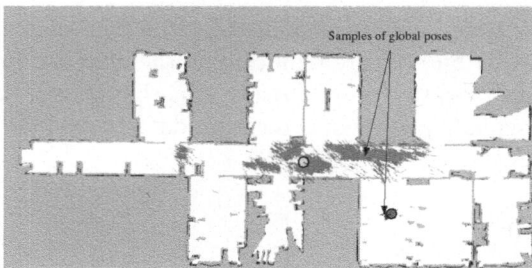

Fig. 5. Dark robot drives into an office and localizes itself globally

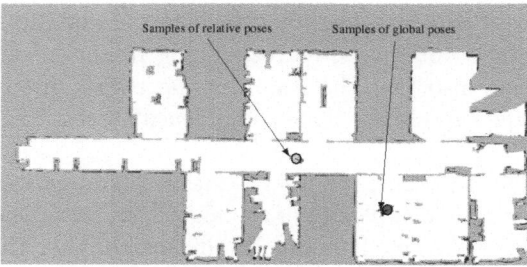

Fig. 6. Bright robot detects the path of dark robot and estimates its relative pose with respect to dark robot

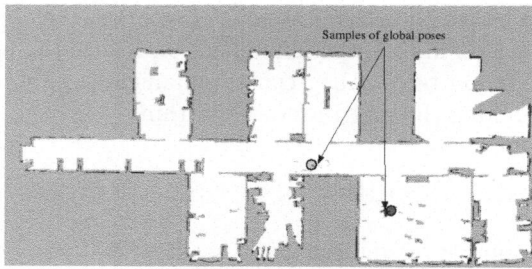

Fig. 7. Bright robot fuses its global pose sample set with that of dark robot

"AllemaniACs" [7]. The software can produce noisy odometry and laser scan data and therefore can very accurately simulate real robots. Figures 4-7 give an example of a situation when a robot (bright one) refines its belief about the global pose using the belief of the other (dark) robot.

We carried out several simulation runs for groups from 2 to 5 robots and for each case computed the average time per robot required to globally localize a robot both by the single-robot MCL and multi-robot MCL. The simulations showed that the average localization time per robot is approximately linearly reduced with an increased number of robots. The robots detected the paths of other robots on average in only two matching steps, however for more symmetric environments the robots may be required to follow longer paths of other robots in order to match the paths with high probability. The linear reduction in the localization time may be explained by the fact that we let the robots start at locations distributed quite uniformly in the map and because the robots could determine their relative poses already at earlier stages, they could use the considerable orthogonality in their perception data to exclude false hypotheses (samples) faster.

5   Conclusion

We presented a new approach to the Monte Carlo localization for multi-robot systems. This approach differs from existing ones in the field in that it does not require the detection and identification of robots by each other for determining their relative poses, which are used to mutually refine the posterior distributions over global poses of the robots. The simulations showed an average reduction in the localization time linear in the number of robots. Still the approach might fail in highly symmetric environments. In such cases it might be reasonable to combine our algorithm with approaches such as [9], where the robots detect each other by cameras. In the near future will also test the algorithm with "real" robots.

As we said in the beginning, the main reason for us to use the scans from other robots instead of simulating them locally is computational. In the future, we also plan to transfer the ideas of this paper to the multi-robot SLAM problem. Here we expect a pay-off that goes beyond savings in computation, as the robots share information about parts of the environment which only some of them have mapped so far.

References

1. A.Doucet, J.F.G. de Freitas, and N.J. Gordon, editors. *Sequential Monte Carlo Methods In Practice*. Springer Verlag, New York, 2001.
2. J. Borenstein, B. Everett, and L. Feng. *Navigating Mobile Robots: Systems and Techniques*. A.K. Peters, Ltd., 1996.
3. W. Burgard, A. B. Cremers, D. Fox, D. Hähnel, G. Lakemeyer, D. Schulz, W. Steiner, and S. Thrun. Experiences with an interactive museum tour-guide robot. *Artificial Intelligence*, 114(1-2), 2000.
4. I. Cox. Blanche - an experiment in guidance and navigation of an autonomous robot vehicle. *IEEE Transactions on Robotics and Automation*, 7(2), 1991.
5. A. Doucet. On sequential simulation-based methods for Bayesian filtering. Technical Report CUED/F-INFENGS/TR 310, Cambridge University, Department of Engineering, Cambridge, UK, 1998.
6. A. Doucet, J.F.G. de Freitas, K. Murphy, and S. Russel. Rao-Blackwellised particle filtering for dynamic Bayesian networks. *Proc. of the Conference on Uncertainty in Artificial Intelligence (UAI)*, 2000.
7. A. Ferrein, C. Fritz, and G. Lakemeyer. Allemaniacs 2004 team description, 2004. http://robocup.rwth-aachen.de.
8. D. Fox, W. Burgard, F. Dellaert, and S. Thrun. Monte Carlo localization: Efficient position estimation for mobile robots. *Proc. of the Sixteenth National Conference on Artificial Intelligence (AAAI-99*, 1999.
9. D. Fox, W. Burgard, W. Kruppa, and S. Thrun. A probabilistic approach to collaborative multi-robot localization. *Autonomous Robotics, Special Issue on Heterogeneous Multi-Robot Systems*, 8(3):325–344, 2000.
10. D. Fox, W. Burgard, and S. Thrun. Markov localization for mobile robots in dynamic environments. *Journal of Artificial Intelligence Research*, (11):391–427, 1999.
11. D. Fox, J. Ko, B. Stewart, and K. Konolige. The revisiting problem in mobile robot map building: A hierarchical Bayesian approach. *In Proc. of the Conference on Uncertainty in Artificial Intelligence*, 2003.
12. A. Howard, M.J. Mataric, and G.S. Sukhatme. Localization for mobile robot teams: A distributed mle approach. *In Proc.of the 8-th International Symposium in Experimental Robotics (ISER'02)*, 2002.

13. A. Jensen and A. la Cour-Harbo. *Ripples in Mathematics: the Discrete Wavelet Transform.* Springer, 2001.
14. L.P. Kaelbling, A.R. Cassandra, and J.A. Kurien. Acting under uncertainty: Discrete Bayesian models for mobile robot navigation. *In Proc. of the IEEE/RSJ Internatinal Conference on Intelligent Robots and Systems (IROS)*, 1996.
15. J. Ko, B. Stewart, D. Fox, K. Konolige, and B. Limketkai. A practical, decision theoretic approach to multi-robot mapping and exploration. *In Proc. of the IEEE/RSJ International Conference on Intelligent Robots and Systems*, 2003.
16. R. Kurazume and S. Hirose. An experimental study of a cooperative positioning system. *Autonomous Robots*, 8(1):43–52, 2000.
17. J. Liu and R. Chen. Sequential Monte Carlo methods for dynamic systems. *Journal of the American Statistical Association*, (93):1032–1044, 2001.
18. F. Lu and E. Milios. Robot pose estimation in unknown environments by matching 2d range scans. *Proc. IEEE Comp. Soc. Conf. on Computer Vision and Pattern Recognition*, pages 935–938, 1994.
19. M. Montemerlo and S. Thrun. Simultaneous localization and mapping with unknown data association using fastslam. *Proc. of the IEEE International Conference on Robotics and Automation (ICRA)*, 2003.
20. I. Nourbaksh, R. Powers, and S. Birchfield. Dervish an office-navigating robot. *AI Magazine*, 16(2), 1995.
21. M. Pitt and N. Shephard. Filtering via simulation: auxiliary particle filter. *Journal of the American Statistical Association*, (94):590–599, 1999.
22. S.I. Roumeliotis and G.A. Bekey. Collective localization: A distributed kalman filter approach. *In Proc. of the IEEE International Conference on Robotics and Automation*, 2:1800–1087, 2000.
23. D. B. Rubin. Using sir algorithm to simulate posterior distributions. In M.H. Bernanrdo, K.M. van De Groot, D.V. Lindley, and A.F.M. Smith, editors, *Bayesian Statistics 3*. Oxford University Press, Oxford, UK, 1998.
24. R. Simmons and S. Koenig. Probabilistic robot navigation in partially observable environments. *In Proc. of the International Joint Conference on Artificial Intelligence (IJCAI)*, 1997.
25. S. Thrun. A probabilistic online mapping for teams of mobile robots. *International Journal of Robotics Research*, 20(5), 2001.
26. S. Thrun, D. Fox, W. Burgard, and F. Dellaert. Robust Monte Carlo localization for mobile robots. *Artificial Intelligence*, 2001.

Modeling Moving Objects in a Dynamically Changing Robot Application

Martin Lauer, Sascha Lange, and Martin Riedmiller

Neuroinformatics Group,
Institute for Computer Science and Institute for Cognitive Science,
University of Osnabrück, 49069 Osnabrück, Germany
{Martin.Lauer, Sascha.Lange, Martin.Riedmiller}@uos.de

Abstract. In this paper we discuss the topic of modeling a moving object in a robot application considering as example the ball movement in a robot soccer environment. The paper focuses on the question of how to estimate the ball velocity reliably even if the ball collides with an obstacle or the observing robot is moving itself. We propose a new estimation algorithm based on a direct estimate using ridge regression and show its performance in reality.

1 Introduction

For many applications of autonomous robots it is important to model moving objects that are in the surroundings of the robot to avoid collisions or enable interaction. E.g. an autonomous car must track the neighboring cars and estimate their velocity and position. In the domain of soccer playing autonomous robots it is important to have knowledge of the current ball position and movement. Hence we need an approach to estimate both reliably even if the ball is kicked by another robot, is hidden behind an obstacle or collides with another object. In this paper we want to focus on this domain and derive an algorithm to detect the ball in an omnidirectional camera image and estimate its position and velocity, even if the robot itself is moving.

In the RoboCup domain [7] of soccer playing robots which serves as testbed for our approach several publications deal with the topic of ball detection in a camera image, e.g. [9], but only a few publications consider the question of how to estimate the ball velocity [8]. Therefore we want to focus on the latter aspect and describe the approach for ball detection in a camera image only informally to give the reader an impression of the preprocessing that is necessary to get observations of the ball.

The paper is organized as follows: section 2 briefly describes a possible way of detecting a RoboCup Midsize ball in a camera image, while section 3 describes the estimation task to obtain the ball velocity. In the subsequent section we show the performance of this approach in practice.

U. Furbach (Ed.): KI 2005, LNAI 3698, pp. 291–303, 2005.

2   Detecting the Ball

The robot that has been used in the experiments is equipped with a catadioptric sensor. An off-the-shelf CCD-camera is mounted onto the top of the robots, pointing upwards to a hyperbolic mirror. There exist various model-based self-calibration methods (e.g. [6]) for this kind of camera assemblies that require the lens to be mounted exactly in the focal point of the mirror while the camera has to be aligned with its symmetry axis. Due to the constant vibrations and huge forces the assembly is exposed to when mounted on top of a mobile robot moving up to $6m/s$ this restriction is impracticable. Therefore we applied a model-free method: the mapping is approximated by a partially linear function established on more than 200 semi-automatically collected correspondence points. While being less precise, within this approach there are no restrictions on the positioning of mirror and camera.

At the moment, we use color as the only feature to detect the ball. Ambiguities that occur especially outside the field are filtered out with the help of the world model. When the ball has been seen in the last frame, a simple search algorithm is started at the old position [1]. Otherwise we search on radial scan lines ordered around the image center for occurrences of red pixels. The algorithm returns the position of the "ball-region" that is further analyzed in the next processing step.

The world model models the ball by the two-dimensional position that one receives, when projecting the center of the ball on the ground (normally, this is the point, where the ball touches the ground). Unfortunately, it is not possible to detect this point directly in the image, since it is always occluded by the ball itself. Therefore we try to approximately detect the center of the ball. The assumption is, that the center of the two dimensional projection of the ball on the projection plane coincides with the projected position of the center of the ball. Although this assumption formally only holds for an ideal pin-hole camera pointed directly to the center of the ball, the error introduced by this simplification is— according to our empirical measurements— in the range of one or two pixels.

To detect the center of gravity of the ball-region we first apply morphologic operators to remove erroneously classified pixels (e.g. spotlights) and to soften the borders. Afterwards the contour and the center of gravity are extracted using an algorithm first proposed in [3].

Then, the image coordinates of the center of gravity are transformed to real world coordinates using the earlier established mapping. But since the point we have extracted from the image was well above the ground, the resulting point always is behind the perpendicular projection of the ball's center (fig. 1). To correct this, we apply the second projection theorem:

$$\frac{g}{h} = \frac{a}{a+b} \Leftrightarrow a = \frac{g \cdot (a+b)}{h} \tag{1}$$

Since $a + b$ is directly measured by the vision system and the radius of the ball could easily be obtained, the only problem is the surface of the hyperbolic

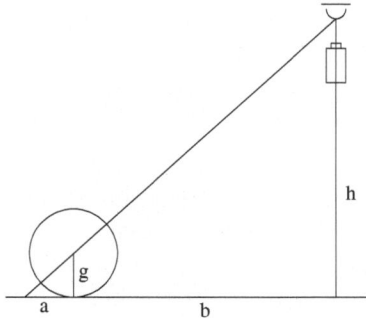

Fig. 1. Application of the intercept theorem to the correction of the ball position. h is the height of the mirror, g the radius of the ball. The image processing algorithm returns a point with distance $a+b$. To correct this point to the perpendicular projection of the ball's center on the floor, we calculate a.

mirror that has no constant "height" h. But because the mirror is rather small in comparison to h and $a + b$, it is possible to simply neglect the curvature of the surface and use a constant height h without any notable loss of precision.

3 Fusion of Single Observations

3.1 Motivation

Up to now we described how the ball can be detected in single camera images. In many situations, this is sufficient for obtaining a meaningful behavior of the robots. But if the ball is rolling or it is temporarily hidden behind an obstacle the robot may be mislead or may exhibit suboptimal behavior. Therefore it is necessary to build a more complex model of the ball movement and hence to be able to predict the ball position and velocity even if the ball is hidden.

Most approaches for the calculation of ball movement that are used up to now in RoboCup use a Kalman filter technique [4] to estimate the ball position and ball velocity [8]. They are based on the assumption of linear ball movement with constant velocity and model the variation of the ball velocity, e.g. by robots pushing the ball or by friction, as random noise. Other approaches sometimes reinit the Kalman filter when a jerky change of the ball movement was observed.

Although Kalman filter based techniques allow to get a crude estimate of the ball velocity they exhibit some shortcomings that lead to imprecision and difficulties in dealing with these techniques:

- the ball movement does not fulfill the Markov condition of stochastically independent noise since friction and slippage are not stochastically independent over time
- jerky changes of the movement due to collisions with robots or due to a robot pushing or kicking the ball are not modeled adequately.

- determining variances and covariances of the motion model and of the sensor probability distributions is difficult and can often be determined only by trial and error. These parameters, especially the covariances, are not very intuitive and therefore determining them is time consuming and needs a lot of expertise.

To avoid these shortcomings, we want to present an alternative way of building a predictive model of ball movement based on an explicit regression model. It is able to increase the precision of the position and velocity estimate. Furthermore, the parameters of the algorithm are more intuitive and can therefore be adjusted more easily.

3.2   Predictive Model for a Free Movement

In many situations the ball is moving freely on the field. In these cases we can assume a linear movement of the ball, i.e. the ball position $p(t)$ at point in time t is given by the equation:

$$p(t) = p_0 + v(t - t_0) \qquad (2)$$

where v is the constant velocity of the ball, t_0 is a certain reference point in time and p_0 is the ball position at t_0. Both, p and v are two-dimensional vectors. The modeling in (2) does not consider the natural deceleration of the ball due to friction, anyhow this modeling is approximately correct for small time intervals.

How to calculate p_0 and v? Instead of using implicit sensor integration approaches like Kalman filtering or Particle filtering [2] we propose the use of a direct approach that recalculates the motion parameters every iteration directly from the latest observations. Hence, we avoid too strong assumptions about the noise level of the observations but we only need to assume that all observations are influenced by noise of the same noise level, no matter how large its variance actually is.

If we have observed the ball at positions $p_1, p_2, \ldots, p_n$ in past at times $t_1, t_2, \ldots, t_n$ we can calculate the parameters of the motion model in such a way that the squared error between observed positions and expected positions according to the motion model is minimized, i.e. we solve the minimization task:

$$\underset{p_0, v}{minimize} \; \frac{1}{2} \sum_{i=1}^{n} ||p_0 + v(t_i - t_0) - p_i||^2 \qquad (3)$$

t_0 can be set arbitrarily, e.g. we can choose $t_0 = t_n$.

Equation (3) exhibits to be a standard linear regression task. By calculating the partial derivatives of (3) and looking for the zeros we obtain the estimates:

$$p_0 = \frac{\sum_{i=1}^{n}(t_i - t_0)^2 \sum_{i=1}^{n} p_i - \sum_{i=1}^{n}(t_i - t_0) \sum_{i=1}^{n}((t_i - t_0)p_i)}{n \sum_{i=1}^{n}(t_i - t_0)^2 - (\sum_{i=1}^{n}(t_i - t_0))^2} \qquad (4)$$

$$v = \frac{n \sum_{i=1}^{n}((t_i - t_0)p_i) - \sum_{i=1}^{n}(t_i - t_0) \sum_{i=1}^{n} p_i}{n \sum_{i=1}^{n}(t_i - t_0)^2 - (\sum_{i=1}^{n}(t_i - t_0))^2} \qquad (5)$$

In contrast to Kalman filtering, the direct estimator does not need any assumption of the noise level by which the observed ball positions are influenced. Hence, we get estimates that are more robust with respect to badly chosen variance parameters.

In the case of very few observations, i.e. if n is small, the estimates from equation (4) and (5) become very noisy. Although the estimates of $\boldsymbol{p}_0$ and $\boldsymbol{v}$ are unbiased, the length of $\boldsymbol{v}$ is expected to be larger than zero even if the ball is not moving, since noisy observations always cause a velocity vector that differs from zero. E.g. if we assume a non-moving ball and estimates for $\boldsymbol{p}_t$ that are independently identically distributed according to a Gaussian the estimates of $\boldsymbol{v}$ will also be distributed Gaussian with mean zero due to (5). But the length of $\boldsymbol{v}$ is not expected to be zero since $||\boldsymbol{v}||^2$ is χ^2-distributed which is a distribution with positive expectation value.

To reduce the effect of noisy observations in the case of small n we can replace the standard linear regression approach shown in equation (3) by a ridge regression with an extra weight decay parameter for the velocity parameter while the position estimate remains unregularized. The minimization task now becomes:

$$\underset{\boldsymbol{p}_0,\boldsymbol{v}}{minimize} \; \frac{1}{2}\sum_{i=1}^{n}||\boldsymbol{p}_0 + \boldsymbol{v}(t_i - t_0) - \boldsymbol{p}_i||^2 + \frac{\lambda}{2}||\boldsymbol{v}||^2 \tag{6}$$

with $\lambda > 0$ chosen appropriately.

In this case, solving (6) leads to:

$$\boldsymbol{p}_0 = \frac{(\lambda + \sum_{i=1}^{n}(t_i - t_0)^2)\sum_{i=1}^{n}\boldsymbol{p}_i - \sum_{i=1}^{n}(t_i - t_0)\sum_{i=1}^{n}((t_i - t_0)\boldsymbol{p}_i)}{n(\lambda + \sum_{i=1}^{n}(t_i - t_0)^2) - (\sum_{i=1}^{n}(t_i - t_0))^2} \tag{7}$$

$$\boldsymbol{v} = \frac{n\sum_{i=1}^{n}((t_i - t_0)\boldsymbol{p}_i) - \sum_{i=1}^{n}(t_i - t_0)\sum_{i=1}^{n}\boldsymbol{p}_i}{n(\lambda + \sum_{i=1}^{n}(t_i - t_0)^2) - (\sum_{i=1}^{n}(t_i - t_0))^2} \tag{8}$$

Comparing (5) and (8) shows the difference: adding $\lambda > 0$ in the denominator leads to a preference of smaller velocities and thus a more reliable estimate in the case of few observations.

The quality of the position and velocity estimate depends on the integration length, i.e. on the number of observations n which are used for the calculation: if n is small the estimates are heavily influenced by noise, if n is large the assumption of the ball rolling with approximately the same velocity is hurt, especially when the ball collides with an obstacle or is kicked by a robot.

3.3   Dealing with Changes of the Ball Movement

The movement of the ball sometimes changes due to (a) the ball is decelerated when it rolls freely across the field or (b) the ball collides with an obstacle or is kicked or pushed by a robot. In the former case the velocity changes only slowly and the assumption of constant velocity is appropriately fulfilled for a moderate number of observations. In the latter case, this assumption is violated anyway. Hence, we have to detect these situations and deal with them in a special way.

Figure 2 shows a situation in which the ball is reflected by an obstacle: the observations A, B and C have been made while the ball was rolling into the direction of the obstacle while the observations D, E and F are made after the collision with the obstacle. Obviously, the entire movement cannot be described by a linear model, but it is possible to describe the movement from A to C and the movement from D to F as two different consecutive linear movements. Thus, we are faced with the problem of detecting the point of change between the first and the second movement.

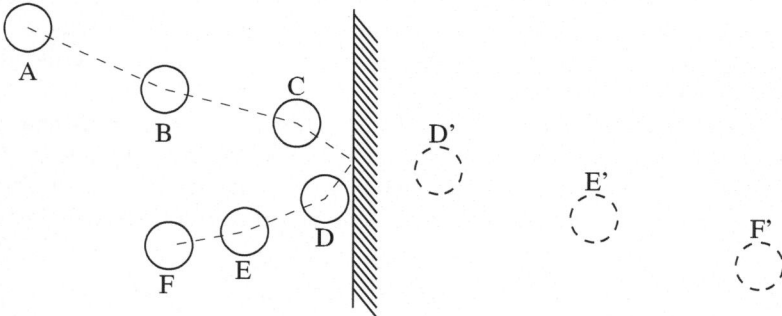

Fig. 2. Collision of the ball with an obstacle: the ball is reflected and changes its direction jerkily. The circles show the positions at which the ball was observed by the robot subsequently (temporal order A-B-C-D-E-F). The points D', E' and F' are predicted ball positions assuming a movement with constant velocity using the observations A,B and C.

Having found such a change point, the current model of the ball movement becomes invalid and has to be replaced by the new one. Therefore we have to recompute the motion model only with observations that have been made after the change point, i.e. we have to remove older observations.

To detect change points in the ball movement we propose a test that always checks whether a new observation fits to the model used so far. If the distance between the predicted ball position and the observed position is larger than a pre-defined threshold we mark this observation as possibly being a change point. But not all observations with large prediction errors are change points. It may happen that due to misdetections of the ball or errors in the robot self-localization an observation may erroneously look like a change point although the ball movement is not interrupted. For that purpose we define a suspicious observation only as change point if the succeeding observation is also suspicious.

In the example given in Figure 2 this would mean: the observations A,B and C match the current motion model and therefore are not at all change points. Comparing observation D with the corresponding prediction D' exhibits an error, which is not large enough to call observation D being suspicious. In contrast, the error between observation E and prediction E' is reasonable and therefore observation E is possibly a change point. When we get observation F the hypothesis of E being a change point is affirmed since the difference between F and the prediction F' is very large. In such a situation we remove old observations from the list of observations and restart calculating a new motion model with the newest observations E and F.

3.4   The Entire Ballfilter

Combining the ideas from section 3.2 and 3.3 leads to a new algorithm that builds an adaptive model of the velocity and position of the moving ball that allows to predict the ball position even if the ball is hidden behind an obstacle.

The algorithm uses a buffer to store observations of the ball. The buffer size varies between a minimum size and a maximum size. The minimum size must be chosen larger than 2 since the predictive model can only be estimated from at least two observations. A good minimum value is 3. The maximal size should be chosen in such a way that the change of the ball velocity due to friction when the ball rolls freely remains small. For cameras that take 30 frames per second, a value of 10–15 is reasonable.

New observations are always inserted into the buffer, old ones are removed. If the predicted ball position and the observed one exhibit a large discrepancy larger than a predefined threshold θ, an observation is marked as being suspicious. If two succeeding observations are suspicious, a change point was detected and the buffer size is reduced to the minimum to forget all old and misleading observations. The value of θ has to be chosen compatible to the noise in the observations. A good choice is a value of $500mm$.

Every time the ball was observed new estimates of ball position and ball velocity are calculated using the observations stored in the buffer. Only in the very beginning when the number of observations is not large enough to estimate both parameters these parameters are not estimated. Figure 3 shows the algorithm step by step.

3.5   Odometer-Based and Self-localization-Based Estimation

The modeling presented in the sections 3.2 and 3.3 is based on observations of the ball position that are represented in a common fixed coordinate system while the camera image always represents information relative to the robot position and robot heading. The robot relative coordinate system moves whenever the robot moves. The robot relative coordinate system and the fixed coordinate system we use are depicted in Figure 4.

If we can rely on an accurate estimate of the robot's pose the best way of calculating estimates of ball position and ball velocity is to perform all calculations in the fixed coordinate system. Then, the observations from the camera image have to be transformed to the fixed coordinate system before applying them in the ball estimator. Accurate estimates of the robot's pose can be calculated using powerful self-localization techniques like [5].

If accurate estimates of the robot's pose are not available we can instead perform the calculations of ball position and velocity in the robot relative coor-

1. start with an empty buffer of observations
2. once obtaining a new observation p_t at time t:
2.1. calculate the prediction $\hat{p}_t$ of the motion model used so far for
     time t.
2.2. if $\|\hat{p}_t - p_t\| > \theta$, mark observation p_t as being suspicious
2.3. if the buffer size is maximal, remove the oldest observation
2.4. insert the new observation p_t into the buffer
2.5. if the newest observations p_t and p_{t-1} are both marked
     suspicious, resize the buffer to minimal size removing the
     oldest observations.
2.6. calculate new estimates p_0 and v using the observations stored
     in the buffer applying equations (7) and (8).
3. wait for new observation and goto 2.

Fig. 3. Sketch of the entire algorithm to estimate the ball position and velocity reliably. The old observations are stored in a buffer of varying size. The minimal size must be larger than 2, the maximal size should be smaller than 20. Good values for minimal and maximal size are 3 and 10.

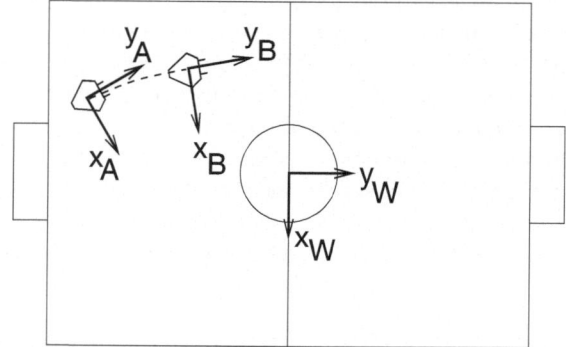

Fig. 4. Coordinate systems used for representation of ball and robot positions: x_W and y_W indicate the coordinate axis of the fixed world coordinate system that is independent of the robot's pose; the robot relative coordinate system depends on the robot's position and heading on the field: seen from the perspective of the robot its x-axis points to the right, its y-axis points to the front. The axis of the robot coordinate system for two different points in time (A and B) are marked with x_A, y_A, x_B and y_B, respectively.

dinate system. In this case, we have to take into account the movement that the robot performs between two observations A and B, i.e. we have to transform all information from the robot relative coordinate system at time A to the robot relative coordinate system at time B (see Figure 4).

A good choice of reference time and reference coordinate system is to set t_0 to the time of the newest observation and also to represent the ball observations in the coordinate system relative to the robots position at time t_0. Hence, obtaining a new ball observation requires to transform the stored observations from the previous coordinate system to the new coordinate system. The calculations needed are simple: if p_A represents a position in the coordinate system of time A and the robot has moved from time A to B by a vector of Δ (relative to the coordinate system of time A) and it has rotated by an angle of ϕ, the position p_A can be transformed into the coordinate system at time B using the equation:

$$p_B = \begin{pmatrix} \cos\phi & \sin\phi \\ -\sin\phi & \cos\phi \end{pmatrix} (p_A - \Delta) \tag{9}$$

To transform the estimated velocity v we only have to consider the rotation:

$$v_B = \begin{pmatrix} \cos\phi & \sin\phi \\ -\sin\phi & \cos\phi \end{pmatrix} v_A \tag{10}$$

The transformation of the stored observations and the position and velocity estimates has to be done directly after receiving a new observation, i.e. between step 2 and 2.1 in Figure 3. The movement of the robot (Δ, ϕ) can be calculated from odometer sensors or exploiting the driving commands sent to the robot and is part of the robot kinematics.

The answer to the question whether an odometer-based or self-localization-based estimation should be preferred depends of the properties of self-localization and sensor signals: on the one hand, self-localization based calculations may be distorted by the noise of the localization approach. On the other hand, odometer-based estimation suffers from the inaccuracy of odometers in the case of slippage. Especially when the robot is blocked by an obstacle and wheelspin occurs an odometer-based ball estimator is significantly distorted.

In our RoboCup team, the "Brainstormers Tribots", we use a very precise self-localization approach [5] and therefore are able to perform all calculation in a fixed coordinate system depicted with x_W, y_W in Figure 4.

3.6   Expired Observations

When the ball is taken off the field or it is far away from the robot or hidden behind an obstacle it cannot be seen by the robot. No new observations are available and the motion model is not updated. Hence, extrapolating the ball position yields an invalid result. Therefore, care has to be taken to assure that the motion estimates and the observations stored in the buffer do not become too old.

This can be guaranteed allowing an observation to be used in the estimation process only for a restricted time, e.g. for 2 seconds. If an observation becomes too old, it is removed from the buffer of stored observations. As well, the time of updating the estimates must be stored. Predictions are declared to be valid only if the time since the latest update is shorter than a pre-defined threshold.

4 Experiments

4.1 Experiments with a Standing Robot

First of all, we want to show the performance of the ball estimator for single trajectories with defined conditions. We made all tests in our laboratories where we have a half-field of the RoboCup-Midsize league. The experiments were performed using the robots of the "Brainstormers Tribots" RoboCup team. We rolled the ball over the field while a robot was observing the ball.

Figure 5 shows the case of the robot standing still. Since the noise of self-localization is almost zero in this case and the optical instruments like camera and mirror are not affected by vibrations the results show the good performance of the ball estimation process. The diagram shows a low-noise estimation of the ball velocity over 1.6 seconds. It is even possible to recognize the decrease of velocity due to friction.

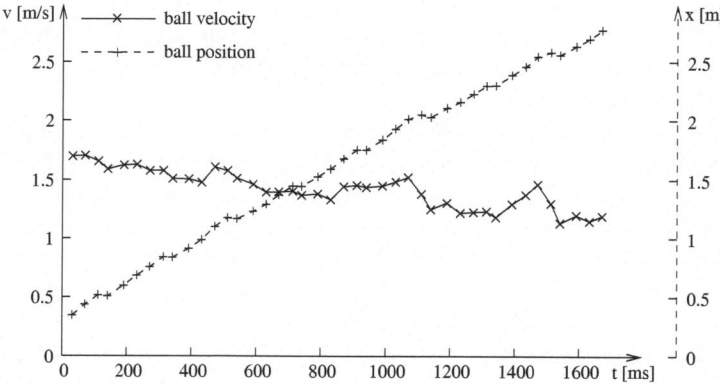

Fig. 5. Ball estimation when the ball was rolling across the field while the robot was standing still. The diagram shows the projection of the ball position (dashed line) onto the x-axis of the field coordinate system and the velocity estimate (solid line), also projected onto the x-axis of the field coordinates. Since the ball rolls freely, the velocity decreases due to friction.

Figure 6 shows the case of the ball colliding with an obstacle. While the ball is moving towards the obstacle for the first 130 milliseconds (negative velocity), it is reflected afterwards and moves to the opposite direction (positive velocity). The ball estimation process recognizes this situation correctly and performs a reset two iterations after the collision happened, i.e. only the three newest observations are left in the observation buffer while the older observations are removed. Otherwise, the large step in the velocity estimate from $-3.5\frac{m}{s}$ to $1\frac{m}{s}$ would not have been possible. Afterwards, the ball velocity is correctly estimated.

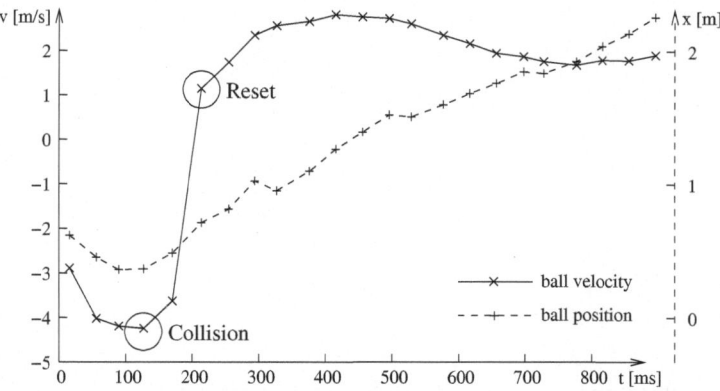

Fig. 6. Case of the ball colliding with an obstacle. The dashed line shows the position of the ball, the solid line the estimate of ball velocity. Both, the position and velocity are projected onto the x-axis of the field coordinate system.

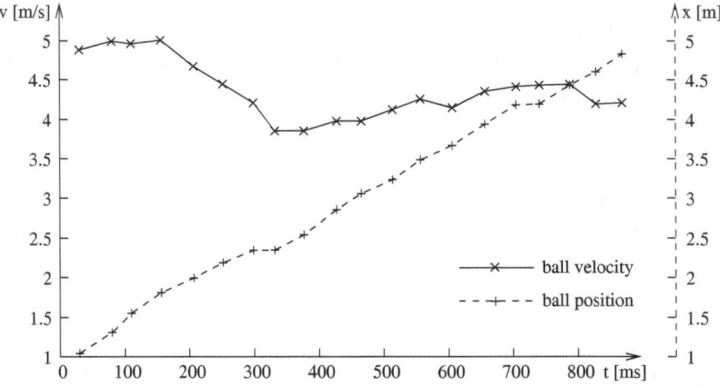

Fig. 7. Case of a moving robot. The dashed line shows the position of the ball, the solid line the estimate of ball velocity. Both, the position and velocity are projected onto the x-axis of the field coordinate system. The robot was moving with $2\frac{m}{s}$ on a circular trajectory.

4.2   Experiments with a Moving Robot

The situation becomes more difficult if the robot itself is moving since the errors of the self-localization and vibrations of the camera and the mirror of the omnidirectional optical system put additional noise onto the estimates. Hence, the quality of estimates is affected.

Figure 7 shows the estimate of a rolling ball when the robot itself was moving with $2\frac{m}{s}$ on a circular trajectory. Despite the robot movement, the estimate of ball velocity remains reliable. Also the direction of the velocity estimate is trust-

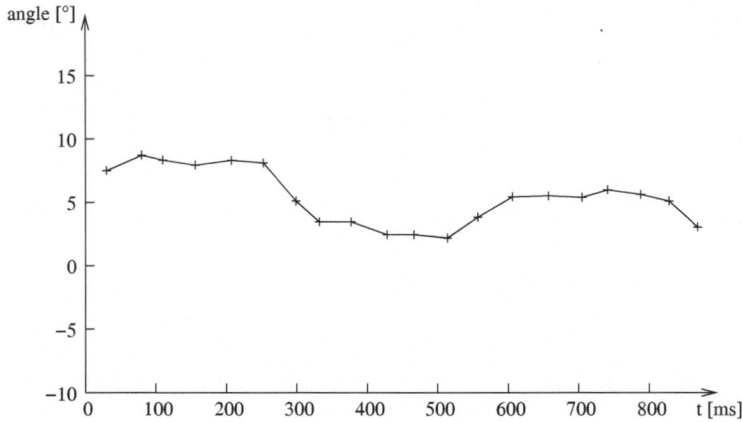

Fig. 8. The estimated direction of ball movement of a rolling ball. The direction varies in the interval from 3 to 9 degree. The robot was moving while estimating the ball velocity.

worthy as indicated in Figure 8: the direction of the velocity vector varies only between 3 and 9 degrees. Thus, the estimates can be used for robot repositioning, e.g. to fetch a moving ball.

4.3   Tournament Experiments

The ball estimation approach has also been tested in a RoboCup tournament. It was used in the German Open 2005 competition in the Midsize league and behaved well. Finally, the team using this approach won the tournament. One of the main reasons was its superior ball handling.

5   Discussion

The approach for ball estimation presented in this paper combines three techniques: (a) modeling a free ball movement linearly and estimating its parameters using ridge regression, (b) recognizing change points of the movement comparing observed and predicted ball positions and (c) considering the movement of the observer using either a precise self-localization approach or odometer values to calculate the observer's movement between two observations.

To avoid unintuitive and therefore difficult to use parameters we developed an estimation process based on regression instead of a Kalman filter. Furthermore, using such a direct approach avoids determining the noise level of the observations.

The comparison of predicted and observed ball positions enables us to recognize when the ball movement changed jerkily. If so, removing old observations allows us to react adequately to such a situation while keeping the newest observations allows to immediately replace the invalidated velocity estimate by a new one.

The experiments presented in this paper show that the new approach works fine in practice. For a standing robot, the estimate is low-noise and reliable while even for a moving observer the estimate of the ball velocity remains valid. Furthermore, jerky changes in the ball movement lead immediately to a reset of the ball filter. Hence, the presented algorithm can successfully be used in autonomous robots like in the RoboCup domain.

Acknowledgments

This work was granted by the German Research Foundation DFG SPP 1125.

References

1. M. Arbatzat, S. Freitag, M. Fricke, R. Hafner, C. Heermann, K. Hegelich, A. Krause, J. Krüger, M. Lauer, M. Lewandowski, A. Merke, H. Müller, M. Riedmiller, J. Schanko, M. Schulte-Hobein, M. Theile, S. Welker, and D. Withopf. Creating a robot soccer team from scratch: the brainstormers tribots. In *Proc. of RoboCup 2003*, Padua, Italy, 2003.
2. Arnaud Doucet. On sequential simulation-based methods for Bayesian filtering. Technical Report CUED/F-INFENG/TR.310, University of Cambridge, 1998.
3. H. Freeman. On the encoding of arbitrary geometric configuration. *IRE Transactions on Electronic Computers*, EC-10(2):260–268, 1961.
4. Arthur Gelb, editor. *Applied Optimal Estimation*. Cambridge, 1974.
5. Martin Lauer, Sascha Lange, and Martin Riedmiller. Calculating the perfect match: an efficient and accurate approach for robot self-localization. In *Robocup-2005*, 2005.
6. Branislav Micusik and Thomas Pajdla. Para-catadioptric camera auto-calibration from epipolar geometry. In *Asian Conference on Computer Vision*, Jeju Island, Korea, 2004.
7. Daniele Nardi, editor. *RoboCup 2004: Robot Soccer World Cup VIII Robot Soccer World Cup VII*. Springer, 2005.
8. Slav Petrov. Computer Vision, Sensorfusion und Verhaltenssteuerung für Fussball-Roboter. Master's thesis, Freie Universität Berlin, Institut für Informatik, 2004.
9. Andre Treptow, Andreas Masselli, and Andreas Zell. Real-time object tracking for soccer-robots without color information. In *Proceedings of the European Conference on Mobile Robotics (ECMR 2003)*, Radziejowice, Poland, 2003.

Heuristic-Based Laser Scan Matching for Outdoor 6D SLAM

Andreas Nüchter[1], Kai Lingemann[1], Joachim Hertzberg[1],
and Hartmut Surmann[2]

[1] University of Osnabrück, Institute of Computer Science,
Knowledge Based Systems Research Group,
Albrechtstr. 28, D-49069 Osnabrück, Germany
{nuechter, lingemann, hertzberg}@informatik.uni-osnabrueck.de
[2] Fraunhofer Institute for Autonomous Intelligent Systems,
Schloss Birlinghoven, D-53754 Sankt Augustin, Germany
hartmut.surmann@ais.fraunhofer.de

Abstract. 6D SLAM (Simultaneous Localization and Mapping) or 6D Concurrent Localization and Mapping of mobile robots considers six dimensions for the robot pose, namely, the x, y and z coordinates and the roll, yaw and pitch angles. Robot motion and localization on natural surfaces, e.g., driving with a mobile robot outdoor, must regard these degrees of freedom. This paper presents a robotic mapping method based on locally consistent 3D laser range scans. Scan matching, combined with a heuristic for closed loop detection and a global relaxation method, results in a highly precise mapping system for outdoor environments. The mobile robot Kurt3D was used to acquire data of the Schloss Birlinghoven campus. The resulting 3D map is compared with ground truth, given by an aerial photograph.

1 Introduction

Automatic environment sensing and modeling is a fundamental scientific issue in robotics, since the presence of maps is essential for many robot tasks. Manual mapping of environments is a hard and tedious job: Thrun et al. report a time of about one week hard work for creating a map of the museum in Bonn for the robot RHINO [25]. Especially mobile systems with 3D laser scanners that automatically perform multiple steps such as scanning, gaging and autonomous driving have the potential to greatly improve mapping. Many application areas benefit from 3D maps, e.g., industrial automation, architecture, agriculture, the construction or maintenance of tunnels and mines and rescue robotic systems.

The robotic mapping problem is that of acquiring a spatial model of a robot's environment. If the robot poses were known, the local sensor inputs of the robot, i.e., local maps, could be registered into a common coordinate system to create a map. Unfortunately, any mobile robot's self localization suffers from imprecision and therefore the structure of the local maps, e.g., of single scans, needs to be

U. Furbach (Ed.): KI 2005, LNAI 3698, pp. 304–319, 2005.

used to create a precise global map. Finally, robot poses in natural outdoor environments involve yaw, pitch, roll angles and elevation, turning pose estimation as well as scan registration into a problem in six mathematical dimensions.

This paper proposes algorithms that allow to digitize large environments and solve the 6D SLAM problem. In previous works we already presented partially our 6D SLAM algorithm [19,23,24]. In [19] we use a global relaxation scan matching algorithm to create a model of an abandoned mine and in [24] we presented our first 3D model containing a closed loop. This paper's main contribution is an octree-based matching heuristic that allows us to match scans with rudimentary starting guesses and to detect closed loops.

1.1 Related Work

SLAM. Depending on the map type, mapping algorithms differ. State of the art for metric maps are probabilistic methods, where the robot has probabilistic motion and uncertain perception models. By integrating of these two distributions with a Bayes filter, e.g., Kalman or particle filter, it is possible to localize the robot. Mapping is often an extension to this estimation problem. Beside the robot pose, positions of landmarks are estimated. Closed loops, i.e., a second encounter of a previously visited area of the environment, play a special role here. Once detected, they enable the algorithms to bound the error by deforming the already mapped area such that a topologically consistent model is created. However, there is no guarantee for a correct model. Several strategies exist for solving SLAM. Thrun reviews in [26] existing techniques, i.e., maximum likelihood estimation [10], expectation maximization [9,27], extended Kalman filter [6] or (sparse extended) information filter [29]. In addition to these methods, FastSLAM [28] that approximates the posterior probabilities, i.e., robot poses, by particles, and the method of Lu Milios on the basis of IDC scan matching [18] exist.

In principle, these probabilistic methods are extendable to 6D. However no reliable feature extraction nor a strategy for reducing the computational costs of multi hypothesis tracking, e.g., FastSLAM, that grows exponentially with the degrees of freedom, has been published to our knowledge.

3D Mapping. Instead of using 3D scanners, which yield consistent 3D scans in the first place, some groups have attempted to build 3D volumetric representations of environments with 2D laser range finders. Thrun et al. [28], Früh et al. [11] and Zhao et al. [31] use two 2D laser scanners finders for acquiring 3D data. One laser scanner is mounted horizontally, the other vertically. The latter one grabs a vertical scan line which is transformed into 3D points based on the current robot pose. Since the vertical scanner is not able to scan sides of objects, Zhao et al. use two additional, vertically mounted 2D scanners, shifted by 45° to reduce occlusions [31]. The horizontal scanner is used to compute the robot pose. The precision of 3D data points depends on that pose and on the precision of the scanner.

A few other groups use highly accurate, expensive 3D laser scanners [1,12,22]. The RESOLV project aimed at modeling interiors for virtual reality and tele-presence [22]. They used a RIEGL laser range finder on robots and the ICP algorithm for scan matching [4]. The AVENUE project develops a robot for modeling urban environments [1], using a CYRAX scanner and a feature-based scan matching approach for registering the 3D scans. Nevertheless, in their recent work they do not use data of the laser scanner in the robot control architecture for localization [12]. The group of M. Hebert has reconstructed environments using the Zoller+Fröhlich laser scanner and aims to build 3D models without initial position estimates, i.e., without odometry information [14].

Recently, different groups employ rotating SICK scanners for acquiring 3D data [15,30]. Wulf et al. let the scanner rotate around the vertical axis. They acquire 3D data while moving, thus the quality of the resulting map crucially depends on the pose estimate that is given by inertial sensors, i.e., gyros [30]. In addition, their SLAM algorithms do not consider all six degrees of freedom.

Other approaches use information of CCD-cameras that provide a view of the robot's environment [5,21]. Nevertheless, cameras are difficult to use in natural environments with changing light conditions. Camera-based approaches to 3D robot vision, e.g., stereo cameras and structure from motion, have difficulties providing reliable navigation and mapping information for a mobile robot in real-time. Thus some groups try to solve 3D modeling by using planar scanner based SLAM methods and cameras, e.g., in [5].

1.2 Hardware Used in Our Experiments

The 3D Laser Range Finder. The 3D laser range finder (Fig. 1) [23] is built on the basis of a SICK 2D range finder by extension with a mount and a small servomotor. The 2D laser range finder is attached in the center of rotation to the mount for achieving a controlled pitch motion with a standard servo.

The area of up to $180°(h) \times 120°(v)$ is scanned with different horizontal (181, 361, 721) and vertical (128, 256, 400, 500) resolutions. A plane with 181 data points is scanned in 13 ms by the 2D laser range finder (rotating mirror device). Planes with more data points, e.g., 361, 721, duplicate or quadruplicate this time. Thus a scan with 181×256 data points needs 3.4 seconds. Scanning the environment with a mobile robot is done in a stop-scan-go fashion.

Fig. 1. Kurt3D in a natural environment. Left to right: Lawn, forest track, pavement.

The Mobile Robot. Kurt3D Outdoor (Fig. 1) is a mobile robot with a size of 45 cm (length) $\times$ 33 cm (width) $\times$ 29 cm (height) and a weight of 22.6 kg. Two 90 W motors are used to power the 6 skid-steered wheels, whereas the front and rear wheels have no tread pattern to enhance rotating. The core of the robot is a Pentium-Centrino-1400 with 768 MB RAM and Linux.

2   Range Image Registration and Robot Relocalization

Multiple 3D scans are necessary to digitalize environments without occlusions. To create a correct and consistent model, the scans have to be merged into one coordinate system. This process is called registration. If the robot carrying the 3D scanner were precisely localized, the registration could be done directly based on the robot pose. However, due to the unprecise robot sensors, self localization is erroneous, so the geometric structure of overlapping 3D scans has to be considered for registration. As a by-product, successful registration of 3D scans relocalizes the robot in 6D, by providing the transformation to be applied to the robot pose estimation at the recent scan point.

The following method registers point sets in a common coordinate system. It is called *Iterative Closest Points (ICP)* algorithm [4]. Given two independently acquired sets of 3D points, M (model set) and D (data set) which correspond to a single shape, we aim to find the transformation consisting of a rotation $\mathbf{R}$ and a translation t which minimizes the following cost function:

$$E(\mathbf{R}, t) = \sum_{i=1}^{|M|} \sum_{j=1}^{|D|} w_{i,j} \left|\left| m_i - (\mathbf{R}d_j + t) \right|\right|^2 . \tag{1}$$

$w_{i,j}$ is assigned 1 if the i-th point of M describes the same point in space as the j-th point of D. Otherwise $w_{i,j}$ is 0. Two things have to be calculated: First, the corresponding points, and second, the transformation $(\mathbf{R}, t)$ that minimizes $E(\mathbf{R}, t)$ on the base of the corresponding points.

The ICP algorithm calculates iteratively the point correspondences. In each iteration step, the algorithm selects the closest points as correspondences and calculates the transformation $(\mathbf{R}, t)$ for minimizing equation (1). The assumption is that in the last iteration step the point correspondences are correct. Besl et al. prove that the method terminates in a minimum [4]. However, this theorem does not hold in our case, since we use a maximum tolerable distance $d_{\max}$ for associating the scan data. Such a threshold is required though, given that 3D scans overlap only partially.

In every iteration, the optimal transformation $(\mathbf{R}, t)$ has to be computed. Eq. (1) can be reduced to

$$E(\mathbf{R}, t) \propto \frac{1}{N} \sum_{i=1}^{N} \left|\left| m_i - (\mathbf{R}d_i + t) \right|\right|^2 , \tag{2}$$

with $N = \sum_{i=1}^{|M|} \sum_{j=1}^{|D|} w_{i,j}$, since the correspondence matrix can be represented by a vector containing the point pairs.

Four direct methods are known to minimize eq. (2) [17]. In earlier work [19,23,24] we used a quaternion based method [4], but the following one, based on singular value decomposition (SVD), is robust and easy to implement, thus we give a brief overview of the SVD-based algorithm. It was first published by Arun, Huang and Blostein [2]. The difficulty of this minimization problem is to enforce the orthonormality of the matrix $\mathbf{R}$. The first step of the computation is to decouple the calculation of the rotation $\mathbf{R}$ from the translation t using the centroids of the points belonging to the matching, i.e.,

$$c_m = \frac{1}{N} \sum_{i=1}^{N} m_i, \qquad c_d = \frac{1}{N} \sum_{i=1}^{N} d_j \qquad (3)$$

and

$$M' = \{m_i' = m_i - c_m\}_{1,\dots,N}, \qquad D' = \{d_i' = d_i - c_d\}_{1,\dots,N}. \qquad (4)$$

After substituting (3) and (4) into the error function, $E(\mathbf{R}, t)$ eq. (2) becomes:

$$E(\mathbf{R}, t) \propto \sum_{i=1}^{N} ||m_i' - \mathbf{R}d_i'||^2 \quad \text{with} \quad t = c_m - \mathbf{R}c_d. \qquad (5)$$

The registration calculates the optimal rotation by $\mathbf{R} = \mathbf{V}\mathbf{U}^T$. Hereby, the matrices $\mathbf{V}$ and $\mathbf{U}$ are derived by the singular value decomposition $\mathbf{H} = \mathbf{U}\mathbf{\Lambda}\mathbf{V}^T$ of a correlation matrix $\mathbf{H}$. This 3×3 matrix $\mathbf{H}$ is given by

$$\mathbf{H} = \sum_{i=1}^{N} d_i' m_i'^T = \begin{pmatrix} S_{xx} & S_{xy} & S_{xz} \\ S_{yx} & S_{yy} & S_{yz} \\ S_{zx} & S_{zy} & S_{zz} \end{pmatrix}, \qquad (6)$$

with $S_{xx} = \sum_{i=1}^{N} m_{ix}' d_{ix}'$, $S_{xy} = \sum_{i=1}^{N} m_{ix}' d_{iy}'$, $\dots$ [2].

We proposed and evaluated algorithms to accelerate ICP, namely point reduction and approximate kd-trees [19,23,24]. They are used here, too.

3   ICP-Based 6D SLAM

3.1   Calculating Heuristic Initial Estimations for ICP Scan Matching

To match two 3D scans with the ICP algorithm it is necessary to have a sufficient starting guess for the second scan pose. In earlier work we used odometry [23] or the planar HAYAI scan matching algorithm [16]. However, the latter cannot be used in arbitrary environments, e.g., the one presented in Fig. 1 (bad asphalt, lawn, woodland, etc.). Since the motion models change with different grounds, odometry alone cannot be used. Here the robot pose is the 6-vector $P = (x, y, z, \theta_x, \theta_y, \theta_z)$ or, equivalently the tuple containing the rotation matrix and translation vector, written as 4×4 OpenGL-style matrix $\mathbf{P}$ [8].[1] The following heuristic computes a sufficiently good initial estimation. It is based on two ideas. First, the transformation found in the previous registration is applied

---

[1] Note the bold-italic (vectors) and bold (matrices) notation. The conversion between vector representations, i.e., Euler angles, and matrix representations is done by algorithms from [8].

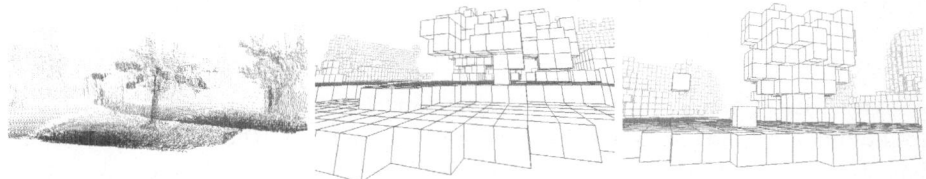

Fig. 2. Left: Two 3D point clouds. Middle: Octree corresponding to the black point cloud. Right: Octree based on the blue points.

to the pose estimation – this implements the assumption that the error model of the pose estimation is locally stable. Second, a pose update is calculated by matching octree representations of the scan point sets rather than the point sets themselves – this is done to speed up calculation:

1. Extrapolate the odometry readings to all six degrees of freedom using previous registration matrices. The change of the robot pose $\Delta \mathbf{P}$ given the odometry information $(x_n, z_n, \theta_{y,n})$, $(x_{n+1}, z_{n+1}, \theta_{y,n+1})$ and the registration matrix $\mathbf{R}(\theta_{x,n}, \theta_{y,n}, \theta_{z,n})$ is calculated by solving:

$$
\begin{pmatrix} x_{n+1} \\ y_{n+1} \\ z_{n+1} \\ \theta_{x,n+1} \\ \theta_{y,n+1} \\ \theta_{z,n+1} \end{pmatrix} = \begin{pmatrix} x_n \\ y_n \\ z_n \\ \theta_{x,n} \\ \theta_{y,n} \\ \theta_{z,n} \end{pmatrix} + \left(\begin{array}{c|c} \mathbf{R}(\theta_{x,n}, \theta_{y,n}, \theta_{z,n}) & 0 \\ \hline & \begin{matrix} 1\ 0\ 0 \\ 0\ 1\ 0 \\ 0\ 0\ 1 \end{matrix} \end{array} \right) \cdot \underbrace{\begin{pmatrix} \Delta x_{n+1} \\ \Delta y_{n+1} \\ \Delta z_{n+1} \\ \Delta \theta_{x,n+1} \\ \Delta \theta_{y,n+1} \\ \Delta \theta_{z,n+1} \end{pmatrix}}_{\Delta P} .
$$

Therefore, calculating $\Delta \mathbf{P}$ requires a matrix inversion. Finally, the 6D pose $\mathbf{P}_{n+1}$ is calculated by

$$
\mathbf{P}_{n+1} = \Delta \mathbf{P} \cdot \mathbf{P}_n
$$

using the poses' matrix representations.

2. Set $\Delta \mathbf{P}_{\text{best}}$ to the 6-vector $(\mathbf{t}, \mathbf{R}(\theta_{x,n}, \theta_{y,n}, \theta_{z,n})) = (\mathbf{0}, \mathbf{R}(\mathbf{0}))$.

3. Generate an octree $\mathfrak{O}_M$ for the nth 3D scan (model set M).

4. Generate an octree $\mathfrak{O}_D$ for the $(n+1)$th 3D scan (data set D).

5. For search depth $t \in [t_{\text{Start}}, \ldots, t_{\text{End}}]$ in the octrees estimate a transformation $\Delta \mathbf{P}_{\text{best}} = (\mathbf{t}, \mathbf{R})$ as follows:
   (a) Calculate a maximal displacement and rotation $\Delta \mathbf{P}_{\text{max}}$ depending on the search depth t and currently best transformation $\Delta \mathbf{P}_{\text{best}}$.
   (b) For all discrete 6-tuples $\Delta \mathbf{P}_i \in [-\Delta \mathbf{P}_{\text{max}}, \Delta \mathbf{P}_{\text{max}}]$ in the domain $\Delta \mathbf{P} = (x, y, z, \theta_x, \theta_y, \theta_z)$ displace $\mathfrak{O}_D$ by $\Delta \mathbf{P}_i \cdot \Delta \mathbf{P} \cdot \mathbf{P}_n$. Evaluate the matching of the two octrees by counting the number of overlapping cubes and save the best transformation as $\Delta \mathbf{P}_{\text{best}}$.

6. Update the scan pose using matrix multiplication, i.e.,

$$\mathbf{P}_{n+1} = \Delta\mathbf{P}_{\text{best}} \cdot \Delta\mathbf{P} \cdot \mathbf{P}_n.$$

Note: Step 5b requires 6 nested loops, but the computational requirements are bounded by the coarse-to-fine strategy inherited from the octree processing. The size of the octree cubes decreases exponentially with increasing t. We start the algorithm with a cube size of 75 cm^3 and stop when the cube size falls below 10 cm^3. Fig. 2 shows two 3D scans and the corresponding octrees. Furthermore, note that the heuristic works best outdoors. Due to the diversity of the environment the match of octree cubes will show a significant maximum, while indoor environments with their many geometry symmetries and similarities, e.g., in a corridor, are in danger of producing many plausible matches.

After an initial starting guess is found, the range image registration from section 2 proceeds and the 3D scans are precisely matched.

3.2   Computing Globally Consistent Scenes

After registration, the scene has to be correct and globally consistent. A straightforward method for aligning several 3D scans is *pairwise matching*, i.e., the new scan is registered against a previous one. Alternatively, an *incremental matching* method is introduced, i.e., the new scan is registered against a so-called *metascan*, which is the union of the previously acquired and registered scans. Each scan matching has a limited precision. Both methods accumulate the registration errors such that the registration of a large number of 3D scans leads to inconsistent scenes and to problems with the robot localization. Closing loop detection and error diffusing avoid these problems and compute consistent scenes.

Closing the Loop. After matching multiple 3D scans, errors have accumulated and loops would normally not be closed. Our algorithm automatically detects a to-be-closed loop by registering the last acquired 3D scan with earlier acquired scans. Hereby we first create a hypothesis based on the maximum laser range and on the robot pose, so that the algorithm does not need to process all previous scans. Then we use the octree based method presented in section 3.1 to revise the hypothesis. Finally, if a registration is possible, the computed error, i.e., the transformation $(\mathbf{R}, \boldsymbol{t})$ is distributed over all 3D scans. The respective part is weighted by the distance covered between the scans, i.e.,

$$c_i = \frac{\text{length of path from start of the loop to scan pose } i}{\text{overall length of path}}$$

1. The translational part is calculated as $\boldsymbol{t}_i = c_i \boldsymbol{t}$.
2. Of the three possibilities of representing rotations, namely, orthonormal matrices, quaternions and Euler angles, quaternions are best suited for our interpolation task. The problem with matrices is to enforce orthonormality and Euler angles show Gimbal locks [8]. A quaternion as used in computer graphics is the 4 vector $\dot{q}$. Given a rotation as matrix $\mathbf{R}$, the corresponding quaternion $\dot{q}$ is calculated as follows:

$$\dot{q} = \begin{pmatrix} q_0 \\ q_x \\ q_y \\ q_z \end{pmatrix} = \begin{pmatrix} \frac{1}{2}\sqrt{\text{trace}\,(\mathbf{R})} \\[4pt] \frac{1}{2}\frac{r_{3,3}-r_{3,2}}{\sqrt{\text{trace}(\mathbf{R})}} \\[4pt] \frac{1}{2}\frac{r_{2,1}-r_{2,3}}{\sqrt{\text{trace}(\mathbf{R})}} \\[4pt] \frac{1}{2}\frac{r_{1,2}-r_{1,1}}{\sqrt{\text{trace}(\mathbf{R})}} \end{pmatrix}, \quad \text{with the elements } r_{i,j} \text{ of } \mathbf{R}.^2 \tag{7}$$

The quaternion describes a rotation by an axis $a \in \mathbb{R}^3$ and an angle θ that are computed by

$$a = \begin{pmatrix} \frac{q_x}{\sqrt{1-q_0^2}} \\[6pt] \frac{q_y}{\sqrt{1-q_0^2}} \\[6pt] \frac{q_z}{\sqrt{1-q_0^2}} \end{pmatrix} \qquad \text{and} \qquad \theta = 2\arccos q_o.$$

The angle θ is distributed over all scans using the factor c_i and the resulting matrix is derived as [8]:

$$\mathbf{R}_i = \begin{pmatrix} \cos(c_i\theta) + a_x^2(1-\cos(c_i\theta)) & a_z\sin(c_i\theta) + a_xa_y(1-\cos(c_i\theta)) \\ -a_z\sin(c_i\theta) + a_xa_y(1-\cos(c_i\theta)) & \cos(c_i\theta) + a_y^2(1-\cos(c_i\theta)) \\ a_y\sin(c_i\theta) + a_xa_z(1-\cos(c_i\theta) & -a_x\sin(c_i\theta) + a_ya_z(1-\cos(c_i\theta)) \end{pmatrix}$$

$$\begin{pmatrix} -a_y\sin(c_i\theta) + a_xa_z(1-\cos(c_i\theta)) \\ -a_x\sin(c_i\theta) + a_ya_z(1-\cos(c_i\theta)) \\ \cos(c_i\theta) + a_z^2(1-\cos(c_i\theta)) \end{pmatrix}. \tag{8}$$

The next step minimizes the global error.

---

^{2} If trace $(\mathbf{R})$ (sum of the diagonal terms) is zero, the above calculation has to be altered: Iff $r_{1,1} > r_{2,2}$ and $r_{1,1} > r_{3,3}$ then,

$$\dot{q} = \begin{pmatrix} \frac{1}{2}\frac{r_{2,3}-r_{3,2}}{\sqrt{1+r_{1,1}-r_{2,2}-r_{3,3}}} \\[6pt] \frac{1}{2}\sqrt{1+r_{1,1}-r_{2,2}-r_{3,3}} \\[6pt] \frac{1}{2}\frac{r_{1,2}+r_{2,1}}{\sqrt{1+r_{1,1}-r_{2,2}-r_{3,3}}} \\[6pt] \frac{1}{2}\frac{r_{3,1}+r_{1,3}}{\sqrt{1+r_{1,1}-r_{2,2}-r_{3,3}}} \end{pmatrix}, \text{if } r_{2,2} > r_{3,3} \;\dot{q} = \begin{pmatrix} \frac{1}{2}\frac{r_{3,1}-r_{1,3}}{\sqrt{1-r_{1,1}+r_{2,2}-r_{3,3}}} \\[6pt] \frac{1}{2}\frac{r_{1,2}+r_{2,1}}{\sqrt{1-r_{1,1}+r_{2,2}-r_{3,3}}} \\[6pt] \frac{1}{2}\sqrt{1-r_{1,1}+r_{2,2}-r_{3,3}} \\[6pt] \frac{1}{2}\frac{r_{2,3}+r_{3,2}}{\sqrt{1-r_{1,1}+r_{2,2}-r_{3,3}}} \end{pmatrix},$$

otherwise the quaternion $\dot{q}$ is calculated as

$$\dot{q} = \begin{pmatrix} \frac{1}{2}\frac{r_{1,2}-r_{2,1}}{\sqrt{1-r_{1,1}-r_{2,2}+r_{3,3}}} \\[6pt] \frac{1}{2}\frac{r_{3,1}+r_{1,3}}{\sqrt{1+r_{1,1}-r_{2,2}-r_{3,3}}} \\[6pt] \frac{1}{2}\frac{r_{2,3}+r_{3,2}}{\sqrt{1-r_{1,1}-r_{2,2}+r_{3,3}}} \\[6pt] \frac{1}{2}\sqrt{1-r_{1,1}-r_{2,2}+r_{3,3}} \end{pmatrix}.$$

Diffusing the Error. Pulli presents a registration method that minimizes the global error and avoids inconsistent scenes [20]. The registration of one scan is followed by registering all neighboring scans such that the global error is distributed. Other matching approaches with global error minimization have been published, e.g., [3,7]. Benjemaa et al. establish point-to-point correspondences first and then use randomized iterative registration on a set of surfaces [3]. Eggert et al. compute motion updates, i.e., a transformation $(\mathbf{R}, \mathbf{t})$, using force-based optimization, with data sets considered as connected by groups of springs [7].

Based on the idea of Pulli we designed the relaxation method *simultaneous matching*[23]. The first scan is the masterscan and determines the coordinate system. It is fixed. The following three steps register all scans and minimize the global error, after a queue is initialized with the first scan of the closed loop:

1. Pop the first 3D scan from the queue as the current one.
2. If the current scan is not the master scan, then a set of neighbors (set of all scans that overlap with the current scan) is calculated. This set of neighbors forms one point set M. The current scan forms the data point set D and is aligned with the ICP algorithms. One scan overlaps with another iff more than p corresponding point pairs exist. In our implementation, $p = 250$.
3. If the current scan changes its location by applying the transformation (translation or rotation) in step 2, then each single scan of the set of neighbors that is not in the queue is added to the end of the queue. If the queue is empty, terminate; else continue at step 1.

In contrast to Pulli's approach, our method is totally automatic and no interactive pairwise alignment has to be done. Furthermore the point pairs are not fixed [20]. The accumulated alignment error is spread over the whole set of acquired 3D scans. This diffuses the alignment error equally over the set of 3D scans [24].

4 Experiment and Results

The following experiment has been made at the campus of Schloss Birlinghoven with Kurt3D. Fig. 3 (left) shows the scan point model of the first scans in top view, based on odometry only. The first part of the robot's run, i.e., driving on asphalt, contains a systematic drift error, but driving on lawn shows more stochastic characteristics. The right part shows the first 62 scans, covering a path length of about 240 m. The heuristic has been applied and the scans have been matched. The open loop is marked with a red rectangle.

At that point, the loop is detected and closed. More 3D scans have then been acquired and added to the map. Fig. 4 (left and right) shows the model with and without global relaxation to visualize its effects. The relaxation is able to align the scans correctly even without explicitly closing the loop. The best visible difference is marked by a red rectangle. The final map in Fig. 4 contains 77 3D scans, each consisting of approx. 100000 data points (275×361). Fig. 5 shows two detailed views, before and after loop closing. The bottom part of Fig. 4 displays an aerial view as ground truth for comparison. Table 1

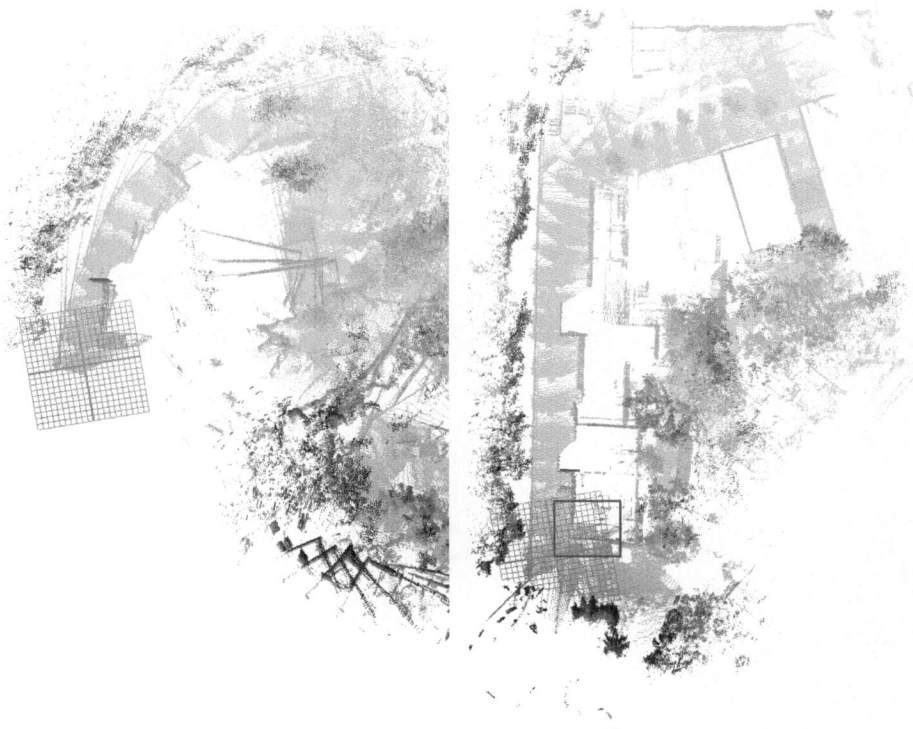

Fig. 3. 3D model of an experiment to digitize part of the campus of Schloss Birlinghoven campus (top view). Left: Registration based on odometry only. Right: Model based on incremental matching right before closing the loop, containing 62 scans each with approx. 100000 3D points. The grid at the bottom denotes an area of $20{\times}20\mathrm{m}^2$ for scale comparison. The 3D scan poses are marked by blue points.

compares distances measured in the photo and in the 3D scene. The lines in the photo have been measured in pixels, whereas real distances, i.e., the (x, z)-values of the points, have been used in the point model. Taking into account that pixel distances in mid-resolution non-calibrated aerial image induce some error in ground truth, the correspondence show that the point model at least approximates reality quite well.

Table 1. Length ratio comparison of measured distances in the aerial photographs with distances in the point model as shown in Fig. 4

1st line	2nd line	ratio in aerial views	ratio in point model	deviation
AB	BC	0.683	0.662	3.1%
AB	BD	0.645	0.670	3.8%
AC	CD	1.131	1.141	0.9%
CD	BD	1.088	1.082	0.5%

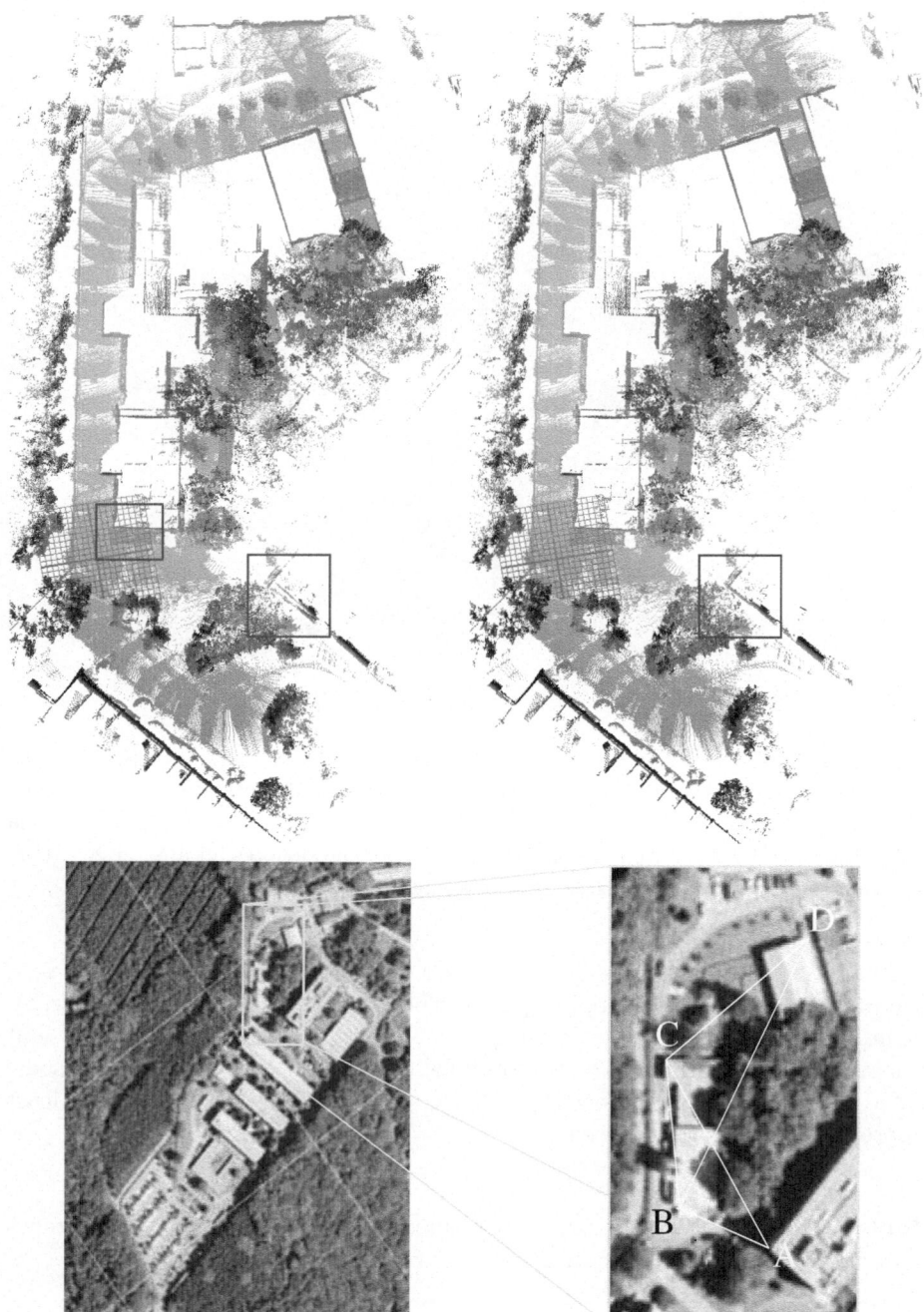

Fig. 4. Top left: Model with loop closing, but without global relaxation. Differences to Fig. 3 right and to the right image are marked. Top right: Final model of 77 scans with loop closing and global relaxation. Bottom: Aerial view of the scene. The points A – D are used as reference points in the comparison in Table 1.

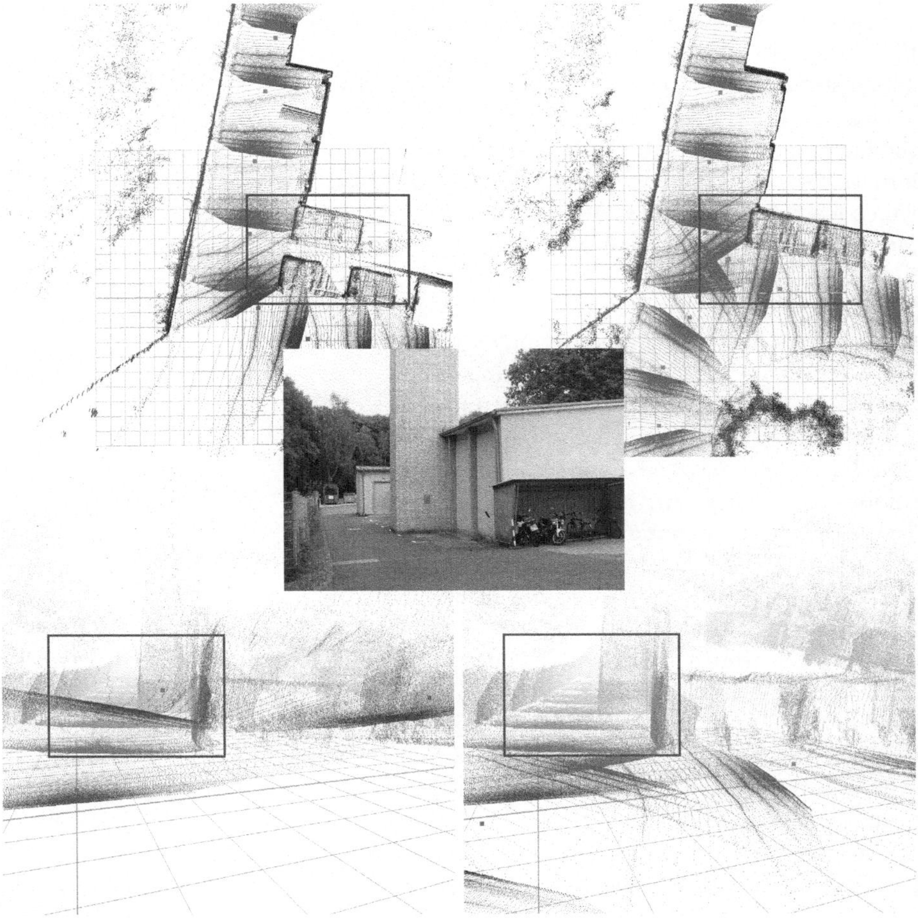

Fig. 5. Detailed view of the 3D model of Fig. 4. Left: Model before loop closing. Right: After loop closing, global relaxation and adding further 3D scans. Top: Top view. Bottom: Front view.

Fig. 6. Detailed views of the resulting 3D model corresponding to robot locations of Fig. 1

Mapping would fail without first calculatingheuristic initial estimations for ICP scan matching, since ICP would likely converge into an incorrect minimum. The resulting 3D map would be some mixture of Fig. 3 (left) and Fig. 4 (right).

Fig. 6 shows three views of the final model. These model views correspond to the locations of Kurt3D in Fig. 1. An updated robot trajectory has been plotted into the scene. Thereby, we assign every 3D scan that part of the trajectory which leads from the previous scan pose to the current one. Since scan matching did align the scans, the trajectory initially has gaps after the alignment (see Fig. 7).

We calculate the transformation $(\mathbf{R}, \boldsymbol{t})$ that maps the last pose of such a trajectory patch to the starting pose of the next patch. This transformation is then used to correct the trajectory patch by distributing the transformation as described in section 3.2. In this way the algorithm computes a continuous trajectory. An animation of the scanned area is available at `http://kos.informatik.uni-osnabrueck.de/6Doutdoor/`. The video shows the scene along the trajectory as viewed from about 1 m above Kurt3D's actual position.

Fig. 7: The trajectory after mapping shows gaps, since the robot poses are corrected at 3D scan poses

The 3D scans were acquired within one hour by tele-operation of Kurt3D. Scan registration and closed loop detection took only about 10 minutes on a Pentium-IV-2800 MHz, while we did run the global relaxation for 2 hours. However, computing the flight-thru-animation took about 3 hours, rendering 9882 frames with OpenGL on consumer hardware.

In addition we used the 3D scan matching algorithm in the context of Robo Cup Rescue 2004. We were able to produce online 3D maps, even though we did not use closed loop detection and global relaxation. Some results are available at `http://kos.informatik.uni-osnabrueck.de/download/Lisbon_RR/`.

5   Discussion and Conclusion

This paper has presented a solution to the SLAM problem considering six degrees of freedom and creating 3D maps of outdoor environments. It is based on ICP scan matching, initial pose estimation using a coarse-to-fine strategy with an octree representation and closing loop detection. Using an aerial photo as ground truth, the 3D map shows very good correspondence with the mapped environment, which was confirmed by a ratio comparison between map features and the respective photo features.

Compared with related approaches from the literature [6,10,26,27,28,29] we do not use a feature representation of the environment. Furthermore our algorithm manages registration without fixed data association. In the data association step, SLAM algorithms decide which features correspond. Wrong correspondences result in unprecise or even inconsistent models. The global scan

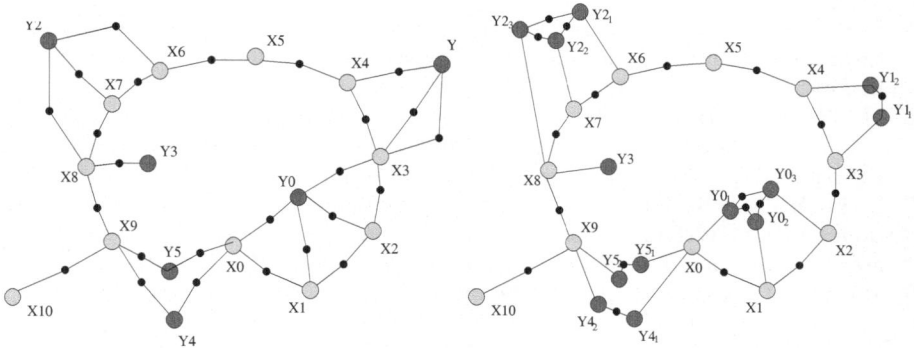

Fig. 8. Abstract comparison of SLAM approaches. Left: Probabilistic methods. The robot poses X_i as well as the positions of the associated landmarks Y_i are given in terms of a probability distribution. Global optimization tries to relax the model, where the landmarks are fixed. Small black dots on lines mark adjustable distances. Right: Our method with absolute measurements Y_i (note there are no black dots between scan poses and scanned landmarks). The poses X_i are adjusted based on scan matching aiming at collapsing the landmark copies Y_{i_k} for all landmarks Y_i. Data association is the search for closest points.

matching based relaxation computes corresponding points, i.e., closest points, in every iteration. Furthermore, we avoid using probabilistic representations to keep the computation time at a minimum. The model optimization is solved in a closed form, i.e., by direct pose transformation. As a result of these efforts, registration and closed loop detection of 77 scans each with ca. 100000 points took only about 10 minutes.

Fig. 8 compares the probabilistic SLAM approaches with ours on an abstract level as presented by Folkesson and Christensen [9]. Robot poses are labeled with X_i whereas the landmarks are the Y_i. Lines with black dots correspond to adjustable connections, e.g., springs, which can be relaxed by the algorithms. In our system, the measurements are fixed and data association is repeatedly done using nearest neighbor search.

Needless to say, a lot of work remains to be done. We plan to further improve the computation time and to use sensor uncertainty models. In addition, semantic labels for sub-structures of the resulting point model will be extracted.

References

1. P. Allen, I. Stamos, A. Gueorguiev, E. Gold, and P. Blaer. AVENUE: Automated Site Modelling in Urban Environments. In *Proc. 3DIM*, Canada, May 2001.
2. K. S. Arun, T. S. Huang, and S. D. Blostein. Least square fitting of two 3-d point sets. *IEEE Transactions on PAMI*, 9(5):698 – 700, 1987.
3. R. Benjemaa and F. Schmitt. Fast Global Registration of 3D Sampled Surfaces Using a Multi-Z-Buffer Technique. In *Proc. 3DIM*, Ottawa, Canada, May 1997.
4. P. Besl and N. McKay. A method for Registration of 3–D Shapes. *IEEE Transactions on PAMI*, 14(2):239 – 256, February 1992.

5. P. Biber, H. Andreasson, T. Duckett, and A. Schilling. 3D Modeling of Indoor Environments by a Mobile Robot with a Laser Scanner and Panoramic Camera. In *Proc. IROS*, Sendai, Japan, September 2004.
6. M. W. M. G. Dissanayake, P. Newman, S. Clark, H. F. Durrant-Whyte, and M. Csorba. A Solution to the Simultaneous Localization and Map Building (SLAM) Problem. *IEEE Transactions on Robotics and Automation*, 17(3):229 – 241, 2001.
7. D. Eggert, A. Fitzgibbon, and R. Fisher. Simultaneous Registration of Multiple Range Views Satisfying Global Consistency Constraints for Use In Reverse Engineering. *Computer Vision and Image Understanding*, 69:253 – 272, March 1998.
8. Matrix FAQ. Version 2, http://skal.planet-d.net/demo/matrixfaq.htm, 1997.
9. J. Folkesson and H. Christensen. Outdoor Exploration and SLAM using a Compressed Filter. In *Proc. ICRA*, pages 419–426, Taipei, Taiwan, September 2003.
10. U. Frese and G. Hirzinger. Simultaneous Localization and Mapping – A Discussion. In *Proc. IJCAI Workshop on Reasoning with Uncertainty in Robotics*, 2001.
11. C. Früh and A. Zakhor. 3D Model Generation for Cities Using Aerial Photographs and Ground Level Laser Scans. In *Proc. CVPR*, Hawaii, USA, December 2001.
12. A. Georgiev and P. K. Allen. Localization Methods for a Mobile Robot in Urban Environments. *IEEE Transaction on RA*, 20(5):851 – 864, October 2004.
13. M. Golfarelli, D. Maio, and S. Rizzi. Correction of dead-reckoning errors in map building for mobile robots. *IEEE Transaction on TRA*, 17(1), May 2001.
14. M. Hebert, M. Deans, D. Huber, B. Nabbe, and N. Vandapel. Progress in 3–D Mapping and Localization. In *Proceedings of the 9th International Symposium on Intelligent Robotic Systems, (SIRS '01)*, Toulouse, France, July 2001.
15. P. Kohlhepp, M. Walther, and P. Steinhaus. Schritthaltende 3D-Kartierung und Lokalisierung für mobile Inspektionsroboter. In *18. Fachgespräche, AMS*, 2003.
16. K. Lingemann, A. Nüchter, J. Hertzberg, and H. Surmann. High-Speed Laser Localization for Mobile Robots. *J. Robotics and Aut. Syst.*, (accepted), 2005.
17. A. Lorusso, D. Eggert, and R. Fisher. A Comparison of Four Algorithms for Estimating 3-D Rigid Transformations. In *Proc. BMVC*, England, 1995.
18. F. Lu and E. Milios. Globally Consistent Range Scan Alignment for Environment Mapping. *Autonomous Robots*, 4(4):333 – 349, October 1997.
19. A. Nüchter, H. Surmann, K. Lingemann, J. Hertzberg, and S. Thrun. 6D SLAM with an Application in autonomous mine mapping. In *Proc. ICRA*, 2004.
20. K. Pulli. Multiview Registration for Large Data Sets. In *Proc. 3DIM*, 1999.
21. S. Se, D. Lowe, and J. Little. Local and Global Localization for Mobile Robots using Visual Landmarks. In *Proc. IROS*, Hawaii, USA, October 2001.
22. V. Sequeira, K. Ng, E. Wolfart, J. Goncalves, and D. Hogg. Automated 3D reconstruction of interiors with multiple scan–views. In *Proc. of SPIE, Electronic Imaging, 11th Annual Symposium*, San Jose, CA, USA, January 1999.
23. H. Surmann, A. Nüchter, and J. Hertzberg. An autonomous mobile robot with a 3D laser range finder for 3D exploration and digitalization of indoor en vironments. *Journal Robotics and Autonomous Systems*, 45(3 – 4):181 – 198, December 2003.
24. H. Surmann, A. Nüchter, K. Lingemann, and J. Hertzberg. 6D SLAM A Preliminary Report on Closing the Loop in Six Dimensions. In *Proc. IAV*, 2004.
25. S. Thrun. Learning metric-topological maps for indoor mobile robot navigation. *Artificial Intelligence*, 99(1):21–71, 1998.
26. S. Thrun. Robotic mapping: A survey. In G. Lakemeyer and B. Nebel, editors, *Exploring Artificial Intelligence in the New Millenium*. Morgan Kaufmann, 2002.
27. S. Thrun, W. Burgard, and D. Fox. A probabilistic approach to concurrent mapping and localization for mobile robots. *Machine Learning and Auton. Rob.*, 1997.

28. S. Thrun, D. Fox, and W. Burgard. A real-time algorithm for mobile robot mapping with application to multi robot and 3D mapping. In *Proc. ICRA*, 2000.
29. S. Thrun, Y. Liu, D. Koller, A. Y. Ng, Z. Ghahramani, and H. F. Durrant-Whyte. Simultaneous localization and mapping with sparse extended information filters. *Machine Learning and Autonomous Robots*, 23(7 – 8):693 – 716, July/August 2004.
30. O. Wulf, K. O. Arras, H. I. Christensen, and B. A. Wagner. 2D Mapping of Cluttered Indoor Environments by Means of 3D Perception. In *Proc. ICRA*, 2004.
31. H. Zhao and R. Shibasaki. Reconstructing Textured CAD Model of Urban Environment Using Vehicle-Borne Laser Range Scanners and Line Cameras. In *Proc. ICVS*, pages 284 – 295, Vancouver, Canada, July 2001.

A Probabilistic Multimodal Sensor Aggregation Scheme Applied for a Mobile Robot*

Erik Schaffernicht, Christian Martin, Andrea Scheidig,
and Horst-Michael Gross

Ilmenau Technical University,
Department of Neuroinformatics and Cognitive Robotics
andrea.scheidig@tu-ilmenau.de

Abstract. Dealing with methods of human-robot interaction and using a real mobile robot, stable methods for people detection and tracking are fundamental features of such a system and require information from different sensory. In this paper, we discuss a new approach for integrating several sensor modalities and we present a multimodal people detection and tracking system and its application using the different sensory systems of our mobile interaction robot HOROS working in a real office environment. These include a laser-range-finder, a sonar system, and a fisheye-based omnidirectional camera. For each of these sensory information, a separate Gaussian probability distribution is generated to model the belief of the observation of a person. These probability distributions are further combined using a flexible probabilistic aggregation scheme. The main advantages of this approach are a simple integration of further sensory channels, even with different update frequencies and the usability in real-world environments. Finally, promising experimental results achieved in a real office environment will be presented.

1 Introduction

Dealing with Human-Robot-Interaction (HRI) in real-world environments, one of the general tasks is the realization of a stable people detection and the respective tracking functions. Depending on the specific application that integrates a person detection, different approaches are possible. Typical approaches use visual cues for face detection, a laser-range-finder for detection of moving objects, like legs, or acoustical cues for sound source detection.

Projects like EMBASSI [1], which aim to detect only the users' faces, usually in front of a stationary station like a PC, typically use visual cues (skin-color-based approaches, sometimes in combination with the detection of edge oriented features). Therefore, these approaches cannot be applied for a mobile robot which has to deal with moving people with faces not always perceivable.

---

* This work is partially supported by TMWFK-Grant #B509-03007 to H.-M. Gross and a HWP-Grant to A. Scheidig.

U. Furbach (Ed.): KI 2005, LNAI 3698, pp. 320–334, 2005.

In [2] a skin-color-based approach for a mobile robot is presented using an extension of particle filters to generate object configurations which represent more then one person in the image.

Other approaches trying to perceive the whole person rather than only her face use laser-range-finders to detect people as moving objects or directly by their legs, e.g. GRACE [3] or TOURBOT [4]. In [5] a approach based on particle representations in joint probabilistic data association filters is presented. Drawbacks of these approaches occur, for instance, in situations where a person stands near a wall and cannot be distinguished, in scenarios with objects yielding leg-like scans, like table-legs or chair-legs, or if the laser-range-finder does not cover 360 degrees of the robot space.

For real-world scenarios, more promising approaches combine more than one sensory channel,like visual cues and the scan of the laser-range-finder. An example for these approaches is the SIG robot [6], which combines visual and auditory cues. People are detected by a face detection system and tracked by using stereo vision and sound source detection. This approach is especially useful for scenarios like face-to-face interaction. Further examples are the EXPO-ROBOTS [7], where people are detected as moving objects by a laser-range-finder (resulting from differences from a given static environment map) firstly. After that, these hypotheses are verified by visual cues. Other projects like BIRON [8] detect people by using the laser-range-finder for detecting leg-profiles and combine these information with visual and auditory cues (anchoring). The essential drawback of these approaches is the sequential processing of the sensory cues. People are detected by laser information only and are subsequently verified by visual cues. These approaches fail, if the laser-range-finder yields no information, for instance, in situations when only the face of a person is perceivable because of leg occlusions.

Therefore, we propose a multimodal approach to realize a robust detection and tracking of people. Compared to other approaches, all used sensory cues are concurrently processed using a probabilistic aggregation scheme, that scales very well with the number of sensors and modalities used in terms of computational complexity. This way people are not only detected by only one feature. They can be detected by their legs and their faces or by only one of this features, respectively. The main advantage of our approach is the simple expandability by integretiing further sensory channels, like sound sources, because of the used aggregation scheme.

The structure of this paper is as follows: first we present the employed different sensory modalities of our mobile robot for people detection and tracking: the omnidirectional camera, the laser-range-finder and the sonar sensors (section 2). Using these modalities, we generate specific probability-based hypotheses about the positions of detected people and combine these probability distributions by covariance intersection (section 3). Respective experimental results are presented in section 4 followed by a short summary and an outlook in section 5.

2 Mobile Interaction Robot Horos

To investigate respective methods, we use the mobile interaction robot Horos as an information system for employees, students, and guests of our institute. The system's task includes that Horos autonomously moves in the institute, detects people as possible interaction partners and interacts with them, for example, to answer questions like the current whereabouts of specific people.

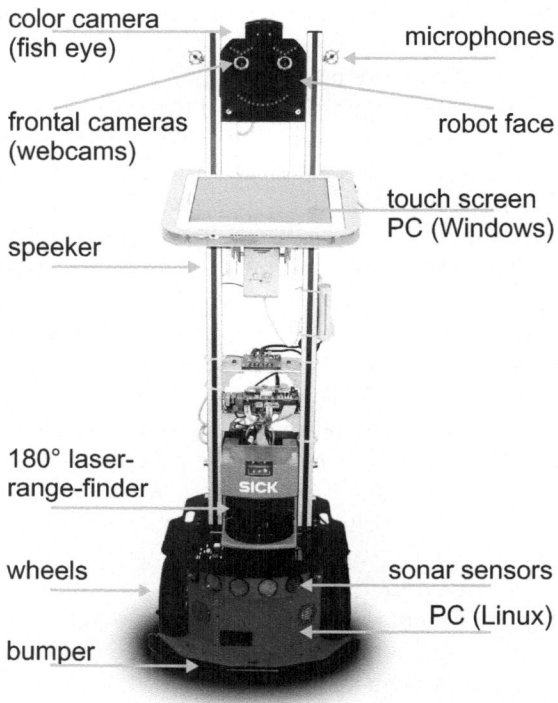

Fig. 1. Sensory and motory modalities of the mobile interaction robot Horos (HOme RObot System)

The hardware platform of Horos is an extended Pioneer-based robot from ActiveMedia. It integrates an on-board PC (Pentium M, 1.6 GHz, 512MB) and is equipped with a laser-range-finder (Sick) and sonar sensors. For the purpose of HRI, this platform was mounted with different interaction-oriented modalities (see Figure 1).

This includes a tablet PC (Pentium M, 1.1 GHz, 256MB) running under Windows XP for pen-based interaction, speech recognition and speech generation. It was further extended by a robot face which includes an omnidirectional fisheye camera and two microphones. Moreover, we integrated two frontal webcams for the visual analysis of dialog-relevant user features (e.g. age, gender, emotions).

Subsequently, only the omnidirectional camera, the laser-range-finder and the sonar sensors are discussed in the context of the people detection and tracking task.

2.1   Laser-Based Information

The laser-range-finder of HOROS is a very precise sensor with a resolution of one degree, perceiving the frontal 180 degree field of HOROS (see Figure 2, left). It is fixed on the robot approximately 30 cm above the ground. Therefore, it can only perceive the legs of people (see Figure 2, right).

Based on the approach presented in [9], we also analyze the scan of the laser-range-finder for leg-pairs using a heuristic method. The measurements are segmented into local groups of similar distance values. Then each segment is checked for different conditions like width, deviation and others that are common for legs. The distance between segments classified as legs is pairwise computed to determine whether this could be a human pair of legs. For each pair found, the distance and direction is extracted.

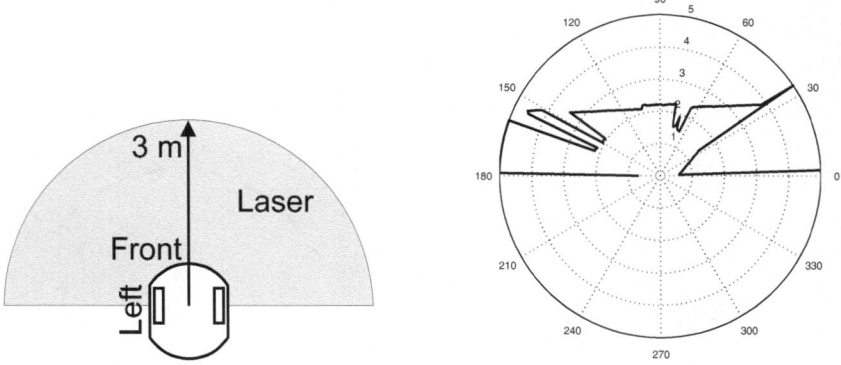

Fig. 2. Left: Top view of the schematic HOROS. The sensory range of the laser-range-finder used to detect people is depicted grey. Right: Real scan of the laser-range-finder, depicted in polar coordinate system. In this situation, the robot is standing in a door directed to the top, perceiving a wall in front of it and the opened door to its right. Further, it senses a pair of legs in front of it (at 70°) and another one to its left (at 155°).

This yields very good results for distances of people which stand less than 3 meters away. In a greater distance legs will be missed due to the limited resolution of the laser-range-finder[1]. The strongest disadvantage of this approach is its false-positive classification detecting table-legs, chair-legs and also waste-paper baskets as legs. Also people standing sideway to the robot or wearing long skirts do not yield appropriate values of the laser-range sensor to detect their legs.

---

[1] At a distance of 3 meters the laser beams have a gap of more than 5cm between each other. In greater distances some legs are missed.

2.2 Sonar Information

HOROS is equipped with 16 sonar sensors, arranged at the Pioneer platform approximately 20 cm above the ground. The sound cones have an aperture angle of about 15 degrees. Because of this, a person detection using the sonar sensors does only work by analyzing the sonar scan for leg profiles (see Figure 3 right).

The disadvantage of these sonar sensors is their high inaccuracy. The measurement depends not only on the distance to an object, but also on the object's material, the direction of the reflecting surface, crosstalk effects when using several sonar sensors, and the absorption of the broadcasted sound. Because of these disadvantages, only distances of at most 2 meters can be considered for person detection using these sonar sensors (see Figure 3, left). This means the sonar sensors yield pretty unreliable and inaccurate values, a fact which has to be considered in the generation of a hypothesis of a person detection. For the purpose of a very simple person detection, we assume that all measurements less than 2 meters could be hypotheses for a person. These hypotheses could be further refined by comparing the position of each hypothesis with a given map of the environment. If the hypothesis would correspond to an obstacle in the map, it could be dismissed. The disadvantage of this refinement is, that people standing near by an obstacle would not be considered as a valid hypothesis.

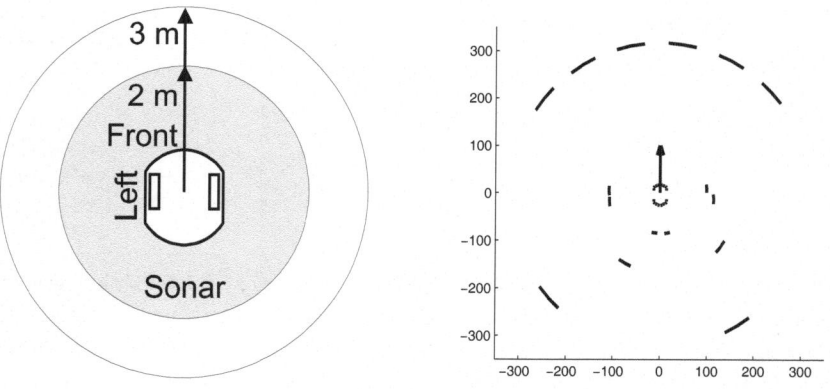

Fig. 3. Left: Top view of the schematic HOROS. The sensoric range of the sonar sensors used to detect people is depicted grey. Right: Real values of the sonar sensors. In this case, the robot is standing in the middle of a floor directed to the top, sensing walls to the left and to the right and a person directly behind it. The distance values in front of the robot (dashed curve) are the result of our range limitation to a maximum of 2 meters.

2.3 Fisheye Camera

As third sensory system, we use an omnidirectional camera with a fisheye lens yielding a 360 degree view around the robot (see Figure 4 left). Because of the task of person detection, the usage of such a camera requires that the position of this camera is lower than the position of the faces. An example of an image resulting from the camera is depicted in Figure 4 (right). To detect people in the omnidirectional camera image, a skin-color-based multi-target-tracker [10] is used. This tracker is based on the condensation algorithm [11] which has been extended, so that the visual tracking of multiple people at the same time is now possible. The particle clouds used to estimate the probability of people in the omnidirectional image will concentrate on the different skin-colored objects. A problem is the possible detection and tracking of non-human skin-color-based objects, like wooden objects or cork pinboards. An essential advantage of this simple approach is, however, that it is much faster than subsampling the whole image trying to find regions of interest and its resistance to minor interferences, like partial occlusions.

A person detection using omnidirectional camera images yields good hypotheses about the direction of a person but not about his distance. Therefore, the integration of the different information from the camera with the data from the laser-range-finder and the sonar sensors should result in a more powerful and robust people detection and tracking system. Subsequently, the developed method for the combination of the several sensory systems will be introduced and discussed.

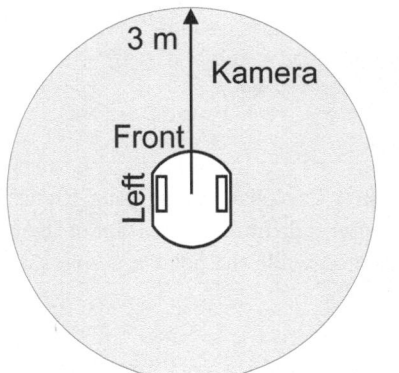 

Fig. 4. Left: Top view of the schematic HOROS. The sensoric range of the omnidirectional camera used to detect people is depicted grey. Right: Real image of the omnidirectional camera with a fisheye lens. A person to be detected is standing in front of the robot and can be seen at the bottom of the image.

3   The Multi-modal Aggregation Scheme

At first, a suitable data representation for the aggregation of the multimodal hypotheses had to be choosen. The possibilities ranged from simple central point representation to probability distributions approximated by particles. The used aggregation scheme is based on Gaussian distributions, see section 3.1. Because of the unknown correlations between the different sensor readings, a *Kalman Filter* based approach was not used to combine these hypotheses. Instead *Covariance Intersection* is applied (section 3.2).

3.1 User Modeling Considering the Different Sensor Information

For the purpose of tracking, the information about detected humans is converted into Gaussian distributions $\phi(\mu, C)$. The mean μ equals the position of the detection in polar coordinates and the covariance matrix C represents the uncertainty about this position. The form of the covariance matrix is sensor-dependent due to the different sensor characteristics described in section 2. Furthermore, the sensors have different error rates of misdetections that have to be taken into account. All computation is done in the defolded cartesian r, φ space, see Figure 5. Examples for the resulting distributions are shown in Fig. 6.

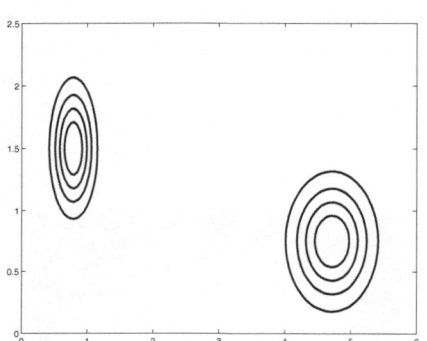 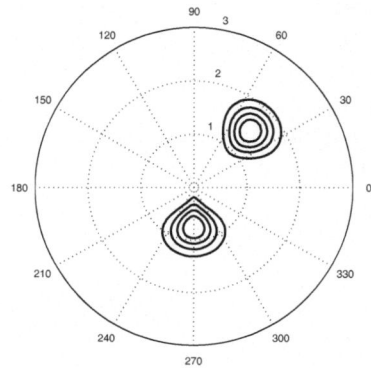

Fig. 5. Left: Two Gaussian hypotheses shown in a Cartesian r, φ system. Right: The same hypotheses in polar r, φ coordinates, the center indicates the position of the robot (Remark: computation is done in the cartesian space, while the polar r, φ space is used for better illustration).

Laser-Based Information: Laser-range-finders yield a very precise measure, hence the corresponding covariances are small and the distribution is narrow (see Figure 6 left). The radial variance is fixed for all possible positions, but the variance of the angular coordinate is distance dependent. A sideways step of a person standing directly in front of the robot changes the angle more than the same movement in a distance of 2 meters. The smaller the distance of the detection the larger the variance has to be. The probability of a misdetection is the lowest of all used sensors, but the laser-range-finder only covers the front area of the robot due to sensor arrangement.

Sonar Information: Information from the sonar tends to be very noisy, imprecise und unreliable. Therefore, the variances are large and the impact on the certainty of a hypothesis is minimal (see Figure 6 middle). Nevertheless, the sonar is included to support people tracking behind the robot. So we are able to form an estimate of the distance in vision-based hypotheses.

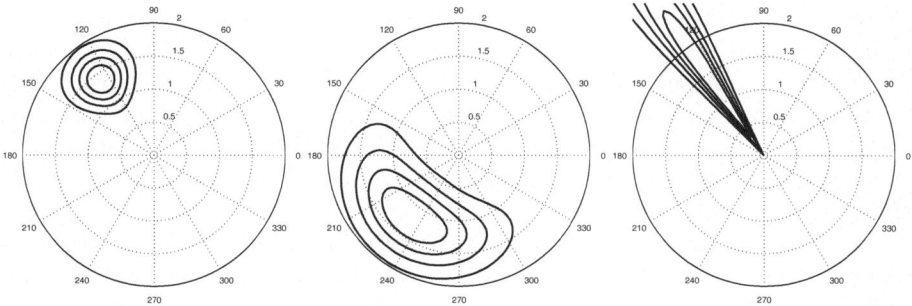

Fig. 6. Examples for generated hypotheses. The center of each plot represents the robot. Left: Hypothesis generated by laser showing a person left in front of the robot. Middle: Sonar-based information showing a hypothesis behind the robot. Right: Camera-based hypothesis without depth information showing the same person as in the left figure.

Fisheye Camera: In contrast to the other sensors, the camera can only provide information about the angle of a detection, but not about the distance of a person. Therefore, for the radial variance of the distance coordinate a very large value was selected, with a fixed mean value (see Figure 6 right). The angular variance is determined by the angular variance of the particle distribution used in the visual skin-color based person tracker (see Section 2.3). The information content of a detection in the image of the fisheye camera is controlled by the position of the detection. In the front area of the robot, the influence is lower, because of the available laser as reliable sensor. Behind the robot, the image is the only source to get information about the presence of a person, the sonar has only supporting character. Thus, the relative weight of a visual hypothesis should be higher behind the robot.

The modeling and integration of additional sensor modalities, like sound localization or other features from the camera image, could be done in a similar way as described above.

3.2   Multi-hypotheses Tracking Using Covariance Intersection

Tracking based on probabilistic methods attempts to improve the estimate x_t of the position of the people at time t. These estimates x_t are part of a local map M that contains all hypotheses around the robot. This map is used to aggregate the sensor hypotheses. Therefore, the movements of the robot $\{u_1, ..., u_t\}$ and the observations about humans $\{z_1, ..., z_t\}$ have to be taken into account. In other words, the posterior $p(x_t|u_1, z_1, ..., u_t, z_t)$ is estimated. This process is assumed to be Markovian. Then the probability can be computed from the previous state probability $p(x_{t-1})$, the last executed action u_t and the current observation z_t. The posterior is simplified to $p(x_t|u_t, z_t)$. After applying the Bayes rule, we get

$$p(x_t|u_t, z_t) \propto p(z_t|x_t)p(x_t|u_t) .$$ (1)

where $p(x_t|u_t)$ can be updated from $p(x_{t-1}|u_{t-1}, z_{t-1})$ using the motion model of the robot and the assumptions about typical movements of people.

A Gaussian mixture $M = \{\mu_i, C_i, w_i | i \in [1, n]\}$ is used to represent the positions of people, where each Gaussian is the estimate for one person. $\phi_i(\mu_i, C_i)$ is a Gaussian centered at μ_i with the covariance matrix C_i. The weight w_i ($0 < w_i \leq 1$) contains information about the contribution of the corresponding Gaussian to the total estimate.

Next, the current sensor readings z_t have to be integrated, after they have been preprocessed as described earlier. If M does not contain any element at time t, all generated hypotheses from z_t are copied to M. Otherwise data association has to be done to determine which elements from z_t and M refer to the same hypothesis. The Mahalanobis distance d_m between two Gaussians $\phi_i \in z_t$ and $\phi_j \in M$ is used as association criterion.

$$\begin{aligned} \mu &= \mu_i - \mu_j \\ C &= C_i + C_j \\ d_m &= \mu C^{-1} \mu^T \end{aligned} \tag{2}$$

This distance is compared to a threshold. As long as there are distances lower than the threshold, the sensor reading i and the hypothesis j with the minimum distance are merged. The problem of merging hypotheses in case two people pass near each other has to be tackled seperately, confer e.g. [12]. The update is done via the *Covariance Intersection* rule (see [13] and [14]), a technique very similar to the *Kalman Filter*. As an advantage of this approach, the unknown correlations between the different sensor readings can be integrated, since this data fusion algorithm does not use any information about the cross-correlation of the inputs. A non-linear convex combination of the means and covariances is computed as follows

$$\begin{aligned} C_{new}^{-1} &= (1 - \omega)C_i^{-1} + \omega C_j^{-1} \; . \\ \mu_{new}^{-1} &= C_{new}\left[(1 - \omega)C_i^{-1}\mu_i + \omega C_j^{-1}\mu_j\right] \; . \end{aligned} \tag{3}$$

The weight ω is chosen to minimize the determinant as

$$\omega = \frac{|C_i|}{|C_i| + |C_j|} \; . \tag{4}$$

The more reliable distribution (that with the smaller determinant of the covariance matrix) is weighted higher in the update. If the current sensor reading is more certain than the current one, the resulting covariance of the hypothesis in M is reduced.

Sensor readings not matching with a hypothesis of M are introduced as new hypothesis in M. The weight w_i is representing the certainty of the corresponding Gaussian. The more sensors support a hypothesis, the higher its weight should be. If the weight passes a threshold, the corresponding hypothesis is considered to be a person. The weight is increased as

$$w_i(t + 1) = w_i(t) + \alpha(1 - w_i(t)) \; , \tag{5}$$

if that hypothesis has been matched with a sensor reading. The time constant $\alpha \in [0, 1]$ is chosen with respect to the certainty of the current sensor (see section 3.1); the more reliable the sensor, the higher the α-weight is. In the case of an unmatched hypothesis, the weight has to be decreased.

$$w_i(t + 1) = w_i(t) - (1 - \theta)\frac{t_{new} - t_{old}}{t_v} \ . \tag{6}$$

The term t_{new} is the current point of time and t_{old} the moment the last sensory input was processed. A person is considered to be lost if t_v seconds passed and no sensor has made a new detection that can be associated with this hypothesis. This temporal control regime is sensor dependent, too. Hypotheses with a weight lower than the threshold θ are deleted.

4   Application

The presented system is in use on the HOROS robot in a real-world office environment. The fact of a changing illumination in different rooms and numerous distractions in form of chairs and tables is quite challenging.

Figure 7 shows a typical aggregation example. In this experiment, the robot was standing in the middle of an office room and did not move. Up to three people were moving around the robot. The enviroment contained several distracting objects, like table legs and skin-colored objects. No sensor modality was able to detect the people correctly. Only aggregation over sensors modalities and time led to the proper result.

The system was able to track multiple people correctly with an accuracy of 93 percent in the experiment. In most cases false negative detections occured behind the robot. The rate of false positive detections is higher, about every forth hypothesis was a misdetection. This is due to the simple cues integrated into the system. But for the intended task of HOROS, the interactive office robot, it is considered to be more important not to miss to many people than finding to many. But there are ways to reduce the amount of false positive detections. Most misdetections are static in the environment, so based on the movement trajectories created by the tracker they can be identified (see section 4.1).

Overall, the presented system improved the performance in the area behind the robot only slightly compared to a simple skin-color tracker. This is, because the sonar-based sensors do not provide many useful information for the tracking task. The main contribution of the sonar sensors is the addition of distance information to existing hypotheses extracted from the fisheye camera and preventing a precipitate extinction of hypotheses in cases of sudden changes in the illumination. In this case, the skin-color tracker will presumably fail, but if the sonar-based information still confirms the presence of the person at the respective position, the hypothesis will not be deleted until the skin-color tracker has recovered.

In the front area of the robot, the system clearly outperforms single sensor-based tracking. Here the influence of the sonar on the result is not observable,

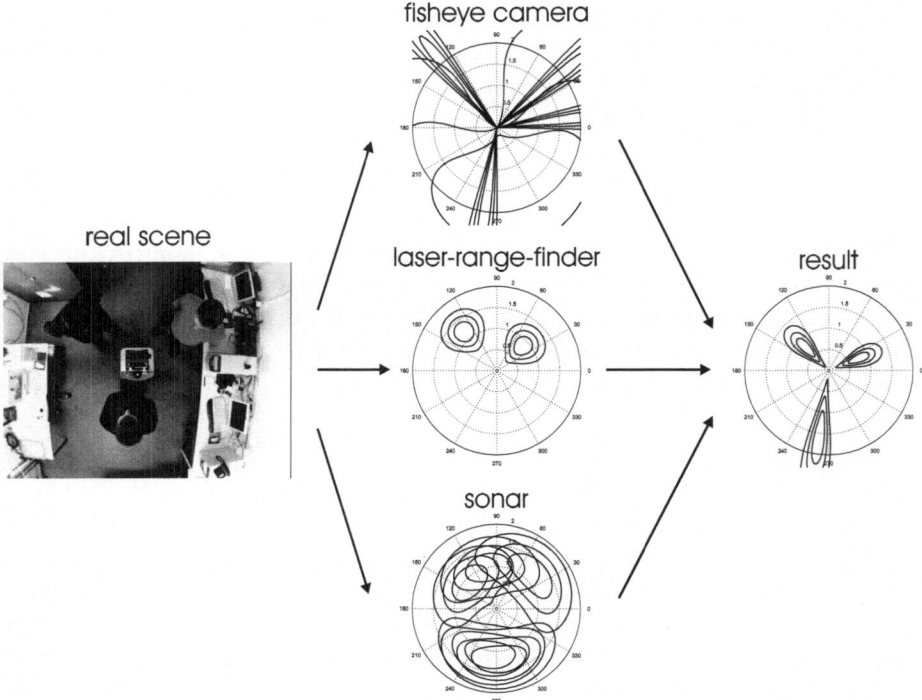

Fig. 7. Aggregation example. The left picture shows the real office scene from a bird's eye view. Three people are surrounding the robot, standing in the middle. The three figures in the middle show the current hypotheses generated by fisheye camera, laser-range-finder, and sonar from top to bottom. No sensor on its own can represent the scene correctly. The final picture displays the aggregated result from the sensors and the previous timestep. This is a correct and sharpened representation of the current situation.

because in most cases the laser-range-finder generates hypotheses more precisely. The laser reduces the deficiency of the skin-color tracker, while the skin-color based information compensates the shortcomings of the laser. These results are observable in Fig. 7. This leads to the assumption that the inclusion of additional sensory systems generating hypotheses about people (e.g. sound source hypotheses) will further improve the performance of this tracking system.

4.1   Trajectory Generation

The system was practically tested in the context of a survey task. HOROS was standing in a hallway in our institute building. His task was to attract attention of people that came by. As soon as the system recognized a person near him, the robot addressed the visitor to come nearer. He then offered to participate in a survey about desired future functionality of HOROS. The people tracking module was used to detect break offs, thus if the user was leaving before finishing

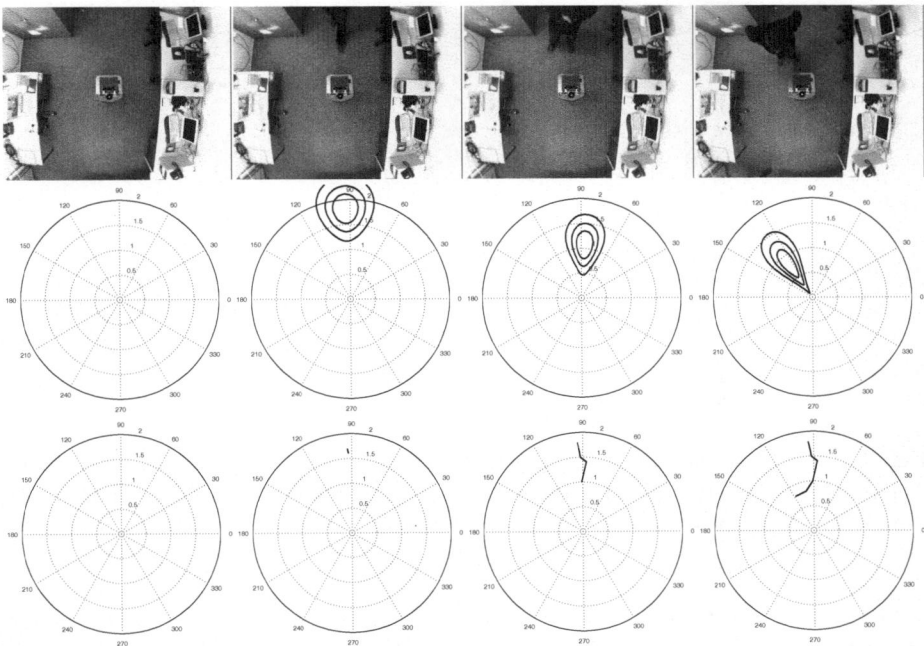

Fig. 8. Trajectory generation. The first row shows the real scene for four discrete time steps from a bird's eye view. In the second row the corresponding results of the tracking system are displayed. In the bottom row the generated trajectories are shown.

the survey. The robot tried to fetch them back and finalize the survey. After the successfull completion of the interaction or a defined time interval with no person coming back, the cycle began again with HOROS waiting for the next interaction partner. The experiment was made in the absence of any visible staff members, so the people could interact more unbiased.

These efforts are repeated from time to time to gather more information, and there is a second, not obvious, intention. The tracking module was used to generate typical movement trajectories of the users. Figure 8 shows the generation of such trajectories. In our future work, we will attempt to classify the path of movement to gain more knowledge about the potential user. In the context of adaptive robot behavior and user models, it is an important issue to assess the interaction partner. The users' movements and the positions relative to the robot are a fundamental step in this direction. If the robot can distinguish between people who are curious, but don't have the heart to step nearer, people who are in a rush and those how just want to interact with the robot, an appropriate reaction can be learned. The use of a multi-person-tracker is a prerequisite, since the experiments show visitors often appearing in groups of two or more people.

Examples for different trajectories are shown in Fig. 9. The most challenging aspects for a classification of trajectories are in our opinion the varying speed of the people and the search for typical movement schemes describing the interest

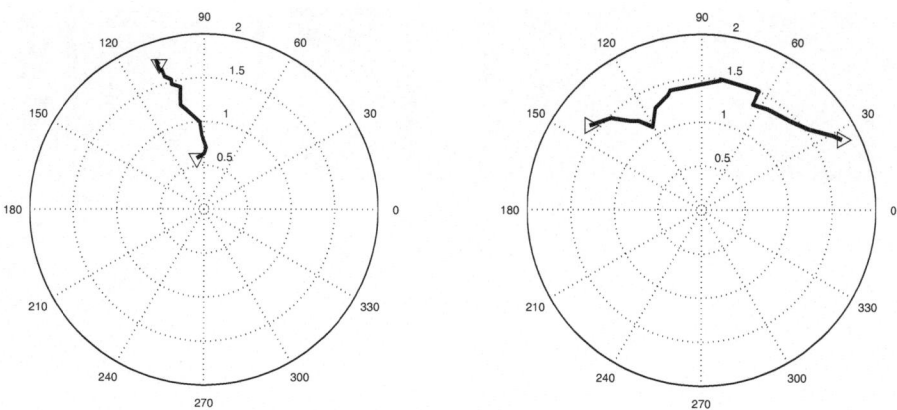

Fig. 9. Left: A trajectory showing a person coming straight towards the robot. Right: The person is crossing from left to right. In doing so, the robot is avoided. The varying time intervals between the movements aren't visible in the figure.

of potential users. Based on the trajectories longtime immovable hypotheses can be discarded with respect to position and interaction status as a false detection.

4.2 Guiding and Following

Another application of the tracking system in the context of our office scenario is the guidance function. The robot can guide visitors from the entrance of the building to the rooms of staff members. If the visitor is leaving sensor range, the robot stops and asks him to return. As soon as the tracker confirms the presence of the user the tour is continued. For this task, a multi-person-tracker is not mandatory, but it allows additional functionality, e.g., a group of people who are unintentionally blocking the robots' path can be detetected and can be asked to clear the path.

The system is able to master the inverted situation, to follow a person, too. This task is not difficult if the user faces the robot. In this case, however, the user has to move backwards, which is unnatural and possibly dangerous. Therefore, the task includes following the person even if the user turns around and no more skin-color is observable by the robot. Without the helpful information of the skin-color tracker, the system successfully follows the user using the hypotheses based on laser and sonar data.

5   Summary and Outlook

We presented a flexible multimodal probability-based approach for detecting and tracking people. It is implemented on our mobile office robot HOROS and is functioning in real-time. Because of the sensor fusion and the probabilistic aggregation, its results are significantly improved compared to a single sensor tracking system. It can be easily extended with other sensors, because there is only the need for a new preprocessing module that produces appropriate Gaussian distributions based on the new sensor readings and an adaption of the weights that model the respective sensor characteristics. The system is able to aggregate data from input modalities with different update frequencies.

In our future work, we will extend the system with additional cues to further increase robustness and reliability for real-world environments. Currently, we are working on the integration of an audio-based speaker localization. In addition, it will be investigated if the face detector by Viola and Jones (see [15] and [16]) can be used for the verification of hypotheses (see [2]) and if it could be integrated into the aggregation scheme itself as an additional cue. Furthermore, we will study the behavior of our system compared to other known approaches and investigate the localization accuracy using labeled data of reference movement trajectories.

References

1. Froeba, B., Kueblbeck, C.: Real-time face detection using edge-orientation matching. In: Audio- and Video-based Biometric Person Authentication (AVBPA'2001). (2001) 78–83
2. Martin, C., Boehme, H.J., Gross, H.M.: Conception and realization of a multisensory interactive mobile office guide. In: IEEE Conference on Systems, Man and Cybernetics. (2004) 5368–5373
3. Simmons, R., Goldberg, D., Goode, A., Montemerlo, M., Roy, N., Sellner, B., Urmson, C., Schultz, A., Abramson, M., Adams, W., Atrash, A., Bugajska, M., Coblenz, M., MacMahon, M., Perzanowski, D., Horswill, I., Zubek, R., Kortenkamp, D., Wolfe, B., Milam, T., Maxwell, B.: Grace: An autonomouse robot for AAAI robot challenge. AAAI Magazine **24** (2003) 51–72
4. Schulz, D., Burgard, W., Fox, D., Cremers, A.: Tracking multiple moving objects with a mobile robot. In: IEEE Conference on Computer Vision and Pattern Recognition (CVPR). (2001) 371–377
5. Schulz, D., Burgard, W., Fox, D., Cremers, A.: People tracking with a mobile robot using sample-based joint probabilistic data association filters. International Journal of Robotics Research (IJRR) **22** (2003) 99–116
6. Nakadai, K., Okuno, H., Kitano, H.: Auditory fovea based speech separation and its application to dialog system. In: IEEE/RSJ International Conference on Intelligent Robots and Systems (IROS-2002). Volume 2. (2002) 1320–1325
7. Siegwart, R., Arras, K.O., Bouabdallah, S., Burnier, D., Froidevaux, G., Greppin, X., Jensen, B., Lorotte, A., Mayor, L., Meisser, M., Philippsen, R., Piguet, R., Ramel, G., Terrien, G., Tomatis, N.: Robox at expo.02: A large scale installation of personal robots. Special issue on Socially Interactive Robots, Robotics and Autonomous Systems **42** (2003) 203–222
8. Fritsch, J., Kleinehagenbrock, M., Lang, S., Fink, G., Sagerer, G.: Audiovisual person tracking with a mobile robot. In Groen, F., Amato, N., Bonarini, A., Yoshida, E., Krse, B., eds.: Int. Conf. on Intelligent Autonomous Systems, IOS Press (2004) 898–906

9. Fritsch, J., Kleinehagenbrock, M., Lang, S., Ploetz, T., Fink, G., Sagerer, G.: Multi-modal anchoring for human-robot-interaction. Robotics and Autonomous Systems, Special issue on Anchoring Symbols to Sensor Data in Single and Multiple Robot Systems **43** (2003) 133–147

10. Wilhelm, T., Boehme, H.J., Gross, H.M.: A multi-modal system for tracking and analyzing faces on a mobile robot. In: Robotics and Autonomous Systems. Volume 48. (2004) 31–40

11. Isard, M., Blake, A.: Condensation - conditional density propagation for visual tracking. International Journal on Computer Vision **29** (1998) 5–28

12. MacCormick, J., Blake, A.: Probabilistic exclusion and partitioned sampling for multiple object tracking. International Journal on Computer Vision **39** (2000) 57–71

13. Julier, S., Uhlmann, J.: A nondivergent estimation algorithm in the presence of unknown correlations. In: Proceedings of the 1997 American Control Conference, IEEE (1997) 2369–2373 vol.4

14. Chen, L., Arambel, P., Mehra, R.: Estimation under unknown correlation: Covariance intersection revisited. IEEE Transactions on Automatic Control **47** (2002) 1879–1882

15. Viola, P., Jones, M.: Fast and robust classication using asymmetric adaboost and a detector cascade. In: NIPS 2001. (2001) 1311–1318

16. Viola, P., Jones, M.: Robust real-time object detection. In: Proc. of IEEE Workshop on Statistical and Computational Theories of Vision. (2001)

Behavior Recognition and Opponent Modeling for Adaptive Table Soccer Playing

Thilo Weigel, Klaus Rechert, and Bernhard Nebel

Institut für Informatik, Universität Freiburg,
79110 Freiburg, Germany
{weigel, rechert, nebel}@informatik.uni-freiburg.de

Abstract. We present an approach for automatically adapting the behavior of an autonomous table soccer robot to a human adversary. Basic actions are recognized as they are performed by the human player, and characteristic action observations are used to establish a model of the opponent. Based on this model, the opponent's playing skills are classified with respect to different levels of expertise and particular offensive and defensive skills are assessed. In response to the knowledge about the opponent, the robot adapts the velocities at which it attacks and defends in order to provide entertaining games for a wide range of human players with different playing skills. Experiments with two different table soccer robots validate our approach.

1 Introduction

For an autonomous system, the best strategy to achieve a certain goal often depends on the behavior of other agents. As the agents usually differ in their behavior patterns, the ability to adapt to these differences dynamically is very beneficial for optimizing the agent's behavior.

Reinforcement learning is a common technique for adapting an agent's policy to the environment [1]. As this is done in a trial-and-error fashion, an agent generally does not derive explicit knowledge about the encountered agents, but rather implicitly learns to act in the most appropriate way. Unfortunately, in realistic environments reinforcement learning approaches are usually too slow to be used for online adaptation.

In contrast, deliberate modification of an agent's behavior on the basis of recognized features of other encountered agents allows to adapt in a much more efficient way. However, this requires to explicitly gather and classify information about other agents' behavior patterns. In game playing, one would like to obtain a model of the opponent which characterizes its playing style, playing skills and general strategy. Based on such a model, the agent then selects a specific strategy which promises to be the optimal response to the opponent.

In this paper, we present an approach for the automated adaptation of a table soccer robot to its human opponent. The basis for the approach is the robust vision-based recognition of basic skills as they are performed by the human players. According to the observations, the opponent is classified with respect to different opponent models. In consequence, the robot adapts its behavior based on the observed playing skills of the opponent.

U. Furbach (Ed.): KI 2005, LNAI 3698, pp. 335–350, 2005.

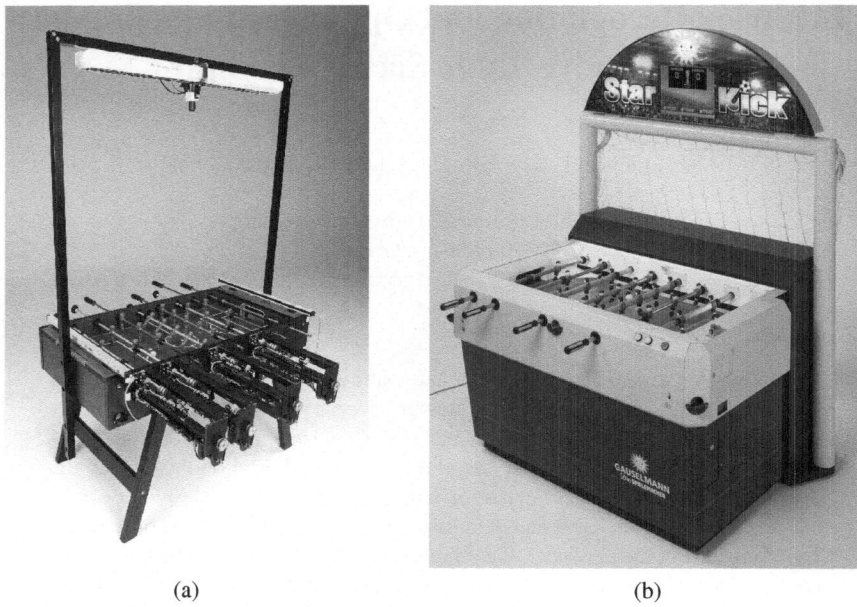

(a)   (b)

Fig. 1. The table soccer robots KiRo (a) and StarKick (b)

The approach is evaluated using two variants of a real table soccer robot. A first prototype version called *KiRo* uses an overhead color camera for observing the players and the ball [2]. A commercial version called *StarKick* uses a black and white camera, which perceives the ball from underneath the table. In many test games StarKick showed to be a competitive challenge even for advanced human players [3]. Figure 1 shows pictures of KiRo and StarKick.

The benefit of adapting to the opponent is twofold. On the one hand, weaknesses of the opponent can be exploited, but on the other hand, the playing level can be adjusted so that the game stays interesting for less experienced players as well.

The rest of this paper is structured as follows. In Section 2, the recognition of basic actions based on vision data is described. Section 3 presents how the observed actions and action parameters are used to derive a model of the opponent's playing skills. In Section 4, we show how the opponent model can be used for adapting the behavior of the table soccer robot. Experimental results are given in Section 5. Section 6 discusses related work and a conclusion is drawn in Section 7.

2   Behavior Recognition

The basis for behavior recogniton are camera images as shown in Figure 2. The images are delivered by the vision system in YUV-format at 50 Hz[1] with a resolution of 384×288 pixels. For StarKick, the possibility to detect the playing figures in the camera images was traded for an infrared-based system, which allows a very robust and reliable

---

[1] This is achieved by processing each half-frame individually.

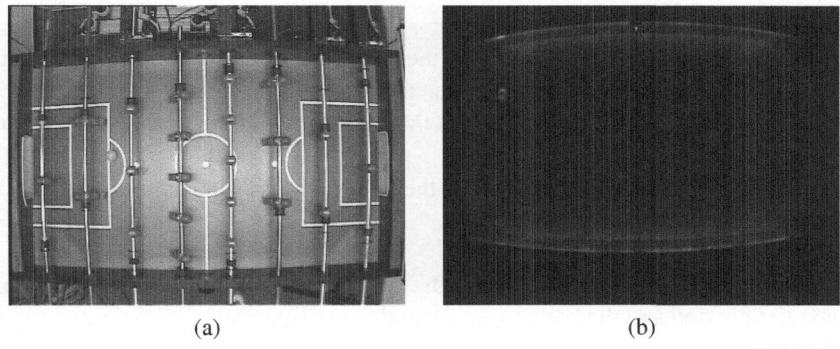

(a)   (b)

Fig. 2. Camera images delivered by KiRo's overhead camera (a) and StarKick's camera underneath the table (b)

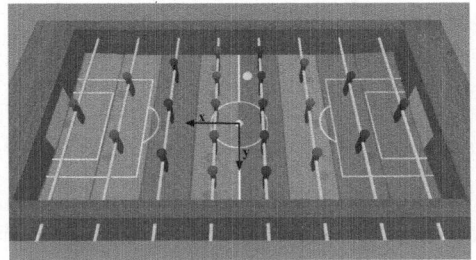

Fig. 3. The influence areas marked with different colors for the red and blue rods

ball detection. Only one's own player positions are available from the motor controller feedback [3]. Even though all player positions can be extracted from KiRo's camera images, it is in general not possible to reliably detect which playing figure caused a certain ball movement: Even at normal game velocities the ball can travel several centimeters between two consecutive camera frames, and due to the limited image resolution, the ball can be recognized only with an accuracy of 3mm at best [4].

For these reasons, we decided to base our approach solely on the observed ball position during a game and the knowledge of the rods' fixed axis locations. The ball's heading and velocity are determined from consecutive ball estimates, which are maintained using a Kalman Filter.

In order to cope with the limited spatial and temporal resolution of the available data, we have chosen a very coarse representation to define and recognize the most common actions during a table soccer game. For each player p and rod r, we introduce a corresponding *influence area* $\mathcal{I}_{p_r}$ in which at least one of the rod's playing figures can manipulate the ball. Figure 3 shows the corresponding coordinate system and the influence areas for the eight rods.

An action is now defined for an entire rod based on where the ball enters its influence area, how it moves inside, and where it leaves the influence area again. We distinguish between the following possibilities for manipulating the ball:

- **BlockBall**
  A rod blocks the ball when it prevents the ball from passing without gaining control over it. This is expressed by the fact that the ball stays inside the influence area only for a short period of time and it leaves the area at the same side where it entered.
- **ControlBall**
  A rod controls the ball when it keeps the ball motionless inside the influence area for a significant period of time.
- **DribbleBall**
  A rod dribbles the ball when it keeps the ball inside the influence area for a significant time period and the ball moves a significant distance.
- **KickBall**
  A rod kicks the ball when it accelerates the ball considerably. A kick can either reverse the ball's trajectory or continue it when the ball arrived from behind. After a dribble or control action, it is always assumed that a kick causes the ball to finally leave the influence area.
- **YieldBall**
  A rod yields the ball when the ball traverses the influence without a significant change in velocity. This action may reflect the intention to let the ball pass from behind, but also may be observed when an intended block failed.

Here, the actions for blocking, kicking and yielding the ball refer to both directions in which the ball can traverse an influence area. However, the direction of the ball is explicitly considered when the opponent's playing skills are assessed.

The above actions can be described more formally by predicates, which are all based on directly observable features. Let $(x_t, y_t, v_t, \vartheta_t)$ be the ball vector at time t with (x_t, y_t) denoting the ball's position, v_t its non-negative velocity and ϑ_t its heading. Let further be $t = 0$ the time the ball enters the influence area $\mathcal{I}_{p_r}$[2]. We can then define the following predicates for the r-th rod of player p:

TimeOut:
$$t > \frac{\Delta_{rod}}{\cos \vartheta_0 v_0}. \tag{1}$$

BallMoved:
$$\frac{1}{t-1} \sum_{i=1}^{t} \sqrt{(x_i - x_{i-1})^2 + (y_i - y_{i-1})^2} > \Delta_d. \tag{2}$$

BallLeft:
$$x_t \notin \mathcal{I}_{p_r}. \tag{3}$$

BallAccelerated:
$$v_t > a v_0 \wedge v_0 < v_{max} \wedge v_t > v_{min}. \tag{4}$$

BallDirectionChanged:
$$\mathrm{sgn}(X_{p_r} - x_0) = \mathrm{sgn}(X_{p_r} - x_t). \tag{5}$$

Here, Δ_{rod} denotes the distance between two neighboring rods. In order to capture when a ball was stopped by a rod, *TimeOut* checks whether the current time t exceeds the time after which the ball is expected to leave the influence area again. A lower and an upper bound limit the timeout to reasonable values. In our experiments, timeouts in the interval $[200msec, 750msec]$ were used. The ball's movements inside an influence area are summed up by the predicate *BallMoved*, which is true when the ball traveled more

---

[2] According to the camera's frame rate, each time step currently corresponds to 20 milliseconds.

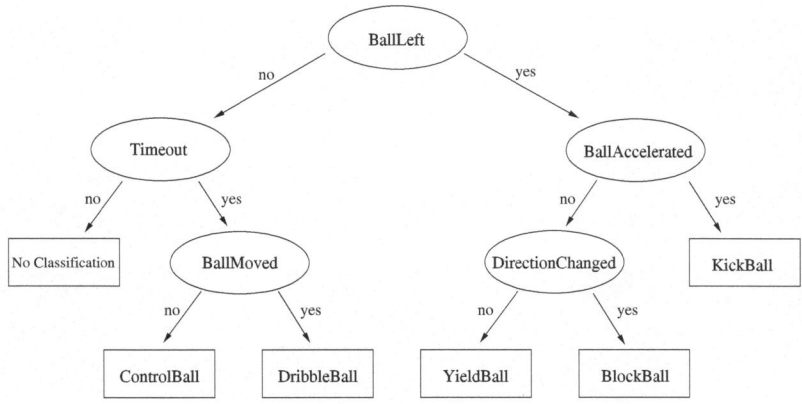

Fig. 4. The decision tree for classifying the basic table soccer actions

than a certain minimum required distance per time step. To detect if the ball has left the influence area, *BallLeft* checks whether the ball coordinates still belong to this area. The predicate *BallAccelerated* determines if the ball's initial velocity has increased considerably. However, the ball is only considered to be deliberately accelerated by a player if the ball's initial velocity has not been too high and and the final velocity is not too low. In our experiments, we achieved good results with $a = 1.35$, $v_{min} = 350 \frac{mm}{s}$ and $v_{max} = 5000 \frac{mm}{s}$. The predicate *BallDirectionChanged* checks if the ball has left the influence area towards the same direction from where it has entered. The distance threshold Δ_d and the acceleration factor a can be determined empirically. However, in the future we plan to learn the values automatically from training data.

The basic actions can now be defined using the above predicates: An action is considered to be recognized if the conjunction of some of the – potentially negated – predicates is true. The decision tree shown in Figure 4 represents these formulas and detects an action for the rod whose influence area currently contains the ball.

The decision tree is evaluated in every cycle. An action can be classified either when the ball has changed to another influence area or when a timeout has occurred. If both *BallLeft* and *Timeout* are false, no action is classified. In case no classification is made, one could in principle try to estimate the action that is currently taking place. However, as an action usually happens within only a few cycles, predictions ahead of time are generally very uncertain.

The predicate *BallAccelerated* reflects if a kick action has occurred. For dribble and control actions the corresponding initial ball velocity is set to zero: $v_0 = 0$. In doing so, after a control or dribble action a kick is always classified as well. Please note that the time t is reset to zero after an action has been recognized, and thus consecutive dribble or control actions may be detected. Such sequences are merged into one single action at a later stage. Figure 5 shows an example series of camera images depicting the blue attacker kicking the ball. As can bee seen, the kicking action takes place within only a few milliseconds.

Since the goalkeeper's movements are limited to the area in front of its goal, it cannot reach the lateral positions of its influence area. In consequence, when the ball

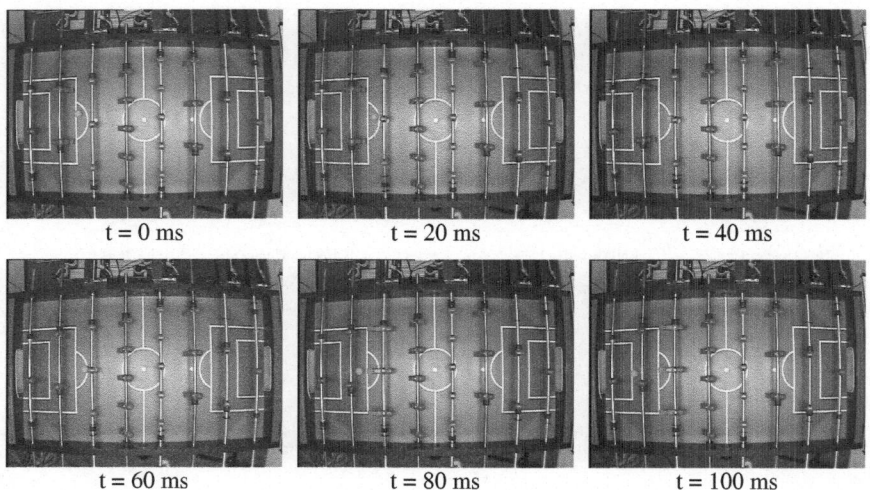

Fig. 5. Camera images in a 100 ms period during a kick action of the blue attacker

bounces back from the goal border, a block action would be assessed even though it is impossible that the goalkeeper has actively blocked the ball. In such cases, the corresponding observations are filtered and no action is classified.

When the ball moves very fast, it may cross entire influence areas inbetween two consecutive camera frames so that no ball observations are available for these areas. As in such situations the corresponding rods obviously let the ball pass, appropriate observations are generated artificially so that *YieldBall* actions are assessed for these rods. However, when a goal has been shot and the ball is thrown in again from the center of the horizontal table border, the ball does not traverse neighboring influence areas, and thus no yield actions should be classified. Such situations can be detected reliably by checking if the ball was last observed in a defender's or goalkeeper's influence area and then (re-)appeared in a rectangle around the regions where the ball can be thrown in. In consequence, the above heuristic is suspended, the ball's initial velocity v_0 is assumed to be zero, and the allowed timeout is initialized with a fixed value. In this way, after a throw-in, always a *KickBall* is assessed, but a *ControlBall* or a *DribbleBall* are only classified when there has been a certain time delay before the kick action.

3 Opponent Modeling

A player's playing skills and general playing style can be assessed based on the observed frequency and characteristics of the recognized basic actions. For this, a set of parameters is observed and evaluated for each occurrence of a basic action. We distinguish between the same actions as presented in Section 2. Each observation is rated with respect to a predefined standard yielding a value in the interval $[0, 1]$. For example, the rating of the ball's velocity v_t after a kick action is mapped to $[0, 1]$ by the following function:

$$r_{kick}^{vel} = \begin{cases} 0, & \text{if } v_t < v_{min}, \\[2mm] \dfrac{v_t - v_{min}}{v_{max} - v_{min}}, & \text{if } v_{min} < v_t < v_{max}, \\[2mm] 1, & \text{else.} \end{cases} \qquad (6)$$

Here, v_{min} and v_{max} denote predefined minimum and maximum velocities, which were set in our case to $300 \frac{mm}{sec}$ and $2000 \frac{mm}{sec}$, respectively.

With r_a^o denoting the rating of an observation o for the action a and $action$ denoting a placeholder for any possible action, the following observation ratings are taken into account:

- r_{action}^{num} – The percentage of occurences of an action with respect to the total number of observed actions so far.
- r_{action}^{vel} – The ball's maximum velocity while an action took place.
- $r_{dribble}^{distance}$ – The distance the ball moved during *DribbleBall* .
- r_{kick}^{gain} – How far the ball was shot by *KickBall* relative to the distance it could be shot until reaching the opponent goal.
- $r_{moveKick}^{dribbleVel}$ – The velocitiy at which the ball was dribbled right before a move kick.

The observations can directly be related to a player's *playing skills*. Other observations, like the time the ball was kept inside an influence area by a control action, would not necessarily reveal how well the opponent plays, but rather show the opponent's *playing style*. Two additional observations can be interpreted as observing two higher-level actions:

- **MoveKick**
  A *MoveKick* is performed when a rod moves the ball sideways with considerable speed and then kicks the ball forward immediately. This action is observed when a fast *DribbleBall* is directly followed by a *KickBall*.
- **ControlledDribble**
  A *ControlledDribble* is performed when a rod controls the ball right after it has dribbled the ball. This action is observed when at least once a *DribbleBall* is immediately followed by a *ControlBall*.

As these actions usually require advanced playing skills, they are helpful for distinguishing between unexperienced and more advanced players.

The ratings of the different observations for one action are now combined to one quality measure for this particular action. As the rating of an action's occurrence frequency does not relate to an individual action, but rather provides an additional global quality measure, it is incorporated at a later stage. In fact, for *Block*, *Control*, *Yield* and *ControlledDribble* actions, only their occurrence frequency is of interest[3]. Therefore, only the following quality measures have to be calculated as the weighted sum of observation ratings:

---

[3] For block actions the velocity at which the ball approaches a rod could be taken as a quality measure. However, the quality would then depend very much on the opponent's kicking skills.

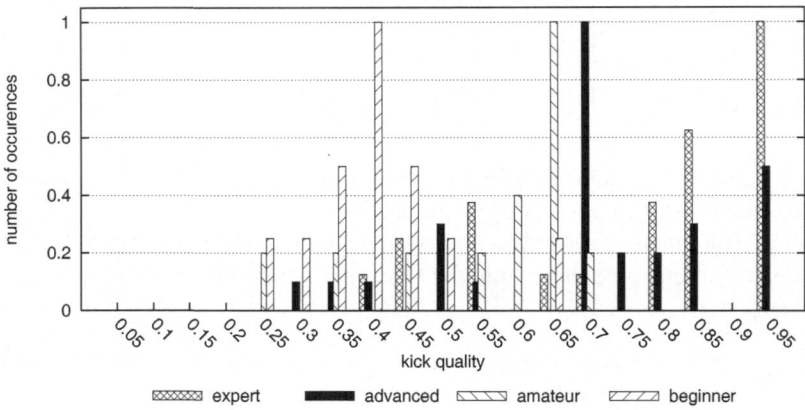

Fig. 6. The normalized number of occurrences of the kick qualities for four human players with different playing skills

$$q_{dribble} = \frac{\alpha_1 r^{vel}_{dribble} + \alpha_2 r^{distance}_{dribble}}{\alpha_1 + \alpha_2}, \tag{7}$$

$$q_{kick} = \frac{\alpha_3 r^{gain}_{kick} + \alpha_4 r^{vel}_{kick}}{\alpha_3 + \alpha_4}, \tag{8}$$

$$q_{moveKick} = r^{dribbleVel}_{moveKick}, \tag{9}$$

where the $\alpha_{\{1...4\}}$ denote some weights.

Usually, an action is carried out by a player with different quality levels during a game. The performance of a player may sometimes be very poor due to mistakes and sometimes be very good just by luck. In order to capture a player's general playing skills, we eliminate such "outliers" and assume that a player's real capabilities are best reflected in the quality measure which occurred the most times. For this, we discretize the quality measures corresponding to one action into *quality levels* and build a histogram over these levels. The maximum of a histogram is then taken as the overall quality measure. Figure 6 shows such a histogram for the kick qualities of four different players and illustrates how better players achieve higher qualities.

The playing skills of an opponent player can now be assessed based on the actions' quality measures and occurrence frequencies. We distinguish between the general skills for blocking, dribbling and kicking the ball:

$$\mathcal{S}_{block} = \frac{\beta_1 r^{num}_{block} + \beta_2 r^{num}_{control} + \beta_3 r^{num}_{dribble} + \beta_4 r^{num}_{kick} + \beta_5 (1 - r^{num}_{yield})}{\beta_1 + \beta_2 + \beta_3 + \beta_4 + \beta_5}, \tag{10}$$

$$\mathcal{S}_{dribble} = \frac{\beta_6 \hat{q}_{dribble} + \beta_7 r^{num}_{controlledDribble} + \beta_8 (1 - r^{num}_{lost})}{\beta_6 + \beta_7 + \beta_8}, \tag{11}$$

$$\mathcal{S}_{kick} = \frac{\beta_9 \hat{q}_{kick} + \beta_{10} \hat{q}_{moveKick} + \beta_{11} r^{num}_{moveKick}}{\beta_9 + \beta_{10} + \beta_{11}}. \tag{12}$$

Here, $\mathcal{S}_s$ denotes a certain skill, $\hat{q}_a$ denotes the maximum quality extracted from the histogram for action a and $\beta_{\{1...11\}}$ are weights. The block skill expresses a player's ability to prevent the ball from rolling towards one's own goal. For this, the occurence percentages of block, control, dribble and kick actions are summarized and weighted against the occurrence percentage of yield actions which let the ball pass contrary to one's own playing direction. The dribble skill combines dribble and controlledDribble actions taking into account the dribble quality and the occurence percentage of controlledDribble actions. Additionally, the number of times the ball was lost is considered. This is the case whenever the velocity of a kick after a dribble or controlledDribble action is rated as zero. Only low ratings r_{lost}^{num} can yield high dribble skills, indicating that most of the time the ball was rather deliberately played than accidentally lost. Since even beginners kick the ball very often while more experienced players may choose to dribble the ball more often, we decided to ignore the occurrence percentage of kick actions when assessing the kicking skills. Only the frequency of moveKick actions, as well as the kick and moveKick quality are taken into account. An opponent model for assessing a player's playing skills can now be defined as a vector that contains the three skill measures:

$$\mathcal{O} = (\mathcal{S}_{block}, \mathcal{S}_{dribble}, \mathcal{S}_{kick})^T. \tag{13}$$

This vector can now be classified with respect to predefined opponent classes. In our case, these classes describe different levels of playing skills and are defined by model vectors with fixed quality measures. In our experiments, we used the following definitions: $\mathcal{O}_{beginner} = (0.3, 0.2, 0.1)^T$, $\mathcal{O}_{amateur} = (0.5, 0.4, 0.3)^T$, $\mathcal{O}_{advanced} = (0.6, 0.5, 0.5)^T$, $\mathcal{O}_{expert} = (0.8, 0.8, 0.8)^T$.

For comparing an observed opponent model M with an opponent class C, the match between M and C is defined as the sum of the squared differences between the corresponding skill measures:

$$match(M, C) = \sum_{a \in \mathcal{A}} (\mathcal{S}_a^C - \mathcal{S}_a^M)^2, \tag{14}$$

where $\mathcal{A} = \{block, dribble, kick\}$.

Classifying a model vector M can now be considered as finding the class C with the minimum deviation to M. Thus, minimizing $match$ yields the desired classification:

$$class(M) = \arg \min_{C \in \mathcal{O}} match(M, C), \tag{15}$$

where $\mathcal{O} = \{\mathcal{O}_{beginner}, \mathcal{O}_{amateur}, \mathcal{O}_{advanced}, \mathcal{O}_{expert}\}$. The function $class(M)$ assesses the general playing skills of an observed adversary player. This information can be used in the following for adapting one's own behavior appropriately.

4 Behavior Adaptation

An agent can utilize the knowledge of the opponent's playing skills in various ways. One goal could be to exploit weaknesses of the opponent for increasing one's own playing performance. Against weaker opponents, a goal could also consist in lowering one's own playing standard to the opponent's level for keeping the game interesting.

Table 1. The factors for the maximum move and turn velocity with respect to the opponent's playing level

	Beginner	Amateur	Advanced	Expert
f_{move}	0.4	0.6	1.0	1.0
f_{turn}	0.2	0.5	0.8	1.0

Since StarKick is capable of beating even advanced human players, its playing level usually needs to be lowered in order to maintain the game entertaining for the average human player. This can be achieved by lowering the velocities at which the robot shoots the ball and at which it moves a rod towards a certain blocking position. While the first weakens the robot's offense, the latter decreases its defensive play.

A more elaborate way of adaptation would consist in employing different types of actions according to the opponent's playing characteristics. Currently, we are working on actions for dribbling and passing the ball. These actions provide a very attractive game to watch but also have a higher risk of losing the ball to the opponent. Therefore, the use of these actions could be made dependent on the risk that an opponent would take too much advantage of them.

At present, we adapt the behavior based on adapting the move and shoot velocities so that the game maintains balanced between the robot and the human opponent. Based on the opponent model proposed in the previous section, there are two alternatives for doing so. Both approaches adjust the velocities by multiplying constant factors $f_{move}, f_{turn} \in [0, 1]$ to the maximum velocity at which a rod moves and turns.

One way to establish these factors is to refer to predefined values according to the assessed adversary playing level. Table 1 shows the mapping from the opponent's playing level to velocity factors which we used in our experiments.

However, instead of only considering the general playing skills, the opponent's defensive and offensive capabilities can also be assessed individually so that one is able to adapt to the opponent's strengths and weaknesses more directly. A second adaptation scheme therefore adjust the maximum move velocity in response to the opponent's offensive skills and adjusts the maximum turn velocity according to the opponent's defensive skill level. For this, an opponent's defensive and offensive skill level is calculated from the model vector as follows:

$$\mathcal{S}_{defense} = \mathcal{S}_{block}, \tag{16}$$

$$\mathcal{S}_{offense} = \frac{1}{2}(\mathcal{S}_{dribble} + \mathcal{S}_{kick}). \tag{17}$$

The maximum velocities are then adapted in direct response to the currently observed opponent's offensive and defensive skill level, i.e. $f_{move} = \mathcal{S}_{offense}$ and $f_{turn} = \mathcal{S}_{defense}$. Of course, the velocity factors could also simply be established depending on the current game score. However, the score might not always be available and the aim of this work is not to balance the game score but to adapt the difficulty of scoring a goal to an appropriate level for the human player.

In order to enable our table soccer robots to adapt online to changing opponents, old action observations should not influence the current opponent model. Therefore,

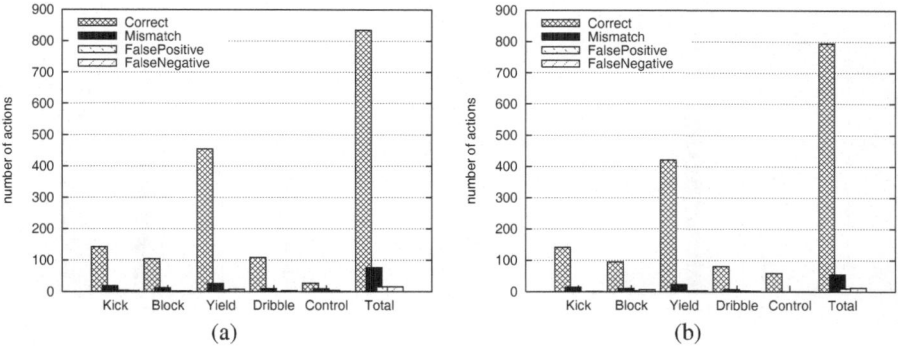

Fig. 7. Action classification results for games against KiRo (a) and games against StarKick (b)

the assessment of the opponent's skills can be limited to the observations which were recorded in a certain time window during the last n seconds[4]. The length of the time window then establishes the trade-off between stableness and reactivity: a long time window is more robust against slight variations in the opponent's playing style. However, a short time window allows to react more quickly when the opponent changed.

5   Results

For the evaluation of the action classification, we recorded twelve log files during games with both KiRo and StarKick. Three different human players – a beginner, an amateur and an advanced human player – played two games on each table. In one game, the robot played with maximum move and turn velocities. In the other game, the robot played very slowly so that the human players had more opportunities to play "their style" without being disturbed. Each log file contains the raw video data recorded at 50 Hz, which sums up to a total amount of 14 GB of data for evaluation. We hand-labeled the log files specifying at which positions the system should assess a certain action. In total, 72.000 frames were analyzed this way. Figure 7 shows the results of comparing the manually created *ground truth* to the output of our system for KiRo (a) and StarKick (b). Always the human opponent and the robot were evaluated. We distinguished between *correct* classifications, *mismatches*, *false positives* and *false negatives*. When a different action than stated in the ground truth was recognized, a mismatch was counted. When an observed action did not appear in the ground truth, a false positive was recorded. Conversely, when the ground truth contained an action where no action was recognized, a false negative was counted. As can be seen, the classification results were similar for the games with KiRo and StarKick. However, due to the better vision hardware, there occurred fewer misclassifications on StarKick. Altogether, 1790 actions were recorded in the ground truth. 1630 actions were classified correctly while 185 actions (including false positives) were misclassified. This corresponds to a recognition rate of 91.1%. As table soccer is a very fast game, the ball frequently passes a rod's influence area without

---

[4] In our experiments, we used a time window of 60 seconds.

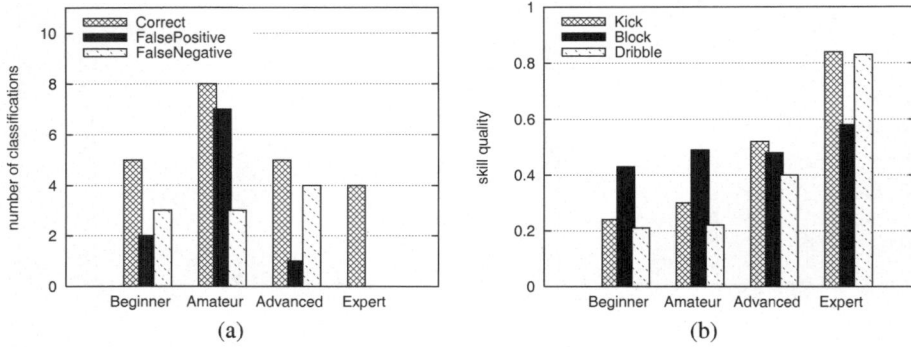

Fig. 8. The classification results for the human players with respect to their playing skills (a), and the assessed skill qualities with respect to the playing skills of the human players (b)

being played. This is reflected in the high occurrence frequency of yield actions. In comparison, the other actions occurred roughly with the same frequency. However, the ball was shot and blocked more frequently than it was controlled and dribbled.

For evaluating the opponent modeling, we took log files from another 32 games, which sums up to 38 GB of data. Each time, a different human player played for about two minutes against StarKick. Each player was automatically classified into one of the opponent classes. The results were then compared to the real playing skills of the human players as assessed by two human observers. Unfortunately, it turned out to be very difficult to make a clear distinction between the playing levels and in many cases a player was assigned a class even though he could have been assigned a neighboring class just as well. Figure 8(a) reveals the classification results for each playing level. As can be seen, some misclassifications occurred. In particular, four advanced players were classified as amateur players, one amateur player was classified as an advanced player, two amateur players were classified as beginners and three beginners were classified as amateurs. In total, 69% of the human players were classified correctly. However, many misclassifications can be ascribed to the inaccurate assessment of the playing levels by the human observers. Figure 8(b) shows the skill measures averaged over the log files corresponding to the same skill level of the human players. Since the skill levels increase with increasing playing skills, the measures are in fact suitable distinguishing features for modeling the opponent. Figure 9 shows individual examples of how theses skill measures develop over time for three human players with different playing skills.

For evaluating the behavior adaptation, an expert human player played twice for eight minutes against StarKick while it automatically adapted its move and turn velocities. In the first game, StarKick adjusted its velocities according to the assessed opponent's defensive and offensive skills. In the second game, StarKick set its velocity with respect to the currently assessed opponent class. In the first two minutes of both games, the human player intentionally played on a poor level like a beginner. In the following two minutes he played roughly on the level of an amateur until he started to play as skillfully as an advanced player for the next two minutes. During the last two minutes he played on a similar level as an expert player.

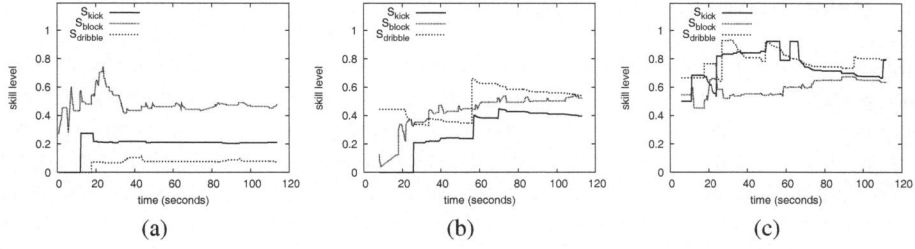

Fig. 9. The skill measures for a human beginner (a), advanced player (b), and expert (c)

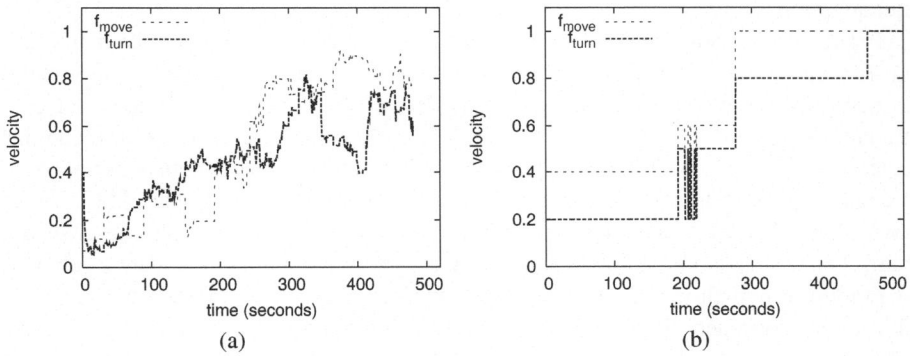

Fig. 10. Evolution over time of the adapted move and turn velocities according to the currently assesed defensive and offensive skills (a) and the currently assessed opponent's playing level (b)

Figure 10(a) shows the velocities which were set during the first game according to the currently assessed defensive and offensive playing skills of the human player. Starting with low velocities, StarKick in fact continuously increased its velocities in order to keep up with the opponent's increasing skill level. However, as StarKick improved its playing level, it got more difficult for the human player to defend against StarKick's strong and frequent kicks. In consequence, after $5\frac{1}{2}$ minutes the human player's defensive play deteriorated for a short while, until in turn the human player adapted to StarKick's increased playing level. Figure 10(b) shows the velocities which were adjusted during the second game with respect to the currently assessed opponent class. The velocities correspond to the fixed values for beginners, amateurs, advanced players, and experts. According to the time window length, there was at least a delay of 60 seconds until StarKick fully adapted to the increased level of play. However, as it is impossible for a human to maintain exactly the same playing level and playing style, and due to the fact, that adaptation is taking place continuously, some variations are possible. While the adaptation to the amateur level took 72 seconds, the adaptation to the advanced level took only 36 seconds. Since it turned out to be very difficult for the human player to continuously play as strong as an expert, the adaptation to the expert level took 108 seconds.

6   Related Work

In many applications it is beneficial to recognize and classify the actions of an interaction partner. Plan recognition attempts to infer plans by logical deduction based on a set of observed actions and an action taxonomy [5]. In this context, Bayesian networks can be used to deal with the uncertainty when a set of observations can be explained by different plans [6]. Lesh *et al.* showed for example how plan recognition can be used in a collaborative email system to reduce the amount of communication required from the user [7]. For all these systems, one assumes that a user behaves according to a fixed plan of a known plan library. In real world dynamic domains like table soccer, this is not necessarily the case. Furthermore, our approach also examines how well an action was carried out and particularly uses such information as quality measures for later adaption.

Especially in sports games, the automated recognition of behaviors is of interest, e.g. for team coaching [8] or commentatory agents [9]. In the RoboCup Simulation League [10] many teams derive explicit models of the opponent with the final goal of adapting a team's own behavior. Similar to our approach, Kaminka *et al.* recognize basic actions based on descriptive predicates. They further use a statistical approach to detect repeating action sequences which are characteristic for a team [11]. Visser and Weland [12] use time series and decision tree learning to induce rules describing a team's behavior. Also Case Based Reasoning is used to predict the opponent's behavior by examining which actions were carried out previously in a similar situation [13]. Akin to our approach, Steffens [14] manually defined feature-based opponent models and improved the team's performance by automatically applying predefined counter strategies. For representing different classes of adversaries, it was proposed to capture the opponent's strategic behavior by accumulating position information in grids and to use decision tree learning for classifying the opponent according to similarities of the accumulated position information [15]. The *ATAC* approach models the opponent's team behavior by probabilistic representations of the opponent's predicted locations. The models have successfully been used to adapt coordinated team plans for setplays [16].

Yet, all these approaches were only evaluated in a simulated environment where the uncertainty in sensing and acting is less severe than in real world environments. In contrast to our work, most of the approaches do not consider the details of an action, but focus on the movements and the positioning of the players. However, in our case the knowledge of how an action is carried out is an important clue for assessing the skills of a player. In that sense, the work of Han and Veloso [17] is particularly related to ours. They formalize and recognize individual basic soccer actions by means of Hidden Markov Models (HMM's) and evaluate their approach in a real world scenario. Unfortunately, their approach cannot be adopted to table soccer because table soccer actions usually do not take long enough time for being captured by a HMM.

7   Conclusion and Outlook

In this paper, we presented an approach for automatically adapting the behavior of an autonomous table soccer robot to a human opponent. Experiments show that it is possible to recognize basic actions in a robust way, that the opponent can reliably be classified with respect to different levels of playing skills, and that the robot adapts its behavior successfully to the playing level of the opponent player.

In the future, we plan to adapt not only the move and shoot velocities but also respond with alternative actions to the opponent's particular playing characteristics. For an even better understanding of the opponent behavior, additionally its playing style could be assessed and general observations about the course of the game, e.g. the ball's distribution over the field could be taken into account. In particular, this could be used to adapt the distribution over mixed strategies. In certain situations mixed (randomized) strategies are the best ones, e.g. when trying to score a goal. If one now observes that the opponent does not play a mixed strategy, but a pure one, one can adapt to it and be more successful. Clearly, such advanced adaptation techniques need the building blocks layed in this paper.

References

1. Sutton, R., Barto, A., eds.: Reinforcement Learning: an Introduction. MIT-Press, Cambridge, Massachusetts (1998)
2. Weigel, T., Nebel, B.: KiRo – An Autonomous Table Soccer Player. [18] 119–127
3. Weigel, T.: KiRo – A Table Soccer Robot Ready for Market. In: Proceedings of the IEEE International Conference on Robotics and Automation (ICRA). (2005) 4277–4282
4. Weigel, T., Zhang, D., Rechert, K., Nebel, B.: Adaptive Vision for Playing Table Soccer. In: Proceedings of the 27th German Conference on Artificial Intelligence, Ulm, Germany (2004) 424–438
5. Kautz, H., Allen, J.: Generalized Plan Recognition. In: Proceedings of the Fifth National Conference on Artificial Intelligence (AAAI), AAAI Press (1986) 32–37
6. Charniak, E., Goldman, R.: A Bayesian Model of Plan Recognition. Artificial Intelligence **64** (1993) 53–79
7. Lesh, N., Rich, C., Sidner, C.: Using Plan Recognition in Human-Computer Collaboration. In: Proceedings of the Seventh International Conference on User Modeling. (1999) 23–32
8. Raines, T., Tambe, M., Marsella, S.: Automated Assistants for Analyzing Team Behaviors. In Veloso, M., Pagello, E., Kitano, H., eds.: RoboCup-99: Robot Soccer World Cup III. Lecture Notes in Artificial Intelligence. Springer-Verlag, Berlin, Heidelberg, New York (2000) 85–102
9. André, E., Binsted, K., Tanaka-Ishii, K., Luke, S., Herzog, H., Rist, T.: Three RoboCup Simulation League Commentator Systems. AI Magazine **21** (2000) 57–66
10. Kitano, H., Asada, M., Kuniyoshi, Y., Noda, I., Osawa, E., Matsubara, H.: RoboCup: A Challenge Problem for AI. AI Magazine **18** (1997) 73–85
11. Kaminka, G., Fidanboylu, M., Chang, A., Veloso, M.: Learning the Sequential Coordinated Behavior of Teams from Observations. [18] 111–125
12. Visser, U., Weland, H.: Using Online Learning to Analyze the Opponent's Behavior. [18] 78–93
13. Wendler, J., Bach, J.: Recognizing and Predicting Agent Behavior with Case Based Reasoning. [19] 729–738
14. Steffens, T.: Feature Based Declarative Opponent Modeling. [19] 125–136
15. Riley, P., Veloso, M.: On Behavior Classification in Adversarial Environments. In Parker, L., Bekey, G., Barhen, J., eds.: Distributed Autonomous Robotic Systems 4. Springer-Verlag, Berlin, Heidelberg, New York (2000) 371–380

16. Riley, P., Veloso, M.: Planning for Distributed Execution through Use of Probabilistic Opponent Models. In: Proceedings of the Sixth International Conference on Artificial Intelligence Planning Systems (AIPS). (2002) 72–81
17. Han, K., Veloso, M.: Automated Robot Behavior Recognition Applied to Robotic Soccer. In Hollerbach, J., Koditschek, D., eds.: Robotics Research: The Ninth International Symposium. Springer-Verlag, London (2000) 199–204
18. Kaminka, G., Lima, P., Rojas, R., eds.: RoboCup-2002: Robot Soccer World Cup VI. Lecture Notes in Artificial Intelligence. Springer-Verlag, Berlin, Heidelberg, New York (2003)
19. Polani, D., Browning, B., Bonarini, A., eds.: RoboCup-2003: Robot Soccer World Cup VII. Lecture Notes in Artificial Intelligence. Springer-Verlag, Berlin, Heidelberg, New York (2004)

Selecting What Is Important: Training Visual Attention

Simone Frintrop[1], Gerriet Backer[2], and Erich Rome[1]

[1] Fraunhofer Institut für Autonome Intelligente Systeme (AIS),
Schloss Birlinghoven, 53754 Sankt Augustin, Germany
[2] Krauss Software GmbH, Tiefe Str. 1, 38162 Cremlingen, Germany

Abstract. We present a new, sophisticated algorithm to select suitable training images for our biologically motivated attention system VOCUS. The system detects regions of interest depending on bottom-up (scene-dependent) and top-down (target-specific) cues. The top-down cues are learned by VOCUS from one or several training images. We show that our algorithm chooses a subset of the training set that outperforms both the selection of one single image as well as simply using all available images for learning. With this algorithm, VOCUS is able to quickly and robustly detect targets in numerous real-world scenes.

1   Introduction and State of the Art

Human visual perception is based on a separation of object recognition into two subtasks [11]: first, a fast parallel pre-selection of scene regions detects object candidates and second, complex recognition restricted to these regions verifies or falsifies the hypothesis. This dichotomy of fast localization processes and complex, robust, but slow identification processes is highly effective: expensive resources are guided towards the most promising and relevant candidates.

In computer vision, the efficient use of resources is equally important. Although an attention system generates a certain overhead in computation, it pays off since reliable object recognition is a complex vision task that is usually computationally expensive. The more general the recognizer – for different shapes, poses, scales, and illuminations – the more important is a pre-selection of regions of interest.

Concerning visual attention, research has so far been focused on localizing the relevant scene parts by evaluating bottom-up, data-driven saliency, featuring research from a psychological [16,19], neuro-biological [2,12] and computational [8,7,1,14] point of view. Koch & Ullman [8] described the first explicit computational architecture for bottom-up visual attention. It is strongly influenced by Treisman's *feature-integration theory* [16] and already contains the main properties of many current visual attention systems, e.g., the one by Itti et al. [7] or [1,14]. These systems use classical linear filter operations for feature extraction, rendering them especially useful for real-world scenes. Another approach is provided by models consisting of a pyramidal neural processing architecture, e.g., the *selective tuning model* by Tsotsos et al. [18].

U. Furbach (Ed.): KI 2005, LNAI 3698, pp. 351–365, 2005.

While much less analyzed, there is strong evidence for top-down influences modifying early processing of visual features due to motivations, emotions, and goals [2]. Only a few computer models of visual attention integrate top-down information into their systems. The earliest approach is the *guided search* model by Wolfe [19], a result of his psychological investigations of human visual search performance. Tsotsos's strongly biologically motivated system considers feature channels separately and uses inhibition for regions of a specified location or those that do not fit the target features [18]. Schill et al. [13] use top-down information from a knowledge-base to select actual fixations from the fixation candidates determined by a bottom-up system. The computation of the bottom-up features is not influenced by the top-down information. Other systems enabling goal-directed search are presented by Hamker [5] and by Navalpakkam et al. [10]; however, Hamker's system does not consider the surroundings of the targets and both do not separate enhancing and inhibiting cues as well as bottom-up and top-down cues. We are not aware of any other well investigated system of top-down visual attention comparable to our approach including the automated learning of features and considering features of both the target and the surroundings.

Our attention system VOCUS was introduced in [4,3], and applied to object detection and recognition in [9,3]. VOCUS performs goal-directed search by extending a well-known bottom-up attention architecture [7] by a top-down part. The bottom-up part computes saliencies in the feature dimensions intensity, orientation, and color independently, weights maps according to the exclusivity of the feature, and finally fuses the saliencies into a single map. The top-down part uses in a search phase learned feature weights to determine which features to enhance and which ones to inhibit. The weighted features contribute to a top-down saliency map highlighting regions with target-relevant features. This map is integrated with the bottom-up map, resulting in a global saliency map and the focus of attention is directed to its most salient region.

The focus of this paper is the learning phase, where relevant feature values for a target are learned using one or several training images. Here, the system automatically determines which features to regard in order to separate the target best from its surroundings.

Learning weights from one single training image yields good results when the target object occurs in all test images in a similar way, i.e., the background color is similar and the object occurs always in a similar orientation. But when the targets occur on different backgrounds and are rendered from different viewpoints, several training images have to be considered for learning. This is important for us as we intend to integrate the system into a robot control architecture enabling the detection of salient regions and goal-directed search in dynamic environments.

So we devised a new, sophisticated algorithm that chooses suitable training images from a training set. It is shown that training on the chosen subset yields much better results than training on a single or on more training images: the subset is a local optimum. Search results for various real-world scenes are presented, showing that VOCUS is robust and applicable to natural scenes.

In section 2, we start by explaining the attention system VOCUS. Section 3 extends the learning of target-specific weights to several training images and introduces our new algorithm that chooses the most suitable images from a training set. In section 4, we present numerous results on real-world images before we conclude in section 5.

2   The Attention System VOCUS

In this section, we present the goal-directed visual attention system VOCUS (Visual Object detection with a CompUtational attention System) (cf. Fig. 1). With visual attention we mean a selective search-optimization mechanism that tunes the visual processing machinery to approach an optimal configuration [17]. VOCUS consists of a bottom-up part computing data-driven saliency and a top-down part enabling goal-directed search. The global saliency is determined from bottom-up and top-down cues. More details to VOCUS are found in [3].

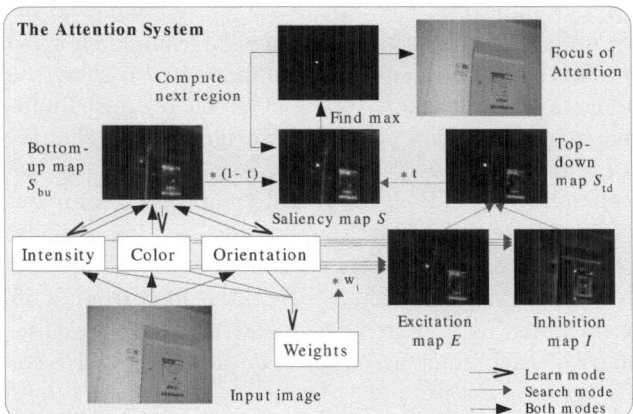

Feature	weights
intensity on/off	0.001
intensity off/on	9.616
orientation 0°	4.839
orientation 45°	9.226
orientation 90°	2.986
orientation 135°	8.374
color green	76.572
color blue	4.709
color red	0.009
color yellow	0.040
conspicuity I	6.038
conspicuity O	5.350
conspicuity C	12.312

Fig. 1. The goal-directed visual attention system with a bottom-up part (left) and a top-down part (right). In learning mode, target weights are learned (blue arrows). These are used in search mode (red arrows). Right: weights for target name plate.

2.1   Bottom-Up Saliency

Feature Computations: The first step for computing bottom-up saliency is to generate image pyramids for each feature to enable computations on different scales. Three features are considered: Intensity, orientation, and color. For the feature intensity, we convert the input image into gray-scale and generate a Gaussian pyramid with 5 scales s_0 to s_4 by successively low-pass filtering and subsampling the input image, i.e., $s_{(i+1)}$ has half the width and height of s_i.

The intensity maps are created by center-surround mechanisms, which compute the intensity differences between image regions and their surroundings. We compute two kinds of maps, the on-center maps I''_{on} for bright regions on dark background, and the off-center maps I''_{off}: Each pixel in these maps is computed by the difference between a center c and a surround σ (I''_{on}) or vice versa (I''_{off}). Here, c is a pixel in one of the scales s_2 to s_4, σ is the average of the surrounding pixels for two different radii. This yields 12 intensity scale maps $I''_{i,s,\sigma}$ with $i \in \{\text{on, off}\}, s \in \{s_2\text{-}s_4\}$, and $\sigma \in \{3,7\}$. The maps for each i are summed up by *inter-scale addition* $\bigoplus$, i.e., all maps are resized to scale 2 and then added up pixel by pixel yielding the intensity feature maps $I'_i = \bigoplus_{s,\sigma} I''_{i,s,\sigma}$.

To obtain the orientation maps, four oriented Gabor pyramids are created, detecting bar-like features of the orientations $\theta = \{0\,°, 45\,°, 90\,°, 135\,°\}$. The maps 2 to 4 of each pyramid are summed up by inter-scale addition yielding 4 orientation feature maps O'_θ.

To compute the color feature maps, the color image is first converted into the uniform CIE LAB color space [6]. It represents colors similar to human perception. The three parameters in the model represent the luminance of the color (L), its position between red and green (A) and its position between yellow and blue (B). From the LAB image, a color image pyramid P_{LAB} is generated, from which four color pyramids P_R, P_G, P_B, and P_Y are computed for the colors red, green, blue, and yellow. The maps of these pyramids show to which degree a color is represented in an image, i.e., the maps in P_R show the brightest values at red regions and the darkest values at green regions. Luminance is already considered in the intensity maps, so we ignore this channel here. The pixel value $P_{R,s}(x,y)$ in map s of pyramid P_R is obtained by the distance between the corresponding pixel $P_{\text{LAB}}(x,y)$ and the prototype for red $r = (r_a, r_b) = (255, 127)$. Since $P_{\text{LAB}}(x,y)$ is of the form (p_a, p_b), this yields: $P_{R,s}(x,y) = ||(p_a, p_b), (r_a, r_b)|| = \sqrt{(p_a - r_a)^2 + (p_b - r_b)^2}$.
On these pyramids, the color contrast is computed by on-center-off-surround differences yielding 24 color scale maps $C''_{\gamma,s,\sigma}$ with $\gamma \in \{\text{red, green, blue, yellow}\}, s \in \{s_2\text{-}s_4\}$, and $\sigma \in \{3,7\}$. The maps of each color are inter-scale added into 4 color feature maps $C'_\gamma = \bigoplus_{s,\sigma} \hat{C}_{\gamma,s,\sigma}$.

Fusing Saliencies: All feature maps of one feature are combined into a conspicuity map yielding one map for each feature: $I = \sum_i \mathcal{W}(I'_i)$, $O = \sum_\theta \mathcal{W}(O'_\theta)$, $C = \sum_\gamma \mathcal{W}(C'_\gamma)$. The bottom-up saliency map S_{bu} is finally determined by fusing the conspicuity maps: $S_{bu} = \mathcal{W}(I) + \mathcal{W}(O) + \mathcal{W}(C)$

The exclusivity weighting $\mathcal{W}$ is a very important strategy since it enables the increase of the impact of relevant maps. Otherwise, a region peaking out in a single feature would be lost in the bulk of maps and no pop-out would be possible. In our context, important maps are those that have few highly salient peaks. For weighting maps according to the number of peaks, each map X is divided by the square root of the number of local maxima m that exceed a threshold t: $\mathcal{W}(X) = X/\sqrt{m} \quad \forall m : m > t$. Furthermore, the maps are normalized after summation relative to the largest value within the summed maps. This yields advantages over the normalization relative to a fixed value (details in [3]).

The Focus of Attention (FOA): To determine the most salient location in S_{bu}, the point of maximal activation is located. Starting from this point, region growing recursively finds all neighbors with similar values within a threshold and the FOA is directed to this region. Finally, the salient region is inhibited in the saliency map by zeroing, enabling the computation of the next FOA.

2.2   Top-Down Saliency

Learning Mode: In learning mode, VOCUS is provided with a training image and coordinates of a rectangle depicting the *region of interest (ROI)* that includes the target. Then, the system computes the bottom-up saliency map and the *most salient region (MSR)* inside the ROI. So, VOCUS is able to decide autonomously what is important in a ROI, concentrating on parts that are most salient and disregarding the background or less salient parts.

Next, weights are determined for the feature and conspicuity maps, indicating how important a feature is for the target (blue arrows in Fig. 1). The weight w_i for map i is the ratio of the mean saliency in the target region $m_{(MSR)}$ and in the background $m_{(image-MSR)}$: $w_i = m_{(MSR)}/m_{(image-MSR)}$. This computation does not only consider which features are the strongest in the target region, it also regards which features separate the region best from the rest of the image. In section 3, we show how the approach is extended to several training images.

Search Mode: In search mode, firstly the bottom-up saliency map is computed. Additionally, we determine a top-down saliency map that competes with the bottom-up map for saliency (red arrows in Fig. 1). The top-down map is composed of an excitation and an inhibition map. The excitation map E is the weighted sum of all feature and conspicuity maps X_i that are important for the learned region, i.e., $w_i > 1$. The inhibition map I shows the features more present in the background than in the target region, i.e., $w_i < 1$:

$$\begin{aligned} E &= \sum_i (w_i * X_i) & \forall i : w_i > 1 \\ I &= \sum_i ((1/w_i) * X_i) & \forall i : w_i < 1 \end{aligned} \tag{1}$$

The top-down saliency map $S_{(td)}$ is obtained by: $S_{(td)} = E - I$. The final saliency map S is composed as a combination of bottom-up and top-down influences. When fusing the maps, it is possible to determine the degree to which each map contributes by weighting the maps with a top-down factor $t \in [0..1]$: $S = (1-t) * S_{(bu)} + t * S_{(td)}$.

With $t = 1$, VOCUS looks only for the specified target. With $t < 1$, also bottom-up cues have an influence and may divert the focus of attention. This is also an important mechanism in human visual attention: a person suddenly entering a room or a deer jumping on the road catch immediately our attention, independently of the task. In humans this *attentional capture* cannot be overridden by top-down search strategies [15]; i.e., for a severely biologically motivated system a top-down factor of 1 should not be allowed. Nevertheless, for a technical system that usually has to solve only one clearly defined task at a time, also a top-down factor of 1 is often useful.

The success of the search is evaluated by the rank of the focus that hits the target, denoted by the **hit number**. For example, if the 2nd focus is on the target, the hit number is 2. The lower the hit number, the better the search performance. If the hit number is 1, the target is immediately detected. In pop-out experiments, the hit number is 1 by definition. If a whole image set is evaluated, we determine the **average hit number**, i.e., the arithmetic mean of the hit numbers of all images. Usually, only a determinate number of fixations is considered so that images with undetected targets are not included in the average. To indicate this, we show in our experimental results the percentage of detected targets (**detection rate**) additionally.

3   Using Several Training Images

Learning weights from one single training image yields good results when the target object occurs in all test images in a similar way, i.e., the background color is similar and the object occurs in nearly identical orientations. These conditions often occur if the objects are fixed elements of the environment. For example, name plates or fire extinguishers usually are placed on the same kind of wall, so the background has always a similar color and intensity. Furthermore, since the object is fixed, its orientation does not vary and it is sensible to learn that fire extinguishers usually have a vertical orientation.

Although the search is already quite successful with weights from a single training image, the results differ somewhat depending on the choice of the training image. This is shown in Tab. 1: a highlighter was searched in a test set of 60 images using the weights from a single training image. The table shows the different results for several training images; the detection rate differs between 95 and 100%.

Table 1. The search for a highligher with the weights w_1 to w_5 learned from 5 training images applied to a test set of 60 images (examples of training and test images in Fig. 3 and 4). The first 10 foci were determined. The performance is shown as the average hit number and the percentage of targets detected within the first 10 foci in parentheses. The performance differs slightly depending on the training image.

Target	# test im	average hit number (and detection rate [%])				
		w_1	w_2	w_3	w_4	w_5
Highlighter	60	1.83 (99%)	1.70 (97%)	1.43 (100%)	1.93 (95%)	1.78 (97%)

Furthermore, the results usually differ slightly depending on the test set the weights are applied to. One training image might fit better to a special image set than to another. To weed out these special cases, it is sensible to take the average weight of at least two training images to enable a more stable performance on arbitrary test sets. For movable objects it is even more important to compute average weights. A highlighter may lie on a dark or on a bright desk and it may have any orientation. Here, it is necessary to learn from several training images which features are stable and which are not.

Table 2. Left: four training examples to learn red bars of horizontal and vertical orientation and on different backgrounds. The target is marked by the yellow rectangle. Right: The learned weights. Column 2–5: the weights for a single training image (vertical bar on bright background (v,b), horizontal on bright (h,b), vertical on dark (v,d), horizontal on dark (h,d)). The highest values are highlighted in bold face. Column 6: average weights. Color is the only stable feature.

Feature	weights for red bar				
	v,b	h,b	v,d	h,d	average
int on/off	0.00	0.01	**8.34**	**9.71**	0.14
int off/on	**14.08**	**10.56**	0.01	0.04	0.42
ori 0°	1.53	**21.43**	0.49	**10.52**	3.61
ori 45°	2.66	1.89	1.99	2.10	2.14
ori 90°	6.62	0.36	5.82	0.32	1.45
ori 135°	2.66	1.89	1.99	2.10	2.14
col green	0.00	0.00	0.00	0.00	0.00
col blue	0.00	0.00	0.01	0.01	0.00
col red	**18.87**	**17.01**	**24.13**	**24.56**	**20.88**
col yellow	16.95	14.87	21.21	21.66	18.45
consp I	7.45	5.56	3.93	4.59	5.23
consp O	4.34	7.99	2.87	5.25	4.78
consp C	4.58	4.08	5.74	5.84	5.00

3.1 Average Weights

To achieve a robust target detection even in changing environments, it is necessary to learn the target properties from several training images. This is done by computing the average weight vector from n training images with the *geometric mean* of the weights for each feature, i.e., the average weight vector $w_{(1,...,n)}$ from n training images is determined by:

$$w_{(1,..,n)} = \sqrt[n]{\prod_{j=1}^{n} w_j}. \tag{2}$$

The geometric mean is more suitable than an arithmetic mean, because the weight values represent relations, so that values like 2 and 0.5 should cancel out each other. If one feature is present in some training images but absent in others, the average values will be close to 1 leading to only a low activation in the top-down map. In Tab. 2 this is shown on the example of searching for red bars: the target occurs in horizontal or vertical orientations and on a dark or bright background; the only stable feature is the red color. This is reflected in the rightmost column of the table which shows the average weights: the weights for intensity and orientation feature maps are almost equal, only weights for the color feature maps show high values. This enables the search for red bars, regardless of the background and the orientation.

3.2   The Algorithm to Choose the Training Images

In the previous example (Tab. 2), four training images were chosen that were claimed to represent the test data. In practice, the problem is: how do we find suitable training images? We could think about the test application and reason about suitable training data or we could just use a bunch of training images that cover many possible contexts presenting the target at different orientations and on different backgrounds. However, this does not guarantee a good training set and, moreover, it depends heavily on the user's experiences and skills. In this section, we introduce an algorithm that chooses the most suitable images from a training set.

Let us first think about how an optimal weight vector could be achieved. Since the average weights do not always improve when more training images are considered, the best performance is usually not achieved by considering all images of a training set T_1. The reason is that training on too similar images results in *overfitting*, e.g., generating too specialized weights. Instead, there exists a subset of T_1, the average weights of which yield the best performance on another image set T_2. The only possibility to find this subset is to test all possible combinations, an effort costing to check 2^n combinations for n training images. Since these computations are too costly even for rather small n, we propose an approximation algorithm that yields a local optimum in performance. Before we introduce the algorithm, we first give a definition:

Definition 1 (Self-test, self-test hit number). *A self-test on image I for a target t means: first, learn the weights* w *for t from image I. Second, apply* w *to I itself. The resulting hit number is the* **self-test hit number.**

The self-test hit number is a good base for comparisons, since the weights of an image itself yield a good chance to discriminate the target from its surrounding. A self-test hit number of 1 indicates that the weights are sufficient to detect the target in similar environments. A larger self-test hit number indicates that there are distractors in the scene that are very similar to the target and that the features of the system are incapable to distinguish between target and distractors. This test is not suitable for deciding whether a training image is useful or not, because if there are distractors in a scene which are too similar according to the given features, there is nothing we can do about it. It might be useful to train on such scenes anyway, because the extracted weights are the best possible solution for these kinds of scenes. Note that a hit number of 2, 3, or even 10 is often still useful since the regions to be investigated by an object classifier are still considerably reduced.

The overall idea of the approximation training algorithm is to first choose one arbitrary image I_1 from the training image set. Then, the weights from I_1 are applied to the whole training set T and the image I_2 is determined on which the hit number is worst. A bad hit number might mean that I_1 was not suitable for this image. Whether this assumption is true can be checked by comparing the hit number with the self-test hit number of I_2. If the latter is better, the assumption was true: I_1 was unsuitable. In this case, I_2 is a good choice to improve the

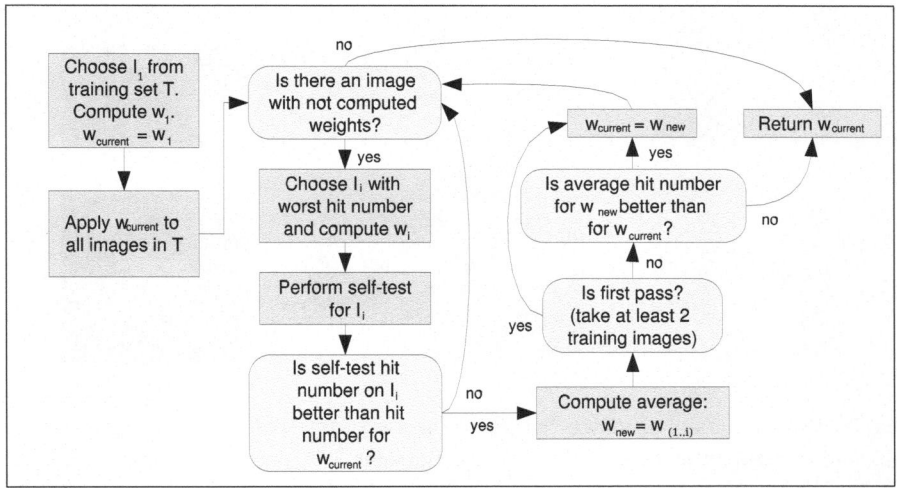

Fig. 2. The algorithm to find the most suitable training images out of an image set and to compute their average weight vector. With this vector, a local optimum in detection quality on the training set is achieved.

weights thus the average weights of I_1 and I_2 are determined. This procedure is continued as long as the average hit rate on the training set improves. A flowchart of the algorithm is shown in Fig. 2.

4   Results

In this section, we show the hit numbers and detection rates of VOCUS when searching for several targets in various real-world scenes. We compare the performance for one and for several training images which were chosen with the presented algorithm. As targets, we used four kinds of objects: two objects which are fixed in our office environment (fire extinguishers and name plates) and two movable objects (a key fob and a highlighter). The highlighter was presented on two different desks, a dark (black) and a bright (wooden) one. For each target, we used a training set of 10 to 54 images and chose suitable training images from the set with the training algorithm in Fig. 2. In Fig. 3, we depict one of the training images for each target as well as the most salient region that VOCUS extracted for learning.

Two experiments demonstrate the power of our algorithm: first, we examine the search performance on four test sets using different numbers of training images. Second, we test VOCUS on a search task in which the environment of the target differs and show that in this case more training images are required.

Experiment 1. Our first experiment examines the effect of different image set sizes on the search quality. We show the search results with the computed vectors first on the training set itself (training phase) and then on a test set (test phase).

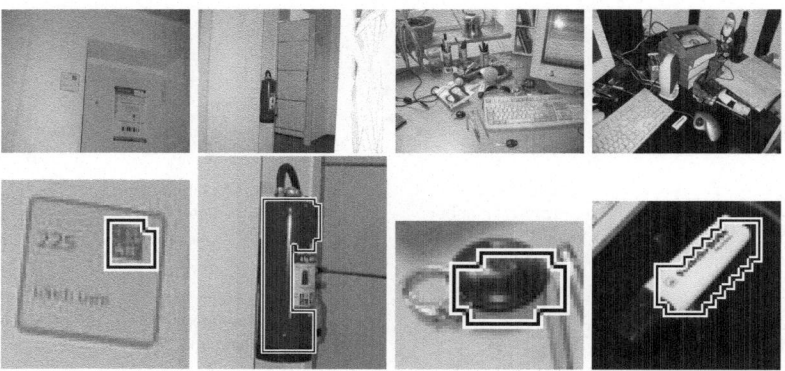

Fig. 3. Top: the training images with the targets (name plate, fire extinguisher, key fob, and a highlighter on the dark desk). Bottom: The part of the image that was marked for learning (region of interest (ROI)) and the contour of the region that was extracted for learning (most salient region (MSR)).

The image sets used for this experiment consisted of images with similar backgrounds for each set (white walls behind the fire extinguisher and the name plate and same desk for the key fob and the highlighter). The images were nevertheless highly complex and include a heavily structured surrounding with many distracting regions.

To compute the weight vectors, we chose the most suitable training images from the training set using the algorithm of Fig. 2. Remember that the algorithm chooses the first image at random and then the images with the worst detection results on the training set; it stops when a local optimum in performance is reached. We document this by presenting the detection results (average hit number and detection rate) for each weight vector that is computed during the iterations of the algorithm (Tab. 3). This corresponds to visualizing the intermediate steps of the algorithm.

It turned out that averaging two training images in most cases outperformed the use of a single image. This is most obvious for the name plate: the detection rate within the first 10 foci increased from 87% to 94% (the detection rate is more important to evaluate the performance than the average focus because a single image that is additionally detected increases the detection rate slightly but decreases the average hit number. That means, a performance of average hit number 2.04 and detection rate of 94% is better than a performance of average hit number 1.97 and detection rate of 93%). Only for the highlighter on the dark desk the detection remains the same. If a single training image yields an equal or better performance than the average from the first two images, we still recommend to use the average because of the risk that the weights from a single training do not generalize well enough on the test set. For three images, the performance does not improve any more, on the contrary, the results are worse for some of the examples (name plate, highlighter bright). Therefore, the algorithm stops with the weights vectors $w_{(1,2)}$ as local optima.

Table 3. Experiment 1: Search performance on training sets with weight vectors from 1, 2, and 3 training images obtained with the algorithm in Fig. 2. The performance is presented as the average hit number on the training set and, in parentheses, the percentage of detected targets within the first 10 foci. The best value is highlighted in bold face. Already two training images yield the local optimum in performance and the algorithm stops. The application of these values to a test data set is presented in Tab. 4.

Target	# train	av. hit number (and detection rate [%])		
	im.	w_1	$w_{(1,2)}$	$w_{(1,...,3)}$
Fire extinguisher	10	1.10 (100%)	**1.00 (100%)**	1.00 (100%)
Key fob	10	1.33 (100%)	**1.00 (100%)**	1.00 (100%)
Name plate	54	1.61 (87%)	**2.04 (94%)**	1.97 (93%)
Highlighter (dark desk)	10	1.50 (100%)	**1.50 (100%)**	1.50 (100%)
Highlighter (bright desk)	10	3.40 (100%)	**2.10 (100%)**	2.40 (100%)

Table 4. Experiment 1: Search performance on test sets with the weight vectors of Tab. 3. The numbers in bold face denote the best performance. Note that the best performance is not always reached at the point that the training proposes: for the highlighter, the best performance is achieved on the dark desk with a single training image and on the bright desk with three training images.

Target	# train	# test	av. hit number (and detection rate [%])		
	im.	im.	w_1	$w_{(1,2)}$	$w_{(1,...,3)}$
Fire extinguisher	10	46	1.14 (100%)	**1.09 (100%)**	1.09 (100%)
Key fob	10	30	1.40 (100%)	**1.23 (100%)**	1.40 (100%)
Name plate	54	238	2.31 (80%)	**2.55 (87%)**	2.28 (86%)
Highlighter (dark desk)	10	30	**1.30 (100%)**	1.37 (100%)	1.37 (100%)
Highlighter (bright desk)	10	30	2.43 (100%)	1.97 (97%)	**2.13 (100%)**

In a second step, we apply these weight vectors to test image sets which were disjoint from the training data (Tab. 4). This shows how the system generalizes on unknown data. As expected the performance is in most cases slightly worse than on the training set, since the weights were chosen to fit the training set. However the detection quality is still very high: fire extinguisher, key fob, and the highlighter on the dark desk are detected in all images (detection rate 100%) and the highlighter on the bright desk is missed only in 3% of the images ($w_{(1,2)}$). In the successful cases, the target is detected in average with the 1st or 2nd focus. The most difficult example is the name plate; here, the target is missed in 13% of the images.

It also revealed that the best performance is not always achieved with the same weights as on the training data: For the highlighter on the dark desk, the best performance is already achieved with the first training image and for the highlighter on the bright desk three training images yield the best performance. This is inevitable since every test set is slightly different and has another com-

Target: name plate

Target: fire extinguisher

Target: highlighter

Target: key fob

Fig. 4. Some of the results from searching the targets of Fig. 3. The FOAs are depicted by red ellipses. After the target was focused, the search was canceled so the number of depicted foci is equal to the number of required fixations. The hardest example is the one in the upper right corner: the poster shows colors similar to the logo of the name plate and diverts the focus so the target is only detected by the 6th focus. In all other depicted examples the target is found with the first or second focus.

bination of weights that fits best for it. Nevertheless, the performance results differ only slightly and the proposed approach yields a good approximation of the optimal performance. Fig. 4 shows some of the test images with some foci of attention.

Table 5. Experiment 2: Search performance on a training set (20 images) with different backgrounds: the highlighter (target) lay on a dark and on a bright desk. The weight vectors are obtained with the algorithm in Fig. 2. The performance is presented as the average hit number on the training set and, in parentheses, the percentage of detected targets within the first 10 foci. The best value is highlighted in bold face. The performance depends on the start image: if the start image is from a bright desk (b), the local optimum is reached for 2 training images. If it is from a dark desk (d), 4 images yield the best performance. The application of these values to a test data set is presented in Tab. 6.

Target	start im.	$w_{i,1}$	$w_{i,(1,2)}$	$w_{i,(1,...,3)}$	$w_{i,(1,...,4)}$	$w_{i,(1,...,5)}$
Highlighter	b	2.45 (100%)	**1.70 (100%)**	1.85 (100%)		
Highlighter	d	2.50 (95%)	1.95 (100%)	1.75 (100%)	**1.55 (100%)**	1.75 (100%)

Table 6. Experiment 2: Search performance on a test set of 60 images with dark and bright backgrounds obtained with the weight vectors of Tab. 5. The numbers in bold face denote the best performance. Note that the best performance is not always reached at the point that the training proposes: for the bright starting image, the best performance is achieved with three training images instead of two.

Target	start im.	$w_{i,1}$	$w_{i,(1,2)}$	$w_{i,(1,...,3)}$	$w_{i,(1,...,4)}$	$w_{i,(1,...,5)}$
Highlighter	b	1.80 (100%)	1.53 (99%)	**1.62 (100%)**		
Highlighter	d	1.83 (99%)	1.58 (100%)	1.55 (100%)	**1.48 (100%)**	1.60 (100%)

Experiment 2. In the previous experiment, the background within each image set was similar. Here we show what happens if a target appears on different backgrounds. To achieve this, we combined the image sets of the highlighter on the dark and on the bright desk into one image set. We expected that here more training images are required to yield a local optimum in performance since the training set is inhomogeneous. It turned out that this is usually true but even here the local optimum is sometimes achieved with two images (cf. Tab. 5). We found that it depends on the starting image how many training images are required until the algorithm stops: when the first image was from the bright desk only two images were needed to yield the local optimum. When it was from the dark desk it took longer until the optimum was reached: the best performance was achieved with the average of four training images. The performance was then better than the performance achieved with two images with a bright-desk starting image.

This results from the fact that the search on the dark desk is considerably easier than on the bright one because of the high contrast of the yellow highlighter to the dark desk. Therefore, weights obtained from the bright desk applied to the dark one yield a good performance but not vice versa. If the starting image is from the bright desk, the images that perform badly are also from the bright desk. After taking the average of two training images, the performance does not improve any more. In contrast, if the starting image is from the dark desk, the bad-performing images are from the bright set and $w_{(1,2)}$ is the average of dark and bright. This is repeated and the average of dark and bright yields a performance that excels the former performance after 4 iterations.

In Tab. 6, the computed weights are applied to a test set of 60 images disjoint from the training set. The detection quality is very high: although the target lay on different backgrounds, it is found in all images and in average with the 1st or 2nd focus. Again, it showed that the optimal performance is not always reached for the same weights as the optimal performance on the training set but the training results yield a good approximation of the optimum. Interestingly, with both kinds of weight vectors (bright and dark starting image) the performance on the test set is better than on the training set. Probably this results from a few difficult example images in the training set which decline the average hit number. Note that despite the different results depending on the start image it is not necessary to attach great importance to the choice of this image since the average hit numbers for both cases are very similar. A randomly chosen image will usually suffice.

5   Conclusion

We have introduced a new algorithm that chooses suitable training images for our attention system VOCUS which detects salient regions in images depending on bottom-up and top-down cues. VOCUS provides a method to quickly localizing object candidates to which computationally expensive recognition algorithms may be applied. Thereby, it provides a basis for robust and fast object recognition in computer vision and robotics.

The presented selection algorithm choses the most suitable training images out of an image set. This selection improved the search performance as shown by experiments involving numerous real-world images. We have shown that the chosen image subset is a local optimum and outperforms the use of a single or all training images. In our experiments, less than five training images were sufficient to yield the local optimum. With this approach, VOCUS detects targets robustly and quickly in various scenes: the target was in average among the first three selected regions.

In future work, we plan to utilize the system for robot control. The attention system will determine salient regions in the robot's environment and search for previously learned objects. Directing the attention to regions of potential interest will be the basic assumption for object detection and manipulation.

Acknowledgements. The authors wish to thank Joachim Hertzberg for supporting our work.

References

1. Backer, G., Mertsching, B. and Bollmann, M. Data- and model-driven Gaze Control for an Active-Vision System. *IEEE Trans. on PAMI* **23(12)** (2001) 1415–1429.
2. Corbetta, M. and Shulman, G. L. Control of goal-directed and stimulus-driven attention in the brain. *Nature Reviews* **3** (3, 2002) 201–215.
3. Frintrop, S. VOCUS: A Visual Attention System for Object Detection and Goal-directed Search. PhD thesis University of Bonn Germany (to appear 2005).
4. Frintrop, S., Backer, G. and Rome, E. Goal-directed Search with a Top-down Modulated Computational Attention System. In: Proc. of DAGM 2005 (accepted) Lecture Notes in Computer Science (LNCS) Springer (2005).
5. Hamker, F. Modeling Attention: From computational neuroscience to computer vision. In: Proc. of WAPCV'04 (2004) 59–66.
6. Hunt, R. W. G. *Measuring colour* Ellis Horwood Limited Chichester, West Sussex, England 1991.
7. Itti, L., Koch, C. and Niebur, E. A Model of Saliency-Based Visual Attention for Rapid Scene Analysis. *IEEE Trans. on PAMI* **20** (11, 1998) 1254–1259.
8. Koch, C. and Ullman, S. Shifts in selective visual attention: towards the underlying neural circuitry. *Human Neurobiology* **4** (4, 1985) 219–227.
9. Mitri, S., Frintrop, S., Pervölz, K., Surmann, H. and Nüchter, A. Robust Object Detection at Regions of Interest with an Application in Ball Recognition. In: Proc. of the Int'l Conf. on Robotics and Automation (ICRA '05) (to appear 2005).
10. Navalpakkam, V., Rebesco, J. and Itti, L. Modeling the influence of task on attention. *Vision Research* **45** (2, 2005) 205–231.
11. Neisser, U. *Cognitive Psychology* Appleton-Century-Crofts New York 1967.
12. Palmer, S. E. *Vision Science, Photons to Phenomenology* The MIT Press 1999.
13. Schill, K., Umkehrer, E., Beinlich, S., Krieger, G. and Zetzsche, C. Scene analysis with saccadic eye movements: Top-down and bottom-up modeling. *Journal of Electronic Imaging* **10** (1, 2001) 152–160.
14. Sun, Y. and Fisher, R. Object-based visual attention for computer vision. *Artificial Intelligence* **146** (1, 2003) 77–123.
15. Theeuwes, J. Top-down search strategies cannot override attentional capture. *Psychonomic Bulletin & Review* **11** (2004) 65–70.
16. Treisman, A. M. and Gelade, G. A feature integration theory of attention. *Cognitive Psychology* **12** (1980) 97–136.
17. Tsotsos, J. K. Complexity, Vision, and Attention. In: Vision and Attention, M. Jenkin and L. R. Harris (Eds.) Springer Verlag 2001 chapter 6.
18. Tsotsos, J. K., Culhane, S. M., Wai, W. Y. K., Lai, Y., Davis, N. and Nuflo, F. Modeling Visual Attention via Selective Tuning. *AI* **78** (1-2, 1995) 507–545.
19. Wolfe, J. Guided Search 2.0: A Revised Model of Visual Search. *Psychonomic Bulletin & Review* **1** (2, 1994) 202–238.

Self-sustained Thought Processes in a Dense Associative Network

Claudius Gros*

Institut für Theoretische Physik, Universität Frankfurt,
Max-von-Laue-Strasse 1, 60438 Frankfurt am Main, Germany

Abstract. Several guiding principles for thought processes are proposed
and a neural-network-type model implementing these principles is presented and studied. We suggest to consider thinking within an associative
network built-up of overlapping memory states. We consider a homogeneous associative network as biological considerations rule out distinct
conjunction units between the information (the memories) stored in the
brain. We therefore propose that memory states have a dual functionality: They represent on one side the stored information and serve, on the
other side, as the associative links in between the different dynamical
states of the network which consists of transient attractors.

We implement these principles within a generalized winners-take-all
neural network with sparse coding and an additional coupling to local reservoirs. We show that this network is capable to generate autonomously a self-sustained time-series of memory states which we identify with a thought process. Each memory state is associatively connected
with its predecessor.

This system shows several emerging features, it is able (a) to recognize external patterns in a noisy background, (b) to focus attention
autonomously and (c) to represent hierarchical memory states with an
internal structure.

1 Introduction

The notion of 'thinking' comes in various flavors. We may associate thinking with
logical reasoning or with a series of associative processes. The latter activity is
performed effortless by the human brain and is at the center of the investigation
we will present here. We will consider in particular associative processes which
may occur also in the absence of any interaction of the brain with the outside
world. It is clear that without any prior stored information - our memories
- this kind of thought process would be semantically empty. We do therefore
investigate the autonomous generation of a time-series of memory-states by a
cognitive system, being it natural or artificial.

We consider consequently thought processes to be characterized by the spontaneous activation of one memory state by another one, leading to a history of

---

* Address at time of submission: Institut für Theoretische Physik, Universität des
Saarlandes, 66041 Saarbrücken, Germany.

U. Furbach (Ed.): KI 2005, LNAI 3698, pp. 366–379, 2005.

memory states. This process should be autonomous and no outside regulative unit should be needed in order to control this dynamical process. In order to make sense, each activated memory state should be closely associated to its predecessor. A history of random memory states would not possibly classify as a true thought process.

A key question in this context is then: When can two or more memory states be considered to be closely associated? Intuitively this is not a problem: Remembering a trip to the forest with our family for a picnic we may associate this activity with a trip to the forest to cut a Christmas tree. These two memory-states are intuitively related. When we store new memory states, like the two trips to the forest in the above example, in our brain, the appropriate associative links need to be generated spontaneously. But how should our brain be capable of finding all possible relations linking this new information associatively with all previously stored memory-states? An exhaustive search would not be feasible, the time needed to perform it would be immense.

This computational problem of embedding new memory states into their relevant semantic context would not occur if no explicit associative links would be needed at all. This can be achieved when considering networks with overlapping memory states. In this case no additional conjunction units describing associative links in between two stored memories are needed. These associative links would be formed by other memories. Any new information learned by the network then acquires naturally meaningful associative links whenever it shares part of its constituent information with other memory states.

We do therefore consider a homogeneous network, with only one kind of constituent building block: the memories themselves. The memory states then show a dual functionality: depending on the initial condition, an associative link in between two or more activity centers could be either a stationary memory state by itself or it could serve to form an association in between two sequential memory states in the course of a thought process.

We propose a generalized neural-network model capable of simulating the here defined kind of thought processes. We do not claim that actual thought processes in biological cybernetic systems (in our brain for instance) will be described accurately by this model. However, the kind of thought processes proposed here seem to be a mandatory requirement if a homogeneous associative network without an external regulative unit wants to acquire true information-processing capabilities. In such kind of networks the information-processing must be self-organized by an autonomous dynamical process. From a functional point of view it is evident that this self-regulated information processing needs to be implemented in biological cognitive systems, like the human brain, in one way or another.

We note that these self-organized association processes work only within dense associative networks, where essentially all activity centers are connected among themselves, forming what one calls in network-theory a 'Giant Strongly Connected-Component' [1]. In a sparse network there would be many unconnected subclusters incapable to communicate autonomously. Therefore we consider here dense and homogeneous associative networks (dHAN) and one might argue that the human brain does fall into this category.

2 Associative and Homogeneous Networks

We consider an associative network with N sites, which we also call activity centers (AC). Each AC represents some specific biologically relevant information, normally in a highly preprocessed form [2]. Examples are ACs for colors, shapes, distances, movements, sounds and so on. Each AC is characterized by the individual activity $x_i \in [0,1]$ $(i = 1, \ldots, N)$. An AC is active when x_i is close to one. We identify ensembles of active ACs with memory states [3].

In addition to the activity levels x_i we introduce for every AC a new variable $\varphi_i \in [0,1]$, which characterizes the level of the individual activity reservoirs. This variable plays, as it will become clear from the discussions further below, a key role in facilitating the self-sustained thought process and distinguishes the dHAN from standard neural networks [4].

We consider here a continuous-time (t) evolution, $x_i = x_i(t)$ and $\varphi_i = \varphi_i(t)$. The differential equations $(\dot{x}_i(t) = \frac{d}{dt}x_i)$

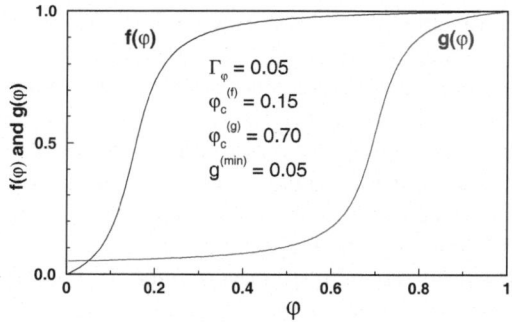

Fig. 1. Illustration of the reservoir-functions $f(\varphi)$ and $g(\varphi)$ as defined by Eq. (4) for $\varphi_c^{(f)} = 0.15$, $\varphi_c^{(g)} = 0.7$, $\Gamma_\varphi = 0.05$ and $g^{(min)} = 0.05$

Table 1. Sets of model-parameters used for the simulations presented here. w/z denote the non-zero matrix elements of the link-matrices $w_{i,j}/z_{i,j}$ entering Eq. (1). The filling/depletion rates for the reservoir $\Gamma_\varphi^{\pm}$ and x_c enter Eq. (3). The critical reservoir-levels for inhibition and activation, $\varphi_c^{(f/g)}$ enter Eq. (4), as well as the width Γ_φ for the reservoir function and the minimal values $f^{(min)} = 0$ and $g^{(min)}$.

	w	z	x_c	Γ_φ^+	Γ_φ^-	$\varphi_c^{(f)}$	$\varphi_c^{(g)}$	Γ_φ	$g^{(min)}$
(a)	0.15	-1.0	0.85	0.004	0.009	0.15	0.7	0.05	0.00
(b)	0.15	-1.0	0.50	0.005	0.020	0.15	0.7	1.00	0.10

$$\dot{x}_i = (1 - x_i)\Theta(r_i)r_i + x_i[1 - \Theta(r_i)]r_i \tag{1}$$

$$r_i = b_i + g(\varphi_i) \sum_{j=1}^{N} w_{i,j} x_j + \sum_{j=1}^{N} z_{i,j} x_j f(\varphi_j) \tag{2}$$

$$\dot{\varphi}_i = \Gamma_\varphi^+ \Theta(x_c - x_i)(1 - \varphi_i) - \Gamma_\varphi^- \Theta(x_i - x_c)\varphi_i . \tag{3}$$

determine the time-evolution of all $x_i(t)$. Here the r_i are growth rates and the b_i the respective biases.[1] We will discuss the role of the bias further below, for the time being we consider $b_i \equiv 0$, if not stated otherwise.[2]

The function $\Theta(r)$ occurring in Eq. (1) is the step function: $\Theta(r) = 1, 0$ for $r > 0$ and $r < 0$ respectively. The dynamics, Eqs. (1) and (3), respects the normalization $x_i \in [0, 1]$ and $\varphi_i \in [0, 1]$ due to the prefactors $(1 - x_i)$, $(1 - \varphi_i)$ and x_i, φ_i for the growth and depletion processes.

The neural-network-type interactions in between the activity centers are given by the matrices $0 \le w_{i,j} \le w$ and $z_{i,j} \le -|z|$ for excitatory and inhibitory connections respectively. The breakdown of the link-matrix in an excitatory and inhibitory sector can be considered as a reflection of the biological observation that excitatory and inhibitory signals are due to neurons and interneurons respectively. Any given connection is either excitatory or inhibitory, but not both at the same time: $w_{i,j} z_{i,j} \equiv 0$, for all pairs (i, j). We do not consider here self-interactions (auto-associations): $w_{i,i} = z_{i,i} \equiv 0$.

We consider here the recurrent case with $w_{i,j} = w_{j,i}$ and $z_{i,j} = z_{j,i}$, but the model works fine also when this symmetry is partially broken. This will happen anyhow dynamically via the reservoir-functions $f(\varphi)$ and $g(\varphi)$. These functions govern the interaction in between the activity levels x_i and the reservoir levels φ_i. They may be chosen as washed-out step functions of a sigmoidal form like

$$g(\varphi) = g^{(min)} + \left(1.0 - g^{(min)}\right) \frac{\text{atan}[(\varphi - \varphi_c^{(g)})/\Gamma_\varphi] - \text{atan}[(0 - \varphi_c^{(g)})/\Gamma_\varphi]}{\text{atan}[(1 - \varphi_c^{(g)})/\Gamma_\varphi] - \text{atan}[(0 - \varphi_c^{(g)})/\Gamma_\varphi]} , \tag{4}$$

with a suitable width Γ_φ. For an illustration see Fig. 1. The effect of the reservoir functions depends on the value of the respective reservoir-levels φ_i, which are governed by Eq. (3).

For $x_i > x_c$ (high activity level) the reservoir-level φ_i decreases with the rate Γ_φ^-. For $x_i < x_c$ (low activity level) the reservoir-level φ_i increases with the rate Γ_φ^+. A low reservoir level will have two effects: The ability to suppress another activity center via an inhibitory link $z_{i,j}$, which will be reduced by $f(\varphi_i) \in [0, 1]$ and the activation by other centers via an excitatory link $w_{i,j}$, which will be reduced by $g(\varphi_i) \in [0, 1]$, see Eq. (2).

The dynamics induced by Eq. (1) leads to a relaxation towards the next stable memory state within a short time-scale of $\Gamma_r^{-1} \approx |w_{i,j}|^{-1} \approx |z_{i,j}|^{-1}$ (for

---

[1] The differential equations (1) and (2) are akin to the Lotka-Volterra equations discussed by Fukai and Tanaka [5].

[2] The time-unit is arbitrary in principle and could be tuned, as well as most of the parameters entering Eqs. (1) and (2), in order to reproduce neurobiologically observed time-scales. For convenience one could take a millisecond for the time unit, or less.

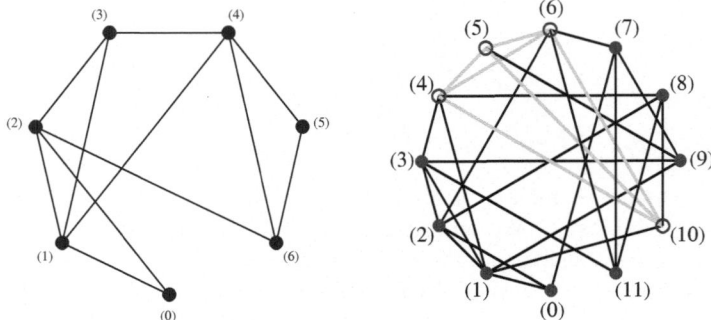

Fig. 2. Two small networks for illustrational purposes. The arrangement of the activity centers (filled blue circles) is arbitrary, here we have chosen a circular arrangement for a good overview. Shown are the excitatory links (the non-zero matrix elements of $w_{i,j}$, black lines). Two activity centers not connected by $w_{i,j}$ are inhibitorily connected via $z_{i,j}$. Left: A seven-center network with five stable memory states: (0,1,2), (1,2,3), (1,3,4), (4,5,6) and (2,6). Right: A 12-center network with 7 2-center memory states, 7 3-center memory states and one 4-center memory state (which is highlighted). It contains a total of 28 links (non-zero matrix-elements of $w_{i,j}$).

the non-zero matrix-elements of the link-matrices). Choosing the rates $\Gamma_\varphi^\pm$ for the reservoir dynamics to be substantially smaller than the relaxation-rates Γ_r we obtain a separation of time-scales for the stabilization of memory states and for the depletion/filling of the activity reservoirs $\varphi_i(t)$ described by Eq. (3). This separation of time-scales is evident in the simulations presented in Fig. 3. For illustrational purposes we present the activity-levels for a (very) small system, the 12-center network illustrated in Fig. 2. The time-scales of the dynamics are however system-size independent. We note, that only a finite number of centers are active at any given time. Before we discuss the dynamics of the thought process more in detail in Sect. 3, we will take a closer look at the nature of the transient attractor stabilized for short time-scales.

2.1 Memory States

We consider here memory states which contain only a finite number, typically between 2 and 7, of constituent activity centers. This is a key difference between the dHAN investigated here and standard neural networks, where a finite fraction of all neurons might be active simultaneously [6].

The stabilization of memory states made up of clusters with a finite number $Z = 2, 3, \ldots$ of activity centers is achieved by an inhibitory background of links:

$$z_{i,j} \leq z < 0, \qquad \forall\,(w_{i,j} = 0, i \neq j) . \tag{5}$$

In Fig. 2 we illustrate a 7-center network. Illustrated in Fig. 2 by the black lines are the excitatory links, i.e. the non-zero matrix-elements of $w_{i,j}$. All pairs (i, j) of activity-centers not connected by a line in Fig. 2 have $z_{i,j} \leq -|z|$. If $|z|$ is big enough, then only those clusters of activity centers are dynamically stable, in which all participating centers are mutually connected.

To see why, we consider an AC (i) outside a Z-center memory state (MS). The site (i) cannot have links (finite $w_{i,j}$) to all of the activity centers (j) making up this memory state. Otherwise (i) would be part of this MS. There are therefore maximally $Z - 1$ positive connections in between (i) and the MS. The dynamical stability of the memory state is guaranteed if the total link-strength between (i) and the MS is not too strong:

$$|z| > \sum_{j \in \text{MS}} w_{i,j}, \qquad |z| > (Z - 1) w , \tag{6}$$

where the second equation holds for the uniform case, $w_{i,j} \equiv w > 0$.

For an illustration of this relation we consider the case $x_3 = x_4 = x_5 = x_6 = 0$ and $x_0 = x_1 = x_2 = 1$ for the 7-center network of Fig. 2. The growth-rate for center (3) is then: $r_3 = 2w - |z|$. For $2w - |z| > 0$ center (3) would start to become active and a spurious state (1,2,3,4) would result. Taking $2w - |z| < 0$ both (0,1,2) and (1,2,3) are stable and viable 3-center memory states.

A 'spurious memory state' of the network would occur if a group of ACs remains active for a prolonged time even though this grouping does not correspond to any stored memory state. No such spurious memory state is dynamically stable when Eq. (6) is fulfilled. For the simulation presented here we have chosen $|z|/w > 6$. This implies that memory states with up to $Z = 7$ activity centers are stable, see Table 1 and Eq. (6).

This kind of encoding of the link-matrices is called a 'winners-take-all' situation[3], since fully interconnected clusters will stimulate each other via positive intra-cluster $w_{i,j}$. There will be at least one negative $z_{i,j}$-link in between an active center of the winning memory state to every out-of-cluster AC, suppressing in this way the competing activity of all out-of-cluster activity centers.

2.2 Hierarchical Memory States

In the above discussion we have considered in part the uniform case $w_{i,j} \equiv w$ for all non-zero excitatory links. In this case, all features making up a memory state are bound together with the same strength. Such a memory state has no internal structure, it is just the reunion of a bunch of semantic nodes with no additional relations in between them. Memory states corresponding to biological relevant objects will however exhibit in general a hierarchical structure [8] [9]. Let us give an example: A memory state denoting a 'boy' may involve a grouping of ACs corresponding to (face), (shirt), (pants), (legs), (red), (green) and so on. This memory state is well defined in our model whenever there are positive links $w_{i,j} > 0$ in between all of them.

There is now the need for additional information like: is 'red' the color of the shirt or of the trousers? That is, there is the need to 'bind' the color red

---

[3] Our winners-take-all setting differs from the so-called 'K-winners-take-all' configuration in which the K most active neurons suppress the activities of all other neurons via an inhibitory background [7].

preferentially to one of the two pieces of clothes. It is possible to encode this internal information into the memory state 'boy' (face,shirt,pants,legs,red,green,..) by appropriate modulation of the internal connections. In order to encode for instance that red is the color of the shirt, one sets the link (red)-(shirt) to be much stronger than the link (red)-(pants) or (red)-(legs). This is perfectly possible and in this way the binding of (red) to (shirt) is achieved. The structure of the memory states defined here for the dHAN is therefore flexible enough to allow for a (internal) hierarchical object representation.

No confusion regarding the colors of the shirt and of the pants arises in the above example when variable link-strengths $w_{i,j}$ are used. Note however, that this is possible only because small and negative links $z_{i,j}$ are not allowed in our model, a key difference to the most commonly used neural-network models. If weak inhibitory links would be present, the boundary of memory states could not be defined precisely. There would be no qualitative difference in between a small negative and a small positive synapsing strength. Furthermore, the stability condition Eq. (6) would break down.

3   Dynamical Thought Processes

In Fig. 3 and Fig. 4 we present an autonomous thought process within a 12-center network with 15 stable memory states, illustrated in Fig. 2. We have chosen a small network here to discuss the properties of the dynamical thought process in detail. The model is however completely scalable and we have performed simulations of networks containing several thousands of sites without any problem on

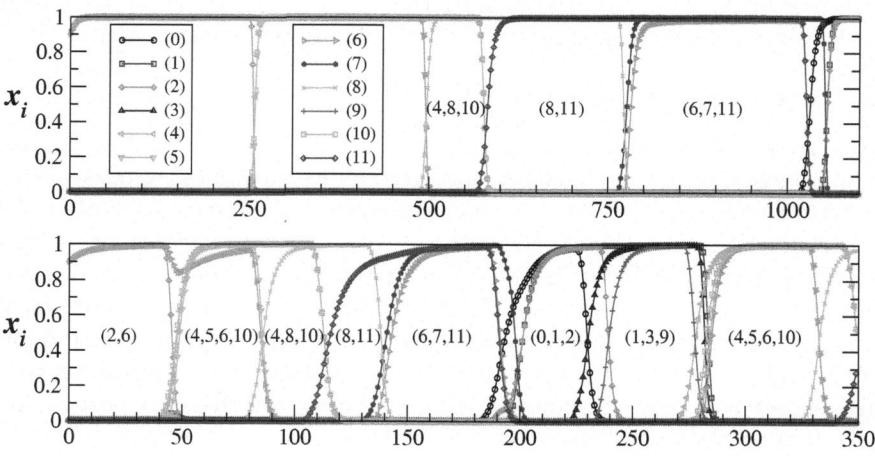

Fig. 3. The activity $x_i(t)$ for the thought process $(2, 6) \rightarrow (4, 5, 6, 10) \rightarrow (4, 8, 10) \rightarrow (8, 11) \rightarrow (6, 7, 11) \rightarrow (0, 1, 2) \rightarrow (1, 3, 9) \rightarrow (4, 5, 6, 10)$ for the 12-site network in Fig. 2. Top/Bottom: Using the parameter sets (a)/(b) of Table 1. Note the different scaling for the respective time-axis.

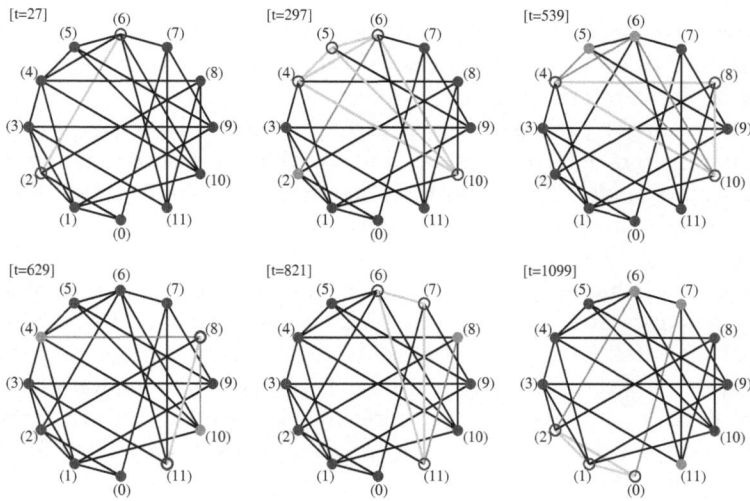

Fig. 4. The thought process $(2,6) \rightarrow (4,5,6,10) \rightarrow (4,8,10) \rightarrow (8,11) \rightarrow (6,7,11) \rightarrow$ $(0,1,2)$ of a 12-site network with 7 2-center, 7 3-center and one 4-center memory state. The non-zero excitatory links $w_{i,j} > 0$ differ from the uniform level w randomly by at most 5%. Compare Fig. 3 for the time-evolution of the variables.

standard computers. The choice of synchronous or asynchronous updating procedures is arbitrary here, due to the continuous-time formulation. No particular care needs to be taken for the integration of the differential equations (1) and (3) as the dynamical process has relaxational properties for the short-time dynamics of both the activities $x_i(t)$ as well as for the reservoir levels $\varphi_i(t)$. The model is numerically robust. The dynamics is numerical stable also for the long-time evolution of the activities $x_i(t)$ which is driven by the respective reservoir levels via the reservoir-functions $f(\varphi)$ and $g(\varphi)$.

The simulations presented in Fig. 3 were performed using the two distinct parameter sets listed in Table 1. For the parameter set (a) we observe very rapid relaxations towards one of the memory-states encoded in the link-matrices of the network, as shown in Fig. 2. The reservoir levels are depleted very slowly and the memory-stable becomes unstable and a different memory-state takes over only after a substantial time has passed. Comparing the thought process shown in Fig. 3 with the network of excitatory links of the dHAN shown in Fig. 2 one can notice that (I) any two subsequent memory states are connected by one or more links and that (II) excitatory links have in fact a dual functionality: To stabilize a transient memory state and to associate one memory state with a subsequent one. The model does therefore realize the two postulates for self-regulated associative thought processes set forth in the introduction.

Biologically speaking it is a 'waste of time' if individual memory states remain active for an exceedingly long interval. The depletion-rate $\Gamma_\varphi^- = 0.009$ for the reservoir-levels is very small for the parameter set (a) listed in Table 1. For the parameter set (b) we have chosen a substantially larger depletion-rate $\Gamma_\varphi^- = 0.02$

together with very smooth reservoir-functions $f(\varphi)$ and $g(\varphi)$ and a finite value for $g^{(min)}$ in order to avoid random drifts for centers active over prolonged periods and fully depleted reservoir levels.

The two thought processes shown in Fig. 3 are for identical link-matrices, only the parameters entering Eqs. (1) and (3) differ. We note that the sequence of memory-state is identical for both parameter sets, the dynamics is stable over a wide range of parameter-values. The history of memory states goes through a cycle, as it is evident for the simulations using the parameter set (b), since the phase-space is finite. The 12-site cluster used in this simulation contains 15 different memory states and the cycle runs over 6 distinct states, a substantial fraction of the total. For larger networks with their very high numbers of stored memory-states [10] the cycle length will be in general very long for any practical purposes. We have, however, not yet performed a systematic study of the cycle-length on the system properties along the same lines usually done for random boolean networks [11].

We note that binary cycles do not occur for the parameters used here. We have chosen $\Gamma_\varphi^+ < \Gamma_\varphi^-$ and the reservoir levels therefore take a while to fill-up again. Active memory states can therefore not reactivate their predecessors, which have necessarily low reservoir levels and small values for $g(\varphi)$. The same temporal asymmetry could be achieved by choosing $w_{i,j} \neq w_{j,i}$. We do not rule-out the use of asymmetric link-matrices for the dHAN, but it is an important property of the dHAN to be able to establish a time direction autonomously. The types of memory-states storable in the dHAN would otherwise be limited to asymmetric states.

4   Details of the Dynamics

In Fig. 5 we present a blowup of the thought process presented in Fig. 3 for the parameter set (b), together with the time-dependence of the growth rates $r_i(t)$. We can clearly observe how competition among the ACs plays a crucial role. For the first transition occurring at $t \approx 78$ sites (1) and (8) compete with each other. Both have two excitatory links with the active cluster (4,5,6,10), see Fig. 2 and Fig. 8. This competition is resolved here by small random differences in between the excitatory links $w_{i,j}$ used here.

The transition occurring at $t \approx 190$ in between the two disjunct memory states (6,7,11) and (0,1,2) takes a substantial amount of time to complete as it goes through the intermediate state (0,7). The memory state (0,7) is actually a valid memory state by itself but is not stabilized here as the AC (7) looses out in the competition with (1) and (2), compare Fig. 5.

Note that in the transition $(4, 8, 10) \rightarrow (8, 11) \rightarrow (6, 7, 11)$ this effect does not occur. The intermediate state (8,11) is stabilized because the reservoir level of (6) had yet not been refilled completely due to a precedent activation. In this case (6) and (7) loose out in competition with (8).

In Fig. 6 we show the results of two simulations on different 12-site clusters with very dense link-matrices $w_{i,j}$ which contain memory-states with up to seven

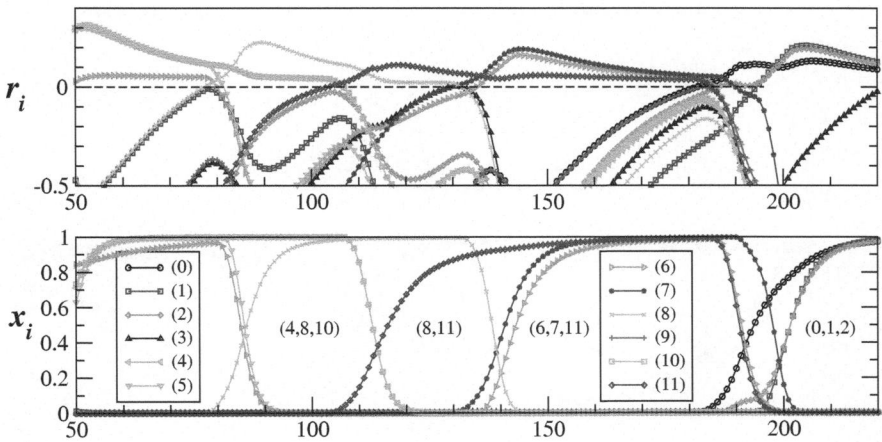

Fig. 5. The growth-rates $r_i(t)$ (top) and the activity $x_i(t)$ (bottom) for the thought process $(4, 5, 6, 10) \to (4, 8, 10) \to (8, 11) \to (6, 7, 11) \to (0, 1, 2)$ for the 12-site network shown in Fig. 2, using the parameter set (b) of Table 1

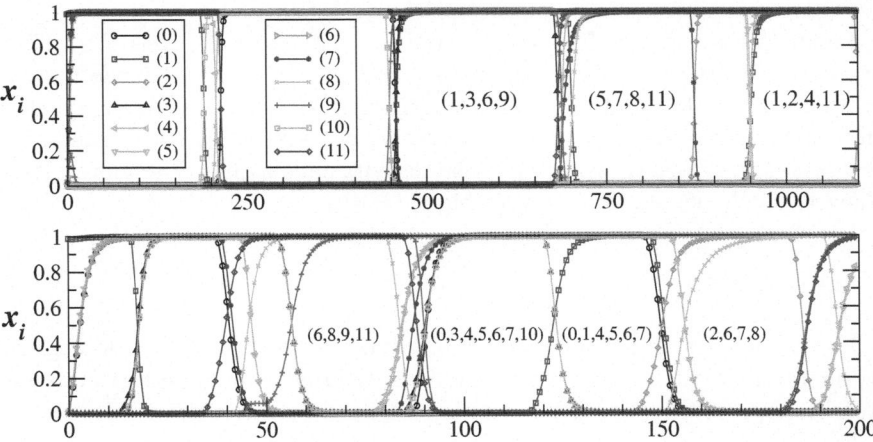

Fig. 6. The activity $x_i(t)$ for two 12-center clusters with a high density of links using the parameter sets (a) and (b) of Table 1 (top/bottom). Top: The thought process is $(1, 4, 6, 7, 11) \to (0, 6, 7, 8, 10) \to (1, 3, 6, 9) \to (5, 7, 8, 11) \to (2, 5, 8, 11) \to (1, 2, 4, 11)$. Bottom: The thought process is $(0, 3, 4, 5, 6, 7, 10) \to (3, 6, 8, 10, 11) \to (6, 8, 9, 11) \to (0, 3, 4, 5, 6, 7, 10) \to (0, 1, 4, 5, 6, 7) \to (2, 6, 7, 8)$.

centers. Simulations for both set of parameters are shown and we observe that the dynamics works perfectly fine. With the appropriate choice Eq. (6) for the link-strength even networks with substantially larger memory states allow for numerical stable simulations.

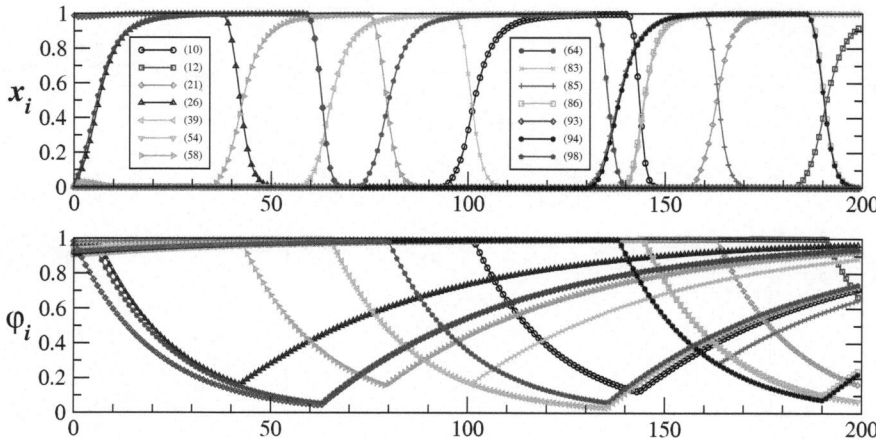

Fig. 7. The activity $x_i(t)$ and the reservoir $\varphi_i(t)$ for a 99-center clusters with $165/143/6$ stable memory states containing $2/3/4$ centers, using the parameter set (b) of Table 1 The thought process is $(26, 64, 93) \rightarrow (58, 64, 93) \rightarrow (39, 58, 83) \rightarrow (39, 83, 98) \rightarrow (10, 39, 98) \rightarrow (54, 85, 86, 94) \rightarrow (21, 54, 86, 94) \rightarrow (12, 21, 54)$

Finally we present in Fig. 7 the activities and reservoir-levels for a 99-site network, using the parameter set (b). Plotted in Fig. 7 are only the activities and the reservoir-levels of the sites active during the interval of observation. We can nicely observe the depletion of the reservoir-levels for the active centers and the somewhat slower recovery once the activity falls again. Similar results are achieved also by simulations of very big networks.

5 Discussion

We have here investigated the autonomous dynamics of the dHAN and neglected any interaction with the outside world. Sensory inputs would add to the growth rates by appropriate time-dependent modulations of the respective bias $b_i(t)$, see Eq. (2). For any sensory input the bias of the involved activity centers would acquire a finite positive value during the interval of the sensory stimulation.

Taking a look at the growth rates plotted in Fig. 5 it becomes immediately clear that the sensory input needs in general a certain critical strength in order to influence the ongoing thought process. The autonomous thought process and the sensory input compete with each other. This kind of competition does not occur in simple attractor neural networks [11], for which any non-zero input induces the network to flow into the nearest accessible attractor. Both for inputs resembling closely a previously stored pattern as well as for random and nonsensical inputs.

The dHAN considered here will, on the other hand, recognize external patterns only when the input is such that it wins the competition with the ongoing thought process, leading to the activation of the corresponding memory state. The strength of the sensory input necessary for this recognition process to be

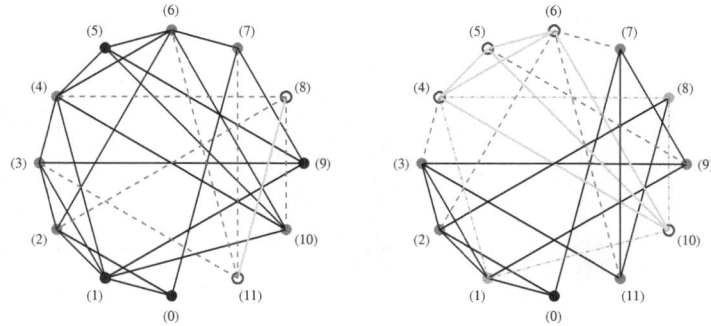

Fig. 8. Dynamical attention-focusing in the 12-site network with the thought process illustrated in Fig. 5, for two different stable memory states. The dashed/dashed-dotted cyan and pink links/circles denote centers linked weakly/strongly (by one/two excitatory links) to the active memory state.

completed successfully depends crucially on the number of links between the current active memory state and the ACs stimulated by the sensory input, as one can see in Fig. 5 and Fig. 8. The sites (1) and (8) have two links to the memory state (4,5,6,10), all other centers have either zero or just one. Any sensory input arriving on sites (1) or (8) would be recognized even for small signal-strength when (4,5,6,10) is active, a sensory input arriving at site (0) would need to be, on the other hand, very large. This property of the dHAN is then equivalent to the capability of focus attention autonomously, an important precondition for object recognition by cognitive systems [12]. Every active memory states carries with it an 'association cloud' and any external input arriving within this area of ACs linked directly to the active memory state will enjoy preferential treatment by the dHAN.

5.1   Biological Considerations

At first sight the model Eq. (1) possesses an unbiological feature. Neglecting the $f(\varphi)$ and $g(\varphi)$ for a moment, the total effective link strength $w_{i,j} + z_{i,j}$ is discontinuous: Either strongly negative ($w_{i,j} = 0$, $z_{i,j} \leq -|z|$), or weakly positive ($0 \leq w_{i,j} \leq w$, $z_{i,j} = 0$), as illustrated in Fig. (9). This property of the link-matrices between the constituent activity centers is crucial for the whole model. It is essential for the stability of the individual memory states, see in Sect. 2.1, and it forms the basis for hierarchical object representations, as discussed in Sect. 2.2. It constitutes a key difference between our and other models of neural networks [11].

The effective link-strength $w_{i,j} + z_{i,j}$ does however not correspond to the bare synapsing-strength in biological neural assemblies. It represents an effective coupling in between local or distributed centers of neural activities and this kind of discontinuous behavior may actually result quite naturally from a simple coupling via intermediate inhibitory interneurons, as illustrated in Fig. (9). When

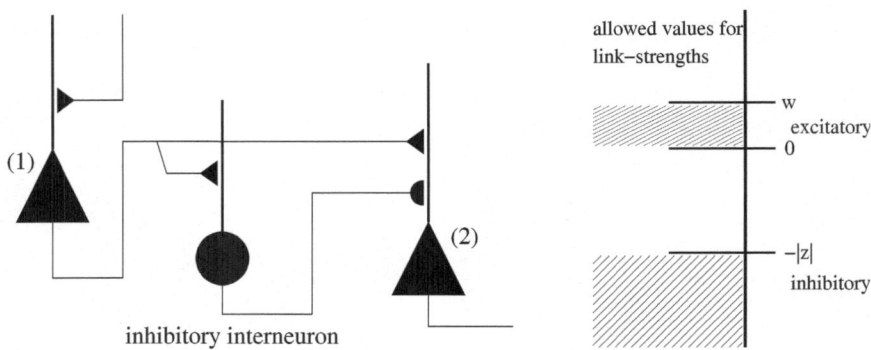

Fig. 9. Illustration of a network of interacting neurons leading to an effective discontinuous inter-neural coupling. Left: The excitatory neuron (1) connects both directly to neuron (2) and indirectly via the inhibitory interneuron. An activation of the interneuron by neuron-(1) may lead to complete inhibition of neuron-(2), masking completely the direct excitatory (1)-(2) link. Right: The resulting range for the allowed link-strengths for the excitatory links $w_{i,j}$ (red shaded region) and for the inhibitory links $z_{i,j}$ (blue shaded region).

the interneuron is active, the effective coupling is strongly inhibitory. When the interneuron is quiet, the coupling is weakly excitatory with Hebbian-type learning. When the interneuron is active, it might as well inhibit the target neuron completely. Integrating out the intermediate inhibitory interneuron leads in this way to an effective discontinuous inter-neuron coupling.

Biologically speaking, one observes that inhibitory synapses are placed all over the target neurons, on the dendrites, the soma and on the axon itself. Excitatory synapses are, however, located mainly on the dendrites. This observation suggests that inhibitory synapses may indeed preempt the postsynaptic excitatory potentials, giving functionally rise to a model like the one proposed here.

6   Conclusions

We have proposed, discussed and implemented a central principle for the self-generated time-series of mental states of a cognitive system - the notion of "duality" for memory states. A cognitive system can make use of the stored information - the memories - only if these memories can be related to each other. We believe that no separate conjunction units are necessary for this job. Memory states and conjunction units responsible for the time-evolution of the thought process are - in our view - just two different aspects of the same coin: Memory states can either be activated, just as normal memory states are supposed to behave, or act as associative links, enabling the dynamics of the thought process.

There may exist a range of possible implementations of this principle, here we have shown that an associative network with overlapping memory states of a generalized neural-network-type will autonomously generate a history of transient memory states when suitable couplings to local reservoir levels are introduced. This self-regulating model exhibits, to a certain extent, autonomous data-processing capabilities with a spontaneous decision-making ability. Here we have been concerned with showing the feasibility of this approach. Further research will be necessary to show that this self-regulating system can carry out specific cognitive tasks.

References

1. Dorogovtsev, S.N., Mendes, J.F.F.: Evolution of Networks. Oxford University Press (2003)
2. Hubel, D., Wiesel, T.: Receptive fields and functional architecture in two nonstriate visual areas (18 and 19) in the cat. J. Neurophysiol. **28** (1965) 229-289
3. McLeod, P., Plunkett, K., Rolls, E.T.: Introduction to connectionist modelling of cognitive processes. Oxford University Press (1998)
4. Müller, B., Reinhardt, J., Strickland, M.T.: Neural Networks, An Introduction. Springer (1995)
5. Fukai, T., Tanaka, S.: A simple neural network exhibiting selective activation of neuronal ensembles: from winner-take-all to winners-share-all. Neural Comp. **9** (1997) 77-97
6. O'Reilly, R.C.: Six principles for biologically based computational models of cortical cognition. Trends Cog. Sci. **2** (1998) 455-462
7. Kwon, T.M., Zervakis, M.: KWTA networks and their application. Multidim. Syst. and Sig. Proccessing **6** (1995) 333-346
8. Riesenhuber, M., Poggio, T.: Are cortical models really bound by the "Binding Problem?. Neuron **24** (1999) 87-93
9. Mel, B., Fiser, J.: Minimizing Binding Errors Using Learned Conjunctive Features. Neural Comp. **12** (2000) 731-762
10. Buhmann, J., Divko, R., Schulten, K.: Associative memory with high information content. Phys. Rev. A **39** (1989) 2689-2692
11. Schuster, H.G.: Complex Adaptive Systems: An Introduction. Scator (2001)
12. Reynolds, J.H., Desimone, R.: The role of neural mechanisms of attention to solve the binding problem. Neuron **24** (1999) 19-29

Why Is the Lucas-Penrose Argument Invalid?

Manfred Kerber

School of Computer Science,
The University of Birmingham,
Birmingham B15 2TT, England
www.cs.bham.ac.uk/~mmk

Abstract. It is difficult to prove that something is not possible in principle. Likewise it is often difficult to refute such arguments. The Lucas-Penrose argument tries to establish that machines can never achieve human-like intelligence. It is built on the fact that any reasoning program which is powerful enough to deal with arithmetic is necessarily incomplete and cannot derive a sentence that can be paraphrased as "This sentence is not provable." Since humans would have the ability to see the truth of this sentence, humans and computers would have obviously different mental capacities. The traditional refutation of the argument typically involves attacking the assumptions of Gödel's theorem, in particular the consistency of human thought. The matter is confused by the prima facie paradoxical fact that Gödel proved the truth of the sentence that "This sentence is not provable."

Adopting Chaitin's adaptation of Gödel's proof which involves the statement that "some mathematical facts are true for no reason! They are true by accident" and comparing it to a much older incompleteness proof, namely the incompleteness of rational numbers, the paradox vanishes and clarifies that the task of establishing arbitrary mathematical truths on numbers by finitary methods is as infeasible to machines as it is to human beings.

1   Introduction

The probably most prominent and most articulate argument why the whole field of AI would be doomed to failure is expressed in the so-called Lucas-Penrose argument, which can be summarized as follows: Since Gödel proved that in each sound formal system – which is strong enough to formulate arithmetic – there exists a formula which cannot be proved by the system (assumed the system is consistent), and since we (human beings) can see that such a formula must be true, human and machine reasoning must inevitably be different in nature, even in the restricted area of mathematical logic. This attributes to human mathematical reasoning a very particular role, which seems to go beyond rational thought. Note that it is not about general human behaviour, and not even about the process of how to find mathematical proofs (which is still only little understood), but just about the checking of (finite) mathematical arguments.

There have been detailed refutations of this argument, actually so many that just a fair review of the most important ones would easily fill a full paper. For this reason I want to mention only the one by Geoffrey LaForte, Pat Hayes and

U. Furbach (Ed.): KI 2005, LNAI 3698, pp. 380–393, 2005.

Ken Ford [10], which gives explanations where Penrose goes wrong, and why his argument does not hold. LaForte et al. discuss in detail arguments which go partly back to Turing, Gödel, and Benacerraf that the assumption under which Gödel's theorem holds for a system, soundness and consistency, are not fulfilled for humans, and that we cannot know whether we are sound or consistent. This, however, would be necessary in order to establish reliably mathematical truths such as Gödel's theorem. In my view, the authors convincingly refute the Lucas-Penrose argument.

So why another paper? Mainly for two reasons, firstly this refutation – as well as the others I am aware of – leaves a paradoxical situation: the sentence which Gödel constructed in his proof, which basically can be described as "This sentence is not provable," is proved, since we know that it is true. But how can it be proved when it is not provable? Or do we not know it and it is not proved? Is it just a confusion of meta and object level? Gödel's result is still puzzling and most interpretations given in order to refute the Lucas-Penrose argument are pretty complicated. Secondly, the arguments of LaForte et al. rely on assumptions about "the nature of science in general, and mathematics in particular." While in my view these arguments are sound, they do neither go to the core of the problem since the Lucas-Penrose argument would be invalid even if these assumptions were false nor are they as easy as they should be (and as Hilbert put it, "You have not really understood a mathematical sentence until you can explain it to the first person you meet on the street in a way that they can understand it.").

The real core of the matter is the fact that there are two classes of mathematical truths, namely those which *can* be finitely proved (Gödel's theorem belongs to this class) and those mathematical truths which *cannot* be finitely proved (Gödel's theorem is about the non-emptiness of this class, but does not belong to it). The latter truths are unknowable in principle to human beings and computers alike (be they real or slightly idealized), since the ones as the others are limited by their finiteness.

In this contribution I want to relate Gödel's incompleteness theorem and the Lucas-Penrose argument (summarized in the next section) via analogy to another incompleteness theorem, namely the incompleteness of the rational numbers, which was discovered 2500 years ago and which is nowadays intuitive (described in section 3). The relation between the two discoveries is closer than one would think on first sight, they have both to do with finite representability. Exactly as most real numbers cannot be finitely represented, it is not possible to prove most true statements in a finite way (see section 4). However, approximations are possible (see section 5). Since the restrictions apply to human beings and machines alike we can conclude (in section 6) that the Lucas-Penrose argument is invalid.

2   Gödel's Incompleteness Theorem and the Lucas-Penrose Argument

Gödel's proof of the incompleteness of logical calculi which are strong enough to deal with arithmetic can be summarized as follows: In a first step Gödel shows that provability can be defined in any logical system that is powerful enough to encode arithmetic. More precisely, if a calculus is given, then it is possible to define a binary predicate symbol "proves[,]", such that for numbers m and n, there is an interpretation in which (semantically) m is a proof of theorem n (for short, proves[m, n]). For instance, "There is no proof of $0 = s(0)$" can be expressed as $\neg\exists m.\text{proves}[m, \lceil 0 = s(0) \rceil]$. $\lceil . \rceil$ is a function which maps a string injectively, concretely here the string "$0 = s(0)$", to a number, the so-called Gödel number, e.g. 170104021703. There are different ways how to produce such a translation and Gödel introduced one in his original paper [6]. In our case, the theorem is translated then to $\neg\exists m.\text{proves}[m, 170104021703]$.

Based on this construction, Gödel enumerates all formulae with one free variable w, which are not provable, as $P_k(w) \leftrightarrow \neg\exists m.\text{proves}[m, \lceil P_w(w) \rceil]$ and by diagonalizing, that is, by identifying k and w:

$$P_k(k) \leftrightarrow \neg\exists m.\text{proves}[m, \lceil P_k(k) \rceil]$$

he constructs a formula $P_k(k)$, which states its own unprovability. That this formula must be true can easily be seen by a contradiction proof: If it were not true, it would be provable, but – assumed that arithmetic is consistent and the calculus correct – only true theorems can be proved, hence it would be true, which is absurd.

As a consequence, Gödel can state that any theory T that is finitely describable and does contain arithmetic is necessarily incomplete. This has led to the following argument by Lucas and Penrose:

- For any theory rich enough to formulate arithmetic and for any calculus there exists a theorem (like $P_k(k)$) which cannot be proved by the corresponding calculus due to Gödel's incompleteness result.
- Since human beings can see the truth of this theorem, human reasoning cannot be based on a fixed calculus.
- Hence human reasoning cannot be mechanized, at least not with currently available techniques.

Lucas first published the argument in [11] and discussed it again in [13]. Penrose has stated it at several places back to [14], a concise description can be found in [15], from which the following quotation is taken:

> "The inescapable conclusion seems to be: Mathematicians are not using a knowably sound calculation procedure in order to ascertain mathematical truth. We deduce that mathematical understanding – the means whereby mathematicians arrive at their conclusions with respect to mathematical truth – cannot be reduced to blind calculation!"

Lucas argues that the argument using the Gödel theorem is more relevant than the corresponding one using Turing's proof of the unsolvability of the halting problem, since *"it raises questions of truth which evidently bear on the nature of mind, whereas Turing's theorem does not; it shows not only that the Gödelian well-formed formula is unprovable-in-the-system, but that it is true. ... But it is very obvious that we have a concept of truth. ... A representation of the human mind which could take no account of truth would be inherently implausible."* [12]

Nota bene, the argument is not about finding a proof of Gödel's theorem in the first place, which is certainly a challenge for computers, but about understanding (checking) it, a challenging problem as well, but one that has been solved. However, Lucas and Penrose will not be convinced by mechanical proofs of Gödel's theorem (as, for instance, described in [2]), since for any such system, limitations would apply which would not apply to human reasoning. But is this true? Do they really not apply to human reasoning? Paradoxical situations – let us include Gödel's discovery in them for a moment – can easily lead to very strange conclusions. If not dealt with great care, we easily manœuvre ourselves into absurd positions as Zeno did on the base of the Achilles-and-the-Tortoise paradox.[1]

What is for instance, about a sentence like the one stated by Anthony Kenny: "John Lucas cannot consistently make this judgement"? Does this not cause similar problems for John Lucas as Gödel's theorem for a machine? Gödel himself concluded that we just cannot prove the consistency of human reasoning, while Alan Turing concluded that we cannot assume humans to be consistent. Both statements are probably true, but that is not the point with respect to the Lucas-Penrose argument. Surely we cannot even assume that our proof checking efforts are flawless.

Any mathematical reasoning about the world – and the world includes human beings and computers – makes idealizations. We can (and do) idealize humans in a way that we assume them to be attentive, never tired, not making any mistakes about proof checking, as we idealize computers as wonderful machines which never crash, are never infected by viruses, never suffer from hardware faults, and never make any mistakes. We can make these idealizations since we can strive towards corresponding ideals and approximate them (albeit never fully achieve them).

---

[1] Zeno's paradox is built on a fictional race between Achilles and a tortoise, in which the tortoise is a bit ahead at the start, at time point t_0. Zeno argues that Achilles has no chance ever to catch up, even him being the fastest runner of all, since when he reaches at time point t_1 the position p_0 where the tortoise was at t_0, it will be already gone there and be at a new position p_1, which Achilles can reach at a time t_2, but then the tortoise has moved on to p_2 and so on. At any point in time t_i the tortoise is ahead of Achilles. Zeno concluded that this proves that motion is impossible. With a deep understanding of infinite series we have reached nowadays, we see where Zeno went wrong. All the time intervals under consideration add up to a finite amount of time and when that has passed, Achilles reaches the tortoise, and after that overtakes it. For many centuries this example was not only funny, but deeply confusing and led to strange conclusions like Zeno's that motion is impossible. We see that special care is advisable when concluding what follows from a paradox.

However, we should not and will not make idealizations in the following, which cannot be approximated. Such an idealization would be to deal with infinitely many objects (digits or proof steps) in finite time and would allow (computers or humans) to deal with infinitely many different cases in a finite amount of time. This could theoretically be done by speeding up and using, for instance, 1 second for the first case, 1/2 second for the second case, 1/4 for the third case, 1/8 for the fourth and so on. This way the whole process of dealing with infinitely many cases would take just 2 seconds.

Lucas and Penrose argue that human beings (and in consequence idealized human beings) can see the validity of Gödel's theorem while idealized computers (and hence real computers) cannot. I will argue that even using such idealizations (and a Turing machine is an idealization of real computers) there are things which idealized computers and idealized human beings can do and cannot do, in principle. Both can understand (and have understood) Gödel's theorem. Gödel's theorem shows that there are truths about arithmetic properties which cannot be recognized by idealized computers and idealized human beings alike. Before we go deeper into this issue let us first step back 2500 years in time, and look at another limitation which holds for humans and machines alike and produced another foundational crisis in mathematics.

3   The Discovery of Irrational Numbers – Another Crisis in Mathematics

Greek mathematicians related the lengths of different line segments and discovered that (for all cases they investigated) they could find integers, such that the relative lengths of any two of them could be expressed by the proportion of two integers. This led to the belief which Pythagoras (570-500 BC) propagated that "All is number." In particular it meant that the proportion of any two lines could be finitely expressed, namely by a pair of two (finite) natural numbers, or as we would say as a rational number (note the dual meaning of rational, that something is a ratio, and that something makes sense).

To the big surprise of the Pythagoreans, not much later they made the discovery that their slogan is not true at all. There are two elementary ancient proofs for it, the one below ascribed to Pythagoras himself, another based on a pentagram ascribed to his student Hippasos of Metapont. It seems not clear who really made the discovery first.

The proof ascribed to Pythagoras is based on a diagonal in a square (it is a strange irony that diagonals play an important role in Pythagoras' incompleteness proof as well as in Gödel's):

Assume a square of length a with a diagonal d as in Fig. 1. By the Pythagorean Theorem we get $d^2 = a^2 + a^2$, that is, $d^2 = 2 \cdot a^2$. Assume the ratio $d : a = p : q$ with p and q minimal positive integers (i.e., p and q are relatively prime). Hence, we have $p^2 = 2 \cdot q^2$, from which we get that p must be even, hence p^2 must be divisible by 4. But then q^2 and in consequence q must be even as well, which is a contradiction, since we assumed p and q to be relatively prime.

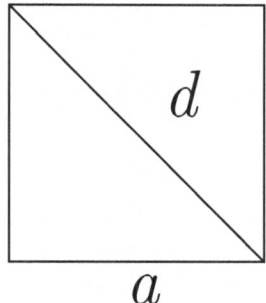

Fig. 1. Square with diagonal

The discovery that the ratio of a and d cannot be represented by two integers caused a foundational crisis in mathematics since one of the axioms, one of the "self-evident truths," on which the Pythagoreans built their mathematics, turned out to lead to a contradiction. Pairs of line segments which cannot be expressed by what we call today a rational number, are called "irrational," the same word we use for things which go beyond rational thought. Why was this dogma so much at the heart of mathematics? Why should the length of a single line in relation to an arbitrarily fixed unit have the ratio of two natural numbers? I think since it was the attempt to describe the continuum of all possible lengths in a finite way. The discovery that it cannot be done as a ratio was an indication that it is not possible to do that in general. However, it is not a proof that it is impossible to describe the length in a finite way altogether. Actually $\sqrt{2}$ (as we represent the ratio of d and a today), is finitely representable as an algebraic number, however, not as the ratio of two integers. A finite representation of $\sqrt{2}$ is, for instance, "the positive zero of the function $f : x \mapsto x^2 - 2$."

We can imagine the shock that the discovery of irrationality caused at the time. It not only destroyed the possibility to represent all lengths as ratios of integers in a very convenient way, but it destroyed also the prospect to finitely represent the corresponding expressions. Mathematics was not far enough developed at the time to see that certain important lengths (actually all the ones we can speak about) can be finitely represented in some other way, and the discovery was not a proof that there are ratios (numbers) which are inherently inexpressible in a finite way.

It took more than 2000 years to gain a deep understanding of what was going on. The final point of this development is the strict construction of the field of real numbers, for instance, in form of Dedekind cuts [4].

Nowadays we distinguish in particular rational, irrational, algebraic, transcendental, computable numbers, and their complements. Real numbers can be represented, for instance, as infinite decimal numbers. Let us look at some examples:

- Integers such as 3 can be represented as
  3.00000000000000000000000000000 ... and defined as $s(s(s(0)))$.
- Rational numbers such as $\frac{1}{3}$ can be represented as
  0.33333333333333333333333333333 ... and defined as $\frac{1}{3} \cdot 3 = 1$.

- Irrational algebraic numbers such as $\sqrt{2}$ can be represented as $1.414211\ldots$ and defined as the positive zero of $f(x) = x^2 - 2$.
- Irrational transcendent numbers such as π can be represented as infinite decimal numbers, e.g., $3.14159265358979323846264338332\ldots$. For special numbers such as π also a finite definition can be given, e.g., $\pi = 4 \cdot \sum_{n=0}^{\infty} \frac{(-1)^n}{2n+1}$ (and effectively constructed from infinite series, although the one given here is not very efficient for practical purposes).
- Computable numbers such as Liouville's transcendent number L can be represented as $1.10100100010000010000001\ldots$. L can be finitely defined by a program that interleaves 1s with an increasing number of 0s.
- True random numbers still can be represented as infinite decimal numbers such as $4.75950161585297612929378858789\ldots$. However, no finite representation is possible at all. True random numbers are all irrational, transcendent, and not computable.

Cantor proved (also by a diagonal argument) that there are more real numbers than natural numbers, that is, that the real numbers are not countable, while rational, algebraic, and computable numbers are all countable.

There are many more numbers (uncountably many) which cannot be finitely represented, but only countably many which can. In particular, most real numbers can only be approximated, while some (and effectively most of the important) real numbers can be finitely described. For instance, the diagonal which caused so much trouble to the Pythagoreans, can be described as $\sqrt{2}$, or as the only positive zero of the function $f(x) = x^2 - 2$. Just adding all those numbers which can be finitely described does not yield a complete structure, however. In order to do so it is necessary to go to the full structure of the real numbers, which include true random numbers or numbers which have a component of randomness (such as a decimal number which is constructed from a random number by interleaving its digits with 1s, e.g., $4.17151915101116111518151219171611121912\ldots$).

While computers are much better at computing with numbers, in principle they suffer from the same limitation as humans do with respect to true random numbers. They can neither store them, nor manipulate them, but only approximate them.

4   Finitary vs. Infinitary – Numbers, Theorems, and Proofs

What has all that to do with logic and proofs? Let us look at the closed interval $[0, 1] = \{x \in \mathbb{R} | 0 \leq x \leq 1\}$ in dual representation (we neglect in the following the problem of the ambiguity of representation with periods of 1s that e.g. $0.0\dot{1}$ is equivalent to $0.1\dot{0}$). In such a representation each number corresponds in a one-to-one form to unary predicates, for instance, the predicate "even" corresponds to the number $2/3$ in dual notation:

$$0. \quad 1 \quad\quad 0 \quad\quad 1 \quad\quad 0 \quad\quad 1 \quad\quad 0 \quad\quad 1 \quad\cdots$$
$$\updownarrow \quad \updownarrow \quad \updownarrow \quad \updownarrow \quad \updownarrow \quad \updownarrow \quad \updownarrow \quad\cdots$$
$$P(0) \ \neg P(1) \ P(2) \ \neg P(3) \ P(4) \ \neg P(5) \ P(6) \ \cdots$$

This one-to-one relation holds for numbers which can be finitely represented (such as the rational number 2/3) and those which cannot be finitely represented (such as random numbers) alike. Since infinite random real numbers are not computable (we will come back to this), this shows that most predicates cannot even be finitely defined, the less we can expect that we can prove their properties in a finite way.

To go beyond incompleteness can of course not be done by adding single elements (be it $\sqrt{2}$ or adding Gödel's $P_k(k)$ formula as an axiom). Rather we have to add all truly infinite elements to it, which are inherently not representable in a finite way (that is, which are not compressible). In the case of rational numbers this leads to the closed field of real numbers, in the case of logic this leads to the introduction of an ω-rule (see e.g., Grzegorczyk, Mostowski, and Ryll-Nardzewski in [7]). In order to understand the ω-rule let us look at proofs in the structure of natural numbers (the argument would hold for other structures as well, of course). The natural numbers are a structure, generated by 0 and the successor function s. They are represented as 0, 1, 2, 3, and so on. How can we prove a property like Goldbach's conjecture that every even number greater than 3 can be written as the sum of two prime numbers using the ω-rule?

The ω-rule has the following form:

$$\frac{P(0) \quad P(1) \quad P(2) \quad \cdots \quad P(n) \quad \cdots}{\forall n \ldotp P(n)}$$

Let $P(n)$ be defined as

$$2 \cdot n \le 3 \vee \exists p \ldotp \exists q \ldotp \text{prime}(p) \wedge \text{prime}(q) \wedge p + q = 2 \cdot n.$$

In order to show $\forall n \ldotp P(n)$ we have to show $P(0)$, $P(1)$, $P(2)$, $P(3)$, $P(4)\ldots$ one after the other. For instance, for $n = 0$ and $n = 1$ we have $2 \cdot n \le 3$, for $n = 2$ we find $p = q = 2$, for $n = 3$, $p = q = 3$, for $n = 4$, $p = 3$ and $q = 5$ and so on. We could go on like that (and computers did go on to big numbers to check the conjecture), but of course we could effectively never write down the full proof this way – assumed there is no counterexample. Likewise we can effectively never write down a single real number in its decimal or dual form. For rational numbers there exists an easy extension of the representation schema which allows for a finite representation by periods: we can write for the number 2/3 in dual notation: $0.\dot{1}\dot{0}$. In the case of proofs there exists a constructive ω-rule which allows for a finite representation in certain cases. We will discuss this in more detail in the next section.

Note that it is important for the argument here not to make the unrealistic idealization of speeding up proofs by halving the time from one step to the next. With such an assumption infinite ω-proofs could be carried through in finite time.

As Grzegorczyk, Mostowski, and Ryll-Nardzewski show, the ω-rule fully captures the concept of structures, like the natural numbers: it simply says, in order to show that a property holds for all natural numbers ($\forall n.P(n)$), prove the property for each natural number $P(0), P(1), P(2), P(3), P(4)\ldots$. Note that the argument as to why the property $P(n)$ above is true may be different for every n, e.g., $P(8)$ since $3 + 13 = 16$ with prime(3) and prime(13), and $P(9)$ since $7 + 11 = 18$ with prime(7) and prime(11). Up to now nobody has seen any pattern in the individual arguments for $P(n)$ which could be generalized. Chaitin's reinterpretation of Gödel's incompleteness proof means that it may be that there is no such pattern. Until somebody has found a (finite) proof for the Goldbach conjecture, or found a counterexample, we cannot rule out that the Goldbach conjecture is true and that there is no finite proof which covers all cases. That is, although the Goldbach conjecture itself can be concisely stated, it may be the case that it is true for reasons which cannot be concisely stated, but which are patternless in the sense that for infinitely many n infinitely many different reasons are needed as to why $P(n)$ is true. In this way, a program of the following form may correspond to an ω-proof (with $m = 2 \cdot n$):

```
m := 2;
bool := true;
repeat
   {m := m+2;
    if
    (not ExistPrimes(m))
    {bool := false}}
until (bool = false)
```

Here `ExistPrimes(m)` is a computable function which tests whether there are numbers p and $m - p$ both smaller than m and both prime numbers.

We do not know whether this program terminates (with m the smallest counterexample to the Goldbach conjecture), or runs forever. The latter would imply the correctness of the Goldbach conjecture. This example also clarifies some relationship between proving and the Halting problem. If the Halting problem were computable, questions such as Goldbach's conjecture could be decided simply by checking of whether a particular program halts or not.

The ω-rule has two advantages: Firstly, it is simple to understand, "If something is true for each single natural number, then it is true for every natural number." Secondly, it provides a complete system (in a similar sense as real numbers are complete using Dedekind cuts or Cauchy sequences). Its major disadvantage is that it does, for most cases, not allow to effectively prove things, but only to approximate the full proof. The transition is one from finitary to infinitary proofs. With this transition it is possible to get a completeness result. Note that this does not contradict Gödel's incompleteness result, since Gödel considers in his incompleteness proof a finite notion of proof only. The encoding of infinite proofs into numbers would not work (at least not by using natural numbers).

The ω-rule is hard to accept as an appropriate rule of proof, since it goes against the main idea of the concept of proof, namely that a proof has to be finite. As Hilbert put it ([9] quoted from [8, p. 381f]), a proof means that we start with "certain formulas, which serve as building blocks for the formal edifice of mathematics, [...] called axioms. A mathematical proof is an array that must be given as such to our perceptual intuition; it consists of inferences according to [a fixed] schema [...] where each of the premisses [...] in the array, either is an axiom or results from an axiom by substitution, or else coincides with the end formula of a previous inference or results from it by substitution. A formula is said to be provable if it is the end formula of a proof." Or as in their simplification Monty Python say in the argument clinic: "An argument is a connected series of statements intended to establish a definite proposition." Although not explicitly stated, the series of statements is assumed to be finite, and mechanically checkable by a proof checker, which can be a relatively simple computer program.

5 Approximating the ω-Rule by Mathematical Induction and the Constructive ω-Rule

Let us first have a closer look at Gödel's theorem again. Its proof is a finite proof that $P_k(k)$ is true. But does $P_k(k)$ not state that it is not finitely provable? No! Here we have to be more precise. Gödel finitely proved that for any fixed finite calculus there exists a true formula $P_k(k)$ which cannot be finitely proved within that calculus. We could well add $P_k(k)$ as an axiom, since it is true in the standard model (we do not discuss here the issue of non-standard models of arithmetic). Then we would get a new finite calculus, which has limitations of its own. This is analogous to the situation in number theory, where the extension of the rational numbers $\mathbb{Q}(\sqrt{2})$, defined as the smallest field which contains the rational numbers and $\sqrt{2}$, (or $\mathbb{Q}(\sqrt{n})$ for a non-square number n) is a finite extension of $\mathbb{Q}$, which allows for a finite description of $\sqrt{2}$, namely just $\sqrt{2}$ is such a description in $\mathbb{Q}(\sqrt{2})$, but $\mathbb{Q}(\sqrt{2})$ is not a complete field and only the transition to the field of real numbers $\mathbb{R}$ makes every Cauchy sequence to converge. The Pythagorean proof shows that rational numbers are not sufficient for all purposes. It turns out that finite representations, in general, are not sufficient. Similarly finite proofs are not sufficient for establishing the truth of arbitrary true formulae. Only the transition to the full ω-rule (exactly as the transition to all real numbers) results in a complete system.

Since the ω-rule is not appropriate for practical proofs in its full generality (unless we make the unrealistic assumption that we can speed up and check infinitely many steps in finite time), we need ways of approximating it. The two most prominent ones are proof by mathematical induction and proof by a constructive variant of the ω-rule.

For numbers we have quite a good understanding what can be represented finitely and we have the important notion of a computable number. For formulae we can define a similar notion of a computable truth, defined by using a constructive version of the ω-rule.

The constructive version of the ω-rule is a computable restriction of the general ω-rule in the sense that each of the $P(n)$ in

$$\frac{P(0) \quad P(1) \quad P(2) \quad \cdots \quad P(n) \quad \cdots}{\forall n{\scriptstyle\bullet}P(n)}$$

can be proved in a uniform way, that is, that there is a generalized proof $\mathsf{proof}(m)$ with a parameter m so that if m is instantiated to a particular n, then $\mathsf{proof}(n)$ is a proof for $P(n)$ for any n. For details, see for instance [1]. Note that we have not such a generalized proof for our outline of a proof of Goldbach's conjecture. For each m we produce a different argument and we would need to check the validity of infinitely many arguments in order to establish the applicability of the ω-rule. If, however, we had a generalized proof, we could check the generalized (finite) proof and apply the constructive ω-rule.

The most prominent way to establish finite arguments for numbers and other structures is mathematical induction. It has been known at least back to Fermat and is a standard way to establish mathematical facts. One such rule is:

$$\frac{P(0) \quad \forall n{\scriptstyle\bullet}P(n) \to P(s(n))}{\forall n{\scriptstyle\bullet}P(n)}$$

An inductive argument of the kind above can be translated to a proof using the constructive ω-rule. If we assume proof_{base} to be a proof for $P(0)$ and for a particular m, $\mathsf{proof}_{step}(m)$ be a proof for $P(m) \to P(s(m))$, then a proof for $P(n)$ for any n can be constructed as: Take the base proof proof_{base}, set m to 0, and establish $P(m)$. Then go through the following loop starting with $m = 0$ and increasing m by 1 until $s(m) = n$: Take the $\mathsf{proof}_{step}(m)$ to establish $P(m) \to P(s(m))$ and apply modus ponens on $P(m)$ and $P(m) \to P(s(m))$ to establish $P(s(m))$. This way, there is a uniform way to establish $P(m)$ for any m, hence the constructive ω-rule is applicable.

That is, we can say that inductive proofs can be transformed to constructive ω-proofs, and any constructive ω-proof is a general ω-proof. While general ω-proofs can be approximated by inductive proofs, and at least as well be approximated by constructive ω-proofs, it is not possible to capture the ω-rule in general by finite means, since the arguments as to why individual expressions such as $P(1)$, $P(2)$, ..., $P(n)$, ... are true, may not be generalizable, or as Gregory J. Chaitin [3] puts it: "... some mathematical facts are true for no reason! They are true by accident!" Taking a cardinality argument, we may go a step further and say: "Most mathematical facts are true for no reason! They are true by accident!" For instance, in the case of Goldbach's conjecture, it is not clear at all that the truth of $P(s(n))$ can be proved from the truth of $P(n)$, or all the previous instances $P(0)$, $P(1)$, $P(2)$, ..., $P(n)$. It may well be that the different cases are all true without any deeper connection and each case necessarily has to be established on its own from first principles.

Table 1. The analogy between numbers and proofs

	Numbers	**Proofs**
concrete proof	proof of irrationality of $\sqrt{2}$	Gödel's proof
generalized incompleteness result	incompleteness of $\mathbb{Q}$	incompleteness of finite calculi
patch to incompleteness	extend $\mathbb{Q}$ to $\mathbb{Q}(\sqrt{2})$	add the Gödel sentence as an axiom
full completion	$\mathbb{R}$, Dedekind cuts, Cauchy sequences	ω-rule
infinite representations	decimal representations of reals	infinitely long proofs
compression	compressed real numbers (e.g., periodic decimals, regular continued fractions	compressed finite proofs (e.g., constructive ω-rule, mathematical induction)
randomness	random numbers	random predicates, random proofs

Finite expressions, be it numbers or proofs, can be represented and fully understood by computers and humans[2]. The completion of logic to infinite systems shows that any reasoning system, be it human or artificial, is even in its idealized form subject to the limitations of finiteness, that is, certain mathematical truths cannot be proved finitely exactly as certain real number cannot be written down finitely. Gödel's proof itself is a finite proof, but it proves that finite proofs are not sufficient to capture truth in the case of systems which are powerful enough to speak about arithmetic.

The analogy between proofs and numbers can be summarized as in Table 1. In order to gain a deeper insight of what can be finitely proved we need a deeper understanding of the nature of proof than the one we have got today. The development of formal calculi corresponds to the understanding of rational numbers, the ω-rule to the understanding of real numbers. In between there is a big scope for finite proofs which go beyond traditional calculi, but avoid infinite arguments. The understanding is, however, not yet so intensively developed as in the case of numbers. We need a classification of proofs which includes the possibility for reformulation and compression of problems as well as abbreviations, extensions to an existing calculus, usage of translations and meta-proofs to establish mathematical facts. All these methods are quite common in the mathematical colloquial language.

---

[2] This is of course another simplification. Mathematical finiteness may go way beyond anything that is practical, and much that is practical for computers is not for humans.

Even when we will have gained a much deeper understanding of the nature of proof, there will still be the limitations which lead to Gödel's incompleteness result. If we take a finitary standpoint then there will be only countably many possible (finite) proofs, but there are uncountably many true statements of arithmetic, that is, too many as that all could have a proof.[3]

6   Conclusion

Pythagoras had a vision, namely that it should be possible that the length of any line segment can be expressed finitely by the ratio of two integers. Likewise Hilbert had a vision, namely that any true mathematical statement can be proved by a finite proof. It looked initially good for the two visions. The Pythagoreans found many examples for which the vision was fulfilled, however, they soon found counterexamples. Likewise Gödel's completeness result [5] was encouraging. But a year later, Gödel proved that Hilbert's vision was not achievable either. The reason why there are problems with the visions, is that there are too many real numbers and there are too many true statements.[4] Exactly as there are random real numbers which cannot be finitely expressed in principle (since they cannot be compressed to a finite representation), there are true statements which cannot be finitely proved in principle (since they are true for no good reason and hence their infinite ω-proofs cannot be compressed into a finite form). The limitations, namely to be able to deal with finite representations only – in computing as well as in reasoning – are limitations which hold for humans and computers alike. Gödel's theorem is about these limitations, but is not subject to them. Exactly as $\sqrt{2}$ is an indication that not all numbers can be finitely represented but it itself actually can, Gödel's theorem is an indication that not all true statements can be finitely proved, but it itself is finitely proved (as well as $P_k(k)$), and has been checked by humans and by computers alike.

Hence Gödel's theorem and the limitations to finiteness to which humans and computers are subject are irrelevant for settling the question whether computers can achieve human-like intelligence or not. Hence the Lucas-Penrose argument is invalid.

---

[3] Note that in full arithmetic we can make in principle uncountably many statements assumed we allow for infinite representations, that is, allow for infinite sets of formulae. For instance, we can represent a number such as π by an infinite set of formulae as: $\Pi = \{3 \leq \pi \leq 4;\ 3.1 \leq \pi \leq 3.2;\ 3.14 \leq \pi \leq 3.15;\ 3.141 \leq \pi \leq 3.142;\ \ldots\}$. If we take all such infinite sets then we get uncountably many true statements, since as for π, we can construct such a set analogously for any other of the uncountably many real numbers. If we allow for finite sets only then we can represent only countably many true statements.

[4] This does not reduce Gödel's theorem to a cardinality argument. However, the cardinality issue makes it plausible. As pointed out in footnote 3, there uncountably many true statements about arithmetic. Gödel's theorem does not explicitly state this – in the same way as Pythagoras did not prove that there are uncountably many numbers. Adding countably many further statements as axioms – such as extending the rational numbers by $\sqrt{n}$ for all n – will not result in a complete system, however.

References

1. Siani Baker, Andrew Ireland, and Alan Smaill. On the use of the constructive omega-rule within automated deduction. DAI Research Paper 560, University of Edinburgh, Edinburgh, United Kingdom, 1991. Also available as: http://www.dai.ed.ac.uk/papers/documents/rp560.html.
2. Alan Bundy, Fausto Giunchiglia, Adolfo Villafiorita, and Toby Walsh. Gödel's incompleteness theorem via abstraction. IRST-Technical Report 9302-15, IRST, Povo, Trento, Italy, February 1993.
3. Gregory J. Chaitin. *The Limits of Mathematics*. Springer-Verlag, Singapore, 1998.
4. Richard Dedekind. *What are numbers and what are they for?* 1888.
5. Kurt Gödel. Die Vollständigkeit der Axiome des logischen Funktionenkalküls. *Monatshefte für Mathematik und Physik*, **37**:349–360, 1930.
6. Kurt Gödel. Über formal unentscheidbare Sätze der Principia Mathematica und verwandter Systeme I. *Monatshefte für Mathematik und Physik*, **38**:173–198, 1931.
7. A. Grzegorczyk, A. Mostowski, and C. Ryll-Nardzewski. The classical and the ω-complete arithmetic. *The Journal of Symbolic Logic*, **23**(2):188–206, 1958.
8. Jean van Heijenoort, editor. *From Frege to Gödel – A Source Book in Mathematical Logic, 1879-1931*. Havard University Press, Cambridge, Massachusetts, USA, 1967.
9. David Hilbert. Über das Unendliche. *Mathematische Annalen*, **95**:161–190, 1926.
10. Geoffrey LaForte, Patrick J. Hayes, and Kenneth M. Ford. Why Gödel's theorem cannot refute computationalism. *Artificial Intelligence*, **104**:265–286, 1998.
11. John R. Lucas. Minds, Machines and Gödel. *Philosophy*, **36**:112–127, 1961.
12. John R. Lucas. A paper read to the Turing Conference at Brighton on April 6th, 1990. http://users.ox.ac.uk/~jrlucas/Godel/brighton.html, 1990.
13. John R. Lucas. Minds, Machines and Gödel: A Retrospect. In Peter Millican and Andy Clark, editors, *Machines and Thought – The Legacy of Alan Turing, Volume 1*, chapter 6, pages 103–124. Oxford University Press, Oxford, United Kingdom, 1996.
14. Roger Penrose. *The Emperor's New Mind*. Oxford University Press, Oxford, United Kingdom, 1989.
15. Roger Penrose. Mathematical intelligence. In Jean Khalfa, editor, *What is Intelligence?*, chapter 5, pages 107–136. Cambridge University Press, Cambridge, United Kingdom, 1994.

On the Road to High-Quality POS-Tagging

Stefan Klatt and Karel Oliva

Austrian Research Institute for Artificial Intelligence*,
Freyung 6/6, A-1010 Vienna, Austria
{stefan.klatt, karel.oliva}@ofai.at

Abstract. In this paper, we present techniques aimed at avoiding typical errors of state-of-the-art POS-taggers and at constructing high-quality POS-taggers with extremely low error rates. Such taggers are very helpful, if not even necessary, for many NLP applications organized in a pipeline architecture. The appropriateness of the suggested solutions is demonstrated in several experiments. Although these experiments were performed only with German data, the proposed modular architecture is applicable for many other languages, too.

1 Introduction

The role of POS-tagging in NLP applications becomes more and more important for at least three reasons: (i) in today's pipeline architecture of NLP analysis systems, POS-tagged data are often used as input for syntactic parsing, (ii) errors made in early stages in such pipeline architectures can rarely be corrected in subsequent stages, (iii) POS-tagged data are also often used as resources for training statistical models for different NLP applications.

Therefore, current state-of-the-art taggers with typical accuracy rates of 95-97% will soon be considered inadequate as components of high-quality NLP systems, since the residual error rate of 5%, combined with an average sentence length of 20 words, means that every sentence contains one tagging error on average[1].

Naturally, one could ask whether a totally unambiguous output of a tagger is needed as the input for the rest of the processing, as the example in (1) shows, in which the word *duck* can either have a verb or a noun reading. On the other hand, it is obvious that eliminating as many inappropriate readings as possible

---

* The Austrian Research Institute for Artificial Intelligence (OFAI) is supported by the Austrian Federal Ministry of Education, Science and Culture and by the Austrian Federal Ministry for Transport, Innovation and Technology. The work reported in this paper was supported by the Grant No. 16614 of the Austrian Science Fund (Fonds zur Förderung der wissenschaftlichen Forschung).

[1] In fact, errors tend to occur together in the same sentence, typically in long sentences. But for NLP applications such as real natural language understanding, it is very important to avoid as many errors as possible.

in secure contexts as in (2) is a goal worth pursuing. Here, since the sequence article+verb is ungrammatical and the word *the* only has the reading of an article, the verb reading of *man* can be eliminated.

(1)   He made her duck/N/V.
(2)   The/ART man/N/V̶ of the year.

Regarding the two processing paradigms in POS-tagging, the statistical one and the linguistic one, only the latter one allows to define what a secure context is. Unfortunately, the construction of linguistic taggers is very demanding. Furthermore they are language dependent, unlike the majority of statistical taggers that only need positive examples in the form of correctly tagged training data, usually created by human annotators[2], as input source for the language independent statistical tagging model.

But these are not the only advantages. Sometimes it is *easier* for a statistical tagger to assign the correct tag than for a linguistic one, as the following example shows. For a linguistic tagger, the context in (3) is insufficient to assign the word *zeitweise* its correct adverbial reading (ADV). Since the corpus frequency of the unigram ADJA (attributive adjective) for *zeitweise* is very rare, a statistical tagger will solve this problem by employing lexical probability.

(3)   Sie hatten zeitweise/ADV/ADJA ehrgeizige/ADJA Pläne/NN.[3]
       They had ambitious plans from time to time.

Comparing the accuracy rates of statistical vs. linguistic taggers for different languages (cf. (Samuelsson and Voutilainen, 1997) for English, (Klatt, 2002) for German) shows that the linguistic ones outperform the statistical ones, or that they help to reduce the error rates of statistical taggers if they are used as the front-end of a hybrid architecture, even if such linguistic front-ends consist of only a few rules (Hajič et al., 2001).

The rest of the paper is organized as follows: In section 2, we discuss several problems that occur in current POS-tagging applications, before we present our solutions of these problems in section 3. The approriateness of the proposed solutions is demonstrated by two experiments in section 4. We conclude with an outlook of further possible improvements in section 5.

2   Problems in Current POS-Tagging

Errors made by current state-of-the-art POS-taggers can be partitioned into three classes: (i) errors resulting from previous stages of processing, mainly insufficient tokenisation and/or incorrect morphological analysis (we summarize

---

[2] Unfortunately not always truly correctly, as our experiments in section 4 show.

[3] In the German examples we use the tags of the STTS (Stuttgart-Tübingen-Tagset (Schiller et al., 1999)), consisting of 54 tags, that was also used in the experiments in section 4.

these as input problems in the following), (ii) errors that can be corrected manually by revoking decisions made on spots which can be recognized (and marked off) as critical during the tagging process, and finally (iii) the residual plain (usual) errors made by the tagger.

2.1 Input Problems

Foreign Material. The language independence of statistical taggers does not help much if they are confronted with texts that contain foreign material (FM) words, as the output of the *TreeTagger* (Schmid, 1999) in (4) shows. Instead of assigning all foreign words inside the brackets their correct FM-tag, the *Tree-Tagger* assigns a mixture of tags very strange from a linguistic point of view.

(4)   von/APPR der/ART Vereinigung/NN für/APPR Kinderlose/NN (/$(National/ADJD Association/NN for/NE the/VVFIN Childless/NE)/$(

The reasons for the incorrect recognition of foreign material can be threefold: (i) The FM-word does not receive any reading by the lexical analysis (cf. subsubsection **Unknown Words**), (ii) The FM-word only receives a native reading by the lexical analysis (cf. subsubsection **Lexical Misassignments**), (iii) The FM-word receives the correct reading among other readings by the lexical analysis (cf. subsubsection **Intra-Tagging Problems**). In order to cope with such situations (and resulting errors), more sophisticated strategies are needed, which we describe in the following.

Unknown Words. While some unknown tokens on the input could indeed be foreign words, the majority is usually part of the native language, from the so-called open word classes (nouns, adjectives, numerals, verbs, and adverbs). These unknown words must be assigned their correct tags in the tagging process[4]. Most of the statistical taggers principally do this in the same way as for known words. Sometimes a suffix analysis is applied to eliminate some of the readings (cf. (Schmid, 1999), (Brants, 2000)). This is a common practice also for linguistic taggers.

Contrary to the majority of statistical taggers that only have a decision window of 2-3 tokens, linguistic taggers can also make use of the whole sentential context and/or linguistic knowledge to employ more sophisticated recognition strategies (and sometimes even wider contexts, e.g. the entire article or corpus, if more appropriate for the task).

Regarding the two unknown words in (5), we can assign the first one a VVFIN-tag by a linguistic context-sensitive rule, since it follows a clause-initial pronoun. Such an assignment would be impossible for a statistical tagger.

For the correct interpretation of the second unknown word a similar powerful linguistic rule as before is not possible. In such a case corpus-based strategies seem to be more appropriate.

---

[4] In the STTS, there are 11 tags representing subclasses of the above mentioned open word classes: NE, NN, ADJA, ADJD, CARD, VVFIN, VVINF, VVPP, VVIZU, VVIMP, ADV.

(5)   Er leaste/? geleaste/? Autos.
    H e leased leased cars.

Lexical Misassignments. A very hard problem for POS-taggers occurs if during the morphological analysis a token receives one or more tag readings, but not the correct one, (e.g. *National* in (4) or the surname *Morgenstern* in (6)). The correct assignment of the latter case can also be done by a named entity recognizer component. The general architecture of our tagger (cf. section 3.2) principally allows the usage of any named entity recognizer marking named entities by SGML-tags. Because of the more innovative ideas of a *far-sighted tokenization* (cf. section 3.1) and an *article-based tagging* (cf. section 3.2), we prefer the application of a partial named entity recognizer that we especially developed for the task of POS-tagging (cf. section 3.2).

(6)   Christian/NE Morgenstern/NN war ein berühmter Dichter.
    Christian Morgenstern was a famous poet.

A simple solution to such misassignments as in (6) is not available, since it would make no sense to mistrust every lexical analysis. Fortunately, most of such misassignments concern FM-words or proper nouns (STTS-tag NE).

Multiword Units. The next problem are multiword units (MWUs). According to the STTS-guidelines each part of a MWU has to be tagged in its original sense, which, however, is problematic for two reasons. (i) This task would require such *original sense* guidelines that do not yet exist (and quite probably never will). (ii) Annotated corpora in this style used for training statistical models could lead to false learning. Hence, in this point we would rather suggest a revision of the guidelines that would allow us to tag MWUs as MWUs [5].

For example, in (7), it seems to be strange to tag *New* as proper noun (NE) and *Yorker* as adjective (ADJA). And in (8), this task looks more like a riddle. Another problem are MWUs as part of a composite expression (cf. (9)), especially if they are surrounded by quotation marks (cf. (10)). But then, we run into another problem – the problem of tokenization.

(7)   New/NE York/NE vs. New/NE Yorker/ADJA
(8)   gang/? und/KON gäbe/?
    part and parcel
(9)   Star Wars-Trilogie
(10)   „Star Wars"-Trilogie

Insufficient Tokenization. The tokenization of (10) by different systems often leads to different representations (cf. (11)-(13)). The last representation is the one used in the NEGRA corpus (Skut et al., 1998) that we do not consider as the best solution (see our solution to this problem in section 3.1).

(11)   „ Star Wars"-Trilogie (3 tokens)
(12)   „ Star Wars " - Trilogie (6 tokens)
(13)   „ Challenge Day " -Büro (5 tokens)

---

[5] Which, contrary to assertions made in the STTS-guidelines (Schiller et al., 1999), should be no technical problem.

However, this is not the only problem that most of the current tokenizers have to deal with but are unable to solve. Information that is relevant for subsequent analysis components is not made transparent. This can lead to repeated redundant processing of the same task. Sometimes it will make the decision more difficult as the next example shows.

In German, the comma is also used as left single quotation mark (inside another pair of quotation marks), see (14). In English texts, we have the same problems with apostrophes as (15) shows. But a space before a comma is not a certain context for a single quotation mark reading, as the spelling error in (16) shows. So, only splitting up the comma from the rest of the token during tokenization is not enough if we do not want to create a very hard or even unsolvable problem for automatic POS-tagging as illustrated in (17): which of the both *commata* has to be tagged as a comma (STTS-tag $,) and which one as a single quotation mark (STTS-tag $()?

(14)   "Das Motto ‚One man, one vote' erschallte."
(15)   "The motto 'One man, one vote' rang out."
(16)   Koblenz,Kaiserslautern and Mainz are cities of Rheinland-Pfalz.
(17)   " Das Motto , One man , one vote ' erschallte . "

2.2   Intra-tagging Problems

The first two problems are those of a limited context window and a uni-directional processing. In the following examples, it is very hard if not impossible for statistical taggers that only use a decision window of 2-3 tokens to assign the word *vergessen* its correct tag reliably. It could be a finite verb (VVFIN), a verb in the infinitive (VVINF) or a past participle (VVPP). Using the whole sentential context, it is usually no problem for a linguistic tagger to locate possible key words (such as auxiliary or modal verbs, or complementizers) to the right or the left (as in (18)-(20)) of the word *vergessen* and to perform the correct assignment.

(18)   Ich glaube, dass sie es wieder einmal vergessen/VVFIN.
        I believe that they forget it once again.
(19)   Er könnte es wieder einmal vergessen/VVINF.
        He could forget it once again.
(20)   Er hat es wieder einmal vergessen/VVPP.
        He has forgotten it once again.

The third problem is the one of an immediate assignment (usually from left to right) instead of applying an easy-first strategy. For example, the readings of a relative pronoun (PRELS) and an article (ART) of the word *das* in (21) can be easily ruled out for linguistically very simple reasons. First, a clause-initial relative pronoun starting with the letter *d* is nearly impossible. And it is strictly impossible if the finite verb does not occur in a clause-final position (right clause bracket). Second, an article cannot precede an unambiguous verb (in German as well as in English).

(21)   Das/~~PRELS~~/~~ART~~/PDS ist/VAFIN nicht schwer!
        That is not difficult!

2.3   Output Problems

These problems are only of interest in off-line applications, especially if the produced texts should be used as training material for several tasks, e.g. statistical tagging.

Statistical taggers are unable to mark off decisions that are problematic from a linguistic point of view. Such are, e.g., the examples in (22) and (23) where even if the whole sentential context is used, it is impossible to assign the correct readings automatically.

(22)   Er war verrückt/ADJD. (The pronoun *Er* = the man)
       He was crazy.
(23)   Er war verrückt/VVPP. (The pronoun *Er* = the table)
       It was moved.

Therefore, marking off uncertain decisions would be of great benefit at least for corpora to be used as training material, since this would often enable a fast and reliable human correction of errors and, consequently, support building better statistical taggers.

3   Proposed Solutions

3.1   Solutions to the Input Problems

Far-Sighted Tokenization. To support the process of tagging from the beginning of text processing, we adopted the ideas in   (Klatt and Bohnet, 2004) to mark off tagging relevant decisions by SGML-tags during tokenization. All such SGML-tags are treated as zero-length tokens in the underlying processing architecture of our tagger (cf. section 3.2).

We also extended the inventory of markers used for annotating text structure units (e.g. for sentences) by markers for headlines, location marks and utterances etc. In (26), we see the output of our tokenizer for the text in (24). The domain between the two quotation marks was analyzed as an utterance consisting of more than one sentence (tag uts), in which the last sentence does not end with a sentence delimiter mark (tag swpm). The domain after the second quotation mark was annotated as a matrix clause (tag mc). Finally, we combined the uts-unit and the mc-unit to a mcuts-unit and annotated this domain with a corresponding tag (mcuts). Contrary to this detailed annotation, (25) shows the insufficient output produced by a lot of current state-of-the-art tokenizers.

A further advantage of this detailed tokenizer annotation concerning the POS-tagging process is that it enables us to tag sentences and headlines by different tagging strategies - this proves reasonable because of reduced sentences often used in headlines.

(24)   "Inflation and inflationary expectations have fallen further. The external
       sector continues to perform well," Central Bank said.

(25)   `<s> " Inflation and inflationary expectations have fallen`
        `further . </s> <s> The external sector continues to perform`
        `well " , Central Bank said . </s>`

(26)   `<mcuts> <uts> " <s> Inflation and inflationary expectations have`
        `fallen further . </s> <swpm> The external sector continues to`
        `perform well </swpm> " </uts> , <mc> Central Bank said .`
        `</mc> </mcuts>`

To solve the problems in (11)-(13) mentioned in section 2.1, such structures are annotated by our tokenizer with the tag `bqm-in-mwu`[6] (cf. (27)), which enables us to transform this word sequence into one MWU in the tagging process and assign to it only one tag (the one of the last word, cf. (28)). The tag `bqm-in-mwu` can then be used for further actions (e.g. deletion of this tag and/or tagging the parts of the MWU in their original sense if necessary).

(27)   `<bqm-in-mwu>`

        „
        Star/NN
        Wars/?
        „

        -
        Trilogie/NN
        `</bqm-in-mwu>`

(28)   `<bqm-in-mwu>`
        „„Star Wars"-Trilogie/NN
        `</bqm-in-mwu>`

Lexical Resources and Their Role in the Tagging Process. For the correct recognition of foreign material (FM) words, we generate special sublexica by a simple bootstrapping algorithm.

Starting with a seed word as the first key word, all new contexts in which the chosen key word occurs in quotation marks are extracted. Next, the frequency distribution of all words in the contexts received so far is investigated. As the next key word, the most frequent unprocessed word is chosen that does not receive any lexical analysis. Furthermore, this new key word is marked as a reliable FM-word candidate. If a former key word candidate received a (native) lexical analysis, it is stored as an uncertain FM-word candidate additionally. The process is iterated until a given parameterizable threshold (in terms of corpus frequency) is reached. In post-processing, the extracted entries are manually checked and stored in two different sublexica that are used by our tagger as described in section 3.2.

If we have to tag entire articles, we also dynamically build a session lexicon, in which we store (i) surname candidates that occur right-adjacent to given first names, (ii) received analysis of unknown words, and (iii) new FM-word candidates (as a result of the tagging process described in the next section) for updating our two FM-lexica.

---

[6] Marking a compound with a possible multiword unit in quotation marks as its first part that is separated from the compound head by a dash.

3.2 Solutions to the Intra-tagging Problems

Usage of *Zero Length* Tokens. As underlying processing architecture, we use a chart as a data structure in which every token is modelled as an edge with length 1 (cf. Figure 1).

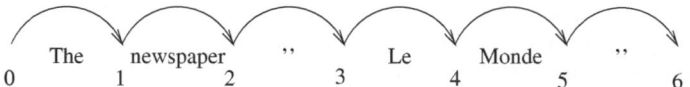

Fig. 1. Usual chart representation

We extend this architecture by a very simple but powerful mechanism: the usage of edges modelling tokens with length 0. If we want, we can use such zero-length tokens as parts of disambiguation rules. If not, we don't have to take care that the occurrence of such tokens could negatively affect other rules, which would be the case for all tokens with the length 1. At the moment, we use quotation marks and all SGML-tags as zero length tokens (cf. Figure 2).

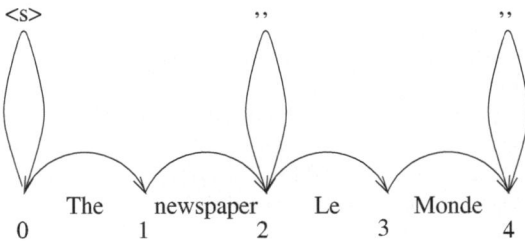

Fig. 2. Chart representation of our tagger

In (30), this enables us to assign the single quotation mark next to the SGML-tag `<sqm>` its correct STTS-tag ($(instead of $,). Without such a marker, the correct assignment would be nearly impossible as discussed before (cf. (17)).

(29)   "Früher sagten sie nein, heute sagen sie ‚Ja, aber'."
       "Earlier they said no, today they say 'Yes, but'.""
(30)   `<ut> ,, <s> Früher sagten sie nein , heute sagen sie`
       `<sqm> ,/$( Ja ,/$, aber ' </sqm> . </s> '' </ut>`

Sentence-Based Linguistic Easy-First Tagging. Using the whole sentential context as decision window, the goal is to eliminate ungrammatical readings in an easy-first manner by linguistic rules. This task is partitioned into several modules of which we describe the ones that cope with the problems mentioned in section 2 in more detail afterwards. But first we present a short look at the overall strategy.

To avoid as many errors as possible (in other words: to achieve a very high recall rate), we have to take special care of the interpretation of unknown and foreign material words. Therefore, we developed three modules: the modules UW_{RB} and UW_{CB} for linguistically rule-based and corpus-based strategies, respectively, to cope with unknown words, and the module FM for the assignment of additional FM-readings (module FM) in the processed sentences.

If we have to process an entire article, we can also duplicate readings of recognized named entities which we found in a secure context somewhere in the text (module NER). Furthermore, we make use of the tokenization tags as part of the tagging process (module TOK).

Later, linguistic rules based on ungrammatical configurations can be applied to filter out impossible tag configurations (module NEG) in the way described in section 2.2. At the moment, we only apply a small set of about 20 bigram rules. Next, nearly ungrammatical configurations are filtered out in the same way using heuristic (i.e. less reliable) linguistic rules (module HEU). In (31) for example, we eliminate the subordinative reading of *um*, since the necessary counterpart (a lexical verb in the *zu*-infinitive) is missing. In (32), we drop the infinitive reading of *vergessen*, because of a missing counterpart in the clause. And in (33), we eliminate the two verbal non-finite readings because of the right adjacent pronoun.

(31)   Er ging um/~~KOUI~~/APPR den See herum.
        He walked around the lake.
(32)   Er hat seinen Koffer im Hotel vergessen/~~VVINF~~/VVPP/VVFIN/ADJD.
        He has forgotten his suitcase in the hotel.
(33)   Hoffentlich vergessen/~~VVINF~~/~~VVPP~~/VVFIN/ADJD sie nicht, ihn einzuladen.
        Hopefully they do not forget to invite him.

Interpretation of Unknown Words. The interpretation of unknown words is bi-parted. First, we use linguistic rules (considering orthographic and morphological properties as well as special contexts) to assign an unknown word its possible readings (module UW_{RB}). Therefore we can easily assign the word *leaste* in (5) its correct finite lexical verb reading, since it is right-adjacent to a clause-initial pronoun (see also the discussion before).

After that we use corpus-based rules for the unresolved cases or to validate the former linguistic assignment (module UW_{CB}). In case of a capitalized or uppercase unknown word, we assign it the two nominal readings NN and NE and optionally an ADJA-reading if it ends with a special adjectival suffix and occurs left-adjacent to a noun or capitalized/uppercase unknown word.

We also assign lowercase words with the same adjective suffix an ADJA-reading, e.g. the word *geleaste* in (5). Additionally, we can also make use of the following corpus-based ADJA-test: If the unknown word occurs more often than a parameterizable threshold left adjacent to a noun or another uppercase written unknown word (for *geleaste* we get a rate of 95% in our corpus), an ADJA-reading is assigned.

In a similar way, we developed corpus-based tests for discriminating plain nouns (NN) from proper nouns (NE), and for distinguishing different verbal as well as adverbial readings.

Application of the Two Foreign Material Lexica. We use FM-words from our lexicon as the starting point for finding other (known and unknown) FM-words in their immediate contexts (module FM). In (34), this enables us to assign *fields* and *fire* an additional FM-tag without considering the context, since we consider *of* as a FM-word that always occurs between two other FM-words.

In the second case, we disambiguate the uncertain FM-words *In* and *a* in a later stage correctly on the assumption that they occur together with two reliable FM-words in a quotation mark context. Such an assignment is impossible for a statistical tagger.

(34)   Sie spielten ,, Fields of/FM fire " und ,, In a big/FM country/FM " .
        They played " Fields of fire " and " In a big country ".

Article-based Tagging Including Partial Named Entity Recognition.
As mentioned before, sometimes a sentential context is not sufficient to avoid tagging errors. Consider the two sentences in (35) and in (36) of one and the same article. In both sentences the word *Schuhmacher* does not receive its correct lexical reading. In the first case, this could be corrected due to the heuristic rule saying that this word is a possible surname because of the preceding (known) first name. In the second case, such a repairing strategy is possible only if the context of the entire article is considered. Therefore, we make use of a corresponding duplication mechanism that assigns every occurrence of such wordforms and their possible derivations relevant additional readings (module NER).

(35)   Toni/NE Schuhmacher/NN wurde zum Fußballer des Jahres gewählt.
        Toni Schuhmacher was voted as football player of the year.
(36)   Schuhmacher gewann diese Auszeichnung erstmalig.
        Schuhmacher/NN won this award for the first time.

Using the *One-Sense-per-Discourse*-constraint (Yarowsky, 1995) these word-forms can thus be disambiguated in respect to the assumed reading, although this will not be a fully reliable decision as the correct tag assignments in the (hypothetical) example (37) show.

(37)   Thomas Mann/NE war ein kluger Mann/NN.
        Thomas Mann was a clever man.

3.3   Solutions to the Output Problems

Additionally to the output of the tagged data, we generate a protocol of problematic tagger decisions that were marked by some of the tagger rules applied. At the moment, we can only process the information of such protocols in off-line decisions (such as the compilation of a nearly error-free corpus that is used as resource for training statistical models) to check uncertain assignments manually.

4   Experiments

4.1   Preparation

For the following two experiments, we compiled a test corpus of the first 21 articles of the manually tagged NEGRA corpus (Skut et al., 1998) consisting of 12735 tokens. For a later comparison with our results, we tagged this corpus with the TNT-Tagger[7] (Brants, 2000), one of the best current statistical taggers for German, receiving a $F_{\beta=1}$-score of 94.24% – or in other words an error rate of 5.76%.

Although the NEGRA corpus was manually annotated, it has the disadvantage that non-sentential discourse units such as headlines are labelled as sentences. Furthermore no article borders and other relevant textual discourse units are annotated. Therefore we tokenized the test corpus by our tokenizer with the result of 31 different tokenizer tag assignments, e.g. 569 sentences, 29 headlines, 35 utterances, 13 tags denoting the author of the article and 7 tags denoting a location at the beginning of the text body (cf. (38)).

```
(38)    <article>
        <hu> Hochheimer Ausstellung erinnert an wichtigen Fund vor
        60 Jahren </hu>
        <su> Der Spiegel aus dem Weinberg </su>
        <su> Auch Kelten waren eitel </su>
        <source> Von Dirk Fuhrig </source>
        <p>
        <loc+dot> HOCHHEIM . </loc+dot>
        <s> Die Selbstbetrachtung im Spiegel liebten offenbar bereits
        die alten Kelten . </s>
        <s> ... </s>
        ... </p>
        </article>
```

4.2 Experiment 1: Application of the Modules LEX, UW$_{RB}$, NEG and HEU

In the first experiment, we used a limited decision window given by the starting and the end tags of the processed structures. After the lexical analysis (module LEX) we applied the modules UW$_{RB}$, NEG and HEU. Table 1 shows the received results. The column *pos* indicates the number of tokens which had to be tagged correctly. The column *unk* contains the number of unknown tokens, and the column *mis* the number of tokens that did not receive the correct reading among one or more tag assignments. The column *corr$_{ua}$* shows the number of tokens that received one unambiguous correct reading, the column *readings* all readings that were considered as correct. The next two columns contain the corresponding recall and precision rates[8]. The last two colums show the corresponding F_β-score and error rates[9].

---

[7] In the experiment, we used the TNT-Tagger version that was also trained on the NEGRA corpus (see http://www.coli.uni-saarland.de/ thorsten/tnt/). In this context, it should be noted that is often argued that the use of a larger training set would produce better results for statistical tagging. Unfortunately, it is often forgotten to remark in the same context that the manual annotation of larger training corpora is also a very time consuming process.

[8] $recall = \frac{pos-(unk+mis)}{pos} * 100\%$ and $precision = \frac{pos}{readings+unk} * 100\%$.

[9] $F_\beta = \frac{(\beta^2+1)*precision*recall}{\beta^2*precision+recall}$ and $err = 100\% - recall$.

Table 1. Results of Experiment 1

Stage	pos	unk	mis	corr$_{ua}$	readings	recall	precision	F$_{\beta=1}$	err
LEX	12217	518	47	7947	18904	95.56	64.38	76.93	4.44
+UW$_{RB}$	12217	15	58	7952	20008	99.43	63.29	77.35	0.57
+NEG	12217	15	58	7978	19912	99.43	63.59	77.57	0.57
+HEU	12217	18	71	11532	14029	99.30	90.01	94.43	0.70

Starting with the lexical analysis (module LEX), we were unable to assign 565 tokens their correct reading. Futhermore we considered 18904 readings of the 12217 tokens that received one or more readings by the lexical analysis as correct.

After the application of the module UW$_{RB}$, we reached the highest recall. Here, we assigned 503 of the 518 unknown words one or more readings; 11 of these words did not receive the correct reading in this process.

Next, we applied the module NEG to filter out ungrammatical readings which only led to a minimally better precision rate.

In the last stage, we applied the module HEU to filter out nearly ungrammatical readings. This resulted in a much better precision rate, but unfortunately we also eliminated 16 correct readings.

Together we eliminated 5963 of 5979 readings after the application of the module UW$_{RB}$ in a correct way resulting in an error rate of 0.27%.

During the evaluation of the modules, we found 194 wrong or highly suspect tag assignments in the Golden Standard (manually tagged NEGRA corpus) of which 160 were corrected by us[10].

The remaining 34 cases were not corrected and were counted as tagging errors, mainly because the STTS-guidelines were not helpful for the determination of the correct reading (see also the discussion in the next section).

4.3 Experiment 2: Additional Application of the Modules UW$_{CB}$, FM, NER and TOK

In this second experiment, we additionally applied the other modules (with a more innovative character) described in section 3. After the application of the first two modules of experiment 1, we applied the modules UW$_{CB}$, FM, NER and TOK followed by the stages NEG and HEU of the first experiment. The results are shown in Table 2.

Using the module +UW$_{CB}$, we were able to assign the correct reading in six cases, e.g. for the verb *deportiert* (engl. *to deport*) the four readings VVPP, VVFIN, VVIMP and ADJD. In nine cases, no reading was given back. Among these words were five FM-words (e.g. *violin*), two verbs (e.g. *wehtue*) and two adjectives (e.g. *funky*).

---

[10] All the experiments, including the one with TnT-Tagger, were re-run after the correction and evaluated with the new version of the corpus.

Table 2. Results of Experiment 2

Stage	pos	unk	mis	corr$_{ua}$	readings	recall	precision	$F_{\beta=1}$	err
LEX+UW$_{RB}$	12217	15	58	7952	20008	99.43	63.29	77.35	0.57
+UW$_{CB}$	12217	9	58	7956	20017	99.47	63.39	77.43	0.53
+FM	12217	9	68	7982	19934	99.40	63.45	77.46	0.60
+NER	12217	9	62	7982	19940	99.44	63.45	77.47	0.56
+TOK	12217	9	62	8025	19903	99.44	63.63	77.60	0.56
+NEG+HEU	12217	12	75	11570	13949	99.32	90.30	94.60	0.68

The application of the module FM led to ten misassignments that are all examples of the highly suspect cases mentioned before. In one article, we found different NEGRA annotations for one and the same word sequence (cf. (39)-(40)). In (41), we indicated our *wrong* annotation.

(39)   The/FM Manson/NE Family/FM
(40)   The/NE Manson/NE Family/NE
(41)   The/FM Manson/FM Family/FM

The application of the module NER enabled us to correctly assign an additional proper name reading (tag NE) to six tokens that received by the module LEX only a NN-tag. The application of the module TOK led to a minimally better precision rate. In one case, we used a location tokenizer-tag to correctly disambiguate the token *HOCHHEIM* (NE) and all derived wordforms of it (e.g. the adjective of origin *Hochheimer*, cf. (38)). The application of the last two modules of experiment 1 did not lead to any changes in respect to experiment 1. Because of the additionally used modules in this experiment we finally received slightly better results than in the first experiment.

At a first glance, these small improvements seem to be irrelevant if we consider the $F_{\beta=1}$-score as important. But for the construction of hiqh-quality NLP taggers (as well as NLP applications based on the results of a tagger component) we consider the avoidance of any single wrong assignments truly crucial.

Finally, it seemed to be no good decision to evaluate the FM module on the NEGRA corpus that includes text material of the early 90's[11]. So it would be interesting to see which impact this module will have on newer corpora that tend to contain more foreign text material than older corpora.

5   Conclusion and Further Work

Comparing our results with the one of the TnT-Tagger ($F_{\beta=1}$-score of 94.24%). we reached a $F_{\beta=1}$-score 94.43% in the first experiment and a $F_{\beta=1}$-score of 94.60% in the second one. This means that if we rate precision and recall equally,

---

[11] In our test corpus, we only find one occurrence of the two highly frequent English words *the* and *of*.

we performed slightly better than one of the best current statistical taggers for German. Although we only receive slightly better results with the application of the modules FM, NER, and TOK, we are convinced that these modules are very important for the creation of high-quality NLP analysis systems.

If we uprate recall against precision, as argued for in section 1 and section 4, the outperformance is considerably higher. We believe that subsequent components in the processing pipeline like parsers benefit much more from the achieved error rate of our tagger (0.68%) than from the achieved $F_{\beta=1}$-score of the TnT-Tagger respective its corresponding error rate of 5.76%. A much higher $F_{\beta=1}$-score could surely be reached by the combination of the two taggers using our tagger as front-end. Unfortunately the TnT-Tagger is not able to function as back-end in such an architecture.

Therefore, we can draw the conclusion that we are on the right track. In the near future, we expect to receive a much better precision rate with the application of the modules in experiment 2, particularly after an expansion of the modules NEG and HEU. In this direction, we plan the application of a new set of constraints motivated by the Topological Field Model (Höhle, 1986) as the next worthwhile step to get nearer to the end of the road to high-quality POS-tagging.

As other major results, we can state that our research proved helpful for locating annotation errors in manually – presumably correct – annotated corpus, and it also showed convincingly that better STTS-guidelines and/or an extension of the STTS-tag inventory are needed.

References

Brants, T.: TnT - a Statistical Part-of-Speech Tagger In *Proceedings of the 6th Applied NLP conference, ANLP-2000*, Seattle, WA (2000).

Hajič, J. and Krbec, P. and Květoň, P. and Oliva, K. and Petkevič, V.: Serial Combination of Rules and Statistics: A Case Study in Czech Tagging In *Proceedings of ACL 2001*, Toulouse (2001).

Höhle, T.: Der Begriff Mittelfeld. Anmerkungen über die Theorie der topologischen Felder In *Weiss, W. and Wiegand E. H. and Reis, M.: Textlinguistik contra Stilistik/Wortschatz und Wörterbuch/Grammatische oder pragmatische Organisation von Rede*, Niemeyer, Tübingen (1986).

Klatt, S. and Bohnet, B.: You don't have to think twice if you carefully tokenize, In *Proceedings of the First International Joint Conference on Natural Language Processing (IJCNLP-04)*, Hainan (2004).

Klatt, S.: Combining a Rule-Based Tagger with a Statistical Tagger for Annotating German Texts In *Busemann, S: KONVENS 2002. 6. Konferenz zur Verarbeitung natürlicher Sprache*, Saarbrücken, Germany (2002).

Samuelsson, C. and Voutilainen, A.: Comparing a Linguistic and a Stochastic Tagger In *Proceedings of the Joint 35th Annual Meeting of the Association for Computational Linguistics* (1997).

Schiller, A. and Teufel, S. and Stöckert, C.: *Guidelines für das Tagging deutscher Textcorpora mit STTS* Technical Report, University of Stuttgart and University of Tübingen (1999).

Schmid, H.: Improvements in part-of-speech tagging with an application to German In *S. Armstrong and K.W. Church and P. Isabelle and S. Manzi and E. Tzoukermann and D. Yarowsky: Natural Language Processing Using Very Large Corpora*, Kluwer, Dordrecht (1999).

Skut, W., Brants, T., Krenn, B. and Uszkoreit, H.: A Linguistically Interpreted Corpus of German Newspaper Text In *Workshop on Recent Advances in Corpus Annotation*, 10th European Summer School in Logic, Language and Information, Saarbrücken, Germany (1998).

Trushkina, J.: Morpho-Syntactic Annotation and Dependency Parsing of German Ph.D. thesis, University of Tübingen (2004).

Yarowsky, D.: Unsupervised Word Sense Disambiguation Rivaling Supervised Methods In *Meeting of the Association for Computational Linguistics* (1995).

Author Index

Lecture Notes in Artificial Intelligence (LNAI)

Vol. 3452: F. Baader, A. Voronkov (Eds.), Logic for Programming, Artificial Intelligence, and Reasoning. XI, 562 pages. 2005.

Vol. 3451: M.-P. Gleizes, A. Omicini, F. Zambonelli (Eds.), Engineering Societies in the Agents World V. XIII, 349 pages. 2005.

Vol. 3446: T. Ishida, L. Gasser, H. Nakashima (Eds.), Massively Multi-Agent Systems I. XI, 349 pages. 2005.

Vol. 3445: G. Chollet, A. Esposito, M. Faundez-Zanuy, M. Marinaro (Eds.), Nonlinear Speech Modeling and Applications. XIII, 433 pages. 2005.

Vol. 3438: H. Christiansen, P.R. Skadhauge, J. Villadsen (Eds.), Constraint Solving and Language Processing. VIII, 205 pages. 2005.

Vol. 3430: S. Tsumoto, T. Yamaguchi, M. Numao, H. Motoda (Eds.), Active Mining. XII, 349 pages. 2005.

Vol. 3419: B. Faltings, A. Petcu, F. Fages, F. Rossi (Eds.), Constraint Satisfaction and Constraint Logic Programming. X, 217 pages. 2005.

Vol. 3416: M. Böhlen, J. Gamper, W. Polasek, M.A. Wimmer (Eds.), E-Government: Towards Electronic Democracy. XIII, 311 pages. 2005.

Vol. 3415: P. Davidsson, B. Logan, K. Takadama (Eds.), Multi-Agent and Multi-Agent-Based Simulation. X, 265 pages. 2005.

Vol. 3403: B. Ganter, R. Godin (Eds.), Formal Concept Analysis. XI, 419 pages. 2005.

Vol. 3398: D.-K. Baik (Ed.), Systems Modeling and Simulation: Theory and Applications. XIV, 733 pages. 2005.

Vol. 3397: T.G. Kim (Ed.), Artificial Intelligence and Simulation. XV, 711 pages. 2005.

Vol. 3396: R.M. van Eijk, M.-P. Huget, F. Dignum (Eds.), Agent Communication. X, 261 pages. 2005.

Vol. 3394: D. Kudenko, D. Kazakov, E. Alonso (Eds.), Adaptive Agents and Multi-Agent Systems II. VIII, 313 pages. 2005.

Vol. 3392: D. Seipel, M. Hanus, U. Geske, O. Bartenstein (Eds.), Applications of Declarative Programming and Knowledge Management. X, 309 pages. 2005.

Vol. 3374: D. Weyns, H. V.D. Parunak, F. Michel (Eds.), Environments for Multi-Agent Systems. X, 279 pages. 2005.

Vol. 3371: M.W. Barley, N. Kasabov (Eds.), Intelligent Agents and Multi-Agent Systems. X, 329 pages. 2005.

Vol. 3369: V. R. Benjamins, P. Casanovas, J. Breuker, A. Gangemi (Eds.), Law and the Semantic Web. XII, 249 pages. 2005.

Vol. 3366: I. Rahwan, P. Moraitis, C. Reed (Eds.), Argumentation in Multi-Agent Systems. XII, 263 pages. 2005.

Vol. 3359: G. Grieser, Y. Tanaka (Eds.), Intuitive Human Interfaces for Organizing and Accessing Intellectual Assets. XIV, 257 pages. 2005.

Vol. 3346: R.H. Bordini, M. Dastani, J. Dix, A.E.F. Seghrouchni (Eds.), Programming Multi-Agent Systems. XIV, 249 pages. 2005.

Vol. 3345: Y. Cai (Ed.), Ambient Intelligence for Scientific Discovery. XII, 311 pages. 2005.

Vol. 3343: C. Freksa, M. Knauff, B. Krieg-Brückner, B. Nebel, T. Barkowsky (Eds.), Spatial Cognition IV. XIII, 519 pages. 2005.

Vol. 3339: G.I. Webb, X. Yu (Eds.), AI 2004: Advances in Artificial Intelligence. XXII, 1272 pages. 2004.

Vol. 3336: D. Karagiannis, U. Reimer (Eds.), Practical Aspects of Knowledge Management. X, 523 pages. 2004.

Vol. 3327: Y. Shi, W. Xu, Z. Chen (Eds.), Data Mining and Knowledge Management. XIII, 263 pages. 2005.

Vol. 3315: C. Lemaître, C.A. Reyes, J.A. González (Eds.), Advances in Artificial Intelligence – IBERAMIA 2004. XX, 987 pages. 2004.

Vol. 3303: J.A. López, E. Benfenati, W. Dubitzky (Eds.), Knowledge Exploration in Life Science Informatics. X, 249 pages. 2004.

Vol. 3301: G. Kern-Isberner, W. Rödder, F. Kulmann (Eds.), Conditionals, Information, and Inference. XII, 219 pages. 2005.

Vol. 3276: D. Nardi, M. Riedmiller, C. Sammut, J. Santos-Victor (Eds.), RoboCup 2004: Robot Soccer World Cup VIII. XVIII, 678 pages. 2005.

Vol. 3275: P. Perner (Ed.), Advances in Data Mining. VIII, 173 pages. 2004.

Vol. 3265: R.E. Frederking, K.B. Taylor (Eds.), Machine Translation: From Real Users to Research. XI, 392 pages. 2004.

Vol. 3264: G. Paliouras, Y. Sakakibara (Eds.), Grammatical Inference: Algorithms and Applications. XI, 291 pages. 2004.

Vol. 3259: J. Dix, J. Leite (Eds.), Computational Logic in Multi-Agent Systems. XII, 251 pages. 2004.

Vol. 3257: E. Motta, N.R. Shadbolt, A. Stutt, N. Gibbins (Eds.), Engineering Knowledge in the Age of the Semantic Web. XVII, 517 pages. 2004.

Vol. 3249: B. Buchberger, J.A. Campbell (Eds.), Artificial Intelligence and Symbolic Computation. X, 285 pages. 2004.

Vol. 3248: K.-Y. Su, J. Tsujii, J.-H. Lee, O.Y. Kwong (Eds.), Natural Language Processing – IJCNLP 2004. XVIII, 817 pages. 2005.

Vol. 3245: E. Suzuki, S. Arikawa (Eds.), Discovery Science. XIV, 430 pages. 2004.

Vol. 3244: S. Ben-David, J. Case, A. Maruoka (Eds.), Algorithmic Learning Theory. XIV, 505 pages. 2004.

Vol. 3238: S. Biundo, T. Frühwirth, G. Palm (Eds.), KI 2004: Advances in Artificial Intelligence. XI, 467 pages. 2004.

Vol. 3230: J.L. Vicedo, P. Martínez-Barco, R. Muñoz, M. Saiz Noeda (Eds.), Advances in Natural Language Processing. XII, 488 pages. 2004.

Vol. 3229: J.J. Alferes, J. Leite (Eds.), Logics in Artificial Intelligence. XIV, 744 pages. 2004.

Vol. 3228: M.G. Hinchey, J.L. Rash, W.F. Truszkowski, C.A. Rouff (Eds.), Formal Approaches to Agent-Based Systems. VIII, 290 pages. 2004.

Vol. 3215: M.G.. Negoita, R.J. Howlett, L.C. Jain (Eds.), Knowledge-Based Intelligent Information and Engineering Systems, Part III. LVII, 906 pages. 2004.